Space Telescopes & Payloads

Space Telescopes & Payloads

Guest Editors

Ariadna Calcines Rosario
John Capone
Sophie Musset
Paula Chadwick

MDPI

Basel • Beijing • Wuhan • Barcelona • Belgrade • Novi Sad • Cluj • Manchester

Guest Editors
Ariadna Calcines Rosario
Department of Physics
Durham University
Durham
UK

John Capone
Optics Branch
NASA GSFC
Greenbelt, MD
USA

Sophie Musset
Applied Physics Laboratory
The Johns Hopkins
University
Laurel, MD
USA

Paula Chadwick
Department of Physics
Durham University
Durham
UK

Editorial Office
MDPI AG
Grosspeteranlage 5
4052 Basel, Switzerland

This is a reprint of the Special Issue, published open access by the journal *Aerospace* (ISSN 2226-4310), freely accessible at: https://www.mdpi.com/journal/aerospace/special_issues/J9BVEV302X.

For citation purposes, cite each article independently as indicated on the article page online and as indicated below:

Lastname, A.A.; Lastname, B.B. Article Title. *Journal Name* **Year**, *Volume Number*, Page Range.

ISBN 978-3-7258-7114-8 (Hbk)
ISBN 978-3-7258-7115-5 (PDF)
https://doi.org/10.3390/books978-3-7258-7115-5

Contents

About the Editors

Ariadna Calcines Rosario

Ariadna Calcines Rosario is an Associate Professor and Head of Optics at the Centre for Advanced Instrumentation of Durham University, UK. She studied at the University of La Laguna, Spain, a degree in Physics-Astrophysics, a Master Degree in Astrophysics – Expert in Astronomical Instrumentation and a PhD in the optical design of the Integral Field Spectrograph of the European Solar Telescope. Additionally, she attended the International School of Advanced Instrumentation and the ESA's International School on Space Optics. Her work includes the design and development of instrumentation for solar and night-time ground-based telescopes and space missions. She is specialised in the design of high resolution integral field spectrographs and image slicers and is interested in Innovation and low TRL technology development. She is currently working in collaboration with UCL to space qualify the image slicer technology in the Extreme Ultra-Violet. She has received multiple awards for her professional career, among them, the Best PhD of Spain prize in astronomical instrumentation and technology development; the Young Female Scientist Talent prize, awarded by the Royal Academy of Sciences of Spain and the ENDECAN award to the scientific and technological talent.

John Capone

John Capone, Ph.D., is an optical engineer at NASA's Goddard Space Flight Center (GSFC), where he serves as the Optical Design Group Lead and the optical lead for a variety of flight projects and proposals. His work focuses on developing innovative solutions to complex optical challenges, particularly in scenarios where commercial software tools fall short of addressing mission requirements. John Capone's contributions have been recognized with awards at both the branch and division levels, including the Instrument Systems and Technology Division (ISTD) Engineering Excellence Award for his pivotal role in the vision system of the Mars Sample Return (MSR) campaign's Capture, Containment, and Return System (CCRS). Recently, he has contributed to the modeling and analysis of the Laser Interferometer Space Antenna (LISA), creating a custom ray-tracing code to achieve the precision required for this groundbreaking mission. John Capone's interests lie in advancing optical system design through the application of custom numerical methods to tackle the unique challenges of spaceflight optical engineering.

Sophie Musset

Sophie Musset is a senior scientist at Johns Hopkins University Applied Physics Laboratory in Maryland, USA. She obtained her Ph.D. in Astrophysics from Paris Observatory in 2016, on the topic of particle acceleration in solar flares, and was awarded a research fellowship of the European Space Agency in 2020. The main topics of her research activities are the fundamental processes of particle acceleration and propagation in the solar atmosphere and solar wind. Her research interests lie in observational diagnostics of solar energetic particles, and the development of community science and machine learning in the analysis of solar observations. She is participating in the development and implementation of space telescopes and missions to understand high-energy phenomena in the solar atmosphere, and specializes in X-ray imagers and spectrometers.

Paula Chadwick

Paula Chadwick graduated with a degree in Physics & Astronomy from Queen Mary, University of London, in 1984. She gained her Ph.D. degree in astroparticle physics in 1987 at Durham University, where she has been based for over 40 years. She is now a Professor and is currently Head of the Physics Department. Her primary interest is in high-energy gamma-ray astronomy from the ground and from space. Having been involved in the construction of ground-based gamma-ray telescopes in Australia, the Canary Islands and Namibia, she is now working on the Cherenkov Telescope Array Observatory and the Southern Wide-field Gamma-ray Observatory. Her scientific interests include indirect detection of dark matter, active galactic nuclei, pulsars and binary systems. Paula has served on many grant-awarding committees in the UK and on several international advisory committees; she is currently a member of the Scientific Advisory Committees of the Astroparticle Physics European Consortium (APPEC) and the Swedish National Space Agency. She has over 400 scientific papers to her name and was awarded the Bragg Medal of the Institute of Physics in 2015.

Preface

Building on the extraordinary success of the James Webb Space Telescope (JWST) and the remarkable technological advancements that have made it possible to deploy a 6.5-meter telescope in space, this Special Issue showcases the latest developments in Space Telescopes and Payloads. With the ongoing exploration of Mars by rovers and the first-ever flight of a helicopter off Earth, as well as the Solar Orbiter's unprecedented imagery of the Sun and Parker Solar Probe touching the solar corona, this is an exciting time in the advancement of space-based instruments central to such groundbreaking missions.

The contributions to this Special Issue present the current and future challenges on Space Telescopes and Payloads that sciences and technology are facing, covering all spectral ranges and scientific cases.

The Guest Editors of this Special Issue (SI) would like to thank all authors for their contributions. Special thanks to August Wang, whose efforts and dedication made this SI possible. Thanks are also extended to the team at MDPI *Aerospace*.

Ariadna Calcines Rosario, John Capone, Sophie Musset, and Paula Chadwick
Guest Editors

aerospace

MDPI

Article

A High-Resolution Mass Spectrometer for the Experimental Study of the Gas Composition in Planetary Environments: First Laboratory Results

Illia Zymak [1,2,*], Ján Žabka [3], Miroslav Polášek [3], Arnaud Sanderink [1,4], Jean-Pierre Lebreton [1], Bertrand Gaubicher [1], Barnabé Cherville [1], Anna Zymaková [2] and Christelle Briois [1]

1 Laboratoire de Physique et Chimie de l'Environnement et de l'Espace (LPC2E), UMR7328 CNRS/Université d'Orléans, 3A, Avenue de la Recherche Scientifique, 45071 Orléans, France
2 The Extreme Light Infrastructure ERIC, ELI Beamlines Facility, Za Radnicí 835, 252 41 Dolní Břežany, Czech Republic
3 J. Heyrovský Institute of Physical Chemistry, Czech Academy of Sciences, Dolejškova 3, 182 23 Prague, Czech Republic
4 Institut für Geologische Wissenschaften, Freie Universität Berlin, Malteserstraße 74-100, D-12249 Berlin, Germany
* Correspondence: illia.zymak@eli-beams.eu

Abstract: A new laboratory Orbitrap^{TM} cell-based mass spectrometer, OLYMPIA (Orbitrap anaLYseur MultiPle IonisAtion), without a C-trap module, has been developed and constructed. The first operation of the Orbitrap^{TM} cell-based device with the continuous ion source and without the C-trap module is reported. OLYMPIA is being developed and used as a workbench platform to test and develop technologies for the next generation of spaceborne mass spectrometers and as a laboratory instrument to perform high-resolution studies of space-relevant chemical processes. This instrument has been used to measure the quantitative composition of $CO/N_2/C_2H_4$ mixtures of the same nominal mass using an electron ionization ion source. The relative abundance of ions has been measured using a short acquisition time (up to 250 ms) with a precision of better than 10% (for most abundant ions) and a mass resolution of 30,000–50,000 (full width at half maximum) over the mass range of m/z 28–86. The achieved mass accuracy of measurements is better than 20 ppm. This performance level is sufficient to resolve and identify the $CO/N_2/C_2H_4$ components of the mixtures. The dynamic range and relative ion abundance measurements have been evaluated using a reference normal isotopic distribution of krypton gas. The measurement accuracy is about 10% for the 4 most abundant isotopes; 6 isotopes are detectable.

Keywords: high-resolution mass spectrometer; Orbitrap; continuous ion source; no C trap; ion optics

1. Introduction

Mass spectrometry is one of the most efficient techniques for in situ experimental studies of the chemical composition of samples of astrochemical relevance. In situ measurements carried out during recent space missions to Saturn and its moons (Cassini-Huygens [1–4]) and the 67P/Churyumov–Gerasimenko comet (Rosetta [5–7]) have provided unique data on the composition of the environment and surface of the Solar System's bodies. These missions unveiled the complexity of the chemical composition of both the atmosphere of Saturn's moon Titan [3,4,8–10] and of comets, which are considered to be relatively primitive small space bodies.

One of the most advanced space mass spectrometers flown onboard space missions, the Rosetta Double Focusing Mass Spectrometer (DFMS) of the Rosetta Orbiter Spectrometer for Ion and Neutral Analysis (ROSINA) module, provided a mass resolution of $m/\Delta m \approx 3000$ at 1% of the peak height (corresponding to ≈9000 at full width at half maximum (FWHM)) with a 1–150 u mass range [11–13], which was sufficient to resolve N_2 from

CO [14]. Detection of prebiotic organic molecules, such as glycine, was also performed, but deconvolution tools and careful examination of the fragmentation pattern were necessary as the resolution was not sufficient for direct and unambiguous identification of complex organic molecules (Altwegg et al., 2016). For this purpose, when the mass resolution of the measured mass spectra is not sufficient, supplementary analytical deconvolution algorithms [15] and complex chemical models [16,17] are required for a comprehensive description of the chemical composition of extraterrestrial environments.

The chemical diversity of the extraterrestrial medium and the limited mass resolution of current state-of-the-art space instruments call for a new generation of spaceborne mass spectrometers and the development of ground-based analytical instruments for additional comparative laboratory studies of space-relevant samples. High-resolution mass spectrometry allows the identification of complex organic compounds without a need for complementary separation techniques, e.g., chromatography [18], spectroscopy [19], or collision-induced dissociation [20]. In recent years, significant technological and scientific advances have been made in the field of high-resolution space mass spectrometry. Several flight instruments and prototypes of high-resolution mass spectrometers are currently under development and are at different space technology readiness levels (TRL). The MAss SPectrometer for Planetary EXploration (MASPEX) (Brockwell et al., 2016) on board NASA's future Europa Clipper mission is characterized by sufficient mass resolution to identify simple hydrocarbons, i.e., $m/\Delta m \approx 7000$ at m/z 2–32 and around 24,000 at m/z 16–114, using the time-of-flight (ToF) technique. Two high-resolution laser ablation space mass spectrometers, the Characterization of Ocean Residues and Life Signatures (CORALS) instrument and the Characterization of Regolith and Trace Economic Resources (CRATER), are being developed as part of an international collaboration between NASA and CNRS laboratories with the support of CNES [21]. They both include the *CosmOrbitrap*, a space-flight prototype of a high-resolution mass analyzer developed by a consortium of six laboratories [22], based on the commercial Orbitrap^{TM} cell (Thermo Fisher Scientific, Bremen, Germany). The *CosmOrbitrap* has a mass resolution that varies from 474,000 to 90,000 for a 9 to 208 m/z range and an acquisition time of 832 ms [22].

In this work, a novel compact high-resolution mass spectrometry (HRMS) laboratory instrument based on the Orbitrap technology, named OLYMPIA (Orbitrap anaLYseur MultiPle IonisAtion) is presented. The instrument is compatible with different types of ion sources that have recently been built in our laboratory. This setup is being used as a proof of concept for laboratory analytical applications and as a workbench platform. In this work, a configuration that uses a continuous electron ionization (EI) ion source for the high-resolution gas analysis of compounds relevant to planetary environments has been tested. Certain decreases in the mass resolution compared to the ultimate values of the best commercially available Orbitrap mass analyzers are expected due to the higher kinetic energy distribution spread of ions and probable diffusion of the gas from the source inside the trap volume for the proposed configuration of the device. Experimental characterization of the mass analyzer is required to facilitate future use of the proposed configuration for laboratory and space applications.

2. Materials and Methods

OLYMPIA (Figure 1) is a HRMS instrument based on the Orbitrap technology for mass analysis proposed and developed by A. Makarov [23]. Its configuration is intended to support the development of future compact and low-resource space instruments without C-traps. The C-trap is a proprietary device used in commercial Orbitrap mass spectrometers for the concentration and increasing monochromaticity of the kinetic energy of ions by collision with the neutral buffer gas prior to injection into an electrostatic trap. It allows achieving ultra-high mass resolution ($m/\Delta m > 10^6$ [24]), but it increases the complexity, mass, and power consumption of the instrument and requires an additional gas reservoir. In contrast, the low-resource design without a C-trap makes OLYMPIA lighter, more transportable, and easier to maintain and operate. It is designed to provide mass resolution adequate for the next generation of space mass spectrometers ($m/\Delta m$ around 50,000) and is equipped with a modular structure ionization source.

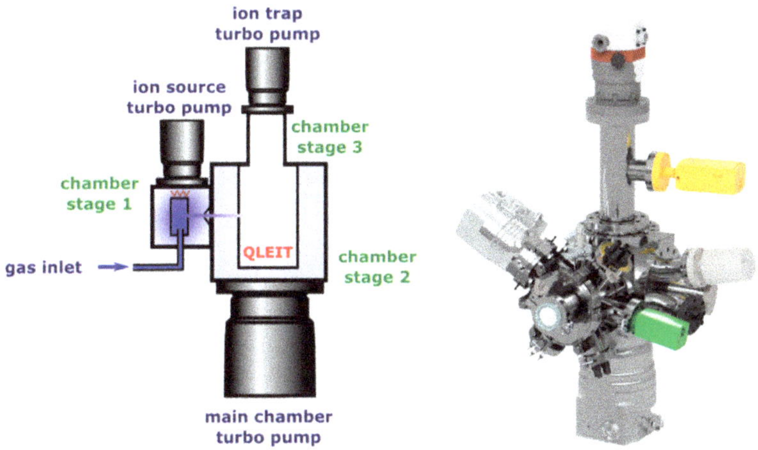

Figure 1. Schematic of the OLYMPIA instrument: principal components of the vacuum system (**left** panel); CAD model of the arrangement of the vacuum system (**right** panel) made of Pfeiffer Vacuum, Edwards Vacuum, and Hositrad Vacuum Technologies commercial off-the-shelf components (publicly available models of turbo pumps, vacuum chambers, ion gauges, flanges, and feedthroughs have been used).

2.1. Ion Source and Ion Optics

In this work, a continuous EI ion source, built in our laboratory, has been used to produce ions from neutral gas samples. It consists of an ionization chamber with a gas inlet capillary, tungsten filaments, and a pusher electrode that deflects ions towards the ion optics. The variable ionization potential values used for the experiment are within the range of 100–150 V. A gas mixing station was used to prepare samples with a defined mixture ratio. The station is connected to a dry-scroll backing pump and a ceramic technology membrane capacitance pressure gauge (Pfeiffer Vacuum CMR 361, Asslar, Germany). The background pressure of the system, 0.8 mbar, limits the purity of the gas mixture, typically prepared at 1000 mbar. The used gauge ensures equal sensitivity to various gases, and the tolerance of the partial pressure of component gases is rated at 0.2% over the selected pressure measurement range. The gas mixture vessel is first filled with the less abundant gas to improve accuracy. The connecting piping is evacuated to a vacuum level better than is measurable with the gauge after each mixture vessel filling. The commercially available pure gas containers used to prepare the mixture have a purity grade several orders of magnitude higher than the resulting mixture.

Similar to the LAb-CosmOrbitrap (Briois et al., 2016, Selliez et al., 2019), ions are directly injected into the trap from the ion source. For this work, a custom electrostatic ion optics module controlled with high voltage (HV) power supplies (iseg Spezialelektronik GmbH, Germany) was used. The ion optics module is composed of in-line electrodes, as shown in Figure 2. The ion optics is used to transport ions extracted from the ion source, to form the focused bunch of ions, and to correct its pointing if it is needed for the optimal injection conditions. The set of ion lenses is directly connected to the Orbitrap mounting plate. The ion beam is focused into the Orbitrap input orifice and can be detected after passing through the Orbitrap cell using an electron multiplier (DeTech 2700, Detector Technologies, Inc., Palmer, AK, United States) placed behind the Orbitrap, opposite the ion optics. The timing parameters of the injection of an ion packet can be optimized using a pulsed ion optics system controlled by an HV switch module (CGC Instruments, Chemnitz,

Germany). In the current experiment, this option has been used as a transition mode between pulsed and continuous ion source modes.

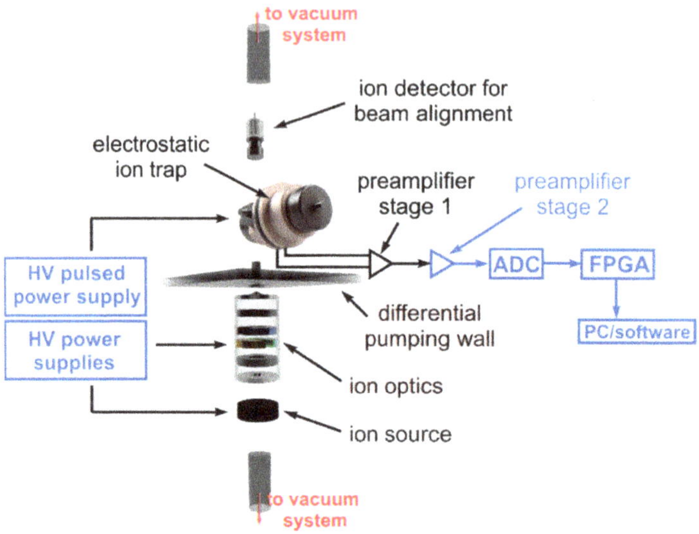

Figure 2. Functional illustration of OLYMPIA. Components inside (resp. outside) the vacuum chamber are indicated in grey (resp. in blue).

2.2. Mass Analyzer, Data Acquisition and Processing

The core part of the instrument is the D30-type commercial Orbitrap^{TM} cell (Figure 2). Similarly to commercial devices based on the Orbitrap cell, in OLYMPIA, after injection into the trap and electrodynamic squeezing [25], ions are dynamically confined inside the trap by the electrostatic field, spinning around the axis with an oscillating movement along the central electrode axis. The frequency of the cyclic motion of ions is defined by the parameters of the trap and the m/z values of the ions [23].

The axial oscillations of the trapped ions induce the current between the receiving electrodes of the Orbitrap cell. This current is detected by the preamplifier. The signal is then converted by an analog-to-digital converter (ADC) and further processed using Fourier transformation algorithms. Finally, the mass-to-charge ratio of ions is evaluated by the frequency of their oscillation [23].

A new customized high-gain differential preamplifier has been developed and constructed by JanasCard Company (Prague, Czech Republic) for the OLYMPIA setup. This preamplifier has been tested to be functional under vacuum conditions. The preamplifier output voltage is digitized by a 16-bit, 20 MHz, Picoscope® ADC unit (Pico Technology, Cambridge, UK).

The time domain data are pre-processed with doubling signal duration by zero-padding to increase the discrete resolution and, thus, the frequency accuracy. Then, a Hann function-shaped window apodization to minimize spectral leakage [26] is applied. The main data processing includes the application of fast Fourier transformation (FFT) using an FFTW implementation of the Cooley–Turkey algorithm [27] to convert the measured time domain signal into a frequency spectrum. The frequency spectrum is then converted to a mass spectrum using the theoretical model of Orbitrap-based mass analyzers [25,28]:

$$f = \frac{1}{2\pi}\sqrt{\frac{k}{m/q}} \tag{1}$$

where f is the frequency of ion axial motion, m is the mass of the ion, k is a calibration factor that characterizes the curvature of the electrostatic field, and q is the electrical charge of the ion. The calibration factor k is calculated for each dataset using the measured frequency and the known exact mass of the calibration ion.

3. Experimental Results

Laboratory measurements have been performed to benchmark the instrument's capabilities to resolve isobaric ions in several gas mixtures. Samples composed of molecular nitrogen (N_2), carbon monoxide (CO), and ethylene (C_2H_4) have been used to determine the resolving power of the mass analyzer at m/z 28. Quantitative measurement capability has been examined with CO and N_2 samples at different mixing ratios. Finally, a pure krypton sample was used to assess the quantification of isotopic abundances.

3.1. N_2/CO/C_2H_4 Gas Mixture

The mass spectra of CO, N_2, and C_2H_4 gases of the same nominal molecular mass (m/z 28), mixed in equal proportions, have been measured (Figure 3). A high resolution ($m/\Delta m > 3000$ at FWHM) is required to resolve CO^+ m/z 27.9949, $N_2{}^+$ m/z 28.0061, and C_2H_4 m/z 28.0313 ions in the spectrum and to measure their respective abundances.

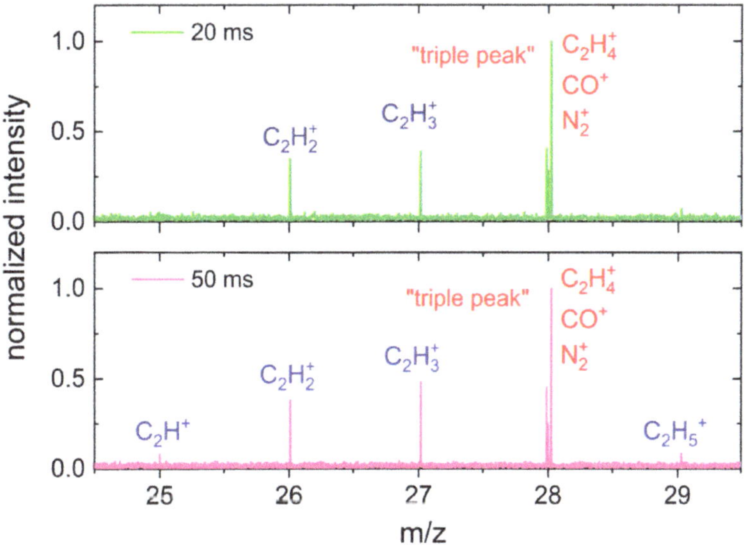

Figure 3. The EI mass spectrum of $N_2/CO/C_2H_4$ gas mixture with a 1:1:1 ratio. Ionization chamber pressure is estimated at 1 mbar. Accuracy of the measured peak mass values is 5–20 ppm compared to NIST atomic weights and isotopic composition data. Spectra are evaluated from the same time domain data measurement by adjusting its duration at the processing phase to remain fixed conditions of the experiment.

The different sampling times from 20 to 500 ms have been used to measure mass spectra for the estimation of the optimal value for the current configuration of OLYMPIA. The maximal reasonable signal duration is about 250 ms due to the relatively high pressure inside the ion trap chamber, approximately 10^{-8} mbar. Longer acquisition reduces the total amplitude of the effective signal and, thus, the signal-to-noise ratio and dynamic range.

The partial mass spectra of the $CO/N_2/C_2H_4$ mixture, are presented in Figure 3. The mass accuracy of the measured peaks varies from 5 to 20 ppm compared to NIST atomic weights and isotopic composition data. A zoom of the region around m/z 28 is shown in Figure 4. Three different peaks corresponding to CO^+, $N_2{}^+$, and $C_2H_4{}^+$ have been successfully resolved. A mass resolution of about 40,000 at m/z 28 has been achieved with a sampling time of less than 250 ms (Figure 4). In Figure 4, a combination of increasing resolution and variation in abundance accuracy is observed for a longer acquisition time.

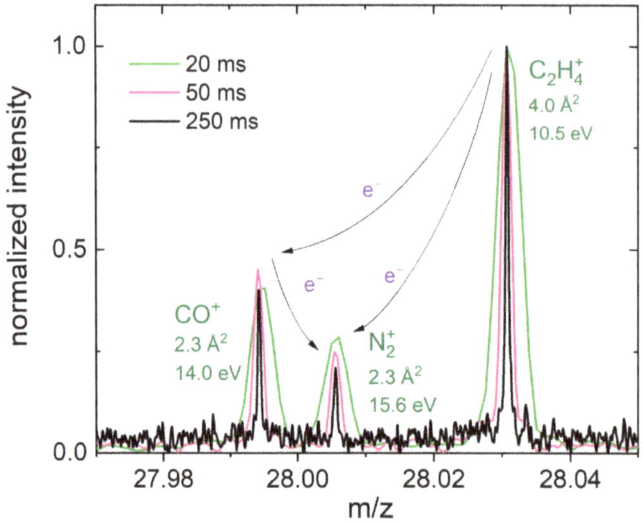

Figure 4. Mass spectra of the $N_2/CO/C_2H_4$ mixture measured with different FFT duration times of the same time domain measurement ($m/\Delta m \approx 6000$ at 20 ms, $m/\Delta m \approx 16{,}000$ at 50 ms, $m/\Delta m \approx 47{,}000$ at 250 ms). The ionization cross-section of neutrals and the ionization energy are given for the three species. Direction of the charge transfer reaction is indicated using the arrows.

The relative abundance of ions, which can be determined from the measured spectrum as peak amplitudes, does not only depend on the initial partial concentration of gases in the mixture. The ionization cross-section (1) and further chemical processes inside the ion source, such as charge transfer (2) and fragmentation (3), have a strong impact on the quantitative composition of the ion bunch injected into the trap.

(1) According to the binary encounter Bethe (BEB) model [29], the electron ionization cross-section, and thus the ionization probability of the C_2H_4 molecules (4.0 Å^2), is almost twice that of the CO and N_2 molecules (2.3 Å^2) (NIST Standard Reference Database 107 data).

(2) After the ionization at low vacuum conditions of the ion source (the pressure is about 1 mbar inside the ionization chamber), charge transfer reactions occur. Cations with lower ionization energies of precursor neutrals (CO—14.0 eV, C_2H_4—10.5 eV, N_2—15.58 eV) [30–32] are produced more effectively.

(3) Ions $C_2H_4{}^+$, their fragments C_2H^+, $C_2H_2{}^+$, or $C_2H_3{}^+$, and secondary reaction products $C_2H_5{}^+$ can be produced from C_2H_4 molecules (Figure 4). Fragmentation of neutrals and reactions of $C_2H_4{}^+$ ions reduce the relative $C_2H_4{}^+$ amplitude.

The combination of these three factors results in a higher abundance of $C_2H_4{}^+$ and a lower abundance of CO^+ and $N_2{}^+$ ions at equal concentrations of neutrals (Figure 5). The exact current of ions extracted from the source and its ratio is not constant and depend on the parameters of the setup, such as the pressure in the source and the accelerating potential of ionizing electrons. Therefore, calibration measurements at relevant conditions are required for quantitative in situ characterization of the sampled gas mixtures.

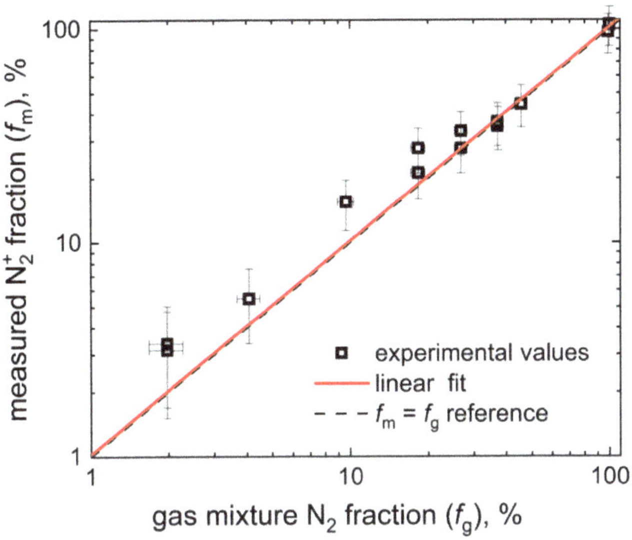

Figure 5. Measured fraction of N_2^+ ions as a function of nitrogen gas ratio in the known mixture. The fit function is calculated for the linear scale data using the least squares method.

3.2. CO/N_2 Gas Mixture

CO and N_2 gases of similar ionization cross-sections mixed in different ratios have been used to examine the quantitative measurement capability of OLYMPIA. These molecules have the same ionization cross-section and so does the probability of the production of ions by electron–neutral interactions, according to the BEB model. Nevertheless, they have different ionization energies, which affect further chemical reactions and the ratio of ions extracted from the source.

Several mixtures with CO fractions ranging from 2% up to 98% in the N_2 gas have been used (Figure 5). The relative intensity of CO^+ in N_2^+ ions was measured with a mass resolution sufficient to identify ions of the same molecular mass ($m/\Delta m \approx 5000$). A trace amount of carbon monoxide gas was found to be detectable in mixtures with a relative abundance of less than 2%.

Measurement of the CO/N_2 mixture showed an estimation of species abundance over two orders of magnitude of the CO concentration. At low N_2 concentrations, the measured N_2^+ to CO^+ ion ratio is slightly higher (but within the range of the statistical error) than the neutral gas ratio in the prepared mixture. The higher number of detected CO^+ ions can also be explained by the higher resulting efficiency of ion production due to the charge transfer processes, compared to Figure 4. The measurable amounts of C^+, O^+, and N^+ fragment ions were not observed.

3.3. Krypton Gas

The sensitivity of the instrument has been verified using a polyisotopic monoatomic gas to improve the amplitude measurement accuracy. The use of one monoatomic gas eliminates the effects of further chemical reactions of formed ions and fragmentation of neutrals. Meanwhile, the isotopic composition of krypton allows to estimate the accuracy of the quantitative characterization of the sample over a broad range of ratios. Krypton gas with 99.998% purity and natural isotopic abundance was selected as a calibrant as it has a well-documented isotopic ratio. It has also been detected in the coma of the 67P/Churyumov–Gerasimenko comet [14], so the study of the possibility of accurate measurement of its isotopic composition is of planetary science interest. A krypton gas container could also be proposed to be used as an onboard calibrant for future space missions that would use an Orbitrap-based spectrometer.

Spectra were measured with a 40 ms acquisition time (Figure 6). All six main isotopes of Krypton, even $^{78}Kr^+$, which is only 0.355% abundant, were successfully detected. The statistical error of the ^{82}Kr, ^{83}Kr, ^{84}Kr, and ^{86}Kr peak amplitude measurements was 9–11%.

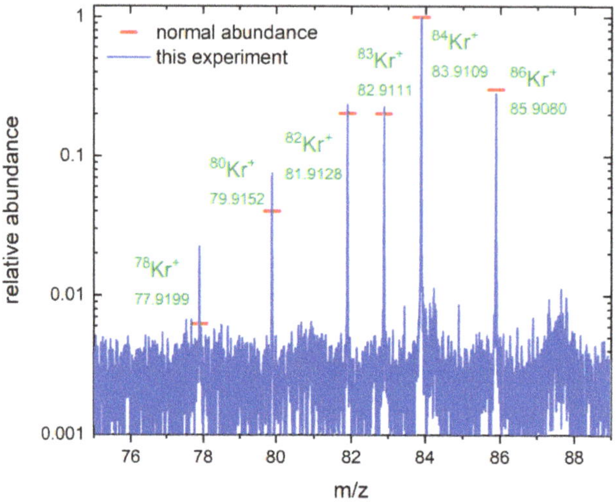

Figure 6. Mass spectra of the krypton gas calibrated and normalized to $^{84}Kr^+$ ion compared to the NIST isotopic composition database (red marks). Experimental m/z values are indicated near labels of isotopes by green numbers (better than 20 ppm mass accuracy).

A higher-than-expected amplitude has been observed in the OLYMPIA data for the least abundant ions of the krypton gas $^{78}Kr^+$ and $^{80}Kr^+$ (Figure 5) and for the CO/N_2 mixture (Figure 6). The measured difference from the normal isotopic distribution for less abundant ions is statistically significant. It indicates an accuracy limit for the measurement of trace amounts of components in the studied samples for the current configuration of the mass analyzer.

4. Outlook for Limited-Resources Space Missions

OLYMPIA is one of the laboratory concepts for future compact Orbitrap-based HRMS instruments for space missions. It specifically targets low-resource missions, including CubeSat-class ones, e.g., the currently developed Space Laboratory for Advanced Variable Instruments and Applications (SLAVIA) space mission dimensioned to 12U size and less than 20 kg mass format. The constraints imposed by the space environment (miniaturization, pressure, low power consumption, limited data transfer, especially for outer solar system bodies) have been considered during the development of this prototype.

For the OLYMPIA setup, the choice was made to not include a C-trap module present in commercial Orbitrap in order to reduce the overall complexity of the space instrument. The feasibility of this configuration had already been proven for pulsed ion sources using the LAb-CosmOrbitrap [22]. In this work, it has been demonstrated that the proposed configuration is compatible with a continuous EI source while ensuring the sufficient performance expected for HRMS.

A mass resolution of 50,000 is desirable to obtain a new level of scientific output and state-of-the-art knowledge for the extraterrestrial medium composition. Selection of the optimal sampling duration for specific tasks of the mission, and target mass resolution, is an important phase of the instrument design. Effective sampling duration is defined by (a) the time stability of power supply designed for space applications (e.g., compact HV source for CubeSats), (b) the pressure in the vicinity of the spacecraft (e.g., in low Earth orbit), and (c) the computing resources required to apply the FFT on board.

Finally, for most space missions, and in particular for low-resource CubeSats, data transfer is highly limited by the transmission capability of the spacecraft. In order to meet the challenges resulting from the proposed space mission application, a trade-off on the optimal signal duration is required to ensure sufficient mass resolution, ion abundance accuracy, and data handling. While the short signal duration decreases the mass resolution, it improves the signal-to-noise ratio and limits computations and data transfer requirements (Figure 4).

5. Conclusions

A new Orbitrap-based HRMS laboratory instrument has been built and characterized. The first operation of the Orbitrap-cell mass analyzer without a C-trap module with the continuous EI ion source has been demonstrated. Its performance has been evaluated by the analysis of gas mixtures composed of chemical species of the same nominal molecular mass (CO, N_2, and C_2H_4). The achieved mass resolution for $^{14}N_2^+$ is more than 40,000 for a 250 ms FFT duration. The experimental resolution value, limited by the pressure inside the Orbitrap cell, is close to the value of about 50,000 (for at least 2–3 points per peak) defined by the FFT theory for this acquisition time and allows to resolve all three chemical species without any additional analytical techniques. With a mixture of CO and N_2 at different mixing ratios and a pure krypton gas sample, the ability of OLYMPIA to measure ion abundances over two orders of magnitude has been assessed with an accuracy of 5–50% from the most abundant to the least abundant fractions. Despite a higher-than-expected measured intensity for less abundant ions (oversensitivity), the error range is less than 10% for the most abundant ions. The dynamic range of the instrument allows the detection of chemical species with a volume ratio of less than 1%. Indeed, the krypton isotope (^{78}Kr) with an abundance of less than 0.4% is clearly detectable in the spectrum with a signal-to-noise ratio of better than 250 for the major isotope.

The performed experiment demonstrates the technical feasibility of space applications of the OLYMPIA setup using a configuration of Orbitrap-based mass spectrometers without a C-trap module (see Outlook for limited-resource space missions). It can be used with continuous or pulsed ion sources, even for applications with high mass resolution requirements. This instrument configuration will reduce mechanical, high-voltage control, and timing module complexity while providing reasonable performances and a mass resolution exceeding that of previous space missions by at least one order of magnitude. The experiment also facilitates the analysis of ions abundant in atmospheres; however, the sensitivity and signal-to-noise ratio of the system should be improved.

With the current mass analyzer configuration, OLYMPIA proves to be an efficient tool, with certain trade-offs related to simplified system design, to study complex gas mixtures relevant to the atmospheres of the Solar System bodies (e.g., Titan), including the Earth, and can be used as a research and development workbench for future space missions.

Author Contributions: I.Z.—design of the experiment, optimization and construction of the experimental setup, data analysis, editing of the manuscript; J.Ž.—design and construction of the experimental setup, data analysis, editing of the manuscript; M.P.—experiment design, data analysis, editing of the manuscript; A.S.—laboratory work, editing of the manuscript; J.-P.L.—data analysis, editing of the manuscript; B.G.—construction of the experimental setup; B.C.—laboratory work; A.Z.—editing of the manuscript; C.B.—group leader, project management, editing of the manuscript. All authors have read and agreed to the published version of the manuscript.

Funding: We acknowledge support from LE STUDIUM Smart Loire Valley Programme/European Union Horizon 2020 research and innovation program under the Marie Sklodowska Curie grant agreement # 665790 and the CNES Research and Technology program. This work was also supported by the Czech Science Foundation (grant No. 21-11931J).

Data Availability Statement: The data presented in this study are available on request from the corresponding author. The data are not publicly available due to use of sensitive information provided by Thermo Fisher Scientific, Bremen, Germany.

Acknowledgments: Authors of the paper gratefully acknowledge Alexander Makarov for his kind support, starting from helping with the conceptual design of a new instrument to the final data analysis. This support motivated the research team to try new methods and approaches.

Conflicts of Interest: We have no conflict of interest to disclose.

References

1. Lebreton, J.-P.; Coustenis, A.; Lunine, J.; Raulin, F.; Owen, T.; Strobel, D. Results from the Huygens probe on Titan. *Astron. Astrophys. Rev.* **2009**, *17*, 149–179. [CrossRef]
2. Woodson, A.K.; Smith, H.T.; Crary, F.J.; Johnson, R.E. Ion composition in Titan's exosphere via the Cassini Plasma Spectrometer I: T40 encounter. *J. Geophys. Res. Space Phys.* **2015**, *120*, 212–234. [CrossRef]
3. Tokar, R.L.; Johnson, R.E.; Thomsen, M.F.; Wilson, R.J.; Young, D.T.; Crary, F.J.; Coates, A.J.; Jones, G.H.; Paty, C.S. Cassini detection of Enceladus' cold water-group plume ionosphere. *Geophys. Res. Lett.* **2009**, *36*, L13203. [CrossRef]
4. Waite, J.H., Jr.; Niemann, H.; Yelle, R.V.; Kasprzak, W.T.; Cravens, T.E.; Luhmann, J.G.; McNutt, R.L.; Ip, W.H.; Gell, D.; De La Haye, V.; et al. Ion Neutral Mass Spectrometer Results from the First Flyby of Titan. *Science* **2005**, *308*, 982–986. [CrossRef]
5. Taylor, M.G.G.T.; Altobelli, N.; Buratti, B.J.; Choukroun, M. The Rosetta mission orbiter science overview: The comet phase. *Philos. Trans. R. Soc. A Math. Phys. Eng. Sci.* **2017**, *375*, 20160262. [CrossRef]
6. Grady, M.; Wright, I.; Engrand, C.; Siljeström, S. The Rosetta Mission and the Chemistry of Organic Species in Comet 67P/Churyumov–Gerasimenko. *Elements* **2018**, *14*, 95–100. [CrossRef]
7. Kissel, J.; Altwegg, K.; Clark, B.C.; Colangeli, L.; Cottin, H.; Czempiel, S.; Eibl, J.; Engrand, C.; Fehringer, H.M.; Feuerbacher, B.; et al. Cosima—High Resolution Time-of-Flight Secondary Ion Mass Spectrometer for the Analysis of Cometary Dust Particles onboard Rosetta. *Space Sci. Rev.* **2007**, *128*, 823–867. [CrossRef]
8. Niemann, H.B.; Atreya, S.K.; Demick, J.E.; Gautier, D.; Haberman, J.A.; Harpold, D.N.; Kasprzak, W.T.; Lunine, J.I.; Owen, T.C.; Raulin, F. Composition of Titan's lower atmosphere and simple surface volatiles as measured by the Cassini-Huygens probe gas chromatograph mass spectrometer experiment. *J. Geophys. Res. Planets* **2010**, *115*, E12006. [CrossRef]
9. Israël, G.; Szopa, C.; Raulin, F.; Cabane, M.; Niemann, H.B.; Atreya, S.K.; Bauer, S.J.; Brun, J.-F.; Chassefière, E.; Coll, P.; et al. Complex organic matter in Titan's atmospheric aerosols from in situ pyrolysis and analysis. *Nature* **2005**, *438*, 796. [CrossRef]
10. Niemann, H.B.; Atreya, S.K.; Bauer, S.J.; Carignan, G.R.; Demick, J.E.; Frost, R.L.; Gautier, D.; Haberman, J.A.; Harpold, D.N.; Hunten, D.M.; et al. The abundances of constituents of Titan's atmosphere from the GCMS instrument on the Huygens probe. *Nature* **2005**, *438*, 779–784. [CrossRef]
11. Balsiger, H.; Altwegg, K.; Bochsler, P.; Eberhardt, P.; Fischer, J.; Graf, S.; Jäckel, A.; Kopp, E.; Langer, U.; Mildner, M.; et al. Rosina—Rosetta Orbiter Spectrometer for Ion and Neutral Analysis. *Space Sci. Rev.* **2007**, *128*, 745–801. [CrossRef]
12. Gasc, S.; Altwegg, K.; Balsiger, H.; Berthelier, J.-J.; Bieler, A.; Calmonte, U.; Fiethe, B.; Fuselier, S.; Galli, A.; Gombosi, T.; et al. Change of outgassing pattern of 67P/Churyumov–Gerasimenko during the March 2016 equinox as seen by ROSINA. *Mon. Not. R. Astron. Soc.* **2017**, *469*, S108–S117. [CrossRef]
13. Fuselier, S.A.; Altwegg, K.; Balsiger, H.; Berthelier, J.J.; Bieler, A.; Briois, C.; Broiles, T.W.; Burch, J.L.; Calmonte, U.; Cessateur, G.; et al. ROSINA/DFMS and IES observations of 67P: Ion-neutral chemistry in the coma of a weakly outgassing comet. *A&A* **2015**, *583*, A2.
14. Rubin, M.; Altwegg, K.; Balsiger, H.; Bar-Nun, A.; Berthelier, J.-J.; Briois, C.; Calmonte, U.; Combi, M.; De Keyser, J.; Fiethe, B.; et al. Krypton isotopes and noble gas abundances in the coma of comet 67P/Churyumov-Gerasimenko. *Sci. Adv.* **2018**, *4*, eaar6297. [CrossRef]
15. De Keyser, J.; Dhooghe, F.; Gibbons, A.; Altwegg, K.; Balsiger, H.; Berthelier, J.-J.; Briois, C.; Calmonte, U.; Cessateur, G.; Equeter, E.; et al. Correcting peak deformation in Rosetta's ROSINA/DFMS mass spectrometer. *Int. J. Mass Spectrom.* **2015**, *393*, 41–51. [CrossRef]
16. Vuitton, V.; Lavvas, P.; Yelle, R.V.; Galand, M.; Wellbrock, A.; Lewis, G.R.; Coates, A.J.; Wahlund, J.-E. Negative ion chemistry in Titan's upper atmosphere. *Planet. Space Sci.* **2009**, *57*, 1558–1572. [CrossRef]
17. Schuhmann, M.; Altwegg, K.; Balsiger, H.; Berthelier, J.-J.; De Keyser, J.; Fuselier, S.A.; Gasc, S.; Gombosi, T.I.; Hänni, N.; Rubin, M.; et al. CHO-Bearing Molecules in Comet 67P/Churyumov-Gerasimenko. *ACS Earth Space Chem.* **2019**, *3*, 1854–1861. [CrossRef]
18. Miller, J.M. Chromatography. *Digit. Encycl. Appl. Phys.* **2009**, 1055–1102. [CrossRef]
19. Goodacre, R.; Shann, B.; Gilbert, R.J.; Timmins, É.M.; McGovern, A.C.; Alsberg, B.K.; Kell, D.B.; Logan, N.A. Detection of the Dipicolinic Acid Biomarker in Bacillus Spores Using Curie-Point Pyrolysis Mass Spectrometry and Fourier Transform Infrared Spectroscopy. *Anal. Chem.* **2000**, *72*, 119–127. [CrossRef]
20. Chai, W.; Piskarev, V.; Lawson, A.M. Negative-Ion Electrospray Mass Spectrometry of Neutral Underivatized Oligosaccharides. *Anal. Chem.* **2001**, *73*, 651–657. [CrossRef]

21. Arevalo, R., Jr.; Selliez, L.; Briois, C.; Carrasco, N.; Thirkell, L.; Cherville, B.; Colin, F.; Gaubicher, B.; Farcy, B.; Li, X.; et al. An Orbitrap-based laser desorption/ablation mass spectrometer designed for spaceflight. *Rapid Commun. Mass Spectrom.* **2018**, *32*, 1875–1886. [CrossRef] [PubMed]
22. Briois, C.; Thissen, R.; Thirkell, L.; Aradj, K.; Bouabdellah, A.; Boukrara, A.; Carrasco, N.; Chalumeau, G.; Chapelon, O.; Colin, F.; et al. Orbitrap mass analyser for in situ characterisation of planetary environments: Performance evaluation of a laboratory prototype. *Planet. Space Sci.* **2016**, *131*, 33–45. [CrossRef]
23. Makarov, A. Electrostatic Axially Harmonic Orbital Trapping: A High-Performance Technique of Mass Analysis. *Anal. Chem.* **2000**, *72*, 1156–1162. [CrossRef] [PubMed]
24. Makarov, A.; Denisov, E.; Lange, O.; Horning, S. Dynamic range of mass accuracy in LTQ orbitrap hybrid mass spectrometer. *J. Am. Soc. Mass Spectrom.* **2006**, *17*, 977–982. [CrossRef]
25. Hu, Q.; Noll, R.J.; Li, H.; Makarov, A.; Hardman, M.; Graham Cooks, R. The Orbitrap: A new mass spectrometer. *J. Mass Spectrom.* **2005**, *40*, 430–443. [CrossRef]
26. Harris, F.J. On the use of windows for harmonic analysis with the discrete Fourier transform. *Proc. IEEE* **1978**, *66*, 51–83. [CrossRef]
27. Frigo, M.; Johnson, S.G. FFTW: An adaptive software architecture for the FFT. In Proceedings of the 1998 IEEE International Conference on Acoustics, Speech and Signal Processing, ICASSP '98 (Cat. No.98CH36181), Seattle, WA, USA, 15 May 1998; Volume 3, pp. 1381–1384.
28. Scigelova, M.; Hornshaw, M.; Giannakopulos, A.; Makarov, A. Fourier Transform Mass Spectrometry. *Mol. Cell. Proteom.* **2011**, *10*, M111.009431. [CrossRef]
29. Hwang, W.; Kim, Y.-K.; Rudd, M.E. New model for electron-impact ionization cross sections of molecules. *J. Chem. Phys.* **1996**, *104*, 2956–2966. [CrossRef]
30. Erman, P.; Karawajczyk, A.; Rachlew-Källne, E.; Strömholm, C.; Larsson, J.; Persson, A.; Zerne, R. Direct determination of the ionization potential of CO by resonantly enhanced multiphoton ionization mass spectroscopy. *Chem. Phys. Lett.* **1993**, *215*, 173–178. [CrossRef]
31. Trickl, T.; Cromwell, E.F.; Lee, Y.T.; Kung, A.H. State-selective ionization of nitrogen in the X 2Σ+gv+ = 0 and v+ = 1 states by two-color (1 + 1) photon excitation near threshold. *J. Chem. Phys.* **1989**, *91*, 6006–6012. [CrossRef]
32. Ohno, K.; Okamura, K.; Yamakado, H.; Hoshino, S.; Takami, T.; Yamauchi, M. Penning Ionization of HCHO, CH2CH2, and CH_2CHCHO by Collision with He(23S) Metastable Atoms. *J. Phys. Chem.* **1995**, *99*, 14247–14253. [CrossRef]

aerospace

MDPI

Article

Data Downloaded via Parachute from a NASA Super-Pressure Balloon

Ellen L. Sirks [1,2,*,†], Richard Massey [1,*,†], Ajay S. Gill [3], Jason Anderson [1], Steven J. Benton [4], Anthony M. Brown [1], Paul Clark [1], Joshua English [1], Spencer W. Everett [5], Aurelien A. Fraisse [4], Hugo Franco [6], John W. Hartley [7], David Harvey [8], Bradley Holder [9], Andrew Hunter [1], Eric M. Huff [5], Andrew Hynous [10], Mathilde Jauzac [1], William C. Jones [4], Nikky Joyce [1], Duncan Kennedy [1], David Lagattuta [1], Jason S.-Y. Leung [3], Lun Li [4,7], Stephen Lishman [1], Thuy Vy T. Luu [4], Jacqueline E. McCleary [11], Johanna M. Nagy [12], C. Barth Netterfield [3,13], Emaad Paracha [13], Robert Purcaru [14], Susan F. Redmond [4], Jason D. Rhodes [5], Andrew Robertson [5], L. Javier Romualdez [7], Sarah Roth [10], Robert Salter [6], Jürgen Schmoll [1], Mohamed M. Shaaban [13,15], Roger Smith [16], Russell Smith [5], Sut Ieng Tam [17] and Georgios N. Vassilakis [11]

1 Department of Physics, Durham University, South Road, Durham DH1 3LE, UK; jason.anderson@durham.ac.uk (J.A.); anthony.brown@durham.ac.uk (A.M.B.); andrew.hunter@durham.ac.uk (A.H.); mathilde.jauzac@durham.ac.uk (M.J.); nikky.s.joyce@durham.ac.uk (N.J.); david.j.lagattuta@durham.ac.uk (D.L.); jurgen.schmoll@durham.ac.uk (J.S.)
2 School of Physics, The University of Sydney and ARC Centre of Excellence for Dark Matter Particle Physics, Sydney, NSW 2006, Australia
3 Dunlap Institute for Astronomy and Astrophysics, University of Toronto, 50 St. George Street, Toronto, ON M5S 3H4, Canada; ajay.gill@mail.utoronto.ca (A.S.G.); leung@astro.utoronto.ca (J.S.-Y.L.); barth.netterfield@utoronto.ca (C.B.N.)
4 Department of Physics, Princeton University, Jadwin Hall, Princeton, NJ 08544, USA; sbenton@princeton.edu (S.J.B.); afraisse@princeton.edu (A.A.F.); wcjones@princeton.edu (W.C.J.); lunl@princeton.edu (L.L.); tluu@princeton.edu (T.V.T.L.); sredmond@princeton.edu (S.F.R.)
5 Jet Propulsion Laboratory, California Institute of Technology, 4800 Oak Grove Drive, Pasadena, CA 91011, USA; spencer.w.everett@jpl.nasa.gov (S.W.E.); eric.m.huff@jpl.nasa.gov (E.M.H.); jason.d.rhodes@jpl.nasa.gov (J.D.R.); andrew.a.robertson@jpl.nasa.gov (A.R.)
6 Columbia Scientific Balloon Facility, 1510 Farm Road 3224, Palestine, TX 75803, USA; hugo.franco@nasa.gov (H.F.); robert.g.salter@nasa.gov (R.S.)
7 StarSpec Technologies Inc., Unit C-5, 1600 Industrial Road, Cambridge, ON N3H 4W5, Canada; john@starspectech.com (J.W.H.); javier@starspectech.com (L.J.R.)
8 Laboratoire d'Astrophysique, EPFL, Observatoire de Sauverny, 1290 Versoix, Switzerland; david.harvey@epfl.ch
9 University of Toronto Institute for Aerospace Studies (UTIAS), 4925 Dufferin Street, Toronto, ON M3H 5T6, Canada; bradley.holder@mail.utoronto.ca
10 NASA Wallops Flight Facility, Balloon Program Office, 34200 Fulton St., Wallops Island, VA 23337, USA; andrew.hynous@nasa.gov (A.H.); sarah.roth@nasa.gov (S.R.)
11 Department of Physics, Northeastern University, 360 Huntington Ave., Boston, MA 02115, USA
12 Department of Physics, Washington University in St. Louis, 1 Brookings Drive, St. Louis, MO 63130, USA; nagy@case.edu
13 Department of Physics, University of Toronto, 60 St. George Street, Toronto, ON M5S 1A7, Canada; emaad.paracha@mail.utoronto.ca (E.P.); m.shaaban@mail.utoronto.ca (M.M.S.)
14 Division of Engineering Science, 40 St. George Street, Room 2110, Toronto, ON M5S 2E4, Canada; robert.purcaru@mail.utoronto.ca
15 Palantir Technologies, 1875 Lawrence St., Denver, CO 80202, USA
16 California Institute of Technology, 1200 E. California Blvd., Pasadena, CA 91125, USA; rsmith@astro.caltech.edu
17 Academia Sinica Institute of Astronomy & Astrophysics, No. 1, Sec. 4, Roosevelt Road, Taipei 10617, Taiwan; sutieng8110@gmail.com
* Correspondence: ellen.sirks@sydney.edu.au (E.L.S.); r.j.massey@durham.ac.uk (R.M.)
† These authors contributed equally to this work.

Abstract: In April 2023, the SUPERBIT telescope was lifted to the Earth's stratosphere by a helium-filled super-pressure balloon to acquire astronomical imaging from above (99.5% of) the Earth's atmosphere. It was launched from New Zealand and then, for 40 days, circumnavigated the globe five times at a latitude 40 to 50 degrees south. Attached to the telescope were four "DRS" (Data Recovery System) capsules containing 5 TB solid state data storage, plus a GNSS receiver, Iridium transmitter, and parachute. Data from the telescope were copied to these, and two were dropped over Argentina. They drifted 61 km horizontally while they descended 32 km, but we predicted their descent vectors within 2.4 km: in this location, the discrepancy appears irreducible below ~2 km because of high speed, gusty winds and local topography. The capsules then reported their own locations within a few metres. We recovered the capsules and successfully retrieved all of SUPERBIT's data despite the telescope itself being later destroyed on landing.

Keywords: balloon instrumentation; data handling; data compression; models and simulations; large detector-systems performance

1. Introduction

The Super-pressure Balloon-borne Imaging Telescope [1] (SUPERBIT) was raised to ~33 km altitude by NASA's 18.8 million cubic foot super-pressure balloon 728NT (https://www.csbf.nasa.gov/map/balloon10/flight728NT.htm). It was launched on 16 April 2023 from Wanaka, New Zealand, was carried east by seasonal polar vortex winds [2], circumnavigated the Earth ~5.5 times, then landed in Argentina on 25 May 2023. During this flight, the telescope acquired deep optical and near-UV imaging of galaxy clusters and other astronomical objects [3]. Early in the mission, data were transmitted to ground stations via Starlink and TDRSS satellite communication links. However, the Starlink connection was lost on 1 May 2023, and TDRSS became unstable on 24 May 2023. The root cause for each failure remains under investigation. All of the data was stored on board, as well as on redundant sets of solid-state drives within drop packages. Several copies of the full dataset were released via parachute prior to the termination of the flight and were safely recovered. The SUPERBIT payload was entirely destroyed upon landing, and although, in this case, the data were eventually recovered from the remains of the onboard drives, the use of parachutes to ensure the safe recovery of the data proved invaluable.

Four Data Recovery System (DRS) capsules were attached to the telescope before launch. As well as a parachute, each included 5 TB solid state data storage, a Global Navigation Satellite System (GNSS) receiver so it could work out its location, and an Iridium Short-Burst Data (SBD) transceiver, so it could report its location to the recovery team [4] (and receive commands to, e.g., turn on or off a beeper). Two DRSs were released, and both were successfully recovered with data intact.

In this article, we describe upgrades (since the first test of a DRS during a SUPERBIT engineering flight in 2019) to their release mechanism, communications with the telescope, and thermal control. We describe the performance of that hardware during operations and the performance of software that we use to predict their decent trajectory. Accurate predictions of the descent vector are essential in choosing a safe landing zone. Neither the trajectory of the main telescope payload nor the parachute is actively controlled, but the timing of the release can be chosen so that the DRS lands in an area that is unpopulated for safety but near a road to facilitate recovery.

Our hardware design (https://github.com/PaulZC/Data_Recovery_System) and software (https://github.com/EllenSirks/pyBalloon) are both open-source. Their success inspired NASA's 'FLOATing DRAGON' Balloon Challenge (https://floatingdragon.nianet.org/). Following this programme, a similar capability to download data or small physical samples will be made available by NASA to all future super-pressure balloon teams to insure against the loss of communications, loss of payload, or destruction of payload upon landing. Our experience suggests that teams should take the offer.

2. Upgraded Hardware Design

Each DRS capsule comprised a Raspberry Pi 3B attached to a custom 300 mm × 125 mm printed circuit board (PCB; see Figure 1). The capsules were powered by switchable 24V DC from the main payload while they were attached, then by two internal 9V lithium batteries after release. These batteries were the only components of the DRS that could be considered potentially hazardous. However, they were compliant with safety test criteria T1–T8 defined in Section 38.3 of [5]. The DRS used a geared pincer mechanism to hold on to a loop of nylon cable tie underneath the main payload. After coordination with local air traffic control, we issued a sequence of three commands (to prevent errors and fault propagation) that used a servo inside the DRS to open the pincer, allowing the DRS to drop away.

Each DRS was enclosed in a 3D-printed CPE hard plastic case and a closed-cell foam shell (see Figure 2). One case was printed in PLA, which also works fine but is harder to handle because it is brittle. The foam shell could absorb impacts and add a degree of waterproofing, which proved useful when one DRS landed in snow. It was manufactured by pouring liquid PU expanding foam mix into a pill-shaped mould a few millimetres less than 6″ in diameter, lined with a chicken roasting bag to create a waterproof outer surface on which contact details can be written [6]. The total mass including the parachute was 1.28 kg.

Figure 1. The DRS is based around this custom PCB with a Raspberry Pi in the middle. An Ethernet cable is plugged into the bottom right of the Pi and attaches to a zero-extraction force connector at the top right corner of the PCB. Two SD card readers are plugged directly into the Pi, and two are attached to USB extension cables to reduce heat production at the sockets. Servo-operated pincer mechanisms can be seen on the right, with the pincer on the front holding on to the main gondola and the release at the back holding the parachute.

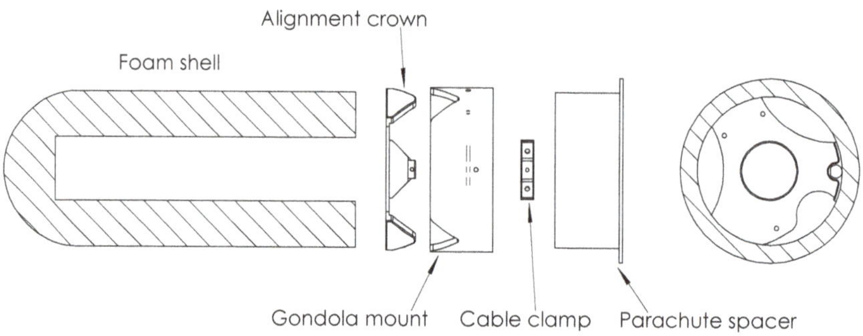

Figure 2. The DRS capsules hang underneath the payload (here, down is shown to the left). They are prevented from swinging or rotating by a 3D-printed 'alignment crown' fixed to their top, which fits inside a mount on the gondola that has an inverted shape.

2.1. Alignment and Attachment to the Main Payload

To prevent the DRS from swinging while hanging under the payload during the 2019 test flight, it was sheathed inside a section of 6″ plastic drainpipe. This worked flawlessly at altitude when two DRS capsules were successfully released. However, during subsequent testing in a thermal vacuum chamber, the foam sometimes expanded and became wedged inside the pipe to drop only when surrounding air pressure increased. To mitigate the risk of a DRS becoming wedged at altitude, we remove the pipe and attached a 3D-printed PETG 'crown' to the top of the DRS (see Figure 2). This fit into an inverted crown that was fixed to the main gondola. The crowns were kept together, and the DRS was held secure by pulling the cable tie taut.

A second pincer mechanism held a 4 foot parachute folded inside a spacer above the DRS. This size parachute yields a terminal velocity $\sim$4 m s^{-1}. This was fast enough to reduce the downrange travel distance (hence, uncertainty in the predicted landing spot, which allowed us to avoid populated areas) while keeping kinetic energy comfortably below a 15 J safety requirement to note cause serious injury or damage.

2.2. Hardwired Ethernet to the Main Payload

It was possible for the DRS to communicate with the main payload via the Raspberry Pi's built-in WiFi. This worked successfully throughout the 2019 test flight. However, the launch of that flight was delayed because of radio interference. After much investigation, it turned out that the interference had nothing to do with WiFi, but it took time to rule out that possibility and to then look for the real source. To militate against future risk of delays and to enable operation on payloads that require radio silence for scientific measurements [7] (e.g., SPIDER;), we use hardwired Ethernet for the 2023 science flight. Inside the DRS, this involved a standard Ethernet cable plugged into the Pi's Ethernet socket. Outside the DRS, two twisted cable pairs were routed to a switch on SUPERBIT, which acted as a DNS server.

The six wires (four for Ethernet and two for power) joined at zero-extraction force connectors (Smiths Interconnect Hypertac® connectors, part numbers D02PB906FSTAH and D02PB906MSTH) that would separate easily when the DRS was released. Two-pin versions of these connectors worked well during the test flight, but we found that versions with more pins did not sit well and occasionally became disconnected during testing. We, therefore, used grub screws to fix half of the solid connector inside the crown and half in the gondola mount so they would be held together until release.

2.3. Thermal Management

During the 2019 engineering flight, each DRS worked for only $\sim$15 min at a time before overheating and shutting down until it naturally cooled. Subsequent testing in a room temperature lab showed two main hot spots (see Figure 3). Assuming a thermal emissivity coefficient $e = 0.95$, the main CPU would reach 42.5 °C while executing a simple Python script (the Raspberry Pi itself was reporting a CPU temperature of 49.4 °C), and the USB sockets would reach 48.9 °C when a 5 GB file was written to an SD card in a small SD card reader. Achieving robust operation during the 2023 science flight required mitigations for both.

To cool the main CPU, we milled a 53 g aluminium heat sink/radiator that sits between the Raspberry Pi and the PCB. It was held against the CPU by a spring glued to the PCB, and contact was ensured with thermal paste. The aluminium block extended through a 56 mm × 14 mm hole in the plastic case, ending flush with that, and was covered with Mylar tape to increase its infrared emissivity. The heat was radiated to space through a hole in the foam shell that was cut by hand (using an indentation in the mould to align it). The ambient temperature in the stratosphere is −40 °C [8]. For what it is worth, we also stuck a thin, off-the-shelf copper radiator to the RAM chip on the back.

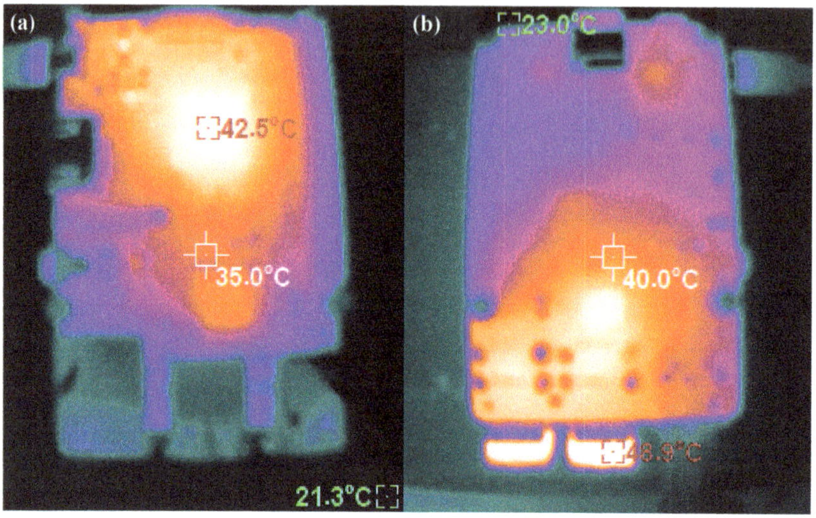

Figure 3. Thermal imaging of the top and bottom of a Raspberry Pi, during testing in a room temperature laboratory. Thermal emission is interpreted as temperature (red, white, green numbers), assuming emissivity $e = 0.95$. (**a**) The main CPU is the hottest component while executing a simple Python script. Nothing is attached to the USB sockets in this image. (**b**) The underside of a Pi attached to a DRS, oriented as in Figure 1. The heat sink and SD card containing the operating system are now cold at the top, but the USB sockets become hot during file transfer to SD cards in small readers mounted in the USB sockets.

To reduce heat production at the USB sockets, we used two 6″ USB extension cables to relocate two SD card readers (they ended up sitting roughly behind the microcontroller).

We also operated each DRS on a 5 min on–20 min off rota: powering up each in turn, copying data, then powering down. During a typical cycle during the night (day), the CPU temperature would increase by 3.0 ± 0.3 °C (3.7 ± 1.0 °C) in 5 min (see Figure 4). On one occasion during night 13, DRS1 was powered for two cycles, and its temperature rose 14.7 °C (from -11.6 °C to 3.1 °C) in 30 min. It continued operating throughout this time and cooled half-way back to the same temperature as the others in approximately 2 h. We do not know why DRS1 ran at -8.2 ± 8.6 °C (4.7 ± 11.2 °C) during the night (day), consistently a few degrees warmer than DRS2 at -11.4 ± 8.1 °C (1.0 ± 10.7 °C) and DRS5 at -11.9 ± 7.4 °C (-1.2 ± 10.6 °C).

Before launch, four DRS capsules were mounted underneath the rest of the payload (see Figure 5). They were oriented with their radiators towards the front of the telescope (which was never allowed to point towards the Sun) and enclosed by a loose skirt of Mylar, the inside of which was aluminised except for a $\sim$45° pleat near the radiator. On two sides, an additional sunshield of foam wrapped in aluminised Mylar also protected the DRSs from incidental knocks. DRS1 was at the back left, DRS2 at the back right, DRS3 at the front right, and DRS5 at the front left (DRS4 was our flight spare). Power and Ethernet cables were routed from each DRS up to SUPERBIT.

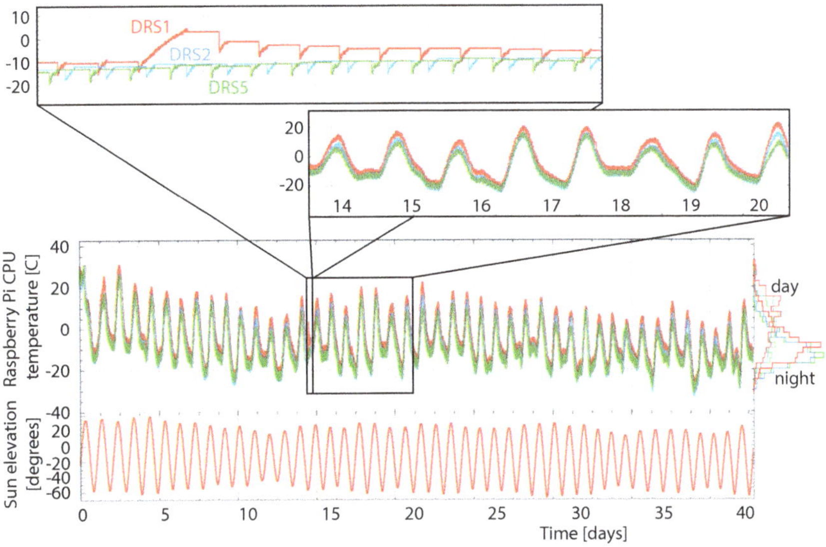

Figure 4. Self-reported CPU temperature of the Raspberry Pi on each DRS during the SUPERBIT mission for DRS1 (red), DRS2 (blue), and DRS5 (green). Temperature principally varied with the 45 diurnal cycles in 40 days (SUPERBIT crossed the international date line 5 times), the length and extent of which varied with SUPERBIT's speed and latitude. The histogram on the right splits times by positive or negative solar elevation. Insets show temperature variations with increased temporal resolution, including, for illustration, the one occasion when DRS1 was left powered on for 30 min. Temperature measurements were only available when a DRS was powered on, and horizontal lines merely show the last reported temperature.

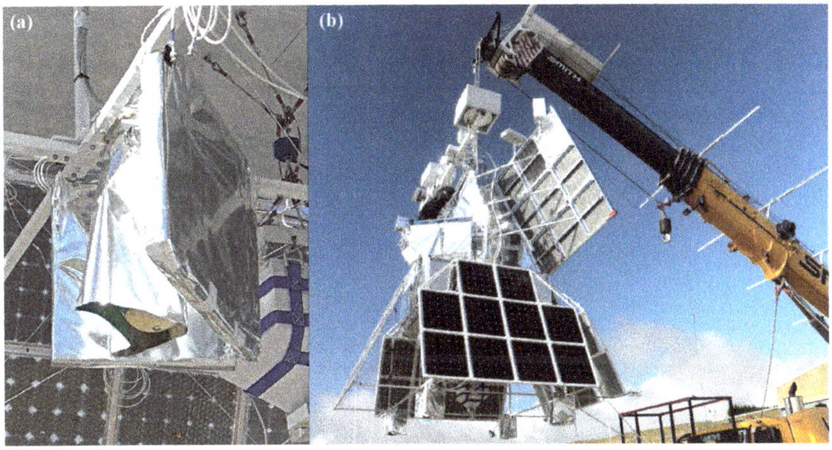

Figure 5. (**a**) Close-up of a DRS in flight configuration. The closed-cell foam shell surrounding the DRS can be seen poking out below a skirt of aluminised Mylar. It is further protected on two sides by sheets of foam insulation also covered by aluminised Mylar. Cables are routed upwards on the mounting frame. (**b**) SUPERBIT suspended on the launch crane. Four DRS capsules can be seen at the bottom, each attached to a corner of the frame holding the solar panels. The blue and white object hanging between them is a ballast hopper. Throughout its mission, the telescope keeps its back (on the right in these photos) oriented towards the Sun.

2.4. Data Storage

Each DRS was fitted with 5 × 1 TB MicroSDXC cards: one in the internal slot that includes the operating system and four in USB card readers. We set these up as independent drives (rather than RAID to further reduce thermal load) and copy data to one of the SD cards at a time, beginning with those mounted in a USB socket. An `rsync` script was not sufficient because the SD cards were independent. Instead, whenever the Pi booted, a custom script compared files on the payload with those on the DRS (matching name and size because some files, such as instrument timestreams, continually grow) and copied new data to an SD card with sufficient space. SUPERBIT's rate of data acquisition was sufficiently slow that we could have duplicated all the data to second SD cards on every DRS but, with so many backups already, we chose not to.

We tested MicroSDXC cards from several manufacturers before the flight. Resilience is more important than speed in this context, and we found considerable variation in cards' abilities to withstand a harsh environment. SanDisk Ultra (A1 UHS-I Class 10 U1) and SanDisk Extreme (A2 UHS-I Class 10 U3 V30) cards worked reliably throughout our tests and the flight. Two different Integral (A2 UHS-I Class 10 U3 V30) cards showed intermittent faults during testing: when used to house the operating system, the Pi would occasionally require power cycling to boot. After a few early problems, this was not reproducible, so we flew two DRSs with SanDisk Ultra SD cards and two with Integral SD cards. All our Lexar (LMSPLAY001T-BNNAG) cards failed during testing, so these were discarded.

2.5. One Failure on Launch

One DRS onboard SUPERBIT stopped responding immediately after launch. It is now impossible to determine why, but we suspect that the zero-extraction force connection became misaligned by the launch shock. It was one of the capsules using an Integral SD card that may have also caused problems, but we power cycled it many times without response. All three other DRS capsules worked reliably for the entire flight or until they were released.

3. Descent and Recovery

We initially intended to release one DRS capsule around mission day 40, 60, 80, and 100, whenever SUPERBIT passed over land, to mitigate the risk of the payload and all its onboard data being lost at sea. SUPERBIT did indeed pass over Chile and Argentina on mission day 40 (25 May 2023). However, in the early hours of the morning, it became apparent that the main payload would need to descend later that day because of reduced communications bandwidth and a weather forecast indicating a southerly trajectory away from further land crossings. After finishing the night's observations and coordinating with local air traffic control, we released (not one but) two DRS capsules over Santa Cruz Province, Argentina. This strategy ensured that independent, redundant sets of data would be available both on board and separated from the main payload.

3.1. Release Time and Location

We attempted to release two DRS capsules nearly simultaneously so they would land close to each other to simplify recovery. Low-latency communications with the payload had been lost, so we were forced to send release commands via Iridium SBD. We transmitted the two commands on independent channels. However, commands are queued and arrive after an unpredictable delay.

Contemporaneous logs recorded on SUPERBIT show that the release commands were received at 12:30:58 and 12:31:25 UT. Taking into account an intentional 30 s delay for the Pi to shut down gracefully, we infer that DRS2 and DRS1 were released at 12:31:28 and 12:31:55 UT, respectively.

The super-pressure balloon (SPB) host logged GNSS locations approximately once per minute. We interpolated between these to infer the location of SUPERBIT at the moments when DRS capsules were released (see Table 1). During the 27 s between release events, SUPERBIT travelled 1.79 km on heading 101°.

Table 1. Contemporaneous logs from onboard SUPERBIT around the time when two DRS capsules were released. Times are UT on 25 May 2023. SPB logs #1–4 recorded accurate GPS locations at known times, which we interpolated to the moments of release for DRS2 and DRS1.

Event	Time (UT)	Latitude [deg]	Longitude [deg]	Altitude [m]
SPB log #1	12:29:33	−47.685	−70.518	32,585
SPB log #2	12:30:37	−47.692	−70.475	32,488
Release of DRS2	12:31:28	−47.696	−70.444	32,439
SPB log #3	12:31:40	−47.697	−70.437	32,428
Release of DRS1	12:31:55	−47.699	−70.421	32,323
SPB log #4	12:32:44	−47.707	−70.367	31,981

3.2. Predicted Descent Trajectories

We predict the DRS descent trajectories and landing sites with `PyBalloon` software [6] that integrates a path followed by a test particle through winds described by Global Forecast System (GFS) weather models. When we did this live, we predicted descent vectors ~61.3 km on heading 107° and a flight time of ~32 min. However, the fast ground speed and high-latency communications meant that our uncertainty in the landing site was dominated by the unpredictable timing of release (2 min uncertainty extended our error ellipses by 8 km). We, therefore, aimed for a large region of open, uninhabited land west of main road 12. These specific circumstances are unusual and unlikely to be repeated.

Knowing with hindsight the precise times of release, we predicted landing sites for DRS1 at −47.8594°, −69.6307° (altitude 878 m), and DRS2 at −47.8557°, −69.6559° (altitude 920 m), with 68% confidence limits of $\sigma_{\parallel} = 2.1$ km in the main direction of motion and $\sigma_{\perp} = 1.9$ km perpendicular to that. With even more hindsight, we were able to interpolate between GFS model wind speeds issued before *and after* the DRS release. These lengthened the model descent vectors by ~1.1 km, to predicted landing sites for DRS1 at −47.8655°, −69.6153 (altitude 863 m) and DRS2 at −47.8600°, −69.6444° (altitude 850 m), with 68% confidence limits $\sigma_{\parallel} = 2.3$ km and $\sigma_{\perp} = 1.9$ km (see Figure 6). Even in the absence of any additional information, the open terrain would have made it possible (albeit time-consuming) to search for bright orange parachutes in an area this size.

3.3. Reported Landing Sites

Unfortunately, neither DRS reported or recorded its location during descent as they were supposed to (and as they did during the 2019 test flight). It is, therefore, impossible to reconstruct their true trajectories. DRS1 spontaneously started reporting back at 14:13:46 UT, after having reached the ground, and DRS2 started reporting back at 14:31:15 UT (intermittently at first and intermittently again overnight). Both DRS capsules reported lines of sight to many GNSS satellites. However, their initial reports included measurements of low battery voltages, and we later found that DRS2 was lying in a patch of snow. We infer that, although the temperature of the Raspberry Pis was successfully managed in space for 40 days (see Section 2.3), the adjacent batteries, which were used only after release, got very cold. We have previously used them in the stratosphere, but not for such extended periods.

Self-reported GNSS coordinates after landing placed DRS1 at −47.88167°, −69.59300° (altitude 786 m) and DRS2 at −47.86368°, −69.67452° (altitude 806 m), after averaging over several readings to provide ~2 m precision. These locations are 66.5 km and 58.3 km horizontally from their release points, respectively, 3.8 km and 1.6 km from the pre-release predicted locations or 2.4 km and 2.3 km from our best-estimate calculations that include weather data available afterwards. The true trajectories appear to have deviated approximately 950 m sideways from the predictions, with one landing 2.2 km earlier than expected and the other travelling 2.1 km farther, both within predicted precision.

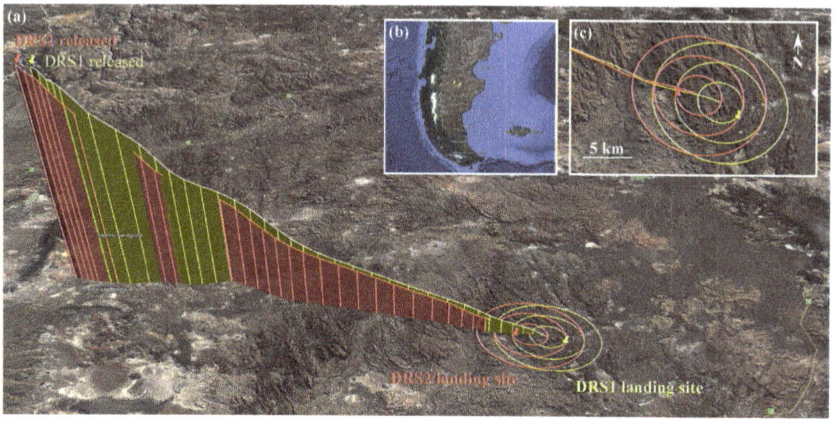

Figure 6. Descent trajectories of DRS1 (yellow) and DRS2 (red) capsules, which were released over southern Argentina on 25 May 2023. Pins mark the known release and landing positions; everything else was modelled using the pyBalloon software [6]. (**a**) View from altitude, looking north. Predicted descent trajectories, spanning 62 km horizontally and 32 km vertically. are shown, with vertical lines every 15 s for the first minute, then every 1 min. (**b**) Location of trajectory within Argentina. (**c**) Zoomed-in map view, comparing the true landing sites to the predicted 1σ, 2σ, and 3σ uncertainty ellipses. The background image was created with Google Earth, using data from SIO, NOAA, U.S. Navy, NGA, GEBCO, and imagery from Landsat/Copernicus.

Our model for both the descent vector and its uncertainty thus appears accurate. The absolute uncertainty on our predictions is larger than previous drops because the high wind speed meant that capsules travelled farther downrange. Indeed, we had to extrapolate uncertainties from most of the test drops used to calibrate our model [6]. But, even in this regime, compared to our predicted landing sites, three of the four coordinate offsets are inside the predicted 68% confidence interval, with one just outside. Compared to our best-guess landing sites, all four offsets are *just* inside. Furthermore, the two near-simultaneous trajectories neatly illustrate the irreducible component of uncertainty caused by small-scale spatial and temporal variations in wind velocity (gusts), as well as local topography, when the predicted trajectory at low altitude is only 21° below the horizontal. The two capsules started only 1.8 km (and 27 s) apart but ended up separated by 6.4 km.

3.4. Successful Recovery

Both DRS capsules were found at their reported locations in rolling hills (see Figure 7) by the search and rescue team for the governor's office of Santa Cruz Province. The bright orange parachutes were visible from a distance, and the beeping was audible. DRS1 was found exactly at its reported location, a few hundred metres from a track and ~24 km from main road 12. DRS2 was found a few metres from its reported location, with evidence of cougar prints in the snow. We surmise that foam and parachute nylon are intriguing but not tasty.

We recovered identical copies of all the data from both released DRS capsules and later from the unreleased DRS and the main data store on SUPERBIT (which also had slightly more telemetry data). However, SUPERBIT had been completely destroyed upon landing, when its parachute failed to detach (perhaps because of similar thermal issues as the DRS capsules; analysis is ongoing), and it was dragged for 3 km through similar terrain, leaving a trail of debris. It is, therefore, remarkable luck that SUPERBIT's solid-state hard drive was later discovered intact. We did not *need* it because data had already been retrieved from the released DRS capsules, but having the original copy enabled us to verify that no data on the SD cards were corrupted.

Figure 7. Landing sites of the DRS capsules in Argentina. (**a**) General terrain and the search and rescue team from the governor's office of Santa Cruz Province. The bright-orange parachutes of DRS1 (**b**) and DRS2 (**c**) were visible from a distance. The white foam shell and release crown were finally also visible; the foam helped to insulate and waterproof DRS2 when it landed on snow.

4. Results

The DRS system comprised both hardware and software. Hardware worked reliably throughout the mission in three of the four capsules. They were power cycled every 15 min for 40 days and booted successfully every time. Data were gradually copied to every DRS as they were acquired and later (after a mean delay of 20 days at altitude plus two days exposed in Argentina) recovered with no bit errors in 219 GB of files. Curiously, one of the 5920 files accumulated 10 bit errors when being copied from the telescope's camera data store to the central data store, but none was further corrupted on either the central data store or the DRS capsules.

The internal batteries in the DRS capsules became very cold at altitude and provided insufficient voltage to power the Iridium transceiver until they warmed up after landing. This meant the true descent trajectories were not recorded, which would have been interesting to compare to our model trajectories. However, the batteries did warm up and worked successfully, even from an exposed, high altitude location on Earth, while covered in snow. The capsules reported their location and were found.

One of the four DRS capsules failed to respond after launch. Following the destruction of the telescope to which it remained attached, it is impossible to determine the cause. We speculate that a cable or connector might have been dislodged by the launch shock, but we also note that the DRS in question used a brand of SD card that had displayed erratic behaviour during testing. Either way, only 75% hardware reliability calls for additional testing before future use to reduce or improve single points of failure.

The software we developed to predict the trajectory of a DRS descending through the Earth's atmosphere worked very well, with accurate predictions and estimates of uncertainty. This was essential to drop the DRS to a safe but accessible landing site. It would also work as backup information to find a DRS that failed to respond after release (although the reliability of our hardware other than during launch meant that this was not needed).

Analysis of the astronomical data collected by SUPERBIT and the DRS capsules will be presented in forthcoming papers.

5. Discussion

The first use of the DRS capsules during a live science mission was a huge success. For a relatively small cost, we insured the scientific returns of SUPERBIT against a loss event *that came true*: high bandwidth communication links failed, then the telescope was destroyed upon landing. We recommend that future balloon missions consider using this or similar systems. Our hardware design and software are open-source and freely available. Further development is continuing at NASA, with plans to offer a facility-class DRS system.

Possible improvements to DRS hardware include (a better way to secure) a low-friction connector so it does not become disconnected during launch. The jerk when a DRS falls away is sufficient that the right amount of rigidity may be provided by a standard Ethernet socket and plug with the retaining lug removed. Power over Ethernet can remove the need for any separate power cables. A related aspect of this is that securing a DRS to the gondola mount currently requires fiddly routing of cables and cable ties inside a cramped case and pulling the ties taut (see Figure 2). More convenient assembly/disassembly will make it easier to perform many more test drops in situ than we achieved.

Thermal management has greatly improved since our test flight in 2019 [6], with the Raspberry Pi operating reliably on demand throughout the mission. The internal batteries got cold, and a (throttled) thermal link from the Pi would have helped them stay ready to work immediately after release. Although our batteries eventually warmed up and worked fine after 40 days cold, we have not tested whether they would also recover as quickly after 100 days cold. It would also have been interesting if the DRS stored GNSS data (at high frequency) during descent, even though it was too cold to power the Iridium transmitter, so the trajectory could be compared to our prediction. This is not possible in the current design concept, which intentionally has only a low-power microcontroller chip and no flash memory available after release, to maximise battery life in case searching for the DRS takes a long time.

Author Contributions: Conceptualization, E.L.S., R.M., L.L., J.D.R., R.S. (Roger Smith) and R.S. (Russell Smith); methodology, A.S.G., A.M.B., P.C., J.E. and R.P.; investigation (DRS hardware), J.A., A.H. (Andrew Hunter), N.J., D.K. and S.L.; investigation (SUPERBIT interface), S.J.B., S.W.E., A.A.F., J.W.H., D.H., B.H., E.M.H., M.J., W.C.J., D.L., J.S.-Y.L., T.V.T.L., J.E.M., J.M.N., C.B.N., E.P., S.F.R., A.R., L.J.R., J.S., M.M.S., S.I.T. and G.N.V.; investigation (SPB interface), H.F., A.H. (Andrew Hynous), S.R. and R.S. (Robert Salter); funding acquisition, R.M.; supervision, R.M., W.C.J. and C.B.N.; writing, E.L.S. and R.M.; software, E.L.S. and P.C. All authors have read and agreed to the published version of the manuscript.

Funding: This research was funded by the Royal Society (grant RGF/EA/180026), the UK Science and Technology Facilities Council (grant ST/V005766/1), and an Impact Acceleration Award to Durham University. Launch and operational support for the 2023 SUPERBIT flight from Wanaka, New Zealand, was provided by NASA. Funding for the development of SUPERBIT was provided by NASA (grant APRA NNX16AF65G), the Canadian Institute for Advanced Research (CIFAR), the Canadian Natural Science and Engineering Research Council (NSERC), and Durham University's astronomy survey fund. This work was done in part at JPL, run under a contract for NASA by Caltech.

Institutional Review Board Statement: Not applicable.

Informed Consent Statement: Not applicable.

Data Availability Statement: Descent trajectories are available from https://github.com/EllenSirks/pyBalloon. Full instructions to build a DRS are available from https://github.com/PaulZC/Data_Recovery_System.

Acknowledgments: We thank the Ministerio de Seguridad de la Provincia Santa Cruz, Argentina, for deploying their search and rescue team to recover the DRS capsules. The search was organised by Ministro de Seguridad Sr. Luca Pratti and Subsecretario de Protección Civil, Sr. Diego Farias. Thank you to Juan Perez, German Alejandro Reynoso, Marcelo Alejandro Pemini. and Alan Gabriel Arbe for going into the field—and, sorry for keeping you awake when we accidentally left one DRS beeping overnight.

Conflicts of Interest: The authors declare no conflicts of interest. The funders had no role in the design of the study; in the collection, analyses, or interpretation of data; in the writing of the manuscript; or in the decision to publish the results.

Abbreviations

The following abbreviations are used in this manuscript:

CPE	Co-polyester, a branched version of PETG
DRS	Data Recovery System
GFS	Global Forecast System (weather model)
GNSS	Global Navigation Satellite System, of which GPS is a subset
GPS	Global positioning system
PCB	Printed circuit board
PETG	Glycol-modified polyethylene terephthalate hard plastic
PLA	Polylactic acid hard plastic
SBD	Short Burst Data (Iridium satellite message)
SPB	Super-pressure (closed, helium-filled) balloon
SUPERBIT	Super-pressure Balloon-borne Imaging Telescope
TDRSS	Tracking and Data Relay Satellite System

References

1. Romualdez, L.J.; Benton, S.J.; Clark, P.; Damaren, C.J.; Eifler, T.; Fraisse, A.A.; Galloway, M.N.; Hartley, J.W.; Jones, W.C.; Li, L.; et al. The design and development of a high-resolution visible-to-near-UV telescope for balloon-borne astronomy: SuperBIT. *arXiv* **2016**, arXiv:1608.02502.
2. Karoly, D. (Ed.) *Meteorology of the Southern Hemisphere*; American Meteorological Society: Boston, MA, USA, 2015.
3. Shaaban, M.M.; Gill, A.S.; McCleary, J.; Massey, R.J.; Benton, S.J.; Brown, A.M.; Damaren, C.J.; Eifler, T.; Fraisse, A.A.; Everett, S.; et al. Weak Lensing in the Blue: A Counter-intuitive Strategy for Stratospheric Observations. *AJ* **2022**, *164*, 245. [CrossRef]
4. Clark, P.; Funk, M.; Funk, B.; Funk, T.; Meadows, R.; Brown, A.; Li, L.; Massey, R.; Netterfield, C. An open source toolkit for the tracking, termination and recovery of high altitude balloon flights and payloads. *J. Instrum.* **2019**, *14*, P04003. [CrossRef]
5. UN. Committee of Experts on the Transport of Dangerous Goods. Recommendations on the Transport of Dangerous Goods. Manual of Tests and Criteria. Technical Report, United Nations, Geneva, 2019. Available online: https://digitallibrary.un.org/record/3846833 (accessed on 29 October 2023).
6. Sirks, E.L.; Clark, P.; Massey, R.J.; Benton, S.J.; Brown, A.M.; Damaren, C.J.; Eifler, T.; Fraisse, A.A.; Frenk, C.; Funk, M.; et al. Download by parachute: Retrieval of assets from high altitude balloons. *J. Instrum.* **2020**, *15*, P05014. [CrossRef]
7. Gualtieri, R.; Filippini, J.P.; Ade, P.A.R.; Amiri, M.; Benton, S.J.; Bergman, A.S.; Bihary, R.; Bock, J.J.; Bond, J.R.; Bryan, S.A.; et al. SPIDER: CMB Polarimetry from the Edge of Space. *J. Low Temp. Phys.* **2018**, *193*, 1112–1121. [CrossRef]
8. Oliver, J.E. Standard atmosphere. In *Climatology*; Springer US: Boston, MA, USA, 1987; pp. 801–803. [CrossRef]

aerospace

MDPI

Article

Performance Analysis of Inter-Satellite and Satellite-Ground Communication: A Report on Flight Data for a Low Earth Orbit CubeSat

Miguel Limón-González [1], Enrique Rafael García-Sánchez [1,*], Héctor Simón Vargas-Martínez [2], Nicolás Quiroz-Hernández [1] and Selene Edith Maya-Rueda [1]

[1] Faculty of Electronics, Benemérita Universidad Autónoma de Puebla, Blvd Valsequillo, Cd Universitaria, Jardines de San Manuel, Puebla 72570, Mexico; miguel.limon@alumno.buap.mx (M.L.-G.); nicolas.quirozh@correo.buap.mx (N.Q.-H.); selene.maya@correo.buap.mx (S.E.M.-R.)

[2] Faculty of Bionics, Aerospace and Mechatronics, Universidad Popular Autónoma del Estado de Puebla, 21 sur #1103, Barrio de Santiago, Puebla 72410, Mexico; hectorsimon.vargas@upaep.mx

* Correspondence: rafael.garciasan@correo.buap.mx

Abstract: Communication between a nanosatellite located in Low Earth Orbit (LEO) and a ground station is limited in regions far from the poles, occurring for only a few minutes on different days and at different times. By utilizing satellite-to-satellite communication, it is possible to transmit and receive information more efficiently, circumventing the restrictions inherent in satellite-ground station links. The objective of this study is to present a comparative report on the results of data transmission through inter-satellite and satellite-to-ground station communication, focusing on a 1U CubeSat nanosatellite (AztechSat-1). This paper discusses the use of the GlobalStar network and a nanosatellite for inter-satellite communication. This paper also discusses the use of proprietary and open-source ground stations for satellite-ground communication. We provide an overview of the GlobalStar network and the associated ground stations involved in this research, along with the results and their subsequent analysis.

Keywords: nanosatellite; satellite communication; GlobalStar; ground station; telemetry; AztechSat-1; satellite network; CubeSat

1. Introduction

Today, small satellites in Low Earth Orbit are typically deployed as constellations to achieve global coverage. To ensure this coverage, hundreds or even thousands of satellites may need to be deployed [1]. Within the classification of small satellites based on weight, nanosatellites (1–10 kg) are one of the most popular categories. One commonly used standard in this category is the CubeSat, which is manufactured in a size of $10 \times 10 \times 10$ cm^3, also known as 1U [2].

Some advantages of manufacturing small satellites are as follows: cost reduction, short development times, launches at an affordable price, commercial applications, and flexibility [3]. Many projects today use commercial off-the-shelf (COTS) components and there are even companies that offer complete systems on the market [4].

Communication between satellites in Geostationary Orbit (GEO) and ground stations usually require large antennas with high transmit power, additionally, they present a high propagation delay due to the long propagation path [5].

Small satellites in LEO offer strategic advantages such as low propagation times, low propagation loss, as well as network robustness. In addition, the power required for data transmission and reception is lower [6].

The concept of satellite constellation is not new and emerged with pioneering companies such as Iridium and GlobalStar twenty years ago. They offered global communication link services that directly competed with cellular networks but had limited success due to high costs [7].

Satellite communication can be classified into two different groups. The first is based on a ground station network architecture, where information is transmitted from the satellite to a ground station and then relayed through existing ground links. This approach requires the deployment of numerous stations for global coverage [8]. The second type is based on a space-based network architecture, which utilizes intersatellite links between satellites and only requires a small number of gateways [9].

In this work, we present a hybrid approach where the GlobalStar constellation is based on the first type of architecture, while the AztechSat-1 nanosatellite is based on both the first and second types of architecture. In the first type, the nanosatellite allows communication with ground stations for radio amateurs, while in the second type, the nanosatellite establishes intersatellite links with the GlobalStar constellation, which subsequently relays the information to its gateways. The information transmitted remains the same in both cases.

There are some initiatives from Space Agencies to promote the use of space using CubeSat-based missions [10]. For example, the ESA Education Programme is composed of many hands-on projects, the Fly Your Satellite! (FYS) program is intended to give university students the opportunity to design, and launch and operate their own CubeSat [11]. On the other hand, NASA's CubeSat Launch Initiative provides universities with the opportunity to launch small satellites into space, enabling students to gain hands-on experience in developing space hardware [12]. It is noteworthy to mention that the AztechSat-1 nanosatellite was part of NASA's ELaNa 25B and 28 mission alongside other experiments such as CryoCube-1, SORTIE, CIRiS, and EdgeCube [13].

1.1. Description of the AztechSat-1 Nanosatellite

The project name, AztechSat-1, pays homage to the Aztec culture and represents the technological challenge it embodies. The AztechSat-1 is a nanosatellite belonging to the 1U CubeSat type category. It was developed by a multidisciplinary group of students from the Universidad Popular Autónoma del Estado de Puebla (UPAEP) in the state of Puebla, Mexico, in collaboration with the Mexican Space Agency (AEM) and the National Aeronautics and Space Administration (NASA). The satellite was launched on 5 December 2019, aboard a Falcon 9 launcher from Cape Canaveral and deployed from the International Space Station (IEE) using the NanoRacks CubeSat Deployer (NRCSD) on 19 February 2020, as shown in Figure 1. It remained operational until its re-entry on 10 December 2021.

Figure 1. AztechSat-1 deployed from the ISS on 19 February 2020 [14].

The primary objective of AztechSat-1, an academic nanosatellite, was to demonstrate communication between a nanosatellite and the GlobalStar constellation at a frequency within the L band (1611.25–1618.725 MHz), aiming to improve data transmission to Earth. Additionally, it aimed to promote amateur radio activities using ultra-high frequency (UHF) frequencies of 435 MHz for upload and 437 MHz for download. Figure 2 shows the mission operation diagram.

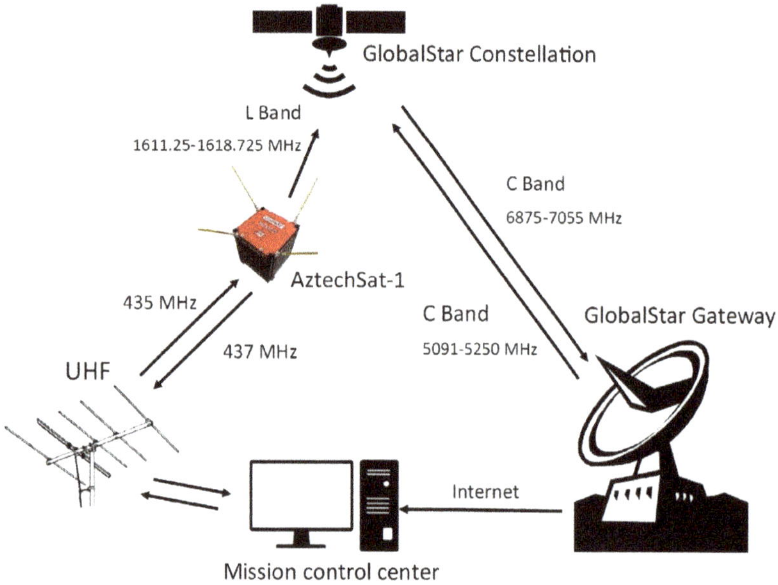

Figure 2. AztechSat-1 mission operation diagram.

The AztechSat-1 nanosatellite is composed of an aluminum structure, an electrical power system (EPS), an on-board computer (OBC), an attitude determination and control system (ADCS), and a radio system for communication (COM) with the ground station in the amateur UHF frequency band. It also includes a communication module for inter-satellite transmission in the L band. The system is equipped with a module that integrates the computer and the radio, as well as an interconnection panel for multipurpose boards. Additionally, it features an omnidirectional antenna system for communication with the ground station and is equipped with a set of 5 solar panels to provide power.

While the system incorporates commercial off-the-shelf (COTS) modules, the payload was specifically designed and manufactured to fulfill the mission objectives. Following its deployment, the nanosatellite powered on and initiated a countdown before transmitting its first packet of data, approximately 40 min later. Subsequently, the antennas were deployed, and the nanosatellite commenced its position control. Initially, the spacecraft was launched at an altitude of approximately 400 km with an inclination angle of 51.6°. The AztechSat-1 orbit simulation is described in Figure 3.

1.2. Intersatellite Communication Description

To achieve the primary mission objective, the AztechSat-1 nanosatellite utilized GlobalStar's Satellite Transmitter Integrated Global Positioning System Receiver (STINGR). This system incorporates a radio transmitter, a GPS receiver, and a patch antenna within a single integrated unit. Operating in simplex transmission mode, the STINGR system can communicate with the GlobalStar constellation. It automatically sends the data to the company's gateways for transmission to the Internet.

For the main parameters of the STINGR communication module, please refer to Table 1. Any microcontroller with serial communication capabilities can use the modem.

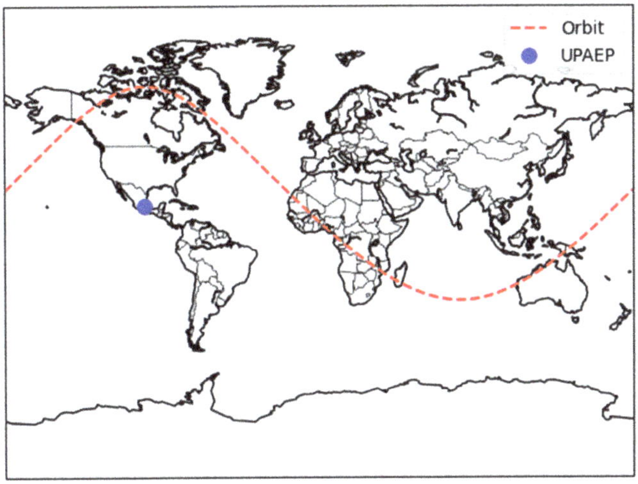

Figure 3. AztechSat-1 orbit simulation.

Table 1. Main parameters of the STINGR communication module [15].

Specifications	STX3	STINGR
Size	28.70 × 20.57 × 4.13 mm	45.05 × 47.37 × 6.3 mm
TX output power	19 dBW	17.5 dBW
TX mode current	390 mA	325 mA
Sleep mode current	7 uA	8 uA
Power Supply	3 to 5 V	3 to 5.5 V
Operating temperature range	−40 °C to +85 °C	−40 °C to +85 °C
Frequency range	1610 to 1620 MHz	1610 to 1620 MHz
Patch antenna included	No	Yes
GPS module included	No	Yes
Price (USD)	$60	$75
Patch antenna price (USD)	$35	$0

1.3. Description of the GlobalStar Satellite Network

The GlobalStar satellite communication system operates in Low Earth Orbit and consists of a constellation of 48 satellites positioned at an approximate altitude of 1414 km. These satellites are distributed among eight planes, each accommodating six satellites, and follow circular orbits with an inclination of approximately 52 degrees. It is worth noting that the GlobalStar constellation is currently undergoing reorganization due to bankruptcy [16].

The original model of the GlobalStar constellation aimed to provide cost-effective and less complex communication solutions by utilizing a hub of ground stations without inter-satellite links (ISLs). The telemetry and command units were planned to be integrated into six host gateways, selected from a total of 200 stations distributed worldwide with redundancy. With an estimated coverage of 50% of the ground stations, each satellite would establish an average of one contact per orbit, which occurs within a period of approximately 10 to 12 min [17].

1.4. Description of the Global Network of SatNOGS Ground Stations

As part of the AztechSat-1 project, a ground station was established to fulfill the secondary objective of promoting activities among radio amateurs. In addition, this study encompasses an analysis of a full stack of open technologies based on open standards. The research uses data that was provided by the Satellite Networked Open Ground Station (SatNOGS) project, which provides access to open-source data contributed by radio amateurs.

Traditionally, when an individual or organization builds a ground station, especially for non-commercial purposes, the system remains inactive for extended periods of time. To maximize the utilization of these resources, projects have been developed to enable their use by multiple users. The SatNOGS project facilitates the connection to a network of open-source ground stations, designed with modularity in mind. This network allows users to access satellite information uploaded by different individuals through a user-friendly web interface. Real-time information from numerous registered ground stations can be obtained through this network. To be compatible with the network, a ground station must have client software installed on an embedded computer. Additionally, a software-defined radio (SDR) should be connected to the computer and properly configured to work with SatNOGS [18].

1.5. State of the Art

The current predominant approach for communication between CubeSats and commercial intersatellite link networks in Low Earth Orbit is based on systems such as Iridium, GlobalStar, and Orbcomm [19].

In a 2008 investigation analyzing Inmarsat, Thuraya, Iridium, and GlobalStar, it was concluded that the GlobalStar network is best suited for nanosatellite payloads with an inclination of 52 degrees at altitudes ranging from 300 to 600 km [20].

TechEdSat 4, launched in 2014, also utilized a communication module for GlobalStar. Although it operated nominally, information regarding its results is unavailable. On August 2016, the secondary payload "exo-brake" was deployed resulting in early orbital reentry [21]. The PicoPanther project has been studying the GlobalStar and Iridium networks since 2008, with its approach remaining controversial until the launch of PhoneSats, particularly PhoneSat 1 Bell [22].

TechEdSat 8, deployed from the International Space Station on 31 January 2019, featured an integrated Near-Earth Network (NEN) S-band radio compatible with Iridium and GlobalStar transmitters. However, it was unable to undergo testing after deployment due to an onboard power system abnormality [23].

In the nanosatellite database (www.nanosats.eu (accessed on 11 September 2023)), there are at least 40 registered CubeSat-type nanosatellites that utilize a GlobalStar-compatible communication module as their primary or secondary downlink band. However, some of these projects were either not released or were subsequently cancelled. Among notable success stories are TSAT, GEARRSSAT, CubeRRT, MakerSat1, and AztechSat-1 [24].

The results of TSAT, a 2U CubeSat, demonstrated the effectiveness of the GlobalStar commercial network without the need for a ground station. The satellite's objective was to validate and characterize the link capacity of the GlobalStar constellation and conduct measurements in an Extremely Low Earth Orbit (ELEO). The team recorded successful simplex link transmissions within the field of view of the patch antenna. TSAT data was received by one or more GlobalStar satellites and transmitted to a gateway. The team received a total of 21,154 packets and noted that the system was capable of transmitting thousands of packets per day on a CubeSat. Although the quantitative capacity was not explored. At the beginning of the mission, TSAT transmitted its maximum number of messages, recording 26 dual-packet messages in 18 min. After ten days, this number was reduced to 7 messages per hour. The system implemented an STX-2 module and was deployed into a circular orbit at 325 km with an inclination of 51.6 degrees [25].

2. Materials and Methods

2.1. Description of the Received Data (SatNOGS and GlobalStar)

For the intersatellite transmission system, a 32-bit Cortex M3 ARM microcontroller was utilized to control the STINGR simplex transmission device, which was equipped with an STX3 modem. This configuration allowed the transmission of packets to a network of LEO satellites using the GlobalStar network. The received data is then forwarded to a web interface where the user interprets the data for further processing.

The device sends a signal to one or more nearby satellites without knowing exactly which satellite will receive the signal. Subsequently, the GlobalStar satellites download the information to the nearest Gateway. Redundant messages are discarded. Each message consists of 9-byte data packets, meaning that the system can transmit only 9 bytes in a single message. Therefore, if the message is larger, multiple packets need to be sent. There is a possibility of message loss when no satellite is within range. As the system operates in simplex mode, a redundancy method is employed to increase the probability of successful message reception.

The information received by the satellite and transmitted to the gateways was presented in a raw log report generated in Excel format. The received packets were placed in different cells using a standardized JSON format. Depending on the configuration 2 or more packets make up a message. A valid message has a length of 157 bytes starting with a marker "<HK>" and ending with "</HK>".

2.2. Description of the Software Developed and Used for the Analysis of the Results

The information sent by the AztechSat-1 and received by the GlobalStar gateways is transmitted in a custom format, needing the development of a computer program to automatically count, categorize, and decode the received messages. AztechSat-1 transmits various parameters such as temperatures of different solar panels, battery voltage, current, and other relevant data. The system data dictionary, providing a comprehensive list of these parameters, can be found in Table 2.

Figure 4 shows the nanosatellite received packets using the GlobalStar link throughout the entire mission. The chart provides a summary of packets categorized by their size in bytes. The transmission rate of messages is grouped into sets of 9-byte packets, with each transmission containing 1 to 16 packets. This allows a maximum of 144 bytes to be transmitted in a single uninterrupted burst. The geographic distribution of the AztechSat-1 packets is shown below in Figure 5 using GlobalStar communication.

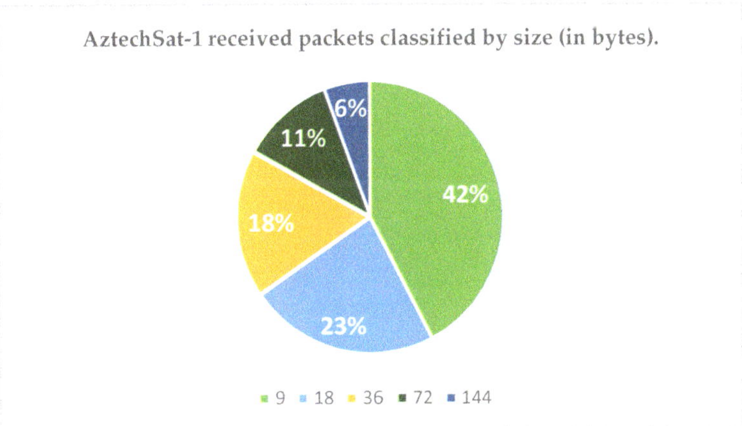

Figure 4. Received packets categorized by size (in bytes).

Table 2. Description of transmitted parameters categorized by subsystem.

Subsystem	Description of Parameters	Unit or Expected Value
Radio	Power amplifier temperature	tenth of °C
	Last Received Signal Strength Indicator	RSSI
	Background RSSI	RSSI
EPS	Power converters voltage—Array [3]	mV
	Battery voltage	mV
	Current in—Array [3]	mA
	Boost converters current	mA
	Battery's output current	mA
	Output current—Array [4]	mA
	Output channels status—Array [4]	0 or 1
	I2C watchdog time left	32-bit number (Seconds)
	GND watchdog time left	32-bit number (Seconds)
	Watchdog reboot number	32-bit number
	Temperature sensor	°C
OBC	Temperature Sensor	°C
	Pulse Width Modulation (PWM) current	Ma
	Last OBC reboot cause	32-bit number
	Magnetometer—X axis	32-bit float number
	Magnetometer—Y axis	32-bit float number
	Magnetometer—Z axis	32-bit float number
	Gyro—X axis	32-bit float number
	Gyro—Y axis	32-bit float number
	Gyro—Z axis	32-bit float number
Payload	Number of Channel	8-bit number
	Number of Bursts	8-bit number
	Minimum Burst Interval: Units of 5 s.	01h to 3Ch (5 to 300 s)
	Maximum Burst Interval: Units of 5 s.	02 h to 78 h (10 to 600 s)
	Status flags ▪ GPS subsystem active. ▪ Device is in active mode. ▪ GPS Subsystem does have a fix. ▪ GPS Subsystem failure. ▪ Simplex Transmitter failure. ▪ Battery low.	8-bit number
	Number of seconds since the device unit last attempted to send a satellite transmission.	16-bit number
	Number of seconds until the device unit attempts to send a satellite transmission.	16-bit number
	Packet size of last or current message.	8-bit number
	Currently waiting on or sending burst number	8-bit number
	Number of seconds until burst transmission number 2	16-bit number
	Number of seconds until burst transmission number 3	16-bit number
	Total messages transmitted in current mode.	16-bit number
	Total Packet transmission count since hard power on.	16-bit number
	STINGR Antenna temperature	°C
	Payload Board temperature	°C
ADCS	Coarse Sun Sensors Value—Array [+Y, +X, −X, −Y, −Z]	16-bit number
	Solar panel's temperature on +Y axis	32-bit float number
	Solar panel's temperature on +X axis	32-bit float number
	Solar panel's temperature on −X axis	32-bit float number
	Solar panel's temperature on −Y axis	32-bit float number
	Solar panel's temperature on −Z axis	32-bit float number
	Bdot status	−2 to 2
	Bdot value from low pass filter slow	32-bit float number
	Bdot value from low pass filter slow2	32-bit float number
	Value of detumbled state	0 or 1

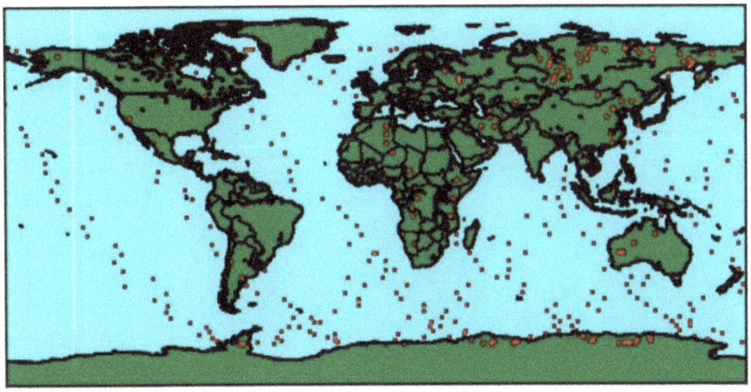

Figure 5. Geographic distribution of received packets from the AztechSat-1 using GlobalStar, the red dots represent regions where packets where received.

For the ground station-based system, the AztechSat-1 nanosatellite utilized a half-duplex radio operating at a UHF frequency that was compatible with the ASM + Golay protocol and offered Gaussian Frequency Shift Keying (GFSK) modulation options. The communication with the onboard computer was facilitated through the I2C communication protocol. The nanosatellite was equipped with a 50-ohm matched omnidirectional antenna, providing a gain of 1.5 dBi at −1 dBi. The use of UHF made it possible to collect data from the university's ground station and the SatNOGS network.

Throughout the operational period of AztechSat-1, SatNOGS received a total of 500 beacon packets. These packets were subsequently processed to extract and interpret the information they contained. On 10 December 2021, the AztechSat-1 nanosatellite reentered Earth's atmosphere, which is evidenced by numerous reports of telemetry available on Twitter and SatNOGS. In Figures 6 and 7 the altitudes and speeds of the last days in orbit of the nanosatellite can be observed.

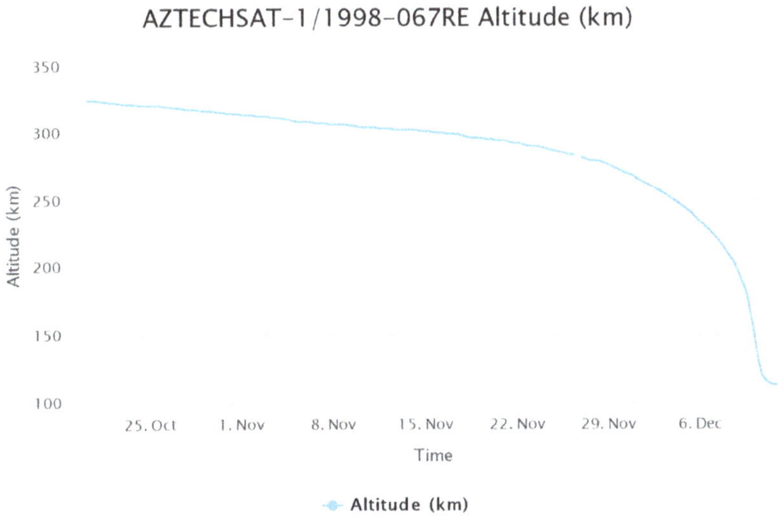

Figure 6. Altitude of the AztechSat-1 during its last days in orbit [26].

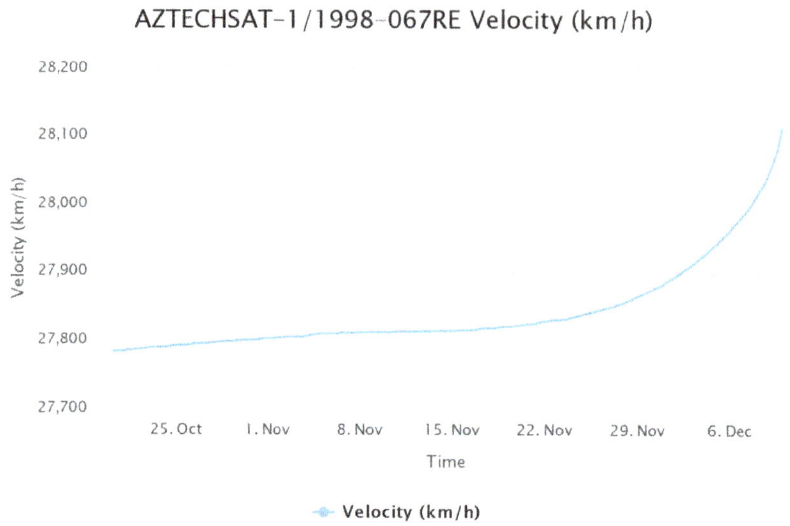

Figure 7. Velocity of the AztechSat-1 during its last days in orbit [26].

2.3. *Strategies for the Use of Statistical Tools, Graphs, among Others for the Presentation of Results*

To download data from SatNOGS, it is necessary to register as a user on the SatNOGS DB page. Once registered, users can request the export of all received frames to a CSV file using the API. This CSV file will contain each received frame along with its corresponding reception time, with each frame listed on a separate line.

3. Results

As previously mentioned, the AztechSat-1 nanosatellite is an academic experiment in the form of a 1U CubeSat. Although it was equipped with an integrated GPS system, its high-power consumption requirements prevented its use during the mission. Consequently, the tracking of the satellite was achieved using NORAD Two-Line Elements (TLEs) shown in Table 3, which served as the primary element for satellite tracking throughout the mission.

Table 3. NORAD Two-Line Element of the AztechSat-1.

Two Line Element (TLE)
1 45261U 98067RE 20274.58306181 +0.00017834 +00000-0 +28188-3 0 9998 2 45261 051.6407 182.8369 0007640 038.2637 321.8894 15.53524114034841

3.1. *Analysis of Data Received from GlobalStar and SatNOGS*

The temperature variations obtained in the month of July 2020 using only the system based on intersatellite communication can be seen in Figures 8 and 9. The location of the sensors is in each of the solar panels for external temperatures and on the onboard computer, electrical power system, and radio boards for internal temperatures.

These figures depict the temperature variations observed during one month of satellite operation. Throughout this period, the satellite experienced maximum and minimum temperatures of 45 °C and −25 °C respectively in the external environment. The internal temperatures ranged from 0 °C to 30 °C, with an abnormal data point of 85 °C on 18 July 2020 possibly indicating an error in reading the sensor or a component failure as shown in Figure 9a.

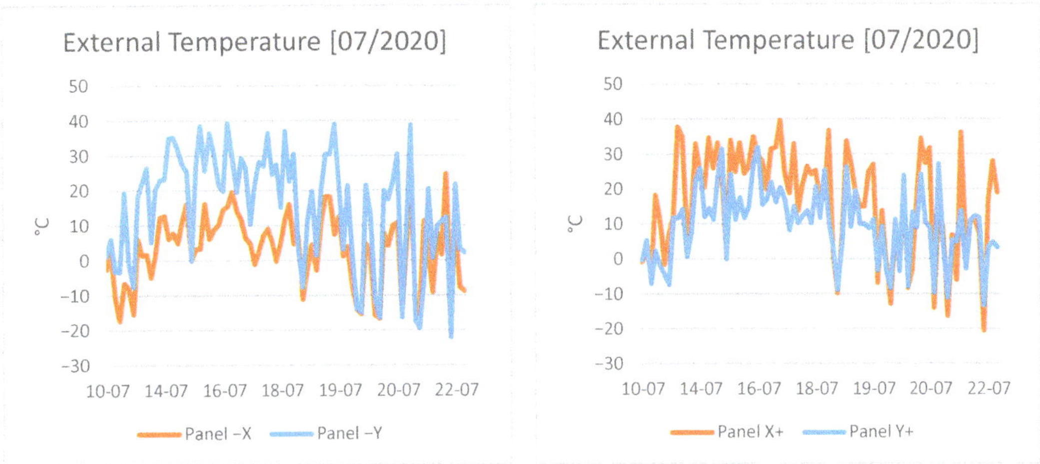

(**a**) (**b**)

Figure 8. Temperature variations on the external panels of the nanosatellite: (**a**) Temperature variations on panels −X and −Y; (**b**) Temperature variations on panels X+ and Y+.

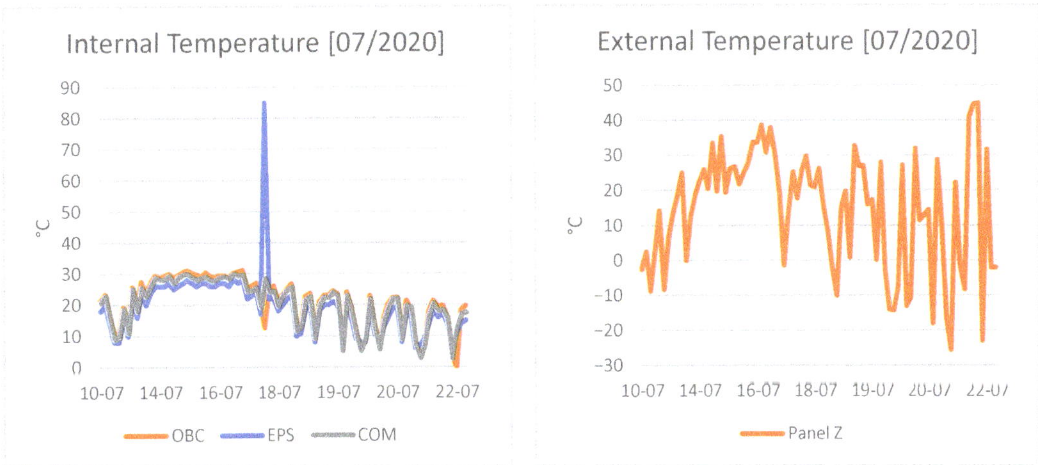

(**a**) (**b**)

Figure 9. Temperature variations on internal and external components: (**a**) Temperature variations on internal boards of the satellite; (**b**) Temperature variations on solar panel Z.

Figure 10 illustrates the relationship between the battery voltage and the input and output current of the EPS system. The behavior of the battery voltage can be observed in comparison to the levels of input and output current. It is evident that the battery charging process occurs when the input current surpasses the output current. Additionally, an anomaly in the charge level is noticeable, potentially correlated with the anomaly depicted in the internal temperature as shown in Figure 9a.

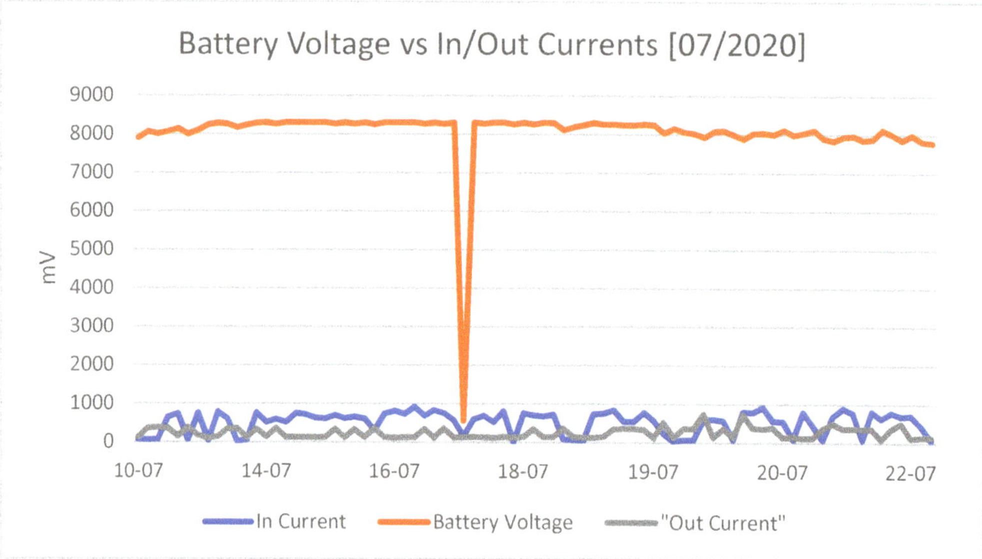

Figure 10. Relationship between battery voltage and input and output Current.

3.2. Analysis of the Results Classified by Inter-Satellite Ground Station

Figure 11 indicates the temperature variations observed in 2021, using the ground station communication system exclusively with data obtained from SatNOGS. The temperature information collected through this system is similar to the obtained through intersatellite link communication.

Figure 11a–c illustrates the exterior temperatures recorded in November 2021, and received via radio frequency (RF) through the SatNOGS platform. Comparing it with the information received via GlobalStar in July 2020, the exterior temperature demonstrates a comparable pattern, with minimums and maximums remaining within the range of −30 °C and 50 °C, respectively.

Figure 11d illustrates the interior temperature recorded in November 2021, this data was received via radio frequency (RF) through the SatNOGS platform. Comparing it with the information received via GlobalStar in July 2020, the internal temperatures demonstrate a comparable pattern, with minimums and maximums that remain between −1 °C and 32 °C. The information on the internal temperatures was recorded using sensors located in the OBC, EPS, and radio.

Figure 12 depicts the behavior of the battery voltage during the month of September 2021. After one and a half years of operation, it can be observed that the charging periods have extended, while the system's current consumption has reduced.

It is important to note that the information transmitted through GlobalStar and received by SatNOGS is similar, highlighting the reliability of telemetry data obtained through both communication methods. However, in the case of satellite-to-satellite communication (via GlobalStar), it turned out to be better because more points of data were retrieved. In the case of GlobalStar communication, different packet length configurations were utilized. The GlobalStar modem has five options for the packet length starting at 9 bytes per packet. The biggest packet length is 144 bytes per packet.

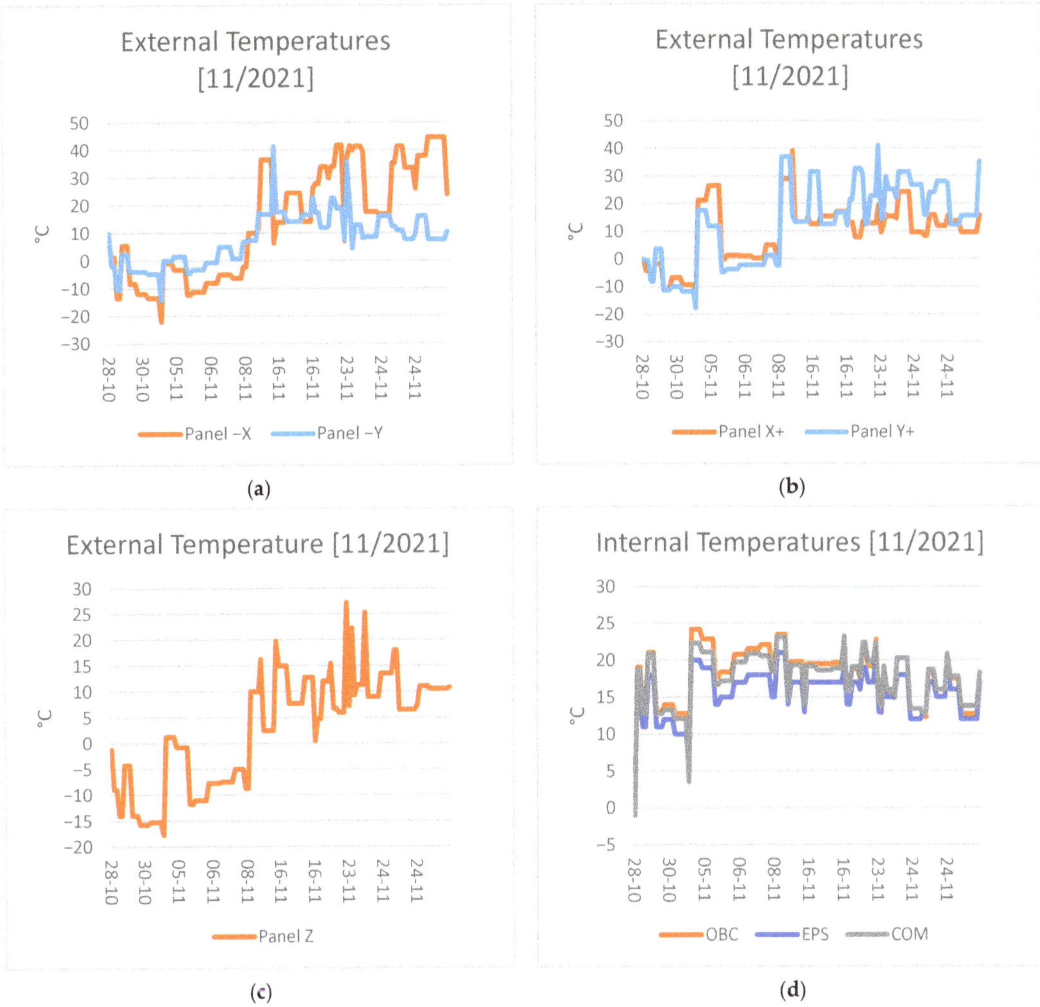

Figure 11. Temperature variations on internal and external components in November 2021: (**a**) Temperature variations on solar panels −X and −Y; (**b**) Temperature variations on solar panels X+ and Y+; (**c**) Temperature variations on solar panels Z.; (**d**) Temperature variations on internal boards of the nanosatellite.

3.3. Analysis of Data Received from GlobalStar and Ground Stations

The obtained information from GlobalStar and different ground stations was compared to assure reliability. To do this comparison we took samples of data collected from the ground stations where the data from GlobalStar were coincident on date and time. Figure 13 depicts charts where the similarity between the data received by RF and GlobalStar is confirmed.

The telemetry data was collected by both ground stations and GlobalStar. The obtained results show more information via ground stations than via GlobalStar satellites. The GlobalStar data indicated a delay in comparison with the ground station data.

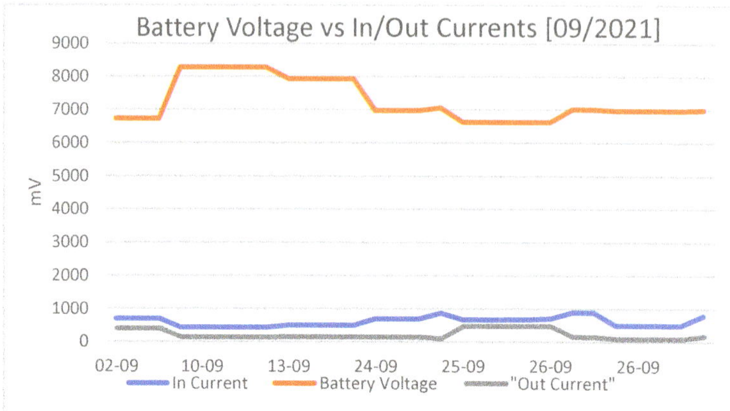

Figure 12. Voltages and currents of the nanosatellite, captured in September 2021.

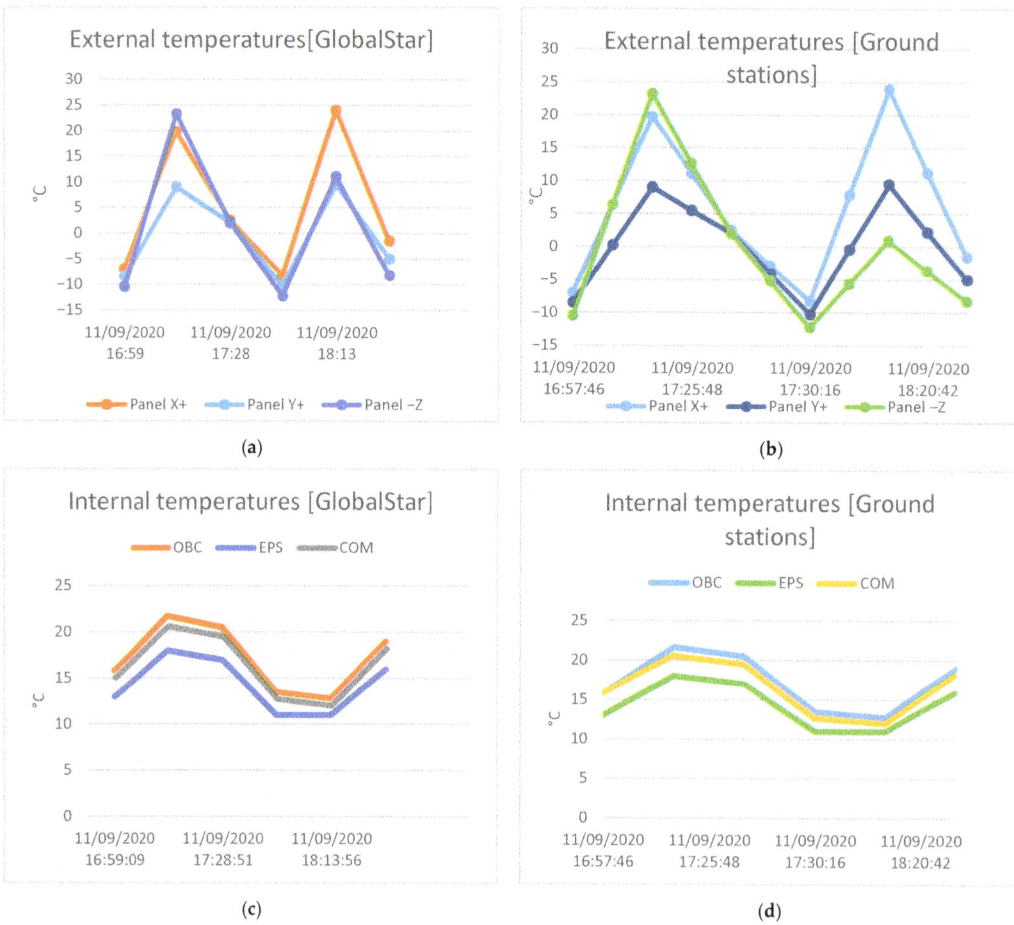

Figure 13. Temperature variations on internal and external components on 11 September 2020 using data from ground-to-satellite link and satellite-to-satellite link: (**a**) Temperature variations on solar

panels −X, −Y and −Z using data received by GlobalStar; (**b**) Temperature variations on solar panels −X, −Y and −Z using data received by ground stations; (**c**) Temperature variations on internal boards of the satellite using data received by GlobalStar; (**d**) Temperature variations on internal boards of the satellite using data received by ground stations.

For satellite-to-satellite communication, different packet length configurations were set up. The GlobalStar modem has five options for the packet length starting at 9 bytes with increments of 9 bytes. The biggest packet length is 144 bytes. The sent message containing the housekeeping data of the nanosatellite has a length of 157 bytes and each packet has 2 bytes with information for message reassembling. This means that for a 9 bytes packet length configuration, 23 packets were needed to reassemble the message; and for a 144 bytes packet length configuration, only 2 packets were needed.

Table 4 presents the result obtained with each packet length configuration. It can be observed that the configuration of 144 bytes per packet has the biggest success rate. Success rate refers to the relationship between received messages and decoded messages. The reason why messages could not be decoded was mainly because fragments were missing and the message could not be reassembled.

Table 4. Results obtained with different configurations of message length.

Message Length	Received Packets	Received Messages (Partial)	Received Messages (Complete)	Decoded Messages (Success)	Success Rate
9	5246	145	0	0	0.00%
18	2857	143	5	4	3.50%
36	2166	142	32	16	22.54%
72	1409	140	59	55	42.14%
144	692	135	60	60	44.44%
Total:	12,370	705	156	135	22.13%

It is important to note that with the smallest packet length configuration (9 bytes per packet), 5246 packets were received, out of which only 145 messages could be identified. However, all these messages were missing fragments, making it impossible to decode them. On the other hand, with the largest message length configuration (144 bytes per packet), 692 packets were received, out of which 135 messages were identified, and only 60 arrived complete.

The smallest packet length configuration resulted in more valid received packets, however, most of the fragments were lost. When the message was increased, the complexity of reassembling the message at the receiver's end was reduced. This is why, despite receiving more valid packets, very few messages were successfully reconstructed, as many fragments were missing. With the largest packet length configuration, we received fewer valid packets but the probability of recovering the whole message was higher.

Table 4 shows the results obtained with each of these configurations. As shown, the reception success rate improves as the message length increases. This suggests that less fragmented transmitted messages are less susceptible to loss of data fragments.

4. Discussion

The use of services with global coverage like GlobalStar as a link to download data from space seems to be a good alternative to other traditional forms of communication such as ground-to-satellite communication. Beyond traditional radiofrequency links, these services including GlobalStar, Starlink, and Iridium, offer unprecedented global reach, low-latency connections, and reliability that are essential for a broad spectrum of space missions. Their cost-effectiveness, scalability, and ease of deployment make them an attractive option for data transfer in remote or challenging space environments.

In this work, we study the idea of enhancing ground-to-satellite communication with satellite-to-satellite links. Our results show a better performance of data covering more geographical areas using intersatellite communication than only using ground station communication.

So far, successful mission and the current literature shows little information in a brief period in similar orbits. The AztechSat-1 project has demonstrated two kinds of satellite communication which have enhanced the received data. Data obtained from GlobalStar has increased the amount of information in comparison to only ground-satellite communication.

In our future work, we will study the performance of a constellation of proprietary nanosatellites with a commercial constellation in an environment that involves real-world constraints. Our next step is a collaborative initiative between the Mexican Space Agency and the National Aeronautics and Space Administration which is currently underway to develop a constellation of at least four nanosatellites for animal telemetry applications. This joint project has been named the AztechSat Constellation. Its primary objective is to establish a marine fauna monitoring system. The project will involve professors and students from various Mexican universities, who will contribute their expertise to the design and integration of nanosatellites for deployment in Low Earth Orbit.

5. Conclusions

The results obtained throughout the operational lifespan of the AztechSat-1 nanosatellite offer valuable insights. Firstly, they serve to validate temperature data obtained from a nanosatellite in a Low Earth Orbit, providing a comprehensive understanding of temperature patterns in such an orbital environment. Additionally, the energy consumption pattern of a communication nanosatellite can be identified.

Moreover, the study highlights the effectiveness of the GlobalStar commercial network in intersatellite links, demonstrating its capability without the need for a ground station. A comparison is made between the quality of data obtained through the GlobalStar network and that acquired using open-source ground stations with the support of radio amateurs, employing the same satellite at different time periods.

The conclusions drawn from both forms of communication are presented. It is noted that packet length demonstrated improved performance in intersatellite communication. Furthermore, the analysis examines whether ground stations provided additional information beyond that gathered by SatNOGS.

Author Contributions: Conceptualization, H.S.V.-M.; writing—original draft preparation, H.S.V.-M., N.Q.-H. and S.E.M.-R.; writing—review & editing, M.L.-G. and E.R.G.-S.; methodology M.L.-G. and E.R.G.-S.; formal analysis M.L.-G. and E.R.G.-S.; project administration, H.S.V.-M.; validation, N.Q.-H. and S.E.M.-R.; supervision, N.Q.-H. and S.E.M.-R. All authors have read and agreed to the published version of the manuscript.

Funding: This study was mainly supported by a sectoral fund for Research, Technological Development, and Innovation in Space Activities AEM-CONACYT. Project number: 292767 under the title "Diseño y construcción del módulo de enlace intersatelital AztechSat-1–GlobalStar", a Council of Science and Technology of the State of Puebla (CONCYTEP) fund. Project number 08 Bis/2018 under the title "Diseño y construcción del Módulo de Enlace Intersatelital "AzTechSat-1"–GlobalStar, an internal budget from UPAEP, a non-profit university located in Puebla, Mexico, since year 2017 to 2019, as a part of a continuing education project for students from the Electronics department for the CubeSat subsystems acquisition, development, and manufacturing.

Data Availability Statement: Not applicable.

Acknowledgments: We would like to express our sincere gratitude to the dedicated students and professors of the Universidad Popular Autónoma del Estado de Puebla for their significant contributions in ensuring the success of the AztechSat-1 project. We are immensely thankful to the Mexican Space Agency for their unwavering support and guidance throughout the development of the nanosatellite. Our heartfelt appreciation also goes to NASA for their invaluable assistance in launching, testing, and sharing their expertise, as well as for their generous commitment of time in addressing our queries. We extend our gratitude to the National Council of Humanities Science and Technology in Mexico (CONAHCYT) for their support in providing scholarships for human resources training and resources to fulfill the project's requirements. We are grateful to the Benemérita Universidad Autónoma de Puebla (BUAP) and the Instituto Nacional de Astrofísica, Óptica y Electrónica (INAOE) for their assistance in conducting various tests and experiments.

Conflicts of Interest: The authors declare no conflict of interest.

References

1. He, G.; Gao, X.; Sun, L.; Zhang, R. A Review of Multibeam Phased Array Antennas as LEO Satellite Constellation Ground Station. *IEEE Access* **2021**, *9*, 147142–147154. [CrossRef]
2. Sinha, R.; Verma, A.; Choudhary, S. General Analysis of Cube Satellite Technology: An Overview. *Int. J. Comput. Appl.* **2019**, *178*, 19–22. [CrossRef]
3. Malisuwan, S.; Kanchanarat, B. Small Satellites for Low-Cost Space Access: Launch, Deployment, Integration, and In-Space Logistics. *Am. J. Ind. Bus. Manag.* **2022**, *12*, 1480–1497. [CrossRef]
4. Yaqoob, M.; Lashab, A.; Vasquez, J.C.; Guerrero, J.M.; Orchard, M.E.; Bintoudi, A.D. A Comprehensive Review on Small Satellite Microgrids. *IEEE Trans. Power Electron.* **2022**, *37*, 12741–12762. [CrossRef]
5. Kolawole, M.O. *Satellite Communication Engineering*; CRC Press: Boca Raton, FL, USA, 2017.
6. ITU-R. *Simulation Methodologies for Determining Statistics of Short-Term Interference between Co-Frequency, Codirectional Non-Geostationary-Satellite Orbit Fixed-Satellite Service Systems in Circular Orbits and Other Non-Geostationary Fixed-Satellite Service Systems in Circular Orbits or Geostationary-Satellite Orbit Fixed-Satellite Service Networks*; International Telecommunication Union: Geneva, Switzerland, 2003.
7. Curzi, G.; Modenini, D.; Tortora, P. Large Constellations of Small Satellites: A Survey of Near Future Challenges and Missions. *Aerospace* **2020**, *7*, 133. [CrossRef]
8. Tanenbaum, A.S. Network Protocols. *ACM Comput. Surv.* **1981**, *13*, 453–489. [CrossRef]
9. Budenske, J.R.; Millikin, K.S.; Bonney, J.C.; Ramanujan, R.S.; Sands, O.S. Space network architecture technologies. In Proceedings of the 2002 IEEE Aerospace Conference Proceedings, Big Sky, MT, USA, 9–16 March 2002; Volume 3, pp. 3–1061.
10. Camps, A. Nanosatellites and Applications to Commercial and Scientific Missions. In *Satellites Missions and Technologies for Geosciences*; IntechOpen: Rijeka, Croatia, 2019; ISBN 978-1-78985-996-6.
11. Marée, H.; Galeone, P.; Kinnaird, A.; Callens, N. *The ESA Education Programme and Its ESA Academy*; University of Leicester: Leicester, UK, 2020; ISBN 978-1-912989-09-6.
12. Crusan, J.; Galica, C. NASA's CubeSat Launch Initiative: Enabling broad access to space. *Acta Astronaut.* **2019**, *157*, 51–60. [CrossRef]
13. National Aeronautics and Space Administration: Past ElaNa CubeSat Launches. CubeSat Launch Initiative. 2021. Available online: https://www.nasa.gov/content/past-elana-cubesatlaunches (accessed on 17 August 2023).
14. Tabor, A. What Is AzTechSat-1? Available online: http://www.nasa.gov/ames/aztechsat-1 (accessed on 18 August 2023).
15. Garcia, J. *Telemetry Subsystem with Higher Access Duration for LEO Satellites*; Morehead State University: Morehead, KY, USA, 2018.
16. Butash, T.; Garland, P.; Evans, B. Non-geostationary satellite orbit communications satellite constellations history. *Int. J. Satell. Commun. Netw.* **2021**, *39*, 1–5. [CrossRef]
17. Smith, D.; Hendrickson, R. Mission control for the 48-satellite Globalstar constellation. In Proceedings of the IEEE MILCOM '95, San Diego, CA, USA, 5–8 November 1995; Volume 2, pp. 828–832.
18. Croissant, K.; White, D.; Álvarez, X.C.; Adamopoulos, V.; Bassa, C.; Boumghar, R.; Brown, H.; Daradimos, I.; Dohmen, P.; Flawinne, J.; et al. *An Updated Overview of the Satellite Networked Open Ground Stations (SatNOGS) Project*; Utah State University: Logan, UT, USA, 2022.
19. Al-Hraishawi, H.; Chougrani, H.; Kisseleff, S.; Lagunas, E.; Chatzinotas, S. A Survey on Nongeostationary Satellite Systems: The Communication Perspective. *IEEE Commun. Surv. Tutor.* **2023**, *25*, 101–132. [CrossRef]
20. Khan, K. Data Communication with a Nano-Satellite Using Satellite Personal Communication Networks (S-Pcns). Master's Thesis, University of Central Florida, Orlando, FL, USA, 2007.
21. Murbach, M.; Alena, R.; Luna, A.G. The TechEdSat/PhoneSat Missions for Small Payload Quick Return. In Proceedings of the Small Satellite Conference, Logan, UT, USA, 6–11 August 2016; Utah State University: Logan, UT, USA, 2016.
22. Rodriguez, C.; Boiardt, H.; Bolooki, S. CubeSat to commercial intersatellite communications: Past, present and future. In Proceedings of the 2016 IEEE Aerospace Conference, Big Sky, MT, USA, 5–12 March 2016; pp. 1–15.

23. Wong, Y.; Schaire, S.; Bundick, S.; Kegege, O.; Altunc, S.; Fetterer, P.; Murbach, M.; Priscal, C.; Perrotto, T.; Celeste, P. NASA Near Earth Network (NEN) DVB-S2 Demonstration Testing for Enhancing Data Rates for CubeSat/SmallSat Missions. In Proceedings of the Annual Conference on Small Satellites, Logan, UT, USA, 3–8 August 2019.
24. Kulu, E. Nanosats Database. Available online: https://www.nanosats.eu/index.html (accessed on 1 August 2023).
25. Voss, H.D.; Dailey, J.F.; Crowley, J.C.; Bennett, B.; White, A.J. TSAT Globalstar ELaNa-5 Extremely Low-Earth Orbit (ELEO) Satellite. In Proceedings of the 28th Annual AIAA/USU Conference on Small Satellites, Logan, UT, USA, 14–16 October 2014; American Institute of Aeronautics and Astronautics: Reston, VA, USA, 2014.
26. Orbit.ing-now.com AZTECHSAT-1/45261/1998-067RE. Available online: https://orbit.ing-now.com/satellite/45261/1998-067re/aztechsat-1/ (accessed on 20 October 2023).

aerospace MDPI

Article

CMAG: A Mission to Study and Monitor the Inner Corona Magnetic Field

David Orozco Suárez [1,2,*], Jose Carlos del Toro Iniesta [1,2], Francisco Javier Bailén Martínez [1,2], María Balaguer Jiménez [1,2], Daniel Álvarez García [1,2], Daniel Serrano [3], Luis F. Peñin [3], Alicia Vázquez-Ramos [4], Luis Ramón Bellot Rubio [1,2], Julia Atienzar [1,2], Isabel Pérez Grande [2,5], Ignacio Torralbo Gimeno [2,5], Esteban Sanchis Kilders [2,6], José Luis Gasent Blesa [2,6], David Hernández Expósito [2,7], Basilio Ruiz Cobo [2,7], Javier Trujillo Bueno [7,8,9], Robertus Erdélyi [10,11,12], Jackie A. Davies [13], Lucie M. Green [14], Sarah A. Matthews [14], David M. Long [15], Michail Mathioudakis [15], Christian Kintziger [16], Jorrit Leenaarts [17], Silvano Fineschi [18] and Eamon Scullion [19]

1 Instituto de Astrofísica de Andalucía (IAA-CSIC), Apdo. de Correos 3004, E-18080 Granada, Spain; jti@iaa.es (J.C.d.T.I.); fbailen@iaa.es (F.J.B.M.); lbellot@iaa.es (L.R.B.R.)
2 Spanish Space Solar Physics Consortium (S^3PC), 18008 Granada, Spain; isabel.perez.grande@upm.es (I.P.G.); ignacio.torralbo@upm.es (I.T.G.); brc@iac.es (B.R.C.)
3 SENER Aerospace, Severo Ochoa 4, E-28760 Tres Cantos, Spain; luis.penin@aeroespacial.sener (L.F.P.)
4 Departamento de Física Teórica y del Cosmos (DFTC), Universidad de Granada (UGR), Campus de Fuentenueva, E-18071 Granada, Spain; aliciavr@ugr.es
5 Instituto de Microgravedad "Ignacio da Riva", Universidad Politécnica de Madrid, Plaza Cardenal Cisneros 3, E-28040 Madrid, Spain
6 Department of Electronic Engineering, Universitat de València Estudi General (UVEG), Avda. de la Universitat s/n, E-46100 Burjassot, Spain
7 Instituto de Astrofísica de Canarias, Vía Láctea, s/n, E-28080 Santa Cruz de Tenerife, Spain; jtb@iac.es
8 Departamento de Astrofísica, Facultad de Física, Universidad de La Laguna, E-38200 Santa Cruz de Tenerife, Spain
9 Consejo Superior de Investigaciones Científicas, 28006 Madrid, Spain
10 Solar Physics & Space Plasma Research Center (SP2RC), School of Mathematics and Statistics, University of Sheffield, Hounsfield Road, Sheffield S3 7RH, UK; robertus@sheffield.ac.uk
11 Department of Astronomy, Eötvös Loránd University, Pázmány Péter sétány 1/A, H-1117 Budapest, Hungary
12 Gyula Bay Zoltan Solar Observatory (GSO), Hungarian Solar Physics Foundation (HSPF), Petőfi tér 3., H-5700 Gyula, Hungary
13 STFC-RAL Space, Harwell Campus, Didcot, Oxford OX11 0QX, UK
14 Mullard Space Science Laboratory, University College London, Dorking, Surrey, Dorking RH5 6NT, UK
15 Astrophysics Research Centre, School of Mathematics and Physics, Queen's University Belfast, University Road, Belfast BT7 1NN, UK; m.mathioudakis@qub.ac.uk (M.M.)
16 Centre Spatial de Liège, STAR Institute, University of Liège (ULiège), B-4000 Liège, Belgium; ckintziger@uliege.be
17 Department of Astronomy, Institute for Solar Physics, Stockholm University, AlbaNova University Centre, SE-106 91 Stockholm, Sweden
18 INAF—Osservatorio Astrofisico di Torino, Via Osservatorio 20, 10025 Pino Torinese, Italy
19 Department of Mathematics, Physics and Electrical Engineering, Northumbria University, Newcastle upon Tyne NE1 8ST, UK; eamon.scullion@northumbria.ac.uk
* Correspondence: orozco@iaa.es

Abstract: Measuring magnetic fields in the inner corona, the interface between the solar chromosphere and outer corona, is of paramount importance if we aim to understand the energetic transformations taking place there, and because it is at the origin of processes that lead to coronal heating, solar wind acceleration, and of most of the phenomena relevant to space weather. However, these measurements are more difficult than mere imaging because polarimetry requires differential photometry. The coronal magnetograph mission (CMAG) has been designed to map the vector magnetic field, line-of-sight velocities, and plane-of-the-sky velocities of the inner corona with unprecedented spatial and temporal resolutions from space. This will be achieved through full vector spectropolarimetric observations using a coronal magnetograph as the sole instrument on board a spacecraft, combined with an external occulter installed on another spacecraft. The two spacecraft will maintain a formation flight distance of 430 m for coronagraphic observations, which requires a 2.5 m occulter disk radius.

The mission will be preferentially located at the Lagrangian L5 point, offering a significant advantage for solar physics and space weather research. Existing ground-based instruments face limitations such as atmospheric turbulence, solar scattered light, and long integration times when performing coronal magnetic field measurements. CMAG overcomes these limitations by performing spectropolarimetric measurements from space with an external occulter and high-image stability maintained over time. It achieves the necessary sensitivity and offers a spatial resolution of 2.5″ and a temporal resolution of approximately one minute, in its nominal mode, covering the range from 1.02 solar radii to 2.5 radii. CMAG relies on proven European technologies and can be adapted to enhance any other solar mission, offering potential significant advancements in coronal physics and space weather modeling and monitoring.

Keywords: corona; magnetic field; instrumentation; polarimeters; radiative transfer

1. Introduction

1.1. State-of-the-Art in Coronal Science

The solar corona, extending from the transition region into the heliosphere, exhibits fascinating characteristics that challenge our current understanding. The temperature of the corona rises rapidly in its lower layers, reaching temperatures of a few million Kelvin, while the gas pressure decreases sharply through the transition region and then decreases smoothly upward. Within the inner corona, complex magnetic fields give rise to highly structured, bright features known as streamers and coronal loops. These structures, influenced by the solar dynamo and activity cycle, remain enigmatic in terms of their magnetic properties.

Coronal loops, in particular, are hotter and denser than the surrounding plasma, yet the mechanisms responsible for their heating and filling remain elusive [1]. The overall high temperature of the global corona itself also poses a significant mystery. Unraveling the processes of coronal heating is considered a paramount challenge in solar physics. The lack of knowledge about the magnetism of the inner corona further complicates our understanding of these phenomena. This region of the corona hosts intricate magnetic field reconnection processes, leading to the generation of flares and energetic particle events across a broad range of scales.

Flares emit radiation across various wavelengths, from radio waves to hard X-rays and even gamma rays. Many large flares are accompanied by coronal mass ejections (CMEs), which are the primary sources of space weather effects on Earth, potentially triggering severe geomagnetic storms [2]. CMEs also generate powerful shock waves as the plasma expands during their formation and propagation through the corona. Additionally, a multitude of magnetohydrodynamic (MHD) waves can be observed propagating upward into the heliosphere within the inner corona [3,4]. It is evident that magnetic fields play a crucial role in governing all aspects of the inner solar corona, driving the solar wind and influencing space weather phenomena. However, the detailed properties of these magnetic fields remain largely unexplored [5].

The CMAG mission offers an opportunity to address key questions related to the coronal heating problem, flare and CME formation, and the generation and energy transport of waves. By providing fundamental data on the vector magnetic field and velocity measurements of the inner corona at unprecedented spatial and temporal resolutions, CMAG would enable substantial advancements in coronal models beyond existing empirical approximations. Moreover, it would contribute to more accurate space weather predictions, ultimately enhancing our understanding of the Sun and its influence on the heliosphere, the Earth's environment, and our technological systems and society.

1.2. Observing the Inner Solar Corona

Observing the inner solar corona presents a significant challenge primarily due to the faintness of the emitted light from the coronal material compared to the overall brightness of the Sun. Consequently, observing the solar corona requires blocking the light emitted by the solar disk. This can be accomplished during a solar eclipse, which only lasts for a few minutes, or through the use of specialized instruments called coronagraphs. However, the solar coronal light becomes intermingled with background signals originating from scattered light, mainly from the occulter but also from the instrument optics. Despite these difficulties, extensive efforts have been made to carry out coronal observations using both ground- and space-based assets.

Instruments for coronal observations include LASCO [6] and UVCS [7], both aboard SoHO [8]; STEREO/SECCHI [9]; METIS [10] and SoloHI [11] on Solar Orbiter [12]; WISPR [13] onboard the Parker Solar Probe [14]; ASPIICS [15] on Proba-3; the High-Altitude Observatory (HAO)/National Center for Atmospheric Research (NCAR) Coronal Multichannel Polarimeter (CoMP) [16] and the upgraded version uCoMP [17]; and CRYO-NIRSP [18] on DKIST [19]. However, while imaging the corona poses challenges, determining its magnetic field is even more difficult due to the need for differential photometry to measure the four Stokes parameters of polarized light. The aforementioned ground-based instruments, uCoMP and CRYO-NIRSP, have capabilities for magnetic field measurements and provide valuable insights into coronal dynamics, which is known to be driven by the conversion of magnetic energy into radiative energy (in the case of flares) and the kinetic energy of material (in the case of coronal mass ejections or CMEs). Nevertheless, these measurements are significantly hampered by atmospheric turbulence, solar scattered light, and long integration times, resulting in a limited understanding of the intricate details of these energy transformations. However, new diagnostics have emerged that can provide insights into the magnetic fields of the inner corona. For instance, a newly developed diagnostic technique that involves atomic radiation modeling of the contribution of a magnetically induced transition to the Fe X 25.7262 nm spectral line intensity was presented in [20–22] and used, later, to obtain the magnetic field in active region loops [23] using Hinode's [24] Extreme-ultraviolet Imaging Spectrometer EIS [25]. More recently, Yang and collaborators [26] presented a method that utilizes spectroscopic measurements and MHD assumptions to indirectly estimate the strength of the coronal magnetic field. These techniques may be very useful in the future, in the context of other space missions such as SPARK [27]. However, what is truly needed is direct observation of the polarization signals associated with the magnetic field (see a review in [28]). In this respect, it is of interest to note that in 2015, the Chromospheric Lyman-alpha Spectropolarimeter (CLASP) mission obtained the first observations of the linear polarization due to scattering processes in the hydrogen Lyman-alpha line of the solar-disk radiation, while in 2019 and 2021, the Chromospheric Layer Spectropolarimeter (CLASP2 and CLASP2.1) missions obtained unprecedented observations of the linear and circular polarization signals caused by scattering processes and the Hanle and Zeeman effects in the near-UV region of the Mg II h and k lines. These suborbital space experiments opened a window for probing the magnetism and geometry of the plasma in the upper chromosphere and transition region of the Sun [28]. The magnetic field in the inner corona can also be determined using measurements at radio wavelengths [29]. These measurements, in addition to providing fundamental information to understand how the magnetic field is structured, can be of great importance for obtaining simultaneous observations of the inner corona.

The CMAG mission aims to overcome the aforementioned limitations by observing the inner corona from 1.02 $R_\odot$ to 2.5 $R_\odot$ with a consistent spatial resolution of 2.5″, a required spectral resolution of less than 40 pm, a cadence of one minute, and a polarimetric sensitivity of 10^{-3} in the polarization continuum in nominal operation. Achieving these ambitious goals is only feasible through space-based observations and the utilization of an externally occulted coronagraph. Figure 1 represents the squared field of view of CMAG superimposed on a simulated K-corona [30] brightness image derived from a magnetohydrodynamic model (Predictive Science Inc. https://www.predsci.com/portal/home.php, accessed on 1 June 2023). The CMAG mission holds the potential to revolutionize our understanding of the inner corona by providing direct and precise measurements of the magnetic field and shed light on the processes underlying coronal dynamics, solar flares, and CMEs.

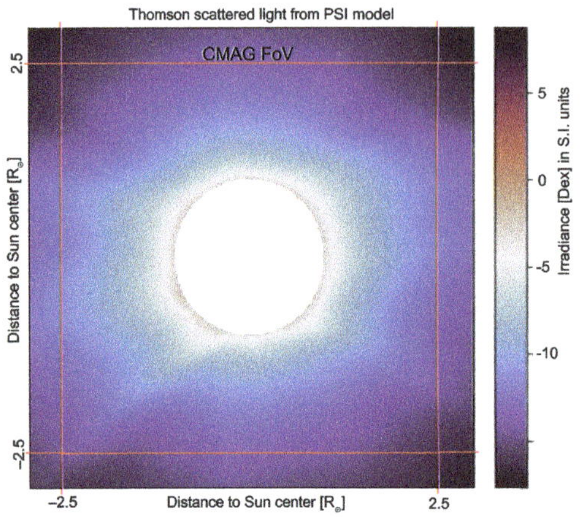

Figure 1. Simulated solar K-corona generated from a magnetohydrodynamic model. The field-of-view (red square) corresponds to 2.5 $R_\odot$, matching that of CMAG.

1.3. Solar Corona Polarimetry

Nowadays, the inference of magnetic fields from spectropolarimetric measurements of the forbidden coronal lines, Fe XIV 530 nm, Fe XI 789 nm, and Fe XIII 1074 nm, has become possible with the interpretation of the Zeeman, scattering, and Hanle effects [31,32]. The Zeeman effect polarizes light from spectral lines due to the presence of magnetic fields in a moderate to strong regime. The Hanle effect, on the other hand, depolarizes light that is already polarized by scattering (or other physical mechanisms) in the presence of weak magnetic fields [33,34]. To measure polarization in coronal lines, high-precision differential photometry is mandatory, and indeed, all four Stokes parameters of solar light must be measured. Quiet coronal line intensities (Stokes I; imaging) are one order of magnitude above the K-corona intensity at 1.10 $R_\odot$ [35]. Linear polarization signals (Stokes Q, U) are weaker, about 1 to 10% of the line core intensity, and hence subject to larger contamination by the already polarized solar corona (the so-called K-corona is linearly polarized by scattering to a large degree). Stokes V is even feebler than linear polarization (a few percent or per mille). Linear polarization is mainly proportional to the degree of anisotropy of the incident light coming from the solar disk, which depends on the height above the Sun's visible surface at where we observe, while the Stokes V amplitude is mainly proportional to the line-of-sight component of the field.

Recent numerical models predict Stokes V amplitudes slightly above 0.1% in coronal loops for the Fe XIII 1074 nm line [32], in line with earlier predictions [36] that suggested amplitudes of about 0.5×10^{-3} for coronal loops and field strengths as low as 2 G. Therefore, measuring circular polarization in the corona is extremely challenging. Simple magnetographic calculations (see Figure 2) suggest signals barely above 0.1 % in the line emission of the quiet corona if the field is stronger than 10 G and when observing at 1.1 $R_\odot$. If one were to measure the magnetic field in locations further away from the solar limb, it would be necessary to focus on specific solar structures with higher density or significantly increase integration times. Hence, observations of magnetic fields in the inner corona are very scarce. Those we currently have are severely limited in terms of spatial and temporal resolution. For example, Lin and collaborators [37] obtained the first observations, requiring a spatial sampling of $20''$ and a temporal integration of over an hour to detect circular polarization signals in the inner corona. Recent measurements suggest much stronger fields, in the range of 50–200 G, for cold and hot coronal loops [23,38], hence predicting larger Stokes V signals in these structures. Coronal electron densities change from 10^9 cm^{-3} at the bottom to 10^6 at 2 $R_\odot$. Therefore, the sensitivity of measurements decreases with distance. However, in loop structures, the density reaches 10^{11} cm^{-3}, and emissivity is larger by one order of magnitude, making the signal-to-noise ratio (S/N) go up, hence increasing polarization sensitivity for equivalent exposure times. Electron density reaches 10^{12} cm^{-3} in flaring loops [1]. When more photons are available, CMAG will achieve the required polarization sensitivity in less time, with an FoV that covers from 1.02 to 2.5 $R_\odot$, and a spatial resolution of 2.5″ at 750 nm. Perhaps the most relevant aspect of the mission is that, in addition to its capabilities, it will take observations from space with unprecedented pointing stability, enabling the measurement of polarization signals in fainter solar structures. Remarkably, no existing or planned mission is able to tackle these challenging measurements, despite the efforts of the solar community over the last few decades. Solar missions with specific instrumentation for performing spectropolarimetry in the corona, such as the solar magnetism explorer (SolmeX) [39] or the Magnetic Imaging of the Outer Solar Atmosphere (MImOSA) [40] missions, have already been proposed. Certainly, the CRYO-NIRSP instrument [18] will likely provide excellent data on the inner corona, although it is expected that the observations will be much more limited in terms of time and have very restricted fields of view compared to CMAG, which aims to observe the entire inner corona. Currently, there are ground-based spectropolarimetric observations of the inner corona, such as those from uCoMP, at a spatial resolution of 6″ and a field of view covering from 1.03 to 2 solar radii. Linear polarization information has already been retrieved although circular polarization still remains elusive so far. It is worth mentioning that on 2 September 2023, the Indian Space Research Organisation launched the Aditya-L1 mission (https://www.isro.gov.in/Aditya_L1-MissionDetails.html accessed on 15 November 2023). The mission features the Visible Emission Line Coronagraph (VELC), developed by the Indian Institute of Astrophysics in Bangalore. The primary objective of VELC is to measure the coronal magnetic field through spectropolarimetric observations of the Fe XIII 1074 nm line.

1.4. Scientific Goals of the CMAG Mission

The mission aims to answer the overarching question, "How does magnetic energy transformation occur in the inner corona, driving coronal heating and all eruptive phenomena relevant to space weather effects on Earth?" This global question breaks down into specific questions described below.

1. What is the structure of the inner coronal magnetic field? Understanding the magnetic field's structuring in coronal loops and streamers and its evolution and relationship with photospheric fields is crucial. CMAG's high spatial resolution (2.5″) and cadence (≈1 min) will determine the strength and direction of the vector magnetic field from close to the surface to the higher layers for providing an empirical basis that help validate and refine current extrapolations and improve coronal magnetic models.
2. What is the coronal origin of the solar wind? The solar wind has distinct components: slow and fast, related to coronal streamers, coronal hole boundaries, and the solar network. CMAG will investigate the magnetic connection between coronal holes, streamers, and the solar wind, aiding the understanding of how the solar wind propagates through the heliosphere and its relationship with coronal waves.
3. How are flares and CMEs generated? Flares and CMEs are phenomena with different effects on the heliosphere and the Earth's environment. CMAG will reveal the inner corona's internal structure and evolution and follow the magnetic field vector during the CME onset and the generation of associated waves. At a resolution of 2.5″, it could also provide a glimpse of the evolution of the magnetic field structure at the reconnection sites.

4. What is the origin of solar coronal heating? Measurements of magnetic field strength will provide a quantitative basis for assessing the origin of non-thermal heat observed in the Sun's inner coronal layers.
5. What is the role of the inner corona in space weather? CMAG's measurements of inner coronal magnetic fields at L5, provided in almost real time, will greatly enhance space weather nowcasting and forecasting capabilities, crucial for predicting radiation and energetic particle impacts on Earth as a consequence of flares and CMEs.

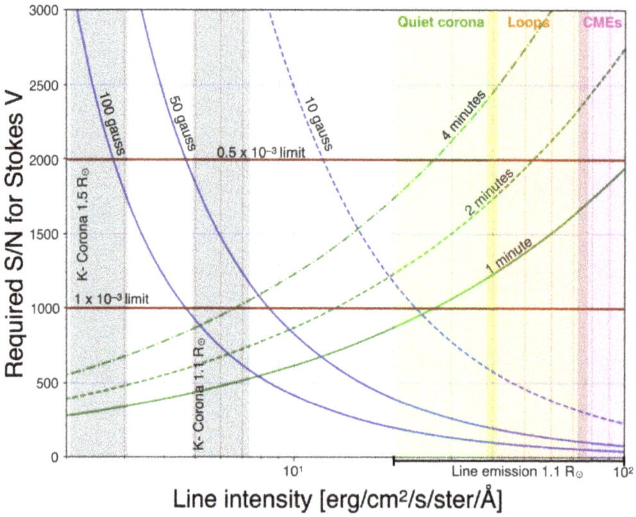

Figure 2. Detectability diagram. Calculations assume the weak field approximation on lines with a Doppler width of 20 km/s [31]. Emission intensities at different regions are represented with shaded areas. Diagram regions below blue lines indicate magnetic field strength detectability for Fe XIII 1074 nm and no atomic alignment. Diagram regions below green lines indicate the reached S/N in Stokes V for the given exposure times with a 10 cm aperture telescope (Table 1) at 1 au from the Sun and a total transmission of 0.4, sensor quantum efficiency of 50%, and a polarimetric efficiency of 0.5.

Table 1. CMAG spacecraft main instrument baseline requirements.

Spatial resolution	2.5″ at 750 nm
Aperture	10 cm
Cadence	≈1 min for S/N = 1000 in Stokes Q, U, and V
Field of view (FoV)	≈2400″ (≈0.66° = 2.5 $R_\odot$)
Targets spectral lines	Fe XIV 530 nm; Fe XI 789 nm; Fe XIII 1074 nm
Spectral resolution	20 pm; 30 pm; 40 pm
Maximum spectral shift over the FoV	25 pm; 30 pm; 50 pm
Detector dimensions	4 k × 4 k pixels
Etalon max. incidence angle [1]	1.28°
Etalon diameter	≈60 mm
Polarization modulator interm. image diameter	≈55 mm
dimensions	<1 m^3
stray light requirements	<0.1% @ 1.02 $R_\odot$

[1] Etalons placed in collimated configuration are often slightly tilted with respect to the incident beam. The etalon tilt induces a wavelength blueshift across the FoV.

1.5. Top Science Requirements

To achieve its science goals, CMAG must collect as many photons as possible to make the required measurements of all four Stokes parameters of the selected spectral line(s) to achieve the required polarimetric performance. The necessary spatial, spectral, and temporal resolutions will be possible through full vector spectropolarimetric observations using a coronal magnetograph as the sole instrument on board a spacecraft, hereafter referred to as the CMAG spacecraft. This will be combined with an external occulter installed on another spacecraft, hereafter referred to as the occulter spacecraft. These two spacecraft will maintain a formation flight distance of 430 m for coronagraphic observations. At that distance, the radius of the occulter has to be around two meters. The observational periods will be of, at least, 2.5 h every twelve hours (this assumes a mission to L5 and a certain solid fuel capacity that may vary). The main scientific and design requirements are listed in Table 1. Since polarimetry demands precise pointing stability (Performance Drift Error, PDE, must be better than 0.125″ in one minute), both spacecraft will adhere to stringent requirements (see next Section).

CMAG's measurements complement those of METIS aboard the Solar Orbiter mission by providing access to the coronal magnetic field. While our goals are similar to those of the (on-ground) large coronagraph of COSMO (the COronal Solar Magnetism Observatory project [41]), we use a much smaller aperture telescope (10 cm vs. 170 cm). Nonetheless, we expect to obtain sharper coronal structure maps due to the absence of atmospheric seeing and a larger throughput, resulting in high S/N observations in less time. Moreover, externally occulted coronagraphy ensures significantly lower scattered light levels (by three orders of magnitude [42]) when the occulter is positioned at 430 m, increasing our polarimetric sensitivity and allowing us to observe much closer to the solar limb (at 1.02 $R_\odot$) where scattered light effects are more detrimental (see Figure 3).

The approved NASA's PUNCH mission [43], expected to be launched in 2025, follows a loose formation flying concept and aims to explore the corona and solar wind by imaging light scattered from free electrons from 6 to 180 solar radii. Missions such as PUNCH will complement the science being carried out now by other missions such as Solar Orbiter, enabling space weather scientists to trace solar phenomena from their origin to their potential impacts on Earth. CMAG's expected benefits will be a significant leap in this endeavor, provided that it is operational in a reasonable time to overlap with the above missions. DKIST will also benefit from pointing its extremely high spatial resolution instrumentation to targets provided by CMAG. Additionally, CMAG is expected to be a full member of the battery of instruments enabling multimessenger solar astronomy, combining its measurements not only with remote-sensing coronagraphs but also with in situ instruments aboard missions such as Solar Orbiter and Parker Solar Probe, ushering in a new era of solar physics [44].

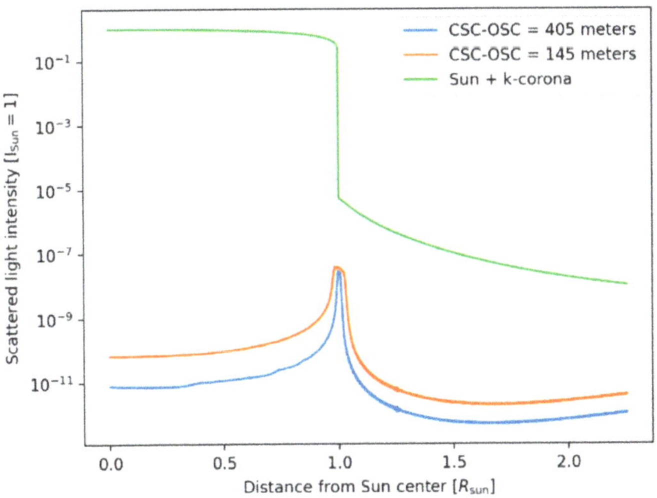

Figure 3. Solar scattered light intensity as a function of the distance to the solar disk center.

2. Mission Concept

2.1. Mission Profile

The mission consists of two spacecraft, the CMAG and the occulter spacecraft, flying in formation at the L5 Lagrangian point for a duration of 7 years. The occulter spacecraft must carry the necessary equipment for enabling formation flying, as well as the occulter disk. The CMAG spacecraft will carry the main scientific payload instrument: a coronal magnetograph. During observations, CMAG will be positioned approximately 430 m behind the occulter, significantly reducing solar disk scattered light by at least one order of magnitude as compared to the Proba-3 mission. Figure 3 shows the relevance of a high intersatellite distance. Unity represents the disk intensity. The regular bright profile without any instrument is plotted in green. The use of an external occulter such as that in Proba-3 at about 140 m from the coronagraph reduces the levels significantly (orange). However, if the occulter is at about 400 m, as in the case of CMAG, the background of scattered light decreases by one more order of magnitude, hence making the observation of the solar corona easier in a unique way. Moreover, the diffraction effects near the limb are much better suppressed, which allows for closer observation to the solar limb, much closer to the transition region of the Sun. This reduction is crucial to extract the polarimetric observations needed to measure the coronal vector magnetic field. Indeed, critical heights of CME onset and prominence instabilities take place from 1.03 to 1.1 $R_\odot$ [45], so those heights close to the solar limb are critical to observe if one wants to follow these events at all their evolutionary stages.

The mission's choice of the Lagrangian L5 point, which positions CMAG at an angle of 60 degrees away from the Sun–Earth line, holds a significant advantage for solar physics and space weather research altogether. This positioning allows us to effectively identify the formation of potentially eruptive structures within the critical region of the corona, often referred to as the "danger" zone for Earth. Moreover, from this unique vantage point, we gain the capability to observe these structures during their gradual ascent in what is known as the "slow-rise phase". This level of observation and insight into solar phenomena is unattainable from the Earth-based perspective and is crucial for enhancing our understanding of solar activities that can impact our space environment. In any case, it is worth noting that the observation of magnetic fields in the inner corona from space remains a significant challenge in solar physics. Therefore, the majority of our (stand-alone) solar objectives could still be achievable from, for example, the Lagrangian L1 point.

CMAG observations can be performed simultaneously and without interference to any other instrumentation aboard the occulter spacecraft. Both spacecraft can be launched together and then transferred to the Lagrangian point. The CMAG spacecraft will travel as a small piggyback of the main primary mission. Both will have the necessary metrology equipment to maintain the right formation flying configuration with respect to one another.

Earth–Sun system Lagrangian points, being in solar orbit, guarantee minimum propellant consumption for maintaining the formation due to their relatively small gravitational perturbations compared to any Earth-bound orbit. Preliminary analyses show that with current technology, mission performance needs can be accomplished. However, in addition, several opportunities have been identified that could potentially improve this further upon design consolidation. A radio-frequency link will be implemented between both spacecrafts providing two key services: on one hand as a data link, on the other hand as a first metrology step, determining relative position thanks to differential RF ranging technology. The main spacecraft, which also carries the occulter, will act as a communication relay to transmit the necessary CMAG data products to the Science Operation Centre. No dedicated communications transceiver between CMAG and the ground segment is necessary, although a low-bit-rate data link with the ground station is foreseen for housekeeping, telemetries, and telecommands. The primary spacecraft, which carries the two-meter radius occulter (making it the main and likely heavier of the two spacecrafts), can also accommodate remote or in situ instrumentation. However, this does not imply that communications with the ground will be established directly from CMAG. It can be devised as a stand-alone mission as well.

The baseline for the CMAG platform is close to a service-like module with a weight of 300 kg and a maximum power consumption of 100 W, excluding any potential structural redesign or additional propellant needs. For the Attitude and Orbit Control System (AOCS), CMAG will carry star trackers and reaction wheels for a fine 3-axis-stabilized pointing mode, combined with coarse rate and Sun sensors for safe operations. A cold gas propulsion system will be used for both relative position control and reaction wheel unloading, ensuring formation flight. The CMAG spacecraft will be able to generate power when outside the occulter shadow and will have active thermal control. The battery capacity will be sized for the foreseen scientific operations, which take place within the shadow cast by the main spacecraft.

A metrology sensor suite is key for mission robustness and for achieving the required performance. Metrology must be accurate enough to meet the mission's scientific requirements and robust enough, at the same time, to guarantee recovering formation flight from a completely lost state after any potential system failure and a smooth handover between the two spacecraft. Four key elements make these operations seamless: the Formation Flying Radio Frequency (FFRF) metrology, based on differential RF ranging from a reduced set of antennas, provides almost omnidirectional coverage with comparatively low accuracy; the Visual-Based System (VBS), placed on the occulter spacecraft and always pointing away from the Sun's direction, can detect a light pattern on CMAG and is capable of robustly providing lateral relative position estimation with relatively high accuracy, as well as coarse longitudinal estimation while maintaining a wide enough field of view to allow for both the Science and the Stand-By Modes (see below); the Fine Longitudinal System (FLS) laser-based metrology that reflects on a corner-cube retroreflector on the occulter spacecraft and provides high-accuracy longitudinal estimation; and the Shadow Position Sensor (SPS), a limb detector that corrects the lateral positioning of the CMAG spacecraft with respect to the occulter shadow. The last two systems guarantee a spacecraft's positioning within a 1 mm error margin in the perpendicular plane (lateral) to the line-of-sight and along it (longitudinal). Relative positioning stability (lateral) is 0.3 mm, while the absolute pointing error and stability are 2.5$''$ and 1$''$ in 1 min, respectively.

2.2. Mission Operations Baseline and Spacecraft Concept

CMAG would be launched in a stacked configuration, piggybacking on the occulter spacecraft. It will be commissioned during the initial stages of the transfer to the Lagrangian point. After arrival, formation flying commissioning and calibration will take place. The CMAG operational phase will comprise three stages (see Figure 4). Science Mode (SM), Stand-By Mode (SBM), and Parking Mode (PM). During SM, CMAG will position the instrument aperture at the necessary location within the occulter shadow and perform station-keeping with high accuracy, using the highest precision elements in the AOCS and metrology chain, and a propulsion system. As the occulter limits the solar panel's power, the shadow time will be limited to the energy storage capacity. Preliminary estimates suggest a duration of approximately 2.5 h until battery depletion.

SBM is needed to optimize mission resources when not performing fine position control. The spacecraft will exit the shadow region to charge the battery while remaining within the coarse metrology's operational range to ensure a swift transition back to fine positioning when needed. This mode also allows for propellant saving since the positioning requirements are much less stringent. The current recharging time estimation is seven hours. A 9.5 h SBM period will then allow for a 12 h observation cycle, making it easier to synchronize CMAG's observations with the ground segment. PM ensures safety against collision for extended periods, although the Lagrangian points already ensure safe and simple operations without collision risk. PM allows CMAG a long-term, passive relative orbit for commissioning, fault recovery, or maintenance. PM can be entered when commanded from the ground. SM and SBM will be fixed and defined by design, autonomously managed by the CMAG spacecraft without operator intervention. The proposed design baselines two observational cycles per day, representing the worst-case scenario, with each cycle providing 2.5 h of scientific operations.

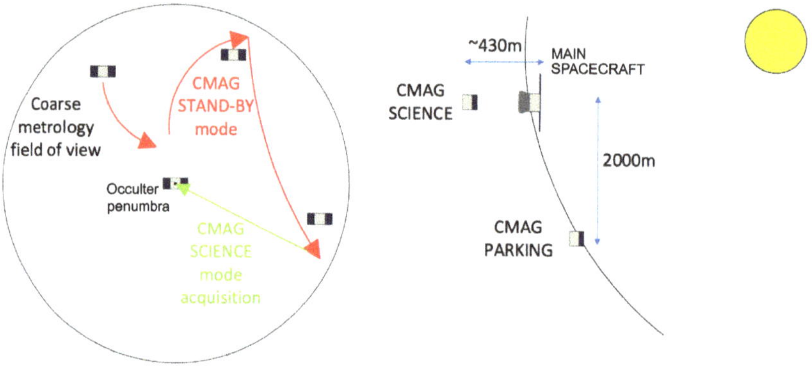

Figure 4. CMAG concept of operations. Autonomous operations (**left**) and non-operational orbit (**right**).

2.3. Payload Concept Design

The proposed CMAG instrument is an imaging spectropolarimeter that measures all four Stokes parameters at given wavelength samples of selected emission spectral lines from the corona: Fe XIV 530 nm, Fe XI 789 nm, and Fe XIII 1074 nm. These spectral lines are among the ones showing the largest intensity above the K-corona. No other mission aims at carrying out vector magnetography and mapping of the LOS and plane-of-the-sky velocities of the highly dynamic inner corona. The baseline design (see Figure 5), with an entrance pupil diameter of 10 cm, combines the first stage of the full disk telescope from the Polarimetric and Helioseismic Imager (PHI) [46] instrument aboard Solar Orbiter with the conceptual design of the TuMag magnetograph [47] for the SUNRISE III mission [48]. Two athermalized doublets focus the Sun on an intermediate focal plane. A field lens placed before that focal plane re-images the entrance pupil to infinity to make the image telecentric. The doublets are optimized to produce the best image of the external occulter where an internal occulter is placed. A polarization modulation package (PMP) based on Liquid Crystal Variable Retarders (LCVRs) carries out the polarization modulation right after a filter wheel containing three prefilters corresponding to the three target spectral lines. A collimator, consisting of a telephoto system to reduce the total track of the instrument, re-images the Sun onto infinity and focuses the pupil on its image focal plane, where the etalon and the Lyot stop are placed. Finally, a camera telephoto system refocuses the image onto the detector with a fixed spatial sampling of 1.25″ (2.5″ resolution). If the size limitations of the instrument allow it, the camera system will provide a telecentric image.

The instrument electronics include: a Data Processing Unit (DPU) for the overall control of the instrument and data processing, communications with the S/C, to carry out the main processing functionalities, and with sufficient storage to keep 2.5 h worth of scientific data; an Analog, Mechanism, and Heater Drivers board (AMHD); a High-Voltage Power Supply (HVPS) for tuning the etalon; a Power Converter Module (PCM); and an Electrical Distribution System (EDS). The DPU is the core of the instrument electronics and has to be tailored to perform data acquisition and accumulation, demodulation, retrieval of the different subproducts (see Section 2.4), and compression. The AMHD will be in charge of synchronizing the PMPs and the etalon, taking care of the thermal stabilization of the more critical elements, and managing housekeeping data for the overall control of the instrument, among other analog tasks. The PCM is the direct interface to the platform and should hence fulfill the necessary requirements and redundancy. The spacecraft will have image stabilization capabilities to guarantee the required 0.125″ PDE stability in one minute. The E-unit will be equipped with the necessary firmware (heritage of the Solar

Orbiter PHI instrument complex on-board data processing) to provide the required data products. On-board preprocessing of payload data will be key for the reduction in the downlink telemetry.

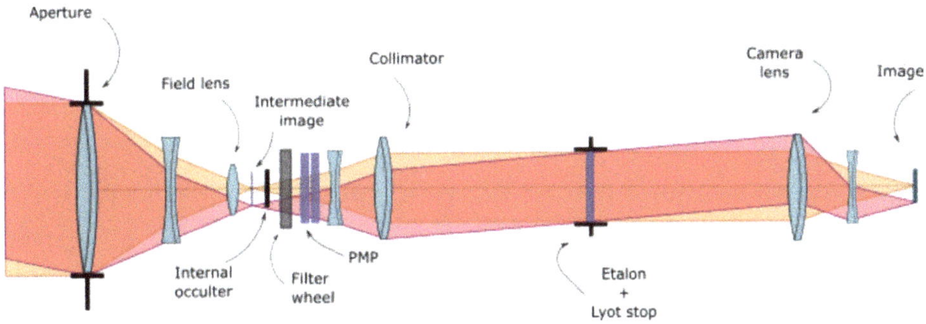

Figure 5. CMAG coronagraph instrument concept.

2.4. CMAG Data Products

The instrument will observe the full Stokes vector in several wavelength samples across the aforementioned coronal lines. To transform the observations into scientific data products, most of them will be processed in the DPU, including dark and flat-fielding corrections, demodulation, normalization, and subsequent transformation to scientific products, minimizing telemetry usage. The expected data products are maps of polarization brightness (K-corona), the line-of-sight plasma velocity component, line Doppler widths, the total line polarization, the (non-disambiguated) orientation of the magnetic field vector, and the amplitude and sign of Stokes V. The latter two products can readily be calculated from the linear polarization signals and from the circular polarization (using the weak field approximation) since the lines lie in the strong limit of the Hanle effect [33,49] in optically thin coronal plasma at locations where circular polarization is above the noise level of 10^{-3}. To increase the polarimetric sensitivity, longer integration times (by adding datasets, reducing spatial resolution, or increasing exposure time) are foreseen in well-defined cases. For instance, when moving away from the solar limb and in the absence of solar structures such as coronal loops, the coronal density drops considerably with distance. Therefore, meeting the signal-to-noise requirement necessitates longer exposure times. In order to infer the full magnetic field vector, downloading raw or preprocessed data is necessary. This is because, to fully interpret optically thin forbidden coronal lines, including collisional excitation and resonance scattering, one needs high computing resources. Other products will be determined on the ground, e.g., coronal density maps can be inferred from the polarization brightness. Intermediate products are also foreseen. For instance, brightness temperatures can also be inferred provided the spectral line profile is available. CMAG will also provide forward modeling tools for the interpretation of the products on the ground and for aiding field extrapolations and modeling of the solar corona.

3. A Mission Opportunity

The CMAG mission was proposed to ESA in February 2022 in response to ESA's call for a "Fast" mission opportunity. In that proposal, the main platform was assumed to be the current ESA's Space Safety Vigil mission, which, with a few configuration modifications and additional equipment, can act as an external occulter and enable formation flying. The combination with the Vigil mission at the L5 Lagrangian vantage point ensures a unique, low-risk, cost-effective, and interdisciplinary scientific return considerably improving current coronal models and space weather predictions with a clear benefit for both the Science and the Space Safety programs of ESA.

The use of Vigil as an external occulter presents an exceptional opportunity for the solar physics and space weather communities, as it allows establishing a connection between photospheric magnetic activity (observed with the Photospheric Magnetic Field Imager instrument, PMI, aboard Vigil [50]) and the corona through CMAG's continuous monitoring of the magnetic field and plasma velocities in the inner corona. Combining CMAG with Vigil's solar instrumentation ensures optimal scientific return and increased space weather forecasting capabilities, which are not expected from any other space agency in the near future, resulting in a significant qualitative leap.

However, CMAG can be adapted to any other space mission, provided it fulfills the science requirements. Modifications to the main mission will be necessary. The main spacecraft needs to carry an occulter disk, which can come from redesigned solar panels or a retractable structure, ensuring the mission's remote sensing payload remains unperturbed. The intersatellite distance, estimated to be 430 m, requires a 2.5 m disk radius. The distance between the two spacecraft, as well as the final external occulter's disk radius, has been defined based on scientific requirements. Additionally, careful consideration has been given to ensure that the mission accommodates the dimensions of the Ariane 6 fairing. The chosen distance is also sufficiently large to prevent any protrusions from potential additional instrumentation on board the occulter spacecraft. This meticulous planning aims to optimize the scientific objectives while ensuring compatibility with the spacecraft's launch configuration. The metrology suite of instruments for formation flight does not necessarily interfere with the main AOCS system design, resulting in an estimated total mass increment of 24 kg, including the occulter disk. CMAG can be used as a science data relay, transmitting data through an intersatellite link to the main mission, which downloads them to the Ground Station. Operationally, CMAG observations will be performed autonomously, providing the necessary data products. It can also be used on demand if the scientific team needs particular observations.

Author Contributions: Conceptualization, D.O.S., J.C.d.T.I., F.J.B.M., M.B.J., D.Á.G., D.S., L.F.P., L.R.B.R., J.A., J.T.B., R.E., J.A.D., L.M.G., S.A.M., I.P.G., I.T.G., E.S.K., J.L.G.B., D.M.L., M.M., C.K., J.L., S.F. and E.S.; methodology, D.O.S., J.C.d.T.I., F.J.B.M., M.B.J., D.Á.G., D.S., L.F.P., L.R.B.R. and J.A.; software, A.V.-R., D.H.E., F.J.B.M. and B.R.C.; investigation, D.O.S., J.C.d.T.I., F.J.B.M., D.Á.G., D.S., L.F.P. and L.R.B.R.; writing—original draft preparation, D.O.S. and J.C.d.T.I.; writing—review and editing, F.J.B.M., M.B.J., D.Á.G., D.S., L.F.P., L.R.B.R., J.A., A.V.-R., I.P.G., I.T.G., E.S.K., J.L.G.B., D.H.E., B.R.C., J.T.B., R.E., J.A.D., L.M.G., S.A.M., D.M.L., M.M., C.K., J.L., S.F. and E.S.; visualization, D.O.S., F.J.B.M., J.A., A.V.-R. and D.S.; supervision, D.O.S., J.C.d.T.I. and M.B.J.; project administration, J.C.d.T.I. and M.B.J.; funding acquisition, D.O.S. and J.C.d.T.I. All authors have read and agreed to the published version of the manuscript.

Funding: This work was funded by AEI/MCIN/10.13039/501100011033/(RTI2018-096886-C5, PID2021-125325OB-C5) and ERDF "A way of making Europe"; "Center of Excellence Severo Ochoa" awards to IAA-CSIC (SEV-2017-0709, CEX2021-001131-S); Plan Andaluz de Investigación, Desarrollo e Innovación (PAIDI 2020) P20_01307; and a Ramón y Cajal fellowship awarded to DOS. DML is grateful to the Science Technology and Facilities Council for the award of an Ernest Rutherford Fellowship (ST/R003246/1).

Data Availability Statement: No new data were created or analyzed in this study. Data sharing is not applicable to this article.

Conflicts of Interest: The authors declare no conflict of interest. The funders had no role in the design of the study; in the collection, analyses, or interpretation of data; in the writing of the manuscript; or in the decision to publish the results.

Abbreviations

The following abbreviations are used in this manuscript:

CMAG	Coronal magnetograph
ESA	European Space Agency
CME	Coronal mass ejection
MHD	Magnetohydrodynamic
LASCO	Large Angle and Spectrometric Coronagraph

UVCS	Ultraviolet Coronagraph Spectrometer
SOHO	Solar and Heliospheric Observatory
STEREO	Solar TErrestrial RElations Observatory
SECCHI	Sun–Earth Connection Coronal and Heliospheric Investigation
METIS	The multiwavelength coronagraph for the Solar Orbiter mission
SoloHI	Solar Orbiter Heliospheric Imager
ASPIICS	Association of Spacecraft for Polarimetric and Imaging Investigation of the Corona of the Sun
CRYO-NIRSP	Cryogenic Near-IR Spectropolarimeter
DKIST	Daniel K. Inouye Solar Telescope
SPARK	Solar Particle Acceleration Radiation and Kinetics
PDE	Performance Drift Error
COSMO	COronal Solar Magnetism Observatory
NASA	National Aeronautics and Space Administration
PUNCH	Polarimeter to Unify the Corona and Heliosphere
AOCS	Attitude and Orbit Control Subsystem
PHI	Polarimetric and Helioseismic Imager
TuMAG	Tunable Magnetograph

References

1. Reale, F. Coronal Loops: Observations and Modeling of Confined Plasma. *Living Rev. Sol. Phys.* **2014**, *11*, 4. [CrossRef] [PubMed]
2. Temmer, M. Space weather: The solar perspective. *Living Rev. Sol. Phys.* **2021**, *18*, 4. [CrossRef]
3. Erdélyi, R.; Fedun, V. Are There Alfvén Waves in the Solar Atmosphere? *Science* **2007**, *318*, 1572. [CrossRef] [PubMed]
4. Mathioudakis, M.; Jess, D.B.; Erdélyi, R. Alfvén Waves in the Solar Atmosphere. *Space Sci. Rev.* **2013**, *175*, 1. [CrossRef]
5. Aschwanden, M.J. *Physics of the Solar Corona. An Introduction with Problems and Solutions*, 2nd ed.; Library of Congress Control Number: 2005937065; Aschwanden, M.J., Ed.; Praxis Publishing Ltd.: Chichester, UK; Springer: New York, NY, USA; Berlin, Germany, 2005; 892p, ISBN 3-540-30765-6.
6. Brueckner, G.E.; Howard, R.A.; Koomen, M.J.; Korendyke, C.M.; Michels, D.J.; Moses, J.D.; Socker, D.G.; Dere, K.P.; Lamy, P.L.; Llebaria, A.; et al. The Large Angle Spectroscopic Coronagraph (LASCO). *Sol. Phys.* **1995**, *162*, 357. [CrossRef]
7. Kohl, J.L.; Esser, R.; Gardner, L.D.; Habbal, S.; Daigneau, P.S.; Dennis, E.F.; Nystrom, G.U.; Panasyuk, A.; Raymond, J.C.; Smith, P.L.; et al. The Ultraviolet Coronagraph Spectrometer for the solar and heliospheric observatory. *Sol. Phys.* **1995**, *162*, 313. [CrossRef]
8. Domingo, V.; Fleck, B.; Poland, A.I. The scientific payload of the space-based Solar and Heliospheric Observatory (SOHO). *Space Sci. Rev.* **1994**, *70*, 7. [CrossRef]
9. Howard, R.A.; Moses, J.D.; Vourlidas, A.; Newmark, J.S.; Socker, D.G.; Plunkett, S.P.; Korendyke, C.M.; Cook, J.W.; Hurley, A.; Davila, J.M.; et al. Sun Earth Connection Coronal and Heliospheric Investigation (SECCHI). *Space Sci. Rev.* **2008**, *136*, 67. [CrossRef]
10. Antonucci, E.; Romoli, M.; Andretta, V.; Fineschi, S.; Heinzel, P.; Moses, J.D.; Naletto, G.; Nicolini, G.; Spadaro, D.; Teriaca, L.; et al. Metis: The Solar Orbiter visible light and ultraviolet coronal imager. *Astron. Astrophys.* **2020**, *642*, A10. [CrossRef]
11. Howard, R.A.; Vourlidas, A.; Colaninno, R.C.; Korendyke, C.M.; Plunkett, S.P.; Carter, M.T.; Wang, D.; Rich, N.; Lynch, S.; Thurn, A.; et al. The Solar Orbiter Heliospheric Imager (SoloHI). *Astron. Astrophys.* **2020**, *642*, A13. [CrossRef]
12. Müller, D.; St Cyr, O.C.; Zouganelis, I.; Gilbert, H.R.; Marsden, R.; Nieves-Chinchilla, T.; Antonucci, E.; Auchère, F.; Berghmans, D.; Horbury, T.S.; et al. The Solar Orbiter mission. *Astron. Astrophys.* **2020**, *642*, A1. [CrossRef]
13. Vourlidas, A.; Howard, R.A.; Plunkett, S.P.; Korendyke, C.M.; Thernisien, A.F.R.; Wang, D.; Rich, N.; Carter, M.T.; Chua, D.H.; Socker, D.G.; et al. The Wide-Field Imager for Solar Probe Plus (WISPR). *Space Sci. Rev.* **2016**, *204*, 83. [CrossRef]
14. Raouafi, N.E.; Matteini, L.; Squire, J.; Badman, S.T.; Velli, M.; Klein, K.G.; Chen, C.H.K.; Matthaeus, W.H.; Szabo, A.; Linton, M.; et al. Parker Solar Probe: Four Years of Discoveries at Solar Cycle Minimum. *Space Sci. Rev.* **2023**, *219*, 8. [CrossRef]
15. Galano, D.; Bemporad, A.; Buckley, S.; Cernica, I.; Dániel, V.; Denis, F.; de Vos, L.; Fineschi, S.; Galy, C.; Graczyk, R.; et al. Development of ASPIICS: A coronagraph based on Proba-3 formation flying mission. *Proc. SPIE* **2018**, *10698*, 106982Y. [CrossRef]
16. Tomczyk, S.; Card, G.L.; Darnell, T.; Elmore, D.F.; Lull, R.; Nelson, P.G.; Streander, K.V.; Burkepile, J.; Casini, R.; Judge, P.G. An Instrument to Measure Coronal Emission Line Polarization. *Sol. Phys.* **2008**, *247*, 411. [CrossRef]
17. Tomczyk, S.; Landi, E. Solar Heliospheric and INterplanetary Environment (SHINE 2019). Available online: https://www.nsf.gov/pubs/2022/nsf22570/nsf22570.htm (accessed on 1 June 2023).
18. Fehlmann, A.; Kuhn, J.R.; Schad, T.A.; Scholl, I.F.; Williams, R.; Agdinaoay, R.; Berst, D.C.; Craig, S.C.; Giebink, C.; Goodrich, B.; et al. The Daniel K. Inouye Solar Telescope (DKIST) Cryogenic Near-Infrared Spectro-Polarimeter. *Sol. Phys.* **2023**, *298*, 5. [CrossRef]
19. Rimmele, T.R.; Warner, M.; Keil, S.L.; Goode, P.R.; Knölker, M.; Kuhn, J.R.; Rosner, R.R.; McMullin, J.P.; Casini, R.; Lin, H.; et al. The Daniel K. Inouye Solar Telescope—Observatory Overview. *Sol. Phys.* **2020**, *295*, 172. [CrossRef]

20. Li, W.; Grumer, J.; Yang, Y.; Brage, T.; Yao, K.; Chen, C.; Watanabe, T.; Jönsson, P.; Lundstedt, H.; Hutton, R.; et al. A Novel Method to Determine Magnetic Fields in Low-Density Plasma Facilitated through Accidental Degeneracy of Quantum States in Fe^{9+}. *Astrophys. J.* **2015**, *807*, 69. [CrossRef]
21. Li, W.; Yang, Y.; Tu, B.; Xiao, J.; Grumer, J.; Brage, T.; Watanabe, T.; Hutton, R.; Zou, Y. Atomic-Level Pseudo-Degeneracy of Atomic Levels Giving Transitions Induced by Magnetic Fields, of Importance for Determining the Field Strengths in the Solar Corona. *Astrophys. J.* **2016**, *826*, 219. [CrossRef]
22. Judge, P.G.; Hutton, R.; Li, W.; Brage, T. On the Fine Structure Splitting of the $3p^43d^4D_{5/2}$ and $3p^43d^4D_{7/2}$ Levels of Fe_x. *Astrophys. J.* **2016**, *833*, 185. [CrossRef]
23. Brooks, D.H.; Warren, H.P.; Landi, E. Measurements of Coronal Magnetic Field Strengths in Solar Active Region Loops. *Astrophys. J.* **2021**, *915*, L24. [CrossRef]
24. Kosugi, T.; Matsuzaki, K.; Sakao, T.; Shimizu, T.; Sone, Y.; Tachikawa, S.; Hashimoto, T.; Minesugi, K.; Ohnishi, A.; Yamada, T.; et al. The *Hinode* (Solar-B) Mission: An Overview. *Sol. Phys.* **2007**, *243*, 3. [CrossRef]
25. Culhane, J.L.; Harra, L.K.; James, A.M.; Al-Janabi, K.; Bradley, L.J.; Chaudry, R.A.; Rees, K.; Tandy, J.A.; Thomas, P.; Whillock, M.C.R.; et al. The EUV Imaging Spectrometer for Hinode. *Sol. Phys.* **2007**, *243*, 19. [CrossRef]
26. Yang, Z.; Bethge, C.; Tian, H.; Tomczyk, S.; Morton, R.; Del Zanna, G.; McIntosh, S.W.; Karak, B.B.; Gibson, S.; Samanta, T.; et al. Global maps of the magnetic field in the solar corona. *Science* **2020**, *369*, 694. [CrossRef]
27. Reid, H.A.S. *The Solar Particle Acceleration Radiation and Kinetics (SPARK) Mission Concept*; University College London: London, UK, 2024. [Manuscript in preparation]
28. Trujillo Bueno, J.; del Pino Alemán, T. Magnetic Field Diagnostics in the Solar Upper Atmosphere. *Annu. Rev. Astron. Astrophys.* **2022**, *60*, 415. [CrossRef]
29. Alissandrakis, C.E.; Gary, D.E. Radio Measurements of the Magnetic Field in the Solar Chromosphere and the Corona. *Front. Astron. Space Sci.* **2021**, *7*, 77. [CrossRef]
30. Computation of the Thomson Scattering of the K-Corona. Available online: https://github.com/aliciavr/thomsonpy (accessed on 1 June 2023).
31. Judge, P.G.; Casini, R.; Tomczyk, S.; Edwards, D.P.; Francis, E. Technical Report, PB2002-102493; NCAR/TN-466-STR, 02. 2001. Available online: https://opensky.ucar.edu/islandora/object/technotes:304 (accessed on 1 June 2023).
32. Schad, T.; Dima, G. Forward Synthesis of Polarized Emission in Target DKIST Coronal Lines Applied to 3D MURaM Coronal Simulations. *Sol. Phys.* **2020**, *295*, 98. [CrossRef]
33. Landi Degl'Innocenti, E.; Landolfi, M. *Polarization in Spectral Lines*; Degl'innocenti, E.L., Landolfi, M., Eds.; University of Fierenze: Firenze, Italy; Arcetri Observatory: Firenze, Italy; Kluwer Academic Publishers: Dordrecht, The Netherland, 2004; Volume 307. [CrossRef]
34. Trujillo Bueno, J. Atomic Polarization and the Hanle Effect. *Adv. Sol.-Polarim.—Theory Obs. Instrum.* **2001**, *236*, 161. [CrossRef]
35. Judge, P.G. Spectral Lines for Polarization Measurements of the Coronal Magnetic Field. I. Theoretical Intensities. *Astrophys. J.* **1998**, *500*, 1009. [CrossRef]
36. Judge, P.G.; Casini, R. Advanced Solar Polarimetry: Theory, Observation, and Instrumentation. *Adv. Sol.-Polarim.—Theory Obs. Instrum.* **2001**, *236*, 503.
37. Lin, H.; Kuhn, J.R.; Coulter, R. Coronal Magnetic Field Measurements. *Astrophys. J.* **2004**, *613*, L177. [CrossRef]
38. Kuridze, D.; Mathioudakis, M.; Morgan, H.; Oliver, R.; Kleint, L.; Zaqarashvili, T.V.; Reid, A.; Koza, J.; Löfdahl, M.G.; Hillberg, T.; et al. Mapping the Magnetic Field of Flare Coronal Loops. *Astrophys. J.* **2019**, *874*, 126. [CrossRef]
39. Peter, H.; Abbo, L.; Andretta, V.; Auchère, F.; Bemporad, A.; Berrilli, F.; Bommier, V.; Braukhane, A.; Casini, R.; Curdt, W.; et al. Solar magnetism eXplorer (SolmeX). *Exp. Astron.* **2012**, *33*, 271. [CrossRef]
40. Peter, H.; Ballester, E.A.; Andretta, V.; Auchère, F.; Belluzzi, L.; Bemporad, A.; Berghmans, D.; Buchlin, E.; Calcines, A.; Chitta, L.P.; et al. Magnetic imaging of the outer solar atmosphere (MImOSA). *Exp. Astron.* **2022**, *54*, 185. [CrossRef]
41. Tomczyk, S.; Burkepile, J.; Casini, R.; Corchado-Albelo, M.; DeLuca, E.; de Toma, G.; de Wijn, A.; Dikpati, M.; Fan, Y.; Farid, S.; et al. COSMO: The COronal Solar Magnetism Observatory. *Bull. AAS* **2023**. [CrossRef]
42. Rougeot, R.; Flamary, R.; Galano, D.; Aime, C. Performance of the hybrid externally occulted *Lyot* solar coronagraph. *Astron. Astrophys.* **2017**, *599*, A2. [CrossRef]
43. Deforest, C.; Killough, R.; Gibson, S.; Henry, A.; Case, T.; Beasley, M.; Laurent, G.; Colaninno, R.; Waltham, N. Polarimeter to UNify the Corona and Heliosphere (PUNCH): Science, Status, and Path to Flight. In Proceedings of the 2022 IEEE Aerospace Conference, Big Sky, MT, USA, 5–12 March 2022; p. 11. [CrossRef]
44. Martinez Pillet, V.; Tritschler, A.; Harra, L.; Andretta, V.; Vourlidas, A.; Raouafi, N.; Alterman, B.L.; Bellot Rubio, L.; Cauzzi, G.; Cranmer, S.R.; et al. Solar Physics in the 2020s: DKIST, Parker Solar Probe, and Solar Orbiter as a Multi-Messenger Constellation. *Proc. Int. Astron. Union* **2020**, *18*, 3–16. [CrossRef]
45. James, A.W.; Williams, D.R.; O'Kane, J. Evolution of the critical torus instability height and coronal mass ejection likelihood in solar active regions. *Astron. Astrophys.* **2022**, *665*, A37. [CrossRef]
46. Solanki, S.K.; del Toro, Iniesta, J.C.; Woch, J.; Gandorfer, A.; Hirzberger, J.; Alvarez-Herrero, A.; Appourchaux, T.; Martínez Pillet, V.; Pérez-Grande, I.; Sanchis Kilders, E.; et al. The Polarimetric and Helioseismic Imager on Solar Orbiter. *Astron. Astrophys.* **2020**, *642*, A11. [CrossRef]

47. Del Toro Iniesta, J.C.; Orozco Suárez, D.; Álvarez-Herrero, A. Instituto de Astrofísica de Andalucía, Granada, Spain, 2024. [Manuscript in preparation].
48. Solanki, S.K. Max Planck Institute for Solar System Research, Göttingen, Germany, 2024. [Manuscript in preparation].
49. Casini, R.; White, S.M.; Judge, P.G. Magnetic Diagnostics of the Solar Corona: Synthesizing Optical and Radio Techniques. *Space Sci. Rev.* **2017**, *210*, 145. [CrossRef]
50. Staub, J.; Fernandez-Rico, G.; Gandorfer, A.; Gizon, L.; Hirzberger, J.; Kraft, S.; Lagg, A.; Schou, J.; Solanki, S.K.; del Toro Iniesta, J.C.; et al. PMI: The Photospheric Magnetic Field Imager. *J. Space Weather Space Clim.* **2020**, *10*, 54. [CrossRef]

aerospace

MDPI

Concept Paper

The Large Imaging Spectrometer for Solar Accelerated Nuclei (LISSAN): A Next-Generation Solar γ-ray Spectroscopic Imaging Instrument Concept

Daniel F. Ryan [1,*], Sophie Musset [2], Hamish A. S. Reid [3], Säm Krucker [1,4], Andrea F. Battaglia [1,5], Eric Bréelle [6], Claude Chapron [6], Hannah Collier [1,5], Joel Dahlin [7], Carsten Denker [8], Ewan Dickson [9], Peter T. Gallagher [10], Iain Hannah [11], Natasha L. S. Jeffrey [12], Jana Kašparová [13], Eduard Kontar [11], Philippe Laurent [6], Shane A. Maloney [10], Paolo Massa [14], Anna Maria Massone [15], Tomasz Mrozek [16], Damien Pailot [6], Melody Pallu [6], Melissa Pesce-Rollins [17], Michele Piana [15], Illya Plotnikov [18], Alexis Rouillard [18], Albert Y. Shih [7], David Smith [19,20], Marek Steslicki [16], Muriel Z. Stiefel [1,5], Alexander Warmuth [8], Meetu Verma [8], Astrid Veronig [9], Nicole Vilmer [21], Christian Vocks [8] and Anna Volpara [15]

1 FHNW School of Engineering, University of Applied Sciences and Arts Northwestern Switzerland, Bahnhofstrasse 6, 5210 Windisch, Switzerland; andrea.battaglia@fhnw.ch (A.F.B.); muriel.stiefel@fhnw.ch (M.Z.S.)
2 European Space Research and Technology Centre (ESTEC), European Space Agency, 2201 Noordwijk, The Netherlands
3 Mullard Space Science Labratory, University College London, Holmbury St. Mary, Dorking RH5 6NT, UK
4 Space Sciences Laboratory, University of California, Berkeley, 7 Gauss Way, Berkeley, CA 94708, USA
5 Institute for Particle Physics and Astrophysics (IPA), Swiss Federal Institute of Technology in Zurich (ETHZ), Wolfgang-Pauli-Strasse 27, 8039 Zurich, Switzerland
6 Astroparticule et Cosmologie, Université Paris Cité, CNRS, CEA, F-75013 Paris, France
7 NASA Goddard Space Flight Center, 8800 Greenbelt Road, Greenbelt, MD 20771, USA; albert.y.shih@nasa.gov (A.Y.S.)
8 Leibniz-Institut für Astrophysik Potsdam (AIP), An der Sternwarte 16, 14482 Potsdam, Germany; awarmuth@aip.de (A.W.); mverma@aip.de (M.V.)
9 Institute of Physics & Kanzelhöhe Observatory, University of Graz, Universitätsplatz 5, 8010 Graz, Austria
10 Dublin Institute of Advanced Studies, 31 Fitzwilliam Place, Dublin D02 XF86, Ireland; peter.gallagher@dias.ie (P.T.G.)
11 School of Physics & Astronomy, University of Glasgow, University Avenue, Glasgow G12 8QQ, UK; iain.hannah@glasgow.ac.uk (I.H.)
12 Department of Mathematics, Physics & Electrical Engineering, Northumbria University, Newcastle upon Tyne NE1 8S, UK
13 Astronomical Institute of the Czech Academy of Sciences (CAS), 251 65 Ondřejov, Czech Republic; jana.kasparova@asu.cas.cz
14 Department of Physics & Astronomy, Western Kentucky University, Bowling Green, KY 42101, USA
15 Methods for Image and Data Analysis (MIDA), Dipartimento di Matematica, Università di Genova, Via Dodecaneso 35, I-16146 Genova, Italy; piana@dima.unige.it (M.P.)
16 Centrum Badań Kosmicznych, PAN, ul. Bartycka 18a, 00-716 Warszawa, Poland; tmrozek@cbk.pan.wroc.pl (T.M.)
17 Istituto Nazionale di Fisica Nucleare (INFN-Pisa), 56127 Pisa, Italy
18 Institut de Recherche en Astrophysique et Planétologie (IRAP), National Center for Space Studies (CNES), Université Toulouse III, 31062 Toulouse, France
19 Physics Department, University of California, Santa Cruz, 1156 High St., Santa Cruz, CA 95064, USA
20 Santa Cruz Institute for Particle Physics, University of California, Santa Cruz, 1156 High St., Santa Cruz, CA 95064, USA
21 Laboratory of Space Studies and Instrumentation in Astrophysics (LESIA), CNRS–UMR 8109, Observatoire de Paris, 5 Place J. Janssen, 92195 Meudon, France; nicole.vilmer@obspm.fr
* Correspondence: daniel.ryan@fhnw.ch

Abstract: Models of particle acceleration in solar eruptive events suggest that roughly equal energy may go into accelerating electrons and ions. However, while previous solar X-ray spectroscopic imagers have transformed our understanding of electron acceleration, only one resolved image of γ-ray emission from solar accelerated ions has ever been produced. This paper outlines a new satellite instrument concept—the large imaging spectrometer for solar accelerated nuclei (LISSAN)—with the capability not only to observe hundreds of events over its lifetime, but also to capture multiple images per event, thereby imaging the dynamics of solar accelerated ions for the first time. LISSAN provides spectroscopic imaging at photon energies of 40 keV–100 MeV on timescales of $\lesssim$10 s with greater sensitivity and imaging capability than its predecessors. This is achieved by deploying high-resolution scintillator detectors and indirect Fourier imaging techniques. LISSAN is suitable for inclusion in a multi-instrument platform such as an ESA M-class mission or as a smaller standalone mission. Without the observations that LISSAN can provide, our understanding of solar particle acceleration, and hence the space weather events with which it is often associated, cannot be complete.

Keywords: particle acceleration; magnetic reconnection; instrumentation; techniques: imaging spectroscopy; corona; coronal mass ejections (CMEs); flares; X-rays; gamma rays

1. Introduction

Particle acceleration is one of the earliest observable consequences of impulsive magnetic energy release, a process that drives many phenomena in the solar corona. Among these are solar eruptive events, the most geo-effective form of space weather. In the standard model of solar eruptive events (Figure 1; see also [1]), the coronal magnetic field becomes highly stressed due to turbulent motions at the "footpoints" below. Here, the magnetic field passes between the lower atmosphere, where the plasma dictates the motion of the field (high-β plasma), to the corona, where the field dictates the motion of the plasma (low-β plasma). Under certain conditions, the excess energy in the field can be suddenly released via magnetic reconnection, which rapidly rearranges the field into a lower stress configuration (transition from blue to green field lines in Figure 1). Depending on the magnetic configuration, the plasma, magnetic field and accelerated particles may escape into the heliosphere as coronal mass ejections (CMEs), "jets" or solar energetic particle (SEP) events. Charged particles accelerated downwards spiral along magnetic field lines towards the footpoints, depositing their energy via Coulomb collisions as they propagate. Because the ambient density in the chromospheric footpoints is orders of magnitude greater than in the corona, most of the particles' energy is deposited there. This rapidly heats and ionises the chromospheric plasma, which ablates back into the corona along the magnetic field lines in a process known as chromospheric evaporation. This produces coronal loops that glow at several million kelvin. The intense broadband radiation produced throughout this process is known as a solar flare.

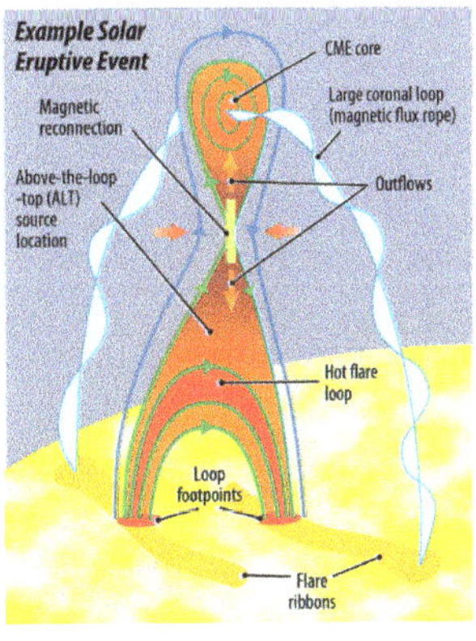

Figure 1. Standard model of a solar eruptive event. Magnetic reconnection releases stored energy in the coronal magnetic field which accelerates charged particles and heats local plasma, causes the growth of flare loop arcades, and can result in the ejection of plasma and the magnetic field into the heliosphere (coronal mass ejection, CME). Figure courtesy of [2,3].

Particles in solar eruptive events can be divided into three populations: accelerated ions, accelerated electrons and hot plasmas. Accelerated ions in the range of 1–100 MeV/nucleon can be detected via various lines in the γ-ray range 1–10 MeV due to nuclear de-excitation, neutron capture, and positron annihilation [4–11]. Accelerated ions with energies greater than 100 MeV/nucleon can be detected via the decay of secondary pions via nuclear reactions with the ambient medium. As the pions decay, the decay products produce a broad-band continuum at photon energies above 10 MeV, with a broad peak around 70 MeV from neutral pion radiation [12,13]. Hot plasmas ($\gtrsim$5 MK) and non-thermally accelerated electrons can be characterised via the hard X-ray (HXR; $\gtrsim$5 keV) bremsstrahlung emission, which is produced as they scatter in the ambient medium. The bremsstrahlung spectrum reflects the velocity distribution of the particles that produced it and can be inverted to calculate the spectrum of the emitting electrons [14,15]. Hot plasmas can therefore be distinguished by their Maxwellian-shaped (thermal) spectrum, which dominates at lower energies ($\lesssim$20 keV), and accelerated electrons can be distinguished by their power-law-shaped (non-thermal) spectrum, which tends to dominate at higher energies ($\gtrsim$15 keV) (e.g., [16–18]). Moreover, unlike other signatures (e.g., radio emission), X-rays can provide measurements of the numbers and energies of accelerated electrons, and hence provide a unique insight into the underlying acceleration process.

The Reuven Ramaty High Energy Solar Spectroscopic Imager (RHESSI; [19]) was the first instrument to simultaneously produce images and spectra of solar eruptive events in HXR on timescales of several seconds. Consequently, its observations have transformed our understanding of electron acceleration and the evolution of solar eruptive events. The Spectrometer/Telescope for Imaging X-rays (STIX; [20]) onboard Solar Orbiter is furthering this understanding by producing similar diagnostics at a higher time cadence from different viewing angles than that of Earth. Despite this, neither RHESSI nor STIX provide sufficient sensitivity and imaging capability in the γ-ray regime to reliably observe ion acceleration. An image of solar γ-ray line emission has only been produced for one solar flare, and the emission centroid has been localised in an additional four [21,22]. Moreover, solar γ-ray imaging has never been performed on timescales sufficient to reveal the spatio-temporal evolution of accelerated ions during a flare. Hence, the role of ion acceleration in solar eruptive events remains largely unknown. This is despite evidence that flare-accelerated ions may carry comparable energy to that of electrons (e.g., [23–25]). Thus, our understanding of solar particle acceleration and the space weather events with which it is often associated cannot be complete until we can reliably observe this missing half of the picture.

In this paper, we outline the large imaging spectrometer for solar accelerated nuclei (LISSAN) satellite instrument concept. LISSAN was originally conceived as part of a suite of instruments for an ESA M-class mission concept (SPARK; [26]). It provides reliable imaging and the spectral characterisation of both electron and ion acceleration in solar eruptive events for the first time. This will enable it to answer several important open questions regarding solar particle acceleration and the initiation of space weather events. We discuss some of the most pressing of these in Section 2 and define the instrument performance required to address them in Section 3. We also lay out LISSAN's expected performance based on its design, which we describe in Section 4. Finally, in Section 5, we summarise the LISSAN instrument concept and its capabilities for revealing new science concerning ion acceleration in solar eruptive events.

2. Unresolved Questions Regarding Ion Acceleration in Solar Eruptive Events

While our understanding of electron acceleration in solar eruptive events has been greatly enhanced in recent decades by instruments like RHESSI and STIX, our understanding of ion acceleration is poorly constrained due to the tiny number of γ-ray spectroscopic imaging observations available. In this section, we discuss three major outstanding questions regarding solar ion acceleration and the observations required to resolve them (Figure 2).

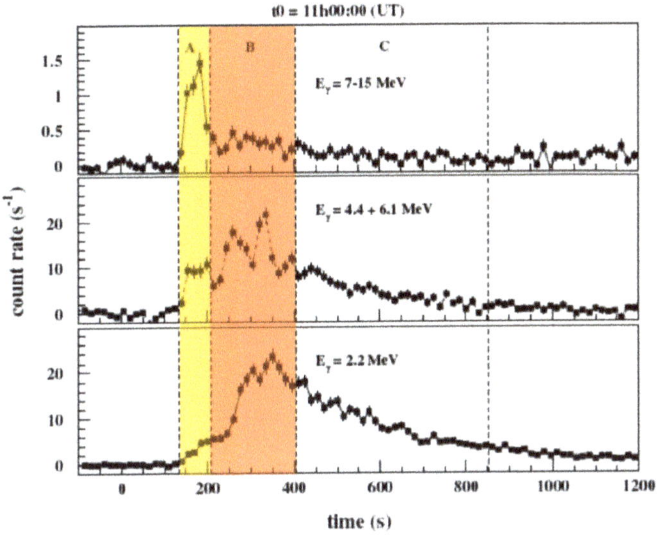

Figure 2. Time histories of the γ-ray emission during the 28 October 2003 X17-class flare due to electron bremsstrahlung (upper panel) and the nuclear line emission (prompt lines: $h\nu$ = 4.4 and 6.1 MeV; delayed neutron capture line 2.223 MeV; INTEGRAL). The dashed lines and shading denote different phases of the flare. Initially (phase A, yellow), impulsive non-thermal bremsstrahlung emission from accelerated electrons dominates. As this ceases, increases in prompt line and delayed neutron capture emission is seen (phase B, orange) before it slowly decays (phase C, white). Figure taken from [27]. For more details, see that study.

2.1. Do Ion and Electron Acceleration in Solar Eruptive Events Share a Close Physical Relationship?

It is still unclear whether there is a close physical link between ion and electron acceleration in solar eruptive events, or whether different mechanisms dominate for the different species at different energies. For example, first-order Fermi acceleration in magnetic islands is expected to operate similarly for both ions and electrons [28], resulting in comparable spectral indices [29]. However, efficient energisation requires pre-heating, for which the responsible mechanisms likely differ between electrons and ions [30,31]. If ion and electron acceleration do share a close physical relationship, we might expect their emission to be well correlated in space and time. However, the few observations to date appear to paint an inconclusive and contradictory picture.

One study [32] examined event-integrated fluences in many events of bremsstrahlung emission above 300 keV from accelerated electrons and 2.223 MeV line emission indirectly produced by protons accelerated above 30 MeV. It found that these signatures correlate well over three orders of magnitude, suggesting that there is a close physical relationship between electron and ion acceleration. However, a detailed study of the 28 October 2003 X17 event showed differences in the spatially integrated time evolution of accelerated electrons and ions during the course of a flare [27]. This found that the emission from accelerated electrons initially dominated (Figure 2, top panel, phase A, yellow) while the γ-ray line emission from accelerated ions became prominent later (Figure 2, middle and lower panels, phase B, orange). Other studies [11,21] have observed differences in the relative locations of electron HXR and ion 2.2 MeV γ-ray line emission in three of the five flares for which γ-ray emission could be localised. Note that "localised" is not the same as "imaged". A single γ-ray image was produced for only one of these events (28 October 2003; Figure 3). For the others, only the emission centroid could be inferred via localisation techniques, but not its size or morphology. Moreover, the spatial uncertainty associated with localisation is greater than when a resolved image can be produced. Nonetheless, the statistically significant separations determined with localisation techniques between electron HXR and ion γ-ray sources may suggest different acceleration mechanisms and/or paths taken by electrons and ions as they precipitate through the solar atmosphere.

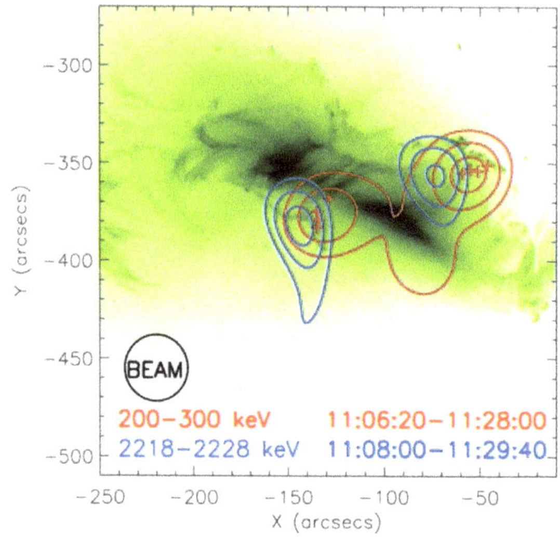

Figure 3. The 50%, 70%, and 90% contours of γ-ray (blue) and HXR (red) images made with RHESSI were overlaid on a TRACE 195 Å image of the 28 October 2003 X17-class flare. The red plus signs indicate the 200–300 keV HXR footpoint locations for successive adjacent intervals of 100, 120, 180, and 240 s. This figure was taken from [22].

To reveal the temporal and spatial connections between accelerated electrons and ions, γ-ray and HXR emission must be imaged simultaneously for a significant number of solar eruptive events. Moreover, they must be imaged on timescales that reveal their evolution during the impulsive phase of flares when electron acceleration is known to be prominent. The one HXR-γ-ray flare image produced to date (Figure 3) is insufficient to achieve these goals.

2.2. How Efficient Is Ion Acceleration in Solar Eruptive Events?

The fraction of particles accelerated out of the ambient Maxwellian velocity distribution and the total energy they contain are essential constraints on acceleration models. Acceleration by magnetic islands [33] and super-Dreicer electric fields in a reconnecting current sheet [34] can accelerate a large fraction of the available electrons and perhaps ions, while mechanisms relying on large-scale sub-Dreicer electric fields cannot [35]. In the few extreme cases where RHESSI observed non-thermal HXR sources near the believed acceleration site above flaring loops, the interpretation of their spectra suggested that, essentially, all electrons in the acceleration region were accelerated in a bulk-energisation process [36,37]. It is unknown how such a process can be sustained given the limited number of electrons available in the acceleration region, although return currents repopulating the region is a possibility [38–42].

In contrast, even less is known about the efficiency of the ion acceleration process. The relative number of accelerated particles of different ion species can be constrained by the shapes and fluences of various γ-ray lines (e.g., alpha/proton ratio; [27]). The relative abundances of heavier accelerated ions can be determined from the relative fluences of broad de-excitation lines produced when heavy energetic ions interact with ambient H and He. Because these lines are broad, they are difficult to distinguish from the underlying continua formed by electrons and nuclei. One study computed both narrow and broad γ-ray line spectra for the most abundant elements and accelerated ions in the solar atmosphere [43]. It compared elemental abundances of a single flare observed with SMM/GRS with similar abundances measured in space for large proton flares and impulsive 3He events. Although the uncertainties were large for most elements, Ref. [43] found that Mg and Fe were significantly enhanced (>3–4σ) relative to C and O, similarly to impulsive 3He-rich events. However, since this study only examined one event, it is unclear whether this result is representative of ion acceleration in flares. Observations of a behind the limb flare with GRANAT/PHEBUS showed that enhancements of heavy ions (Ne-Fe) increased with time over the course of the flare, before finally reaching the highest values observed in solar energetic particle events by in situ measurements [44].

The relative importance as a function of time of the narrow (proton and alpha-particle) and broad (heavier accelerated species) lines would be key evidence for the progress of the energisation process. However, observations to date have not been able to support sufficiently comprehensive studies to determine this.

2.3. *How Are the Most Energetic Particles in the Solar Atmosphere Accelerated?*

Studies of small numbers of events examining the γ-ray lines (1–10 MeV) and pion continuum (>10 MeV) suggest that the accelerated ion spectrum is not a simple power law extending from the non-relativistic (1–100 MeV/nucleon) to relativistic (>few hundred MeV/nucleon) regimes [45–49]. These studies were performed with observations from different instruments, and so uncertainties in cross-calibration mean that their results are not conclusive. They nonetheless raise the question of whether the most energetic particles are accelerated via a different mechanism.

The longevity of some pion emission presents another major challenge to our understanding of how most energetic solar particles are accelerated (e.g., [50]). Once again, the limited observations to date paint an inconclusive and contradictory picture. Fermi-LAT observed emission for more than two hours in 21 solar flares, almost half of the pion-producing flares observed in the first 13 years of the mission. This raises the question of why such high-energy particles are present in the solar atmosphere, despite the apparent absence of other flare signatures [51]. The fact that most accelerated electrons are seen within the first several minutes of a flare suggests that the long-duration pion emission is caused by an acceleration mechanism with no close physical link to that of the electrons. The often quoted alternative is acceleration at a CME-driven shock. One study [52] analysed 30 solar eruptive events with long-duration γ-ray emission and their associations with CMEs and relativistic ions observed near Earth. They concluded that the γ-ray emission is most likely due to back-precipitation from a CME-driven shock. However, another analysis of 14 events [53] found no correlation between the observed interplanetary ions and the numbers required to produce the observed γ-ray emission, casting doubt on the CME scenario. In addition, it has been shown that, due to strong magnetic mirroring, the fraction of ions accelerated at a CME that would back-precipitate to the chromosphere should be in the order of 0.5–1% [54], significantly smaller than that derived from observations [53]. The 20 January 2005 flare demonstrated the possibility that very high-energy ions are accelerated in the flare process, rather than in the interplanetary medium [55–57]. Meanwhile, the Fermi-LAT observations of the 7 March 2012 flare suggest that the emission above 100 MeV migrated from the active region towards the western limb during the first 10 h. This supports the idea that particle acceleration could be occurring at an extended source [58,59]. However, Fermi-LAT's angular resolution is too coarse to resolve the γ-ray sources and the inferred source locations are associated with large uncertainties.

Unravelling the relative roles of flare and interplanetary processes in accelerating high-energy ions is crucial to understanding long-duration events and their potential space weather consequences. However, it is clear that higher sensitivity observations in the HXR and γ-ray regimes are required to sufficiently examine the timing and spectra of electron bremsstrahlung and pion decay radiation. Moreover, this must be performed for a significant number of events, as flare and interplanetary processes may dominate in different scenarios.

3. LISSAN Instrument Requirements

To address the scientific questions outlined in Section 2, LISSAN must meet the performance requirements listed in the middle column of Table 1. A lower energy limit of 40 keV is required to observe the HXR emission from accelerated electrons. A higher energy limit of 80 MeV is required to measure the broadly peaked pion decay emission, which peaks at 70 MeV. To distinguish emission in different parts of the coronal structure, a spatial resolution of $10''$ is required. To detect γ-ray line emission in an active region, and possibly pion decay sources away from the main site of energy release, a field of view with a diameter of $\geq 10'$ is required. To follow the spatio-temporal evolution of the γ-ray line emission, an imaging cadence of 10 s is required. To characterise the alpha/proton ratio via the shapes of the C and O nuclear lines 4.4 and 6.1 MeV ($\sim$1.5–4% FWHM; [60]), an energy resolution of 1.5% FWHM is required at 6.1 MeV. To detect the γ-ray emission from M-class flares, and characterise the temporal evolution of the γ-ray emission during intense events, effective area for imaging of 75 cm^2 and an effective area for spectroscopy of 400 cm^2 are required at 2.2 MeV.

The right column of Table 1 gives LISSAN's expected performance based on its design (Section 4) which meets or exceeds the requirements in all categories. LISSAN represents a significant advance over any previous satellite-based solar γ-ray spectroscopic imager. For example, LISSAN's imaging effective area is 25 times greater than RHESSI's at 2.2 MeV and its effective area for spectroscopy at 2.2 MeV is 20 times greater. This will enable hundreds of solar γ-ray events to be imaged and for multiple images to be produced during individual flares. This is a drastic improvement over the single resolved γ-ray flare image produced to date. LISSAN's improved imaging time cadence will help reveal how accelerated ions evolve in space and time, and how their importance relative to electrons changes during flares. The greater number of events observed will include a range of flare magnitudes, which will help reveal the different conditions that affect the ion acceleration process(es). Despite being crucial to understanding how solar eruptive events occur and evolve, such observations have never before been possible.

Table 1. *Middle column*: Performance required to address the scientific questions outlined in Section 2. *Right column*: LISSAN's predicted performance based on the design outlined in Section 4. In all cases, LISSAN's predicted performance meets or exceeds the requirements.

LISSAN Parameter	Performance Requirement	Predicted Performance
Energy range—low	40 keV	40 keV
Energy range—high	80 MeV	100 MeV
Imaging effective area (2.2 MeV)	75 cm^2	100 cm^2
Spectral effective area (2.2 MeV)	400 cm^2	440 cm^2
Sensitivity (2.2 MeV)	50 photons/cm^2	50 photons/cm^2
Sensitivity (6.1 MeV)	5 photons/cm^2	5 photons/cm^2
Imaging time resolution	10 s	1 s
Angular resolution	$<10''$	$8''$
Field of view	$>10'$ diameter	$13'$ diameter
Energy resolution (6.1 MeV)	1.5% dE/E	1.5% dE/E
Largest observable flare	>X5	>X5

4. LISSAN Design

In this section, we first provide a high-level overview of LISSAN's design that enables it to provide the observational capabilities described in Section 3. We then describe LISSAN's indirect imaging system in more detail before moving onto a more technical discussion of LISSAN's main components. We finish with a brief discussion of LISSAN's spacecraft interface requirements and environmental constraints.

4.1. Instrument Concept

LISSAN reveals the dynamics of accelerated ions and electrons via spectroscopic imaging on timescales of 10 s in a photon energy range of 40 keV–100 MeV. Its basic design is outlined in Figure 4. LISSAN employs an indirect Fourier imaging technique which is valid across its entire spectral range (Section 4.2). Spatial information is encoded into 15 moiré patterns by 15 pairs of slightly offset grids (bigrids) separated by a fixed distance. The moiré patterns each represent a "spatial frequency" at a certain orientation on the sky and can be combined on the ground into images via various Fourier reconstruction techniques. Each bigrid is aligned with a pixelated GAGG crystal scintillator detector which records the time, location, and energy of each incoming photon. This enables both the moiré pattern and spectrum to be reproduced via the same imaging system, hence providing simultaneous imaging and spectroscopy. This imaging concept was pioneered and successfully employed by STIX [20,61]. In addition to its 15 imaging subcollimators (bigrids plus detector), LISSAN has an additional five gridless subcollimators. Four are used for enhanced sensitivity spectroscopy, as the absence of bigrids amplifies the photon throughput to the detector by a factor of 4 at the cost of imaging information. The fifth is used for monitoring background.

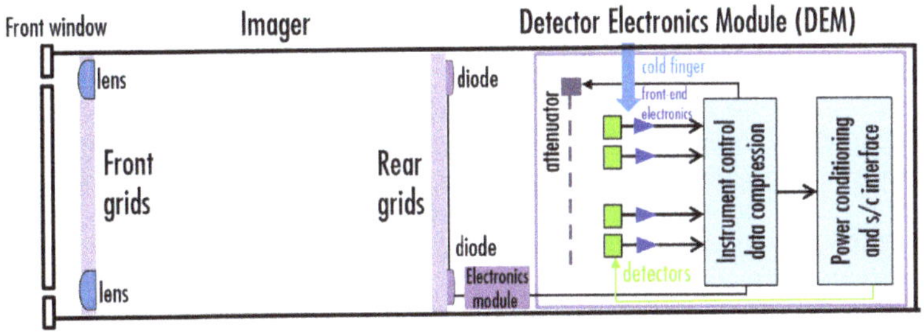

Figure 4. Block diagram showing the main components of LISSAN's design.

Incident photon flux can be moderated by a moveable attenuator that automatically responds to count rate thresholds. This avoids issues that corrupt the measured spectrum such as pile-up, which occurs when the rate at which photons are detected approaches the maximum rate at which the detector system can count. Due to the higher penetrating power of more energetic photons, the attenuator preferentially blocks lower energy X-rays. Thus, lower count rates can be achieved without impacting LISSAN's ability to observe γ-rays from accelerated ions. Moreover, since solar X-ray spectra increase exponentially at lower energies, enough X-rays can still be observed to characterise electron acceleration.

The data processing unit (DPU) performs low-level processing and the compression of data received from the detector system's electronics. It also performs instrument control functions and interfaces with the spacecraft.

Detection of faint solar X-ray and γ-ray signals requires a low background flux. This depends on many factors, including the satellite orbit and phase of the solar cycle. For an orbit at L1 during solar maximum, approximately 2 protons $cm^{-2}\,s^{-1}$ are expected to be incident on the satellite due to isotropic cosmic rays. These will be shielded from LISSAN's detectors by the surrounding spacecraft, as well as by tungsten passive shielding around the detector plane. A more reliable background estimate can be computed via a Monte Carlo Geant4 simulation of the whole system. A plastic scintillator can also be placed below the detectors to tag high-energy particles (GeV cosmic rays), and produce an anti-coincidence signal to the main detectors. Calculations for previous missions, including INTEGRAL, suggest that active shielding may enable a reduction in the background by a factor of 5–10.

LISSAN's imaging axis is defined by the position of the front grids relative to the rear ones. LISSAN's solar aspect system monitors this orientation relative to the Sun, thereby enabling the coordinates of the reconstructed images to be determined.

LISSAN's design enables it to achieve 40 times greater sensitivity than RHESSI at 6.1 MeV (5 counts/cm^2) and at the 2.2 MeV neutron capture line (50 counts/cm^2). This will enable it to detect, image, and characterise emission from accelerated ions in a significant number of flares for the first time. Moreover, LISSAN's design is scalable. Therefore, if mass, volume, and power budgets allow, LISSAN can be augmented with additional subcollimators. These can be used to enhance LISSAN's imaging capabilities by producing more moiré patterns, and/or further boost sensitivity for spectroscopy. Alternatively, if mass and funding budgets are tight, thinner grids could be used that are partially or fully transparent to the highest energy γ-rays. This would reduce LISSAN's imaging capability at these energies (e.g., the C and O lines), but spectroscopic sensitivity would be enhanced by the greater transmission through the grids. Thinner grids would affect lower energy imaging far less—e.g., the Ne, Mg, Fe, and Si lines—and so still provide valuable information on the locations of accelerated ions. LISSAN would therefore still be able to address many of the science questions outlined in Section 2.

4.2. Indirect Imaging System

Photons cannot be focused easily at the energies observed by LISSAN, and so an indirect imaging technique is required [62]. LISSAN's imaging system is composed of 15 imaging subcollimators, each composed of a detector and a pair of grids (bigrids) of parallel slits and slats. Each pair of grids is separated by 3 m and have slightly mismatched pitches (slit/slat size). This casts a moiré pattern onto the detector as photons pass through the slits and are absorbed by the slats (Figure 5). The moiré pattern represents a Fourier imaging component, also known as a complex visibility. This is a spatial sampling on the plane of the sky whose angular resolution is determined by the grid pitch and its direction by the bigrid orientation. The visibility's phase (shift of the moiré pattern's peak across the detector) and amplitude (difference between the moiré pattern's peak and trough intensities) are sensitive to both the extent and position of the source. For instance, any source located off-axis will result in a shifted pattern. The visibilities can be recombined into images using reconstruction algorithms similar to a 2D Fourier transform.

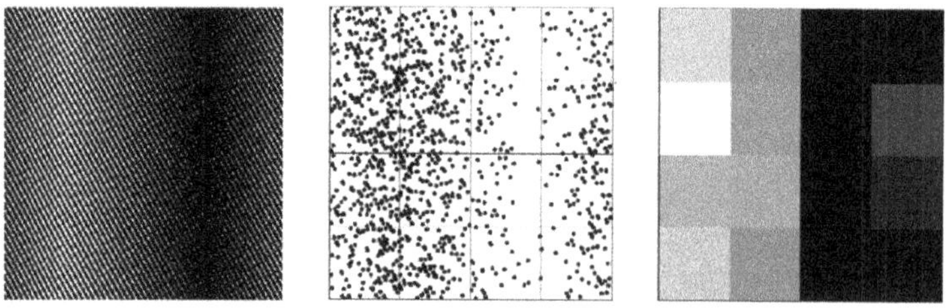

Figure 5. Left: moiré pattern created by the transmission through the grids in one of the LISSAN collimators by a point source on the Sun 100″ from the line of sight (black areas mark areas where the transmission is blocked by the grids). Middle: corresponding counts reaching the LISSAN detector. Right: relative flux registered in each GAGG crystal of the detector array: the highest flux is in white and the lower flux in black. Differences between pixels in a single column are due to counting statistics.

The minimum angular resolution is dictated by the subcollimator with the finest bigrid pitch. However, coarser pitches are also needed to capture the larger scale structure of the source. Moreover, since each subcollimator produces a visibility sensitive in only one direction, it is desirable to have multiple subcollimators with the same or similar resolutions at different orientations to better capture the asymmetric structure. The LISSAN design proposed in this paper therefore includes 5 different bigrid pitches, each arranged at 3 different orientations, resulting in the previously quoted 15 imaging subcollimators. The finest bigrid pitch is 230 µm (115 µm slit width) corresponding to a minimum angular resolution of 8″ FWHM. The distribution of subcollimator resolutions and orientations can be visualised by plotting their corresponding visibilities in Fourier space, known as the (u,v) plane. Altering the pitch and orientation of the bigrids causes the (u,v) points to shift. One possible visibility distribution is shown in the left panel of Figure 6. The resultant point spread function (PSF) is shown in the middle panel. The right panel shows the 50–84 keV image that LISSAN would have produced with this visibility distribution for the M9.7 flare observed by STIX on 31 March 2022. Two flare footpoints produced by HXR bremsstrahlung from precipitating accelerated electrons are clearly visible. While Figure 6 clearly shows that this visibility distribution can produce scientifically useful images, a more optimal distribution may exist. This can be determined by a trade study during the design phase of a LISSAN mission.

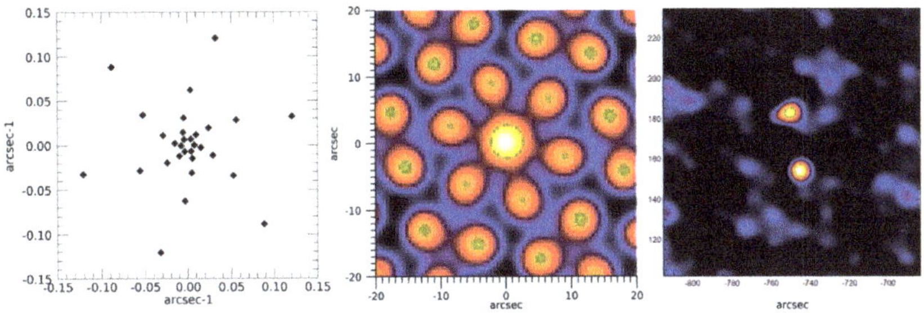

Figure 6. Left: One possible (u,v) plane coverage for LISSAN; Middle: associated point spread function (dirty map of a point source on-axis: this image contains both the X-ray source and instrumental artefacts to be removed with the adequate cleaning algorithms); Right: simulation of a LISSAN image of the 50–84 keV emission from the two hard X-ray footpoints during the M9.7 flare on 31 March 2022, which was observed by STIX on Solar Orbiter. This image was obtained by running the CLEAN image reconstruction algorithm (e.g., [63] and Section 2.5 in [61]) on the dirty image.

The field of view (FOV) of the imaging system is defined by the grid pitch and thickness. The finest pitch of 230 µm and 3 cm grid thickness proposed here results in an FOV with a diameter of 13 arcmin. This is greater than the 10 arcmin required to image the γ-ray line emission in an active region and possible pion decay source away from the main flare site (middle column, Table 1). The thickness of the bigrids is determined by the highest energy at which imaging is required. This is because grids of a certain thickness become transparent to γ-rays above a certain energy, and so moiré patterns are no longer formed.

4.3. Main Instrument Components

4.3.1. Grids

The LISSAN grids are square, with lengths of 15 cm for the front grids and 11 cm for the rear ones. The pitches range from 230 µm to 3720 µm (slit widths 115–1860 µm) and their thickness is 3 cm. The grids can be made of tungsten or a tungsten–copper alloy. Manufacturing grids with such an extreme thickness-to-pitch aspect ratio is challenging, and all conventional manufacturing techniques have drawbacks. Machining can fail due to the length of the channels compared to the thin wall thickness of the grids. Laser ablation is often only suitable for relatively shallow depths, while electrochemical machining usually requires a tool that is expensive to manufacture. A feasible alternative is modern 3D printing. We identified two options:

1. Laser powder bed fusion (cf. [64]): for tungsten, the components for nuclear fusion, medical imaging, and space applications have been manufactured with this technique by the Fraunhofer Institute for Casting, Composite and Processing Technology (IGCV).

2. Three-dimensional screen printing (cf. [65]): Anti-scatter grids with fine pitches and large-aspect ratios for medical imaging were developed in tungsten–copper by the Fraunhofer Institute for Manufacturing Technology and Advanced Materials (IFAM).

Examples of grids manufactured with these techniques are shown in Figure 7. While the manufacturing technologies are mature, grids with such fine pitches have not yet been developed for space applications at the time of writing. Issues currently remain with microcracks in the laser-based printing technique that are susceptible to vibrations during launch. Therefore, further development is required to bring these components to TRL 6. However, mitigation strategies such as switching to a tungsten–copper alloy are already known. The only performance penalty associated with this measure would be a reduction in the field of view from 13″ to 10″, as the material's lower density would require thicker grids to achieve the same absorption. However, this still fulfils the field of view requirement in Table 1.

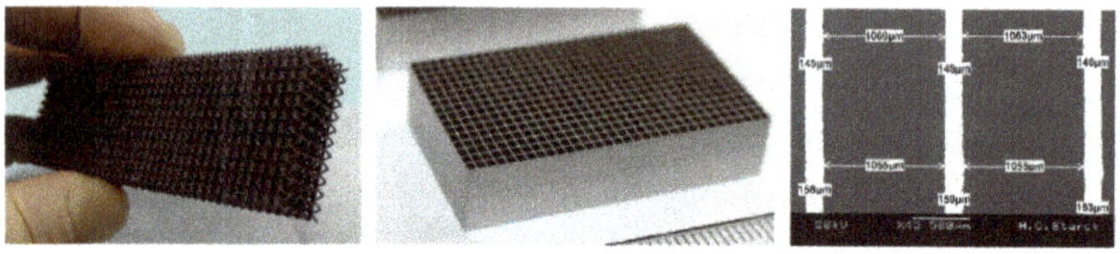

Figure 7. Examples of 3D-printed tungsten components developed by Fraunhofer institutes. (**Left**): tungsten grid structure for nuclear fusion applications, manufactured by laser powder bed fusion (IGCV). (**Middle**): tungsten anti-scatter grid for medical imaging manufactured by screen printing (IFAM). (**Right**): Measured dimensions of anti-scatter grid (IFAM).

4.3.2. Detectors

The LISSAN detectors provide a spectral resolution that is better than 1.5% FWHM at 6.1 MeV which is sufficient to measure the Doppler profiles of the C and O lines at 4.4 and 6.1 MeV (Table 1). Each detector is composed of a 4×4 array of gadolinium aluminium gallium garnet (GAGG) crystal pixels, which provide a total active area of 100 cm^2. GAGG:Ce are new crystals which are fast (50–100 ns), bright (30–50 photons/keV), and non-hygroscopic. They perform at moderate cooling temperatures compared to Germanium-based detectors and have a high quantum efficiency in the γ-ray regime. The detection chain is composed of:

1. GAGG crystals that scintillate when X-rays and γ-rays are absorbed;
2. Silicon photomultipliers (SiPMs) and their front-end electronics (FEE; analog/digital front–end ASIC and voltage regulation) that record the scintillation;
3. Fully digital back-end electronics (BEEs).

A single pixel is composed of a 2.5×2.5 cm^2 GAGG crystal with a thickness of 10 cm for good high-energy photon absorption. They are surrounded by a 3 M Vikuiti reflecting foil. Each pixel is readout at both ends by SiPM matrices, each composed of 4×4 pixels (6 mm), resulting in a total matrix size comparable to the ends of the GAGG pixel. The analog signals from each 16-pixel matrix are combined before being sent to the FEE, which are cooled to 0 °C.

The LISSAN detector units are covered by a Polyether Ether Ketone (PEEK) cover, transparent to γ-rays, to shield the detectors against visible photons. Each GAGG pixel produces two SiPM signals (giving a total of 32 analog signals) which are readout by two IDEAS/APOCAT fast 16-channel ASICs. The APOCAT ASIC, conceived by the IDEAS company and inherited from SIPHRA ASIC, is radiation-hardened and optimised for SiPM readout and has been used by several space missions. The two APOCAT ASICs provide digital information on the detected events, SiPM HV voltage ($\sim$50 V), temperature probes to monitor SiPM temperatures, and voltage and current probes to verify the health of the detection unit. The APOCAT ASIC also enables high-voltage fine-tuning to optimise the detector performance and uniformity between detector units.

The detector plane is composed of 20 LISSAN detection units, each associated with a subcollimator (15 imaging and 5 gridless). The total detection area is 2000 cm^2. After full functional and performance validation, these detection units are mounted in a main structure of aluminium alloy or AlBemet. All the detection units' information is sent to the BEE card, which is based on two nanoExplore NG-LARGE FPGA hardened components. The FPGA then produces photon event lists, which are sent to the data processing unit (DPU). The FPGA can produce higher-level products such as spectra, light curves, and images, instead of, or as well as, the event lists depending on the telemetry budget.

APC and LESIA are currently developing a new detector system funded by CNES known as the fast gamma-ray spectrometer (FGS) which is similar to that required by LISSAN. It is also based on new GAGG scintillators readout by SiPMs and controlled by the IDEAS/APOCAT fast ASIC (Figure 8). FGS differs from the LISSAN detection unit design only in its crystal size: $2 \times 2 \times 1$ cm^3 for FGS compared to $2.5 \times 2.5 \times 10$ cm^3 for LISSAN. LISSAN can therefore leverage the FGS development, which is expected to bring this technology to TRL 6 in 2024. It is important to note that, although the FGS system will soon be space-qualified, its main components, i.e., GAGG crystals, SiPMs, and the IDEAS APOCAT ASIC have already flown in space. Although SiPM dark current has been observed in space missions to increase with the radiation dose, this effect is not expected to be significant for LISSAN as the SiPMs are shielded by their surroundings. We anticipate a fluence of less than 10^9 protons cm^{-2} (67 MeV equivalent) during a three-year mission. Radiation measurements made with CNES at the Louvain-la-Neuve cyclotron have shown that the Hamamatsu SiPM will operate sufficiently with this dosage.

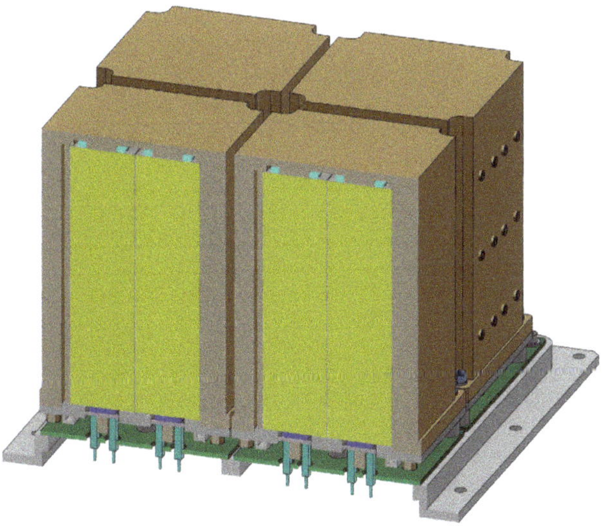

Figure 8. CAD drawing of LISSAN's detector system design arranged in four 2×2 pixel arrays (4×4 pixels in total). The artificial cutaways in the two nearest arrays reveal the yellow GAGG:Ce crystals (pixels) surrounded by 3M Vikuiti reflecting foil. Each pixel is read out by SiPMs at their top and bottom and are connected to the APOCAT ASICs below.

4.3.3. Solar Aspect and Twist Monitoring System

LISSAN's solar aspect system determines the imaging axis by monitoring the relative position of the front and rear grids, including their shift and twist, in relation to the Sun. This will be achieved by mounting two aspect system units in two corners of the instrument, as shown in Figure 4. Each unit is composed of a small optical lens in the front collimator assembly and a photodetector in the rear. The position of Sun-centre can then be determined from the image produced. This system provides an absolute pointing precision to better than 4″. A similar system is currently being successfully used by STIX. An alternative is to directly image the Sun with a CCD or CMOS detector. The pros and cons of each system can be determined by a trade study during the initial design phases of a LISSAN mission, but both are viable.

4.3.4. Data Processing Unit (DPU)

Figure 9 shows the main components of the DPU subsystem. The central element is an Aeroflex GR712RC chip implementing a Leon3FT processor. The CPU is clocked by the 20 MHz oscillator in order to minimise the power consumption. The DPU is equipped with 8 MBytes of Aeroflex SRAM operating memory and a further 2 MBytes of Aeroflex MRAM memory for storing the application software. The MRAM has very good radiation immunity. The size of the SRAM and MRAM can be increased depending on LISSAN's requirements. The DPU's start-up (boot) software is stored in PROM memory from Aeroflex. This also has good radiation immunity and is supplied by a 3.3 V voltage source. Communication with the spacecraft is achieved via SpaceWire using Aeroflex drivers and receivers UT54LVDS031LVUCC and UT54LVDS032LVUCC. The Leon3FT processor is widely used by ESA space projects and several space-qualified real-time operating systems can run on it, e.g., RTEMS OS.

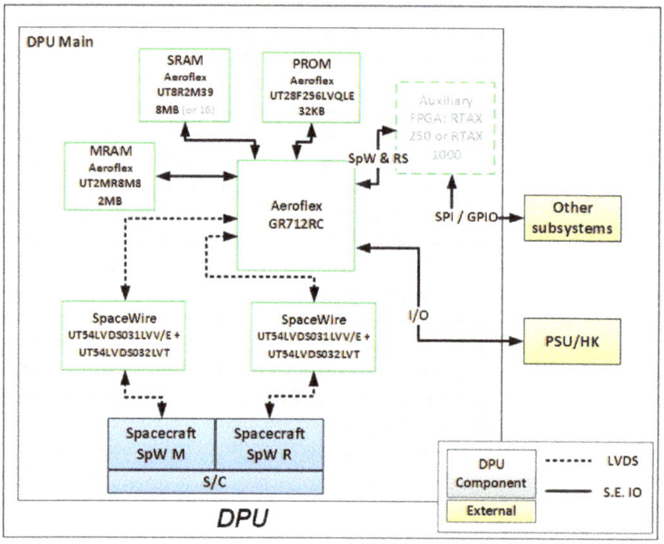

Figure 9. LISSAN DPU subsystem.

The LISSAN DPU is based on GR712RC, which is used in different space applications. A similar design concept was used by the SWI instrument onboard JUICE, and a similar architecture was used in PROBA3. An alternative concept is based on RTAX2000S/SL, which was used by STIX.

4.4. Spacecraft Interface Requirements and Environmental Constraints

The resources required by LISSAN are summarised in Table 2. The mass is driven by the mass of the individual grids (∼3 kg each) and detector units (∼6 kg each). The dimensions are 350 cm × 75 cm × 75 cm to accommodate the grids and the separation between them needed to achieve the described imaging performance. The data rate will be highly dependent on the solar activity. Each photon count in each detector segment is individually recorded as a 4-byte data record. During the largest flare (>X10), the data volume transmitted from the DPU to the onboard storage could reach 10 GB in ~10 min. Such events are rare, and the average data rate is expected to be at least 2 orders of magnitude lower. Onboard storage will be used to store large data volumes if needed. Binning the data in energy and time on request, as is performed for STIX, can also be considered if the telemetry is limited. The strongest thermal requirement is driven by the FEE operating temperature (~0 °C). This can be achieved with a cold finger and radiators. To locate solar targets in its FOV, LISSAN requires an absolute pointing accuracy of 30″. To maintain the angular resolution of the instrument, a pointing stability of 0.2″ over a second is required. L1 and LEO equatorial orbits both provide an environment with sufficiently low background radiation for LISSAN to achieve its required sensitivity.

Table 2. Resources required for LISSAN.

Resource	Requirement
Mass	370 kg
Volume	1.96 m^3
Power (peak)	125 W
Operating temperature (FEE)	0 °C

5. Summary

LISSAN provides solar X-ray and γ-ray imaging spectroscopy in the range of 40 keV–100 MeV on timescales in the order of 10 s. Its observations will answer many open scientific questions regarding electron and ion acceleration in the solar corona, whose underlying physical processes also drive many other phenomena throughout the universe. Its ability to unveil the acceleration of ions is particularly valuable because of their anticipated energetic importance in solar eruptive events and the fact that the spatio-temporal regime has largely been unexplored. Only one resolved γ-ray image of a solar flare has previously been achieved. Most of the technology required by LISSAN is already TRL 6 or greater, while other technologies are currently being brought to this level by ongoing funded efforts. LISSAN is appropriate for deployment on a multi-instrument ESA M-class mission or as a smaller standalone mission. See [26] for one such multi-instrument mission concept.

Author Contributions: The writing of this paper was led by D.F.R. and S.K. The LISSAN concept was developed as part of the SPARK mission concept which was coordinated by H.A.S.R., S.M. and D.F.R. All other authors contributed by providing scientific or instrumentation expertise on the development of the LISSAN concept and/or the paper itself. All authors have read and agreed to the published version of the manuscript.

Funding: This research received no external funding.

Institutional Review Board Statement: Not applicable.

Informed Consent Statement: Not applicable.

Data Availability Statement: Not applicable.

Acknowledgments: Parts of the science case for LISSAN builds upon those developed for the FOXSI SMEX [2] and FIERCE MIDEX [3] mission proposals. We acknowledge the work by those teams that helped make this paper possible.

Conflicts of Interest: The authors declare no conflict of interest.

Abbreviations

The following abbreviations are used in this manuscript:

APC	Laboratoire Astroparticule & Cosmologie
ASIC	Application-specific integrated circuit
BEE	(Detector unit) back-end electronics

CCD	Charge-coupled device
CME	Coronal mass ejection
CMOS	Complementary metal–oxide–semiconductor
CNES	Centre National de la Recherche Scientifique
CPU	Central processing unit
DPU	Data processing unit
ESA	European Space Agency
EUV	Extreme ultraviolet
FEE	(Detector unit) front-end electronics
Fermi-LAT	Fermi large-area telescope
FGS	Fast Gamma-ray spectrometer
FOV	Field of view
FPGA	Field-programmable gate array
FWHM	Full-width half-maximum
GAGG(CE)	Gadolinium aluminium gallium garnet (Cerium)
HXR	Hard X-rays ($\gtrsim$5 keV)
IGCV	Fraunhofer Institute for Casting, Composite and Processing Technology
IFAM	Fraunhofer Institute for Manufacturing 294 Technology and Advanced Materials
L1	Lagrange point 1
LEO	Low Earth orbit
LESIA	Laboratoire d'études spatiales et d'instrumentation en astrophysique
LISSAN	Large Imaging Spectrometer for Solar Accelerated Nuclei
MRAM	Magnetoresistive random-access memory
PEEK	Polyether ether ketone
PROM	Programmable read-only memory
PSF	Point spread function
RHESSI	Reuven Ramaty high-energy solar spectroscopic Imager
SEPs	Solar energetic particles
SiPM	Silicon photomultiplier
SMM/GRS	Solar maximum mission/gamma-ray spectrometer
SRAM	Static random-access memory
STIX	Spectrometer/telescope for imaging X-rays
TRL	Technology readiness level

References

1. Shibata, K. New observational facts about solar flares from YOHKOH studies—Evidence of magnetic reconnection and a unified model of flares. *Adv. Space Res.* **1996**, *17*, 9–18. [CrossRef]
2. Christe, S.; Alaoui, M.; Allred, J.; Battaglia, M.; Baumgartner, W.; Buitrago-Casas, J.C.; Chen, B.; Chen, T.; Dennis, B.; Drake, J.; et al. The Focusing Optics X-ray Solar Imager (FOXSI). *Bull. Am. Astron. Soc.* **2023**, *55*, 065. [CrossRef]
3. Shih, A.Y.; Glesener, L.; Krucker, S.; Guidoni, S.; Christe, S.; Reeves, K.; Gburek, S.; Caspi, A.; Alaoui, M.; Allred, J.; et al. Combined Next-Generation X-ray and EUV Observations with the FIERCE Mission Concept. In Proceedings of the American Meteorological Society 100th Annual Meeting (AMS), Boston, MA, USA, 12–16 January 2020. [CrossRef]
4. Chupp, E.L. High-Energy Neutral Radiations from the Sun. *Annu. Rev. Astron. Astrophys.* **1984**, *22*, 359–387. [CrossRef]
5. Ramaty, R. Nuclear processes in solar flares. *Phys. Sun* **1986**, *2*, 291–323.
6. Chupp, E.L. Evolution of our understanding of solar flare particle acceleration: (1942–1995). *AIP Conf. Proc.* **1996**, *374*, 3–31. [CrossRef]
7. Trottet, G.; Vilmer, N. The Production of Flare-Accelerated Particles at the Sun. In *European Meeting on Solar Physics*; Simnett, G.M., Alissandrakis, C.E., Vlahos, L., Eds.; Springer: Berlin/Heidelberg, Germany, 1997; Volume 489, p. 219. [CrossRef]
8. Share, G.H.; Murphy, R.J. Gamma Ray Spectroscopy in the Pre-HESSI Era. *High Energy Sol. Phys. Workshop* **2000**, *206*, 377.
9. Vilmer, N.; MacKinnon, A.L.; Trottet, G.; Barat, C. High energy particles accelerated during the large solar flare of 1990 May 24: X/γ-rayobservations. *Astron. Astrophys.* **2003**, *412*, 865–874. [CrossRef]
10. Share, G.H.; Murphy, R.J. Gamma Radiation from Flare-Accelerated Particles Impacting the Sun. *Geophys. Monogr. Ser.* **2006**, *165*, 177. [CrossRef]
11. Vilmer, N.; MacKinnon, A.L.; Hurford, G.J. Properties of Energetic Ions in the Solar Atmosphere from γ-ray and Neutron Observations. *Space Sci. Rev.* **2011**, *159*, 167–224. [CrossRef]

12. Stecker, F.W. The Cosmic γ-Ray Spectrum from Secondary Particle Production in Cosmic-Ray Interactions. *Astrophys. Space Sci.* **1970**, *6*, 377–389. [CrossRef]
13. Murphy, R.J.; Dermer, C.D.; Ramaty, R. High-Energy Processes in Solar Flares. *Astrophys. J. Suppl. Ser.* **1987**, *63*, 721. [CrossRef]
14. Brown, J.C. The Deduction of Energy Spectra of Non-Thermal Electrons in Flares from the Observed Dynamic Spectra of Hard X-ray Bursts. *Sol. Phys.* **1971**, *18*, 489–502. [CrossRef]
15. Piana, M.; Massone, A.M.; Kontar, E.P.; Emslie, A.G.; Brown, J.C.; Schwartz, R.A. Regularized Electron Flux Spectra in the 23 July 2002 Solar Flare. *Astrophys. J. Lett.* **2003**, *595*, L127–L130. [CrossRef]
16. Holman, G.D.; Aschwanden, M.J.; Aurass, H.; Battaglia, M.; Grigis, P.C.; Kontar, E.P.; Liu, W.; Saint-Hilaire, P.; Zharkova, V.V. Implications of X-ray Observations for Electron Acceleration and Propagation in Solar Flares. *Space Sci. Rev.* **2011**, *159*, 107–166. [CrossRef]
17. Kontar, E.P.; Brown, J.C.; Emslie, A.G.; Hajdas, W.; Holman, G.D.; Hurford, G.J.; Kašparová, J.; Mallik, P.C.V.; Massone, A.M.; McConnell, M.L.; et al. Deducing Electron Properties from Hard X-ray Observations. *Space Sci. Rev.* **2011**, *159*, 301–355. [CrossRef]
18. Benz, A.O. Flare Observations. *Living Rev. Sol. Phys.* **2017**, *14*, 2. [CrossRef]
19. Lin, R.P.; Dennis, B.R.; Hurford, G.J.; Smith, D.M.; Zehnder, A.; Harvey, P.R.; Curtis, D.W.; Pankow, D.; Turin, P.; Bester, M.; et al. The Reuven Ramaty High-Energy Solar Spectroscopic Imager (RHESSI). *Sol. Phys.* **2002**, *210*, 3–32. [CrossRef]
20. Krucker, S.; Hurford, G.J.; Grimm, O.; Kögl, S.; Gröbelbauer, H.P.; Etesi, L.; Casadei, D.; Csillaghy, A.; Benz, A.O.; Arnold, N.G.; et al. The Spectrometer/Telescope for Imaging X-rays (STIX). *Astron. Astrophys.* **2020**, *642*, A15. [CrossRef]
21. Hurford, G.J.; Schwartz, R.A.; Krucker, S.; Lin, R.P.; Smith, D.M.; Vilmer, N. First Gamma-Ray Images of a Solar Flare. *Astrophys. J. Lett.* **2003**, *595*, L77–L80. [CrossRef]
22. Hurford, G.J.; Krucker, S.; Lin, R.P.; Schwartz, R.A.; Share, G.H.; Smith, D.M. Gamma-Ray Imaging of the 2003 October/November Solar Flares. *Astrophys. J. Lett.* **2006**, *644*, L93–L96. [CrossRef]
23. Ramaty, R.; Mandzhavidze, N. High energy processes in solar flares. *AIP Conf. Proc.* **2000**, *522*, 401–410. [CrossRef]
24. Emslie, A.G.; Dennis, B.R.; Shih, A.Y.; Chamberlin, P.C.; Mewaldt, R.A.; Moore, C.S.; Share, G.H.; Vourlidas, A.; Welsch, B.T. Global Energetics of Thirty-eight Large Solar Eruptive Events. *Astrophys. J.* **2012**, *759*, 71. [CrossRef]
25. Aschwanden, M.J.; Caspi, A.; Cohen, C.M.S.; Holman, G.; Jing, J.; Kretzschmar, M.; Kontar, E.P.; McTiernan, J.M.; Mewaldt, R.A.; O'Flannagain, A.; et al. Global Energetics of Solar Flares. V. Energy Closure in Flares and Coronal Mass Ejections. *Astrophys. J.* **2017**, *836*, 17. [CrossRef]
26. Reid, H.A.S.; Musset, S.; Ryan, D.F.; Calcines Rosario, A.; Dudík, J.; Auchère, F.; Dahlin, J.; Hayes, L.A.; Kerr, G.S.; Nakariakov, V. M.; et al. The Solar Particle Acceleration Radiation and Kinetics (SPARK) mission concept. *Aerospace* **2023**, submitted.
27. Kiener, J.; Gros, M.; Tatischeff, V.; Weidenspointner, G. Properties of the energetic particle distributions during the 28 October 2003 solar flare from INTEGRAL/SPI observations. *Astron. Astrophys.* **2006**, *445*, 725–733. [CrossRef]
28. Dahlin, J.T.; Drake, J.F.; Swisdak, M. The mechanisms of electron heating and acceleration during magnetic reconnection. *Phys. Plasmas* **2014**, *21*, 092304. [CrossRef]
29. Drake, J.F.; Swisdak, M.; Cassak, P.A.; Phan, T.D. On the 3-D structure and dissipation of reconnection-driven flow bursts. *Geophys. Res. Lett.* **2014**, *41*, 3710–3716. [CrossRef]
30. Dahlin, J.T.; Drake, J.F.; Swisdak, M. Parallel electric fields are inefficient drivers of energetic electrons in magnetic reconnection. *Phys. Plasmas* **2016**, *23*, 120704. [CrossRef]
31. Dahlin, J.T.; Drake, J.F.; Swisdak, M. The role of three-dimensional transport in driving enhanced electron acceleration during magnetic reconnection. *Phys. Plasmas* **2017**, *24*, 092110. [CrossRef]
32. Shih, A.Y.; Lin, R.P.; Smith, D.M. RHESSI Observations of the Proportional Acceleration of Relativistic >0.3 MeV Electrons and >30 MeV Protons in Solar Flares. *Astrophys. J. Lett.* **2009**, *698*, L152–L157. [CrossRef]
33. Drake, J.F.; Swisdak, M.; Che, H.; Shay, M.A. Electron acceleration from contracting magnetic islands during reconnection. *Nature* **2006**, *443*, 553–556. [CrossRef] [PubMed]
34. Litvinenko, Y.E.; Craig, I.J.D. Flare Energy Release by Flux Pile-up Magnetic Reconnection in a Turbulent Current Sheet. *Astrophys. J.* **2000**, *544*, 1101–1107. [CrossRef]
35. Holman, G.D.; Benka, S.G. A Hybrid Thermal/Nonthermal Model for the Energetic Emissions from Solar Flares. *Astrophys. J. Lett.* **1992**, *400*, L79. [CrossRef]
36. Krucker, S.; Hudson, H.S.; Glesener, L.; White, S.M.; Masuda, S.; Wuelser, J.P.; Lin, R.P. Measurements of the Coronal Acceleration Region of a Solar Flare. *Astrophys. J.* **2010**, *714*, 1108–1119. [CrossRef]
37. Oka, M.; Krucker, S.; Hudson, H.S.; Saint-Hilaire, P. Electron Energy Partition in the Above-the-looptop Solar Hard X-ray Sources. *Astrophys. J.* **2015**, *799*, 129. [CrossRef]
38. Knight, J.W.; Sturrock, P.A. Reverse current in solar flares. *Astrophys. J.* **1977**, *218*, 306–310. [CrossRef]
39. Emslie, A.G. The effect of reverse currents on the dynamics of nonthermal electron beams in solar flares and on their emitted X-ray bremsstrahlung. *Astrophys. J.* **1980**, *235*, 1055–1065. [CrossRef]
40. van den Oord, G.H.J. The electrodynamics of beam/return current systems in the solar corona. *Astron. Astrophys.* **1990**, *234*, 496–518.
41. Zharkova, V.V.; Gordovskyy, M. The Effect of the Electric Field Induced by Precipitating Electron Beams on Hard X-ray Photon and Mean Electron Spectra. *Astrophys. J.* **2006**, *651*, 553–565. [CrossRef]

42. Alaoui, M.; Holman, G.D. Understanding Breaks in Flare X-ray Spectra: Evaluation of a Cospatial Collisional Return-current Model. *Astrophys. J.* **2017**, *851*, 78. [CrossRef]
43. Murphy, R.J.; Ramaty, R.; Kozlovsky, B.; Reames, D.V. Solar Abundances from Gamma-ray Spectroscopy: Comparisons with Energetic Particle, Photospheric, and Coronal Abundances. *Astrophys. J.* **1991**, *371*, 793. [CrossRef]
44. Trottet, G.; Barat, C.; Ramaty, R.; Vilmer, N.; Dezalay, J.P.; Kuznetsov, A.; Mandzhavidze, N.; Sunyaev, R.; Talon, R.; Terekhov, O. Thin target γ-ray line production during the 1991 June 1 flare. *AIP Conf. Proc.* **1996**, *374*, 153–161. [CrossRef]
45. Alexander, W.M.; Tanner, W.G.; McDonald, R.A.; Schaub, G.E.; Stephenson, S.L.; McDonnell, J.A.M.; Maag, C.R. The Status of Measurement Technologies Concerning Micrometer and Submicrometer Space Articulate Matter Capture, Recovery, Velocity, and Trajectory. In Proceedings of the Workshop on Particle Capture, Recovery and Velocity/Trajectory Measurement Technologies, Houston, TX, USA, 27–28 September 1993; p. 11.
46. Dunphy, P.P.; Chupp, E.L.; Bertsch, D.L.; Schneid, E.J.; Gottesman, S.R.; Kanbach, G. Gamma-Rays and Neutrons as a Probe of Flare Proton Spectra: The Solar Flare of 11 June 1991. *Sol. Phys.* **1999**, *187*, 45–57. :1005143603547 [CrossRef]
47. Kocharov, L.G.; Lee, J.W.; Zirin, H.; Kovaltsov, G.A.; Usoskin, I.G.; Pyle, K.R.; Shea, M.A.; Smart, D.F. Neutron and electromagnetic emissions during the 24 May 1990 solar flare. *Sol. Phys.* **1994**, *155*, 149–170. [CrossRef]
48. Kocharov, L.; Debrunner, H.; Kovaltsov, G.; Lockwood, J.; McConnell, M.; Nieminen, P.; Rank, G.; Ryan, J.; Schoenfelder, V. Deduced spectrum of interacting protons accelerated after the impulsive phase of the 15 June 1991 solar flare. *Astron. Astrophys.* **1998**, *340*, 257–264.
49. Vilmer, N.; Krucker, S.; Trottet, G.; Lin, R.P. Hard X-ray and metric/decimetric spatially resolved observations of the 10 April 2002 solar flare. *Adv. Space Res.* **2003**, *32*, 2509–2515. [CrossRef]
50. Kanbach, G.; Bertsch, D.L.; Fichtel, C.E.; Hartman, R.C.; Hunter, S.D.; Kniffen, D.A.; Kwok, P.W.; Lin, Y.C.; Mattox, J.R.; Mayer-Hasselwander, H.A. Detection of a long-duration solar gamma-ray flare on 11 June 1991 with EGRET on COMPTON-GRO. *Astron. Astrophys. Suppl. Ser.* **1993**, *97*, 349–353.
51. Ryan, J.M.; Lockwood, J.A.; Debrunner, H. Solar Energetic Particles. *Space Sci. Rev.* **2000**, *93*, 35–53. :1026580008909 [CrossRef]
52. Share, G.H.; Murphy, R.J.; White, S.M.; Tolbert, A.K.; Dennis, B.R.; Schwartz, R.A.; Smart, D.F.; Shea, M.A. Characteristics of Late-phase >100 MeV Gamma-Ray Emission in Solar Eruptive Events. *Astrophys. J.* **2018**, *869*, 182. [CrossRef]
53. de Nolfo, G.A.; Bruno, A.; Ryan, J.M.; Dalla, S.; Giacalone, J.; Richardson, I.G.; Christian, E.R.; Stochaj, S.J.; Bazilevskaya, G.A.; Boezio, M.; et al. Comparing Long-duration Gamma-Ray Flares and High-energy Solar Energetic Particles. *Astrophys. J.* **2019**, *879*, 90. [CrossRef]
54. Hutchinson, A.; Dalla, S.; Laitinen, T.; de Nolfo, G.A.; Bruno, A.; Ryan, J.M.; Waterfall, C.O.G. Energetic proton back-precipitation onto the solar atmosphere in relation to long-duration gamma-ray flares. *Astron. Astrophys.* **2022**, *658*, A23. [CrossRef]
55. Grechnev, V.V.; Kurt, V.G.; Chertok, I.M.; Uralov, A.M.; Nakajima, H.; Altyntsev, A.T.; Belov, A.V.; Yushkov, B.Y.; Kuznetsov, S.N.; Kashapova, L.K.; et al. An Extreme Solar Event of 20 January 2005: Properties of the Flare and the Origin of Energetic Particles. *Sol. Phys.* **2008**, *252*, 149–177. [CrossRef]
56. McCracken, K.G.; Moraal, H.; Stoker, P.H. Investigation of the multiple-component structure of the 20 January 2005 cosmic ray ground level enhancement. *J. Geophys. Res. Space Phys.* **2008**, *113*, A12101. [CrossRef]
57. Masson, S.; Pariat, E.; Aulanier, G.; Schrijver, C.J. The Nature of Flare Ribbons in Coronal Null-Point Topology. *Astrophys. J.* **2009**, *700*, 559–578. [CrossRef]
58. Ajello, M.; Albert, A.; Allafort, A.; Baldini, L.; Barbiellini, G.; Bastieri, D.; Bellazzini, R.; Bissaldi, E.; Bonamente, E.; Brandt, T.J.; et al. Impulsive and Long Duration High-energy Gamma-ray Emission from the Very Bright 7 March 2012 Solar Flares. *Astrophys. J.* **2014**, *789*, 20. [CrossRef]
59. Ajello, M.; Baldini, L.; Bastieri, D.; Bellazzini, R.; Berretta, A.; Bissaldi, E.; Blandford, R.D.; Bonino, R.; Bruel, P.; Buson, S.; et al. First Fermi-LAT Solar Flare Catalog. *Astrophys. J. Suppl. Ser.* **2021**, *252*, 13, [CrossRef]
60. Chen, W.; Gan, W.Q. A Spectroscopic Method Based on the Shapes of Nuclear Deexcitation γ-ray Lines in Solar Flares. *Astrophys. J.* **2020**, *895*, 8. [CrossRef]
61. Massa, P.; Battaglia, A.F.; Volpara, A.; Collier, H.; Hurford, G.J.; Kuhar, M.; Perracchione, E.; Garbarino, S.; Massone, A.M.; Benvenuto, F.; et al. First Hard X-ray Imaging Results by Solar Orbiter STIX. *Sol. Phys.* **2022**, *297*, 93, [CrossRef]
62. Hurford, G.J. X-ray imaging with collimators, masks and grids. *ISSI Sci. Rep. Ser.* **2010**, *9*, 223–234.
63. Högbom, J.A. Aperture Synthesis with a Non-Regular Distribution of Interferometer Baselines. *Astron. Astrophys. Suppl. Ser.* **1974**, *15*, 417.
64. Müller, A.v.; Dorow-Gerspach, D.; Balden, M.; Binder, M.; Buschmann, B.; Curzadd, B.; Loewenhoff, T.; Neu, R.; Schlick, G.; You, J.H. Progress in additive manufacturing of pure tungsten for plasma-facing component applications. *J. Nucl. Mater.* **2022**, *566*, 153760. [CrossRef]
65. Studnitzky, T.; Reuter, K.; Stiebritz, G.; Wirth, S. 3D Screen Printing of Anti-Scatter Grids—Part Development and Industrialization. In Proceedings of the EuroPM 2019, Maastricht, The Netherlands, 14–16 October 2019; Euro Powder Metallurgy Conference Series.

aerospace

MDPI

Article

Optical Design of a Miniaturised Solar Magnetograph for Space Applications

Ariadna Calcines Rosario [1,*], Lucie M. Green [2], Alan Smith [2] and David M. Long [3]

1 Centre for Advanced Instrumentation, Department of Physics, Durham University, Durham DH1 3LE, UK
2 Mullard Space Science Laboratory, University College London, Holmbury St. Mary, Dorking, Surrey RH5 6NT, UK
3 Astrophysics Research Centre, School of Mathematics and Physics, Queen's University Belfast, University Road, Belfast BT7 1NN, UK
* Correspondence: ariadna.calcines@durham.ac.uk

Abstract: Measuring the Sun's magnetic field is a key component of monitoring solar activity and forecasting space weather. The main goal of the research presented in this paper is to investigate the possibility of reducing the dimensions and weight of a solar magnetograph while preserving its optical quality. This article presents a range of different designs, along with their advantages and disadvantages, and an analysis of the optical performance of each. All proposed designs are based on the magneto-optical filter (MOF) technique. As a result of the design study, a miniaturised solar magnetograph is proposed with an ultra-compact layout. The dimensions are 345 mm × 54 mm × 54 mm, and the optical quality is almost at the diffraction limit. The design has an entrance focal ratio of F/17.65, with a plate scale of 83.58 arcsec/mm at the telescope image focal plane, and produces a magnification of 0.79. The field of view is 1920 arcsec in diameter, equivalent to ±0.27 degrees, sufficient to cover the entire solar disk.

Keywords: solar magnetograph; space weather; space sciences; solar instrumentation; miniaturised optics

1. Introduction

The Sun is a centrally important star. Its relatively close proximity means that it can be spatially resolved, providing information that is vital for understanding not only our Sun but stars in general [1]. The Sun is an active star, in that it emits a wind of magnetised plasma that flows over and effects all Solar System objects and it sporadically ejects vast clouds of magnetised plasma (known as coronal mass ejections), bursts of electromagntic radiation (known as solar flares), and high-energy particles that affect Solar System bodies. The variations that these forms of activity create in the near-Earth space environment are collectively known as space weather [2]. Space weather can have deleterious impacts on modern life through disruptions to electricity transmission, high-frequency radio communication, and navigation systems.

To understand the physics of the Sun, and the origins of solar activity and space weather, it is crucial to measure and monitor the solar magnetic field. So-called non-potential magnetic fields harbour energy that powers the various forms of solar activity [3], and certain magnetic field configurations form in the Sun's atmosphere prior to the occurrence of flares and coronal mass ejections [4]. In addition, measurements of the magnetic field in the photosphere are used as the boundary condition for space weather tools used to model the solar wind conditions and the propagation of CMEs to the Earth, such as WSA-ENLIL [5] or EUHFORIA [6].

To this end, space and ground-based instrumentation monitoring the Sun has, for decades, included magnetographs that measure the polarisation state of the light of the Sun to determine the magnetic field strength and direction. Most of these instruments exploit the Zeeman effect in which a magnetic field induces both energy-level splitting and polarization of the emitted light. Magnetographs that use this technique make measurements centred on a magnetically sensitive spectral line produced in the lower solar atmosphere. However, the capability to determine the magnetic field in the solar corona via the Hanle effect is growing (e.g., the Daniel K. Inouye Solar Telescope Visible Spectro-Polarimeter [7]).

Current designs for ground and space-based solar magnetographs utilising the Zeeman effect make use of both filtergraph (e.g., the Helioseismic and Magnetic Imager Telescope on the Solar Dynamics Observatory [8]) and spectrograph techniques (the Solar Optical Telescope onboard Hinode [9]) to provide information on the magnetic field. The former allows for relatively high-cadence full-disc magnetograms to be made but with fewer measurements taken across the selected spectral line. The latter provides partial field of view magnetograms that take longer to construct but which have significantly increased spectral resolution and therefore provide higher precision. Space-based magnetographs have an additional design constraint in that they must operate within limited mass, size, and power budgets. The Michelson Doppler Imager instrument onboard the SOHO spacecraft has a total mass of 56 kg [10], the HMI instrument onboard the Solar Dynamics Observatory has a mass of 73 kg, and the Polarimetric and Helioseismic Imager instrument onboard Solar Orbiter has a mass of 35 kg [11]. With a view to the future, smaller and more lightweight magnetographs can be more readily employed and offer the practicality of inclusion on a wider range of missions. A truly compact and lightweight magnetograph opens up the possibility of multiple flight opportunities, including in a constellation that could provide near full-Sun coverage.

One solution to creating a small and lightweight magnetograph is to adapt a technique that has been used in ground-based telescopes. That is, to utilise magneto-optical filters (MOFs; [12]). MOFs contain an elemental vapour immersed in a strong magnetic field, and so only photons close to the atomic resonance frequency pass through, thereby creating a narrow band filter. For example, a potassium vapour filters all photons except those originating in the mid-photosphere, whilst a sodium vapour filters all except those originating in the lower chromosphere. MOFs have a high throughput and wavelength stability and reject the continuum light of the Sun, only allowing through light from the red and blue wings of the Zeeman broadened/split selected spectral line, allowing only these photons to be analysed. Ground-based telescopes having successfully implemented the MOF technology include, for example, MOTH at the Mees Solar Observatory in Hawaii, MOTH II, located at the South Pole in Antarctica [13], and the Tor Vergata Synoptic Solar Telescope (TSST) in La Palma, Spain [14,15]. Previous studies have demonstrated that the use of a magneto-optical filter could lead to a space-based magnetograph design of reduced size and mass (e.g., [16–18]). The study of [18] conceived a space-based magnetograph of mass 14 kg with dimensions 850 × 150 × 150 mm. In this paper, we build on the previous work to make steps towards a conceptual design of a compact (800 mm × 200 mm × 200 mm) lightweight (<15 kg) magnetograph. We search for an optimum optical configuration in order to compress the size of the instrument whilst preserving instrument performance.

2. Magneto-Optical Filter Technique

A magneto-optical filter (MOF) is a narrow band filter that uses heated cells containing a vapour, such as sodium, immersed in a static magnetic field of about 2 kGauss. The vapour cells are typically a few centimetres in each dimension. For potassium, the temperature should be greater than 373.15 K, typically 393.15 K. The temperature of a cell is kept to ±1 K and the cells are placed in vacuum chambers to prevent convection around them. Overall, two vapour cells are needed (labelled as "MOF" and the wing selector, WS, in Figure 1), along with a series of quarter and half-wave plates and two linear polarisers (P1 and P2), as shown in Figure 1. The axes of P1 and P2 are orthogonal to each other and so act as crossed polarisers. The wavelength of the MOF is not sensitive to the incidence angle of the light.

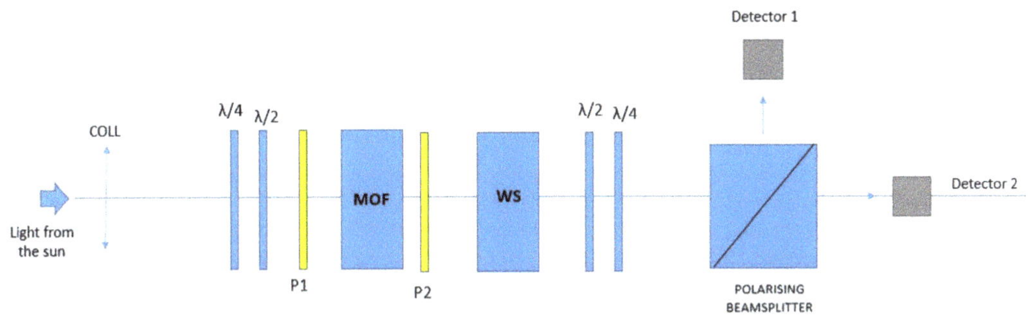

Figure 1. Schematic of a magnetograph layout that employs magneto-optical filters. The first magneto-optical filter in the optical path is labelled as the MOF and the second is labelled as the wing selector (WS). Also included are quarter-wave plates ($\lambda/4$) and half-wave plates ($\lambda/2$) to enable the passage of selected polarisation states. P1 and P2 are two linear polarisers whose axes are orthogonal to each other (crossed polarisers). Based on Barajas, Garcia, and Vo (2011).

Light from the Sun enters the optics and is collimated. A quarter-wave plate ($\lambda/4$) at 45 degrees turns circularly polarised light into linearly polarised. After the quarter-wave plate, a half-wave plate ($\lambda/2$) is used, whose orientation can be varied from 0 degrees to 45 degrees with respect to the axis of the linear polariser P1 (i.e., allows through linearly polarised light). The strong longitudinal magnetic field that is applied to the elemental vapour in the first MOF cell broadens/splits the spectral line and rotates the plane of polarisation of the light (via Faraday rotation). After the MOF cell, a second polariser, P2, is crossed with the input polariser so that ultimately only light of orthogonal linear polarisation passes through. This part of the magnetograph is sometimes called the "pre-filter" or the "filter selection MOF" (see, for example, [13]).

The wing selector converts the linearly polarised light into opposite circularly polarised light in the blue and red wings of the spectral line. Which wing is transmitted is controlled by the orientation of the quarter-wave plate so that full solar disk images can be made in the red and blue wings of the spectral line. A quarter-wave plate optically separates the circular polarisations into orthogonal linear polarisations. After the quarter-wave plate, the two orthogonal linear polarisations are separated using a polarising beam splitter, sending each one to a different detector in this configuration. It would be possible to have both on the same detector using a different polarising beam splitter.

The main factors found in existing solutions based on the MOF technique that pose limits on the spatial dimensions are:

- The optical components required for the instrument.
- The characteristics of the light required at each component location (i.e., collimated beam, polarisation state). This requires the use of optics, which enlarges the magnetograph layout.
- The cell dimensions, including its housing and electronics.
- The detectors, including electronics and cables.

Additional limitations come from the optimisation of the instrument performance. Two cells are required, with different magnetic fields. The magnetic field of the second cell (WS) shall be higher than that of the first cell. For both MOFs, the light needs to be linearly polarised—ideally collimated beams and with intermediate pupil images within each cell. For a better spectral response, collimated beams are also required for the half-wave plates, quarter-wave plates, and linear polarisers. To achieve this, some optical components are needed to generate the collimated beams and intermediate pupil images. All these components contribute to the total weight and also enlarge the dimensions of the instrument.

3. Layouts to Reduce Mass and Volume

Here, we explore four ideas that aim to reduce both the volume and the mass of a solar magnetograph, with a view on providing the basis of a design for a compact and lightweight space-based instrument after having evaluated a range of options.

3.1. Option 1: Re-Imaging System before the Magneto-Optical Filter

A reduction in the size, and therefore mass, of the magneto-optical cells is possible if the diameter of the light beam is reduced. This can be achieved using a re-imaging system after the telescope image focal plane. This approach would also reduce the diameter of all optical components and lead to a reduction in the size of the mechanical mounts and an overall smaller mass and volume for the instrument. However, it would be necessary to evaluate the focal lengths required to achieve good optical performance for the re-imaging system. In order to minimise volume, the focal lengths would need to be as short as possible. A disadvantage of this approach would be a reduction of throughput associated with the addition of the optical surfaces for the re-imaging system. This would be $\geq$0.96. A refractive solution (lenses) might be the most compact one if the focal lengths required are not too large since the angles in a reflective solution would increase the dimensions of the instrument in the direction perpendicular to that of the light propagation.

As shown in Figure 2, although the addition of the relay reduces the size of the beam and thus the size of all optical components, this advantage is negated since the focal lengths of the lenses required increase the total length of the magnetograph. For this reason, it was not considered in the final miniaturised magnetograph layout (see Section 4.3).

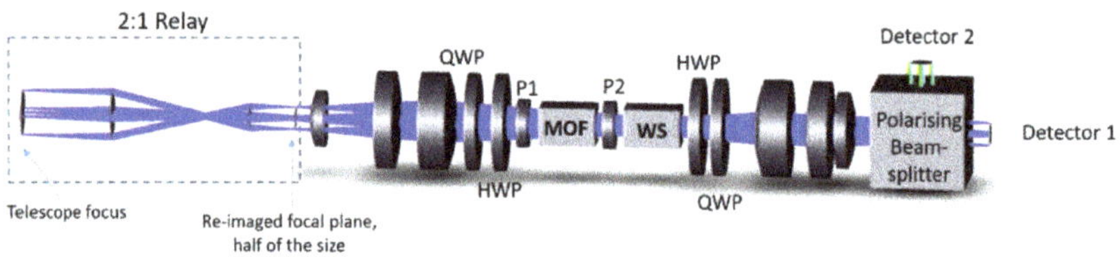

Figure 2. Optical design of a magnetograph using a 2:1 re-imaging system at the telescope focus. The total dimensions are 522.02 mm × 62.99 mm.

3.2. Option 2: One Detector Instead of Two

In the layout of Figure 1, a polarising beam splitter separates the two orthogonal linear polarisations so that each one could be observed on a different detector. However, an alternative polarising beam splitter, such as a Wollaston prism, could be chosen to accommodate both polarisations within the same detector. Using one detector instead of two would reduce volume and mass.

A Wollaston prism separates light into two linearly polarised beams with orthogonal polarisation. The beams will be polarised according to the optical axis of the two right-angle prisms, which are cemented orthogonal prisms of birefringent material, such as calcite. At the exit, two beams are obtained, associated with the ordinary and extraordinary rays. These beams diverge due to the differences in the indexes of refraction. The angle of divergence is a function of the prism's wedge angle and the wavelength. In commercial Wollaston prisms, this angle can vary from 1 degree to 45 degrees.

3.3. Option 3: One Magneto-Optical Filter in Double Pass

The MOF is a significant contributor to the mass and volume of the magnetograph. For both cells, a collimated beam with incoming linear polarisation is required, ideally with a pupil image inside the cells. To meet this, for both cells, some optical components are required between them. Two cells are also needed, with the magnetic field of the second cell, the Wing Selector (WS), higher than the magnetic field of the first cell because its Zeeman component must be superimposed to the transmission peaks of the Macaluso–Corbino effect. The alternative here proposed is to have only one MOF with the magnetic field required for the WS and use it in double pass, as shown in Figure 3.

This would reduce volume and mass since only one cell would be needed and the optical components to generate a pupil image in a collimated beam with linear polarisation would not need to be duplicated, with the space reduction associated with the separation between these components. The size required for the cell in this layout is 64 mm side by 30 mm thickness. The dimensions for the layout with the two detectors is 397 mm × 125.5 mm. The points to be evaluated in this option are the effect of having illumination in both directions for a defined direction of the magnetic field. In one pass, the beam would be in favour of this direction, and in the double pass in opposite direction. Would it be possible to set the system so that the second pass is the one in the direction of the magnetic field for the Macaluso–Corbino effect since it is required to have a higher magnetic field for the WS? Two flat mirrors orthogonal to each other and thus not modifying the polarisation state would be used to revert the beam path back to the MOF for a second pass.

The gap between the mirrors separates the input and output axes, which are parallel. The MOF shall be large enough to accept this separation between the input and output beams without vignetting. This idea could be combined with the previous ones using a reimaging system to reduce the beam size, for example, to a half, and thus be able to have the beams associated with the two directions inside the original cell size.

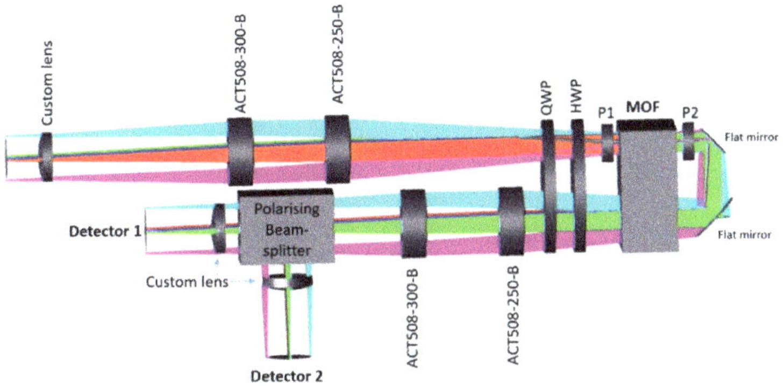

Figure 3. Compact magnetograph layout with one cell used in double pass and two detectors to measure the two polarisation components. The colours represent different fields.

3.4. Option 4: Application of Slicers for Magnetographs

Image slicers are typically used to slice a field of view. This application is discarded for magnetographs since an image of the full disk is required. However, slicers can be used at different locations and with different functionalities, for example a pupil slicer or a slicer at the image focal plane to arrange the image distribution. A pupil slicer (see Figure 4) would produce two beams with the same field of view. By adding a pupil slicer, each arm would correspond to a polarisation state (Figure 5), in which this separation would be performed before the detector, thus controlling the output direction towards the detector and adding flexibility in the layout. If this were combined with the previous ideas, using a re-imaging system after the telescope focus, the beam size would already be reduced. Thereby, a pupil slicer would produce two beams that could fit within the MOFs (with their original size) and also within one detector. A lateral beam displacement beam splitter could be used as a pupil slicer producing the two parallel beams.

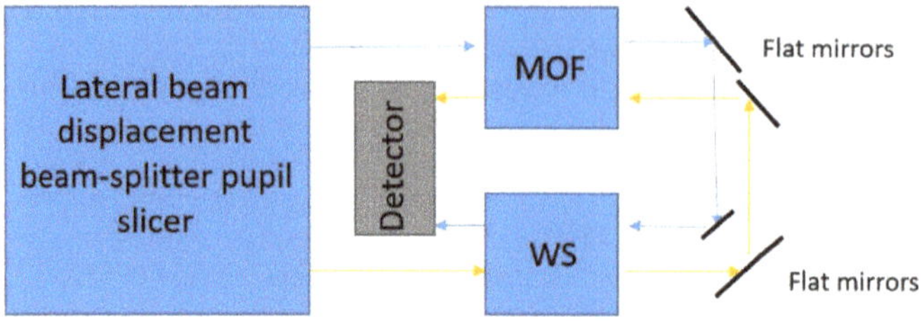

Figure 4. Sketch for a compact magnetograph using a pupil slicer generating two output parallel beams, two cells, and only one detector.

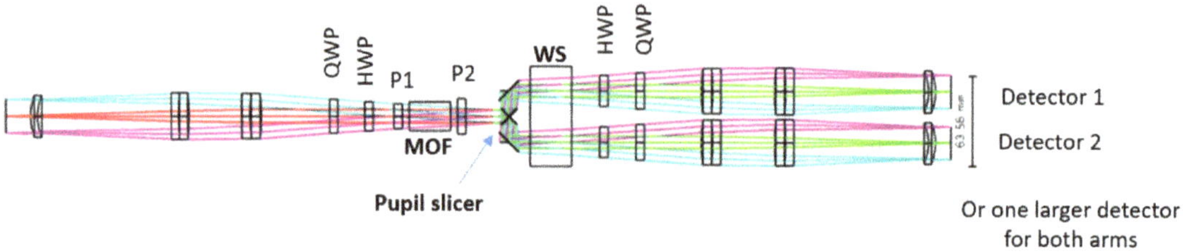

Figure 5. Magnetograph layout using a pupil slicer. The colours represent different fields.

4. Design of a Miniaturised Solar Magnetograph

4.1. Specifications

The specifications, shown in Table 1, are drawn from the HMI instrument's vector telescope on board the Solar Dynamics Observatory and correspond to a solar magnetograph for operational space weather forecasting and space science investigations. The mass of existing solar space magnetographs ranges between 73 and 35 kg. The goal of this research was to reduce the mass to the minimum possible, with a target of 20 kg for the optics. The goal dimensions for the magnetograph are 800 mm × 200 mm × 200 mm.

Table 1. Specifications for a miniaturised solar magnetograph for space applications.

Specifications	
Field of view	1920 arcsec diameter at 1 AU
Telescope diameter	140 mm
Telescope effective focal length	2468 mm
Telescope F/D	F/17.65
Plate scale at the telescope focus	83.58 arcsec/mm
Wavelength	589 nm (Na D line)
Goal dimensions for the magnetograph	800 mm × 200 mm × 200 mm

4.2. Optical Design Options

In order to evaluate the optical performance of the magnetograph, without any additional aberration that would be introduced by the telescope, a paraxial lens has been used to generate a beam and telescope focal plane based on the specifications of Table 1. The first optical design (Figure 6) was a 1:1 telecentric system, using 100 mm focal length lenses for collimator and cameras, which allows enough space for the waveplates (QWP and HWP) and polarisers (P1, P2). An intermediate pupil image was reimagined in the middle of each MOF cell (MOF and WS). This enlarges the instrument size, since another pupil image is required for the second cell (WS). A layout based on this idea generates two intermediate pupil images and an intermediate focus. The space from the first camera lens to the second collimator lens is unused. The length of this first design was 832 mm, slightly larger than the specification. Alternative solutions were investigated to reduce the total length using real lenses, such as the layout of Figure 7, whose length is reduced to 634 mm × 61.5 mm height and provides a pupil image inside each cell. In order to measure the two polarisation components simultaneously, a polarising beam splitter and a second detector were added, as shown in Figure 8, with an increase in the length of the instrument, which is now 650 mm. Most of the optical components for this design are commercial in order to demonstrate feasibility. For space applications, it will be necessary to select space-qualified optical materials.

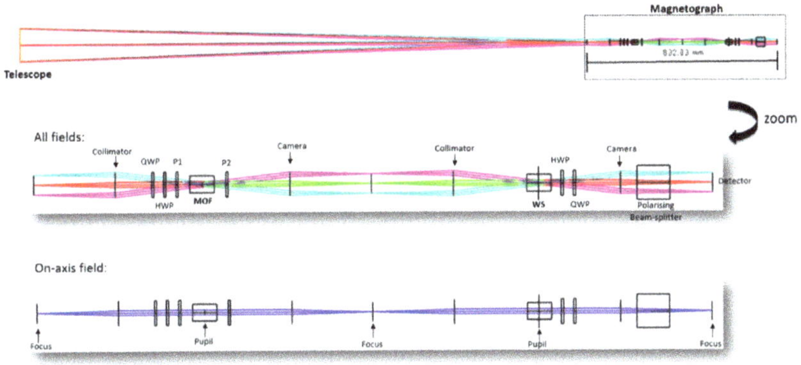

Figure 6. Design of a paraxial 1:1 telecentric magnetograph providing a pupil image for each cell and an intermediate focus. The layout on the top includes a paraxial telescope to produce a beam with the specifications defined in Table 1. The layout in the middle includes all fields, and that at the bottom only the on-axis field. The colours represent different fields.

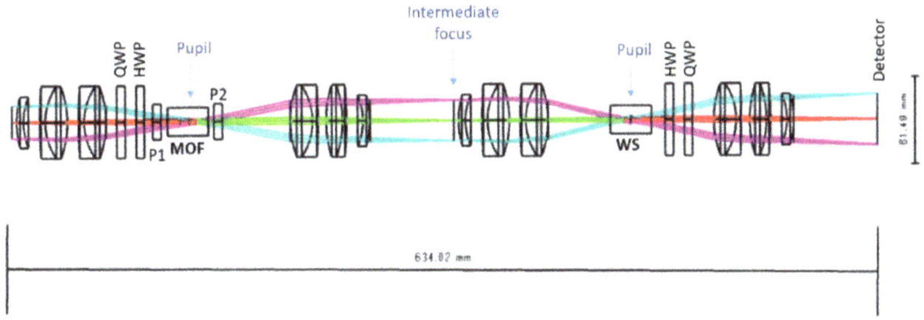

Figure 7. Magnetograph design with a pupil image inside each cell and an intermediate focus available. The total length is 634 mm. The colours represent different fields.

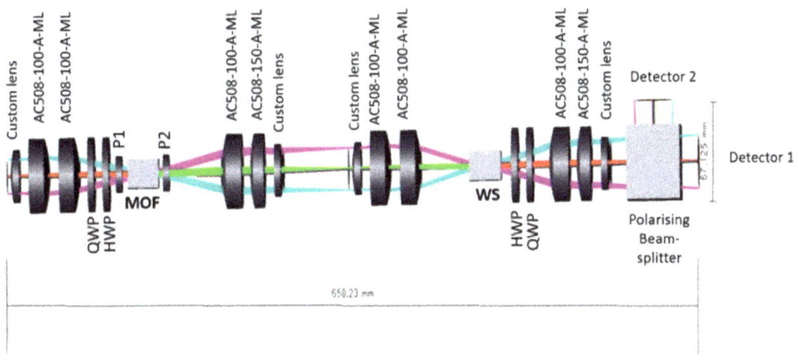

Figure 8. Design of a compact magnetograph using mainly commercial components.The colours represent different fields.

Producing only one intermediate pupil, where one of the cells could be centred at and having the second cell nearby or, alternatively, placing both cells close to that pupil, the number of lenses required will be reduced, optimising the efficiency. An example is illustrated in Figure 9 using only one detector to show the concept and in Figure 10 with two detectors to measure the two orthogonal linearly polarised states of the light simultaneously.

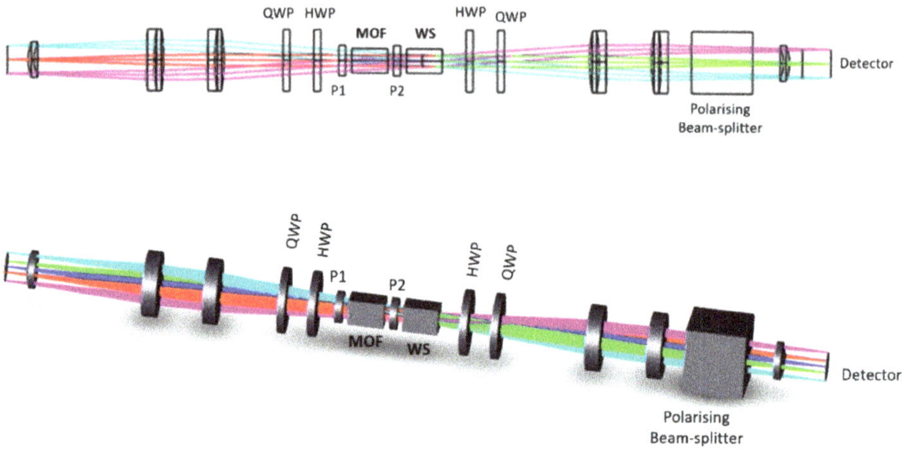

Figure 9. Optical design of a magnetograph with one intermediate pupil located within the wing selector (WS) cell, near the MOF. The colours represent different fields.

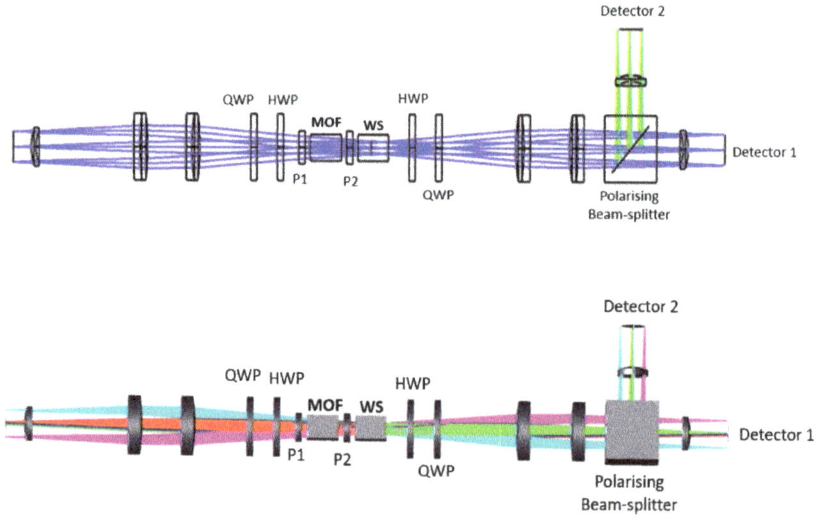

Figure 10. Optical design of a magnetograph with one intermediate pupil located within the wing selector (WS) cell, near the MOF, and using two detectors to measure the two orthogonal linearly polarised states of the light simultaneously. The colours represent different fields.

4.3. Final Design and Optical Quality

The studied layout alternatives investigated the feasibility of proposing a magnetograph within the specified envelope (800 mm $\times$ 200 mm $\times$ 200 mm). Further optimisations led to the final design proposed, reducing the length of the instrument and improving its optical performance.

Our first proposal, shown in Figure 11, is an ultra-compact layout within 378 mm × 70 mm × 70 mm. These dimensions are defined for the optics, without including detector sizes and mechanical mounts.

This design has an entrance focal ratio of F/17.65, with a plate scale of 83.58 arcsec/mm at the telescope image focal plane and an instrument final focal ratio of F/18.45, thus producing a magnification of 1.04. The field of view is 1920 arcsec in diameter, equivalent to ±0.27 degrees, thus meeting all specifications of Table 1.

The optical components are specified in Figure 12. A custom lens with radii 42.33 mm and 33.58 mm is placed at 8 mm from the focus. Thirty millimetres away, two achromatic doublets from THORLABS are used. The first one, with a focal length of 180 mm, is AC508-180-A-ML and the second one, separated by 10 mm, has a focal length of 100 mm and is AC508-100-A-ML. The quarter and half-wave plates are from Edmund Optics. Six millimetres from the half-wave plate, a linear polariser, P1, is used. Both linear polarisers, P1 and P2, are from THORLABS. The size of the cells (MOF, WS) represented in this layout is 20 mm side by 30 mm length. An intermediate pupil image is produced within the first cell (MOF). The two achromatic lenses after the second HWP are from THORLABS, the first one, AC508-100-A-ML, with a focal length of 100 mm and an achromatic doublet, AC508-180-A-ML, with a focal length of 180 mm. The combination of these lenses enables a compact layout. Before the polarising beam splitter, a custom lens, with radii 44.41 mm and 55.37 mm, compensates for aberrations. The polarising beam-splitter cube, with 50.8 mm side, is from THORLABS. The use of mainly commercial optical components was decided in order to build the proposed miniaturised magnetograph and test its performance in the lab.

The proposed solution is very compact, below the specifications for dimensions, and presents a great optical quality almost at the diffraction limit, which is a challenge when considering wide fields of view, short focal lengths, and commercial optical components. The main characteristics are presented in Table 2. Although the proposed design is compact and lightweight, it is not expected to require high-accuracy tolerances. The input and output beams are slow, enabling a depth of focus within which the optical quality is still acceptable. This relaxes tolerances in positioning and angles of incidence. Alignment is more critical in previously considered layouts that share some components, such as the detector or the cell, in order to avoid beam overlapping. No critical aspects have been identified in the proposed layout, however, a detailed tolerance analysis will be performed in later phases, including sensitivity analysis.

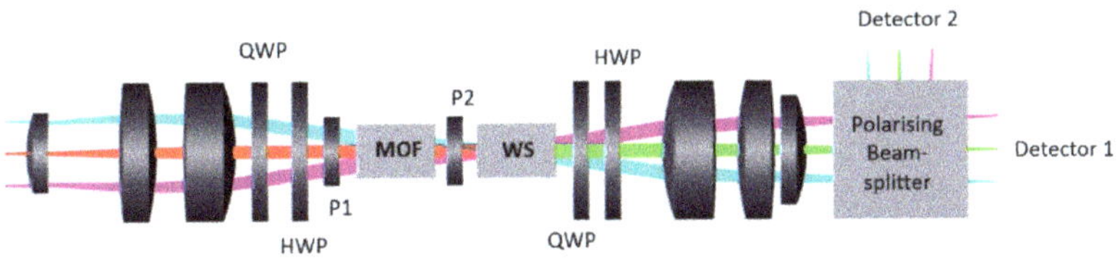

Figure 11. Optical design of a miniaturised solar magnetograph with dimensions 378 mm × 70 mm. The colours represent different fields.

The size of the image on the detector has a diameter of 24.17 mm. If a detector such as that of MDI/SOHO is considered, this would have 1024 × 1024 square pixels of 21 µm side, which leads to a linear size of 21.50 mm × 21.50 mm. The image size can be accommodated to the detector size by modifying the last commercial lens, AC508-180-A-ML to AC508-100-A-ML, which has a shorter focal length. This also produces a more compact layout (see Figure 13), with dimensions 345 mm × 53.9 mm, and the image size has a diameter of 18.43 mm (see Figure 14). Despite being very compact, this layout offers an excellent optical quality, equivalent to that of Figure 15.

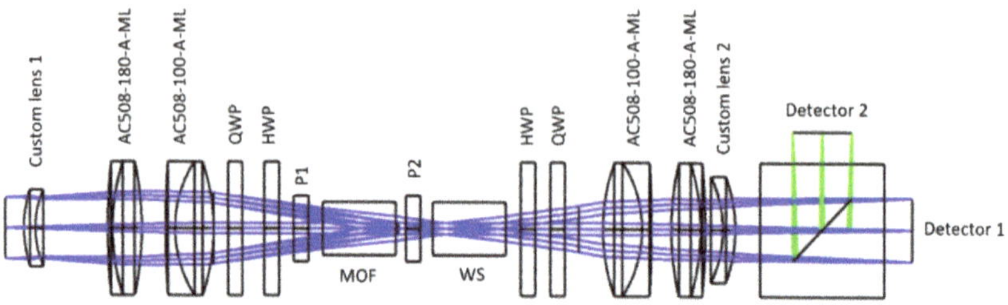

Figure 12. Optical components for the design of a miniaturised solar magnetograph.

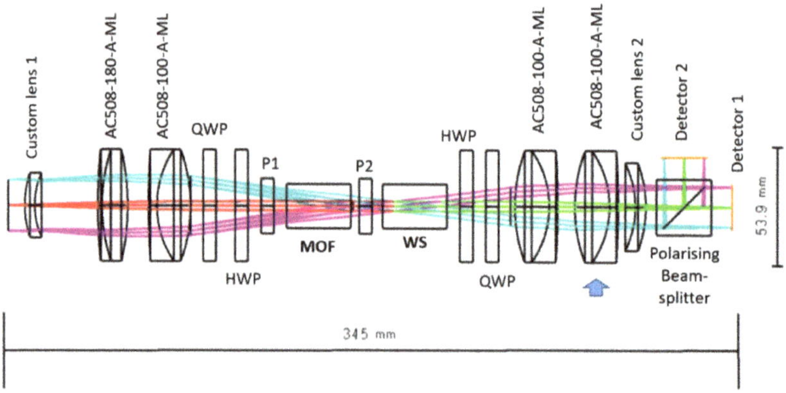

Figure 13. Compact magnetograph design with dimensions 345 mm × 53.9 mm. The colours represent different fields.

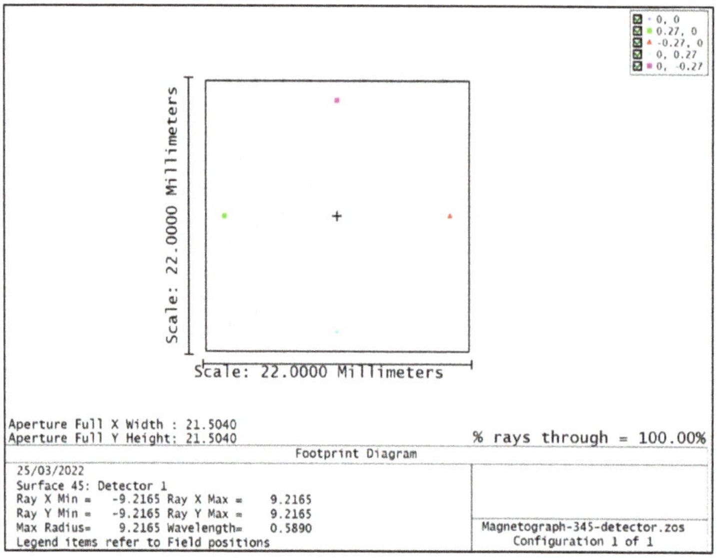

Figure 14. Beam footprint at the detector position showing that with the new design, the image is contained within the detector dimensions.

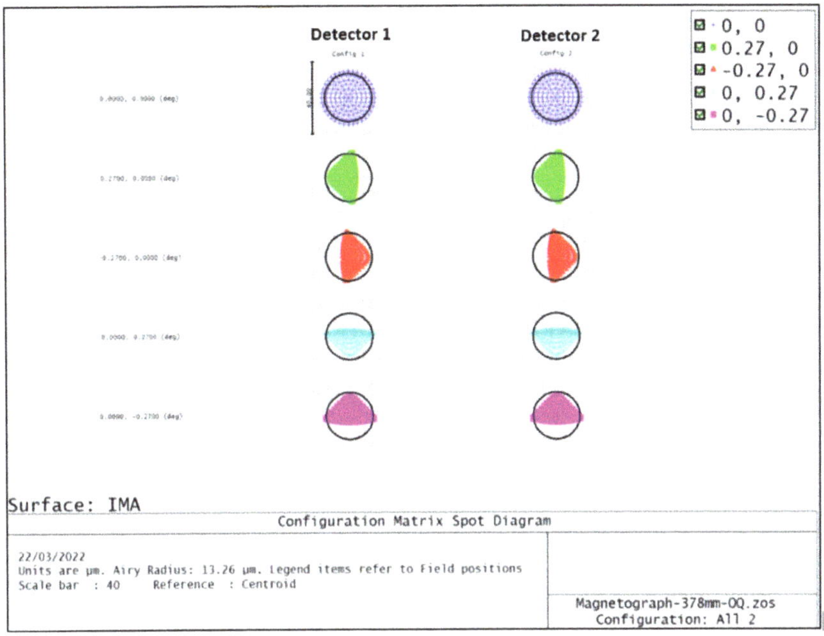

Figure 15. Optical quality obtained at detector 1 (column on the left) and detector 2 (column on the right). For each detector, five field points were defined. The one on the top corresponds to the on-axis field (0,0) and the rest are associated with the values in the legend on the top right, which are defined in degrees. The circle is the Airy disc showing the diffraction limit, within which most of the rays are contained.

A summary of the main investigated optical designs is presented in Table 3, all of them meeting the specifications of Table 1.

Table 2. Characteristics of the proposed layout.

Characteristics of the Proposed Layout	
Dimensions	345 mm × 54 mm × 54 mm
Optical quality	almost diffraction limited
Magnetograph magnification	0.79
Image size	18.43 mm diameter
arcsec/pixel	2.187
Number of intermediate pupils	one

Table 3. Summary of the main investigated layouts.

Description	Length	Layout
1:1 telecentric, two pupils, one per cell	634 mm	Figure 8
Only one intermediate pupil, 24.17 mm image diameter	378 mm	Figure 12
One pupil, 18.43 mm image diameter	345 mm	Figure 13

5. Conclusions

The main goal of this research was to investigate the feasibility of more compact and lightweight solar magnetographs for space applications. Several novel ideas were evaluated to reduce dimensions. The conclusion of this study is that it is possible to design a miniaturised solar magnetograph with diffraction-limited optical quality contained within an envelope of 345 mm × 54 mm × 54 mm. The weight of the optics has been significantly reduced with respect to existing solutions. This design has an entrance focal ratio of F/17.65, with a plate scale of 83.58 arcsec/mm at the telescope image focal plane, and it produces a magnification of 0.79. The field of view is 1920 arcsec in diameter, equivalent to ±0.27 degrees. For a low-earth satellite in a near-equatorial orbit, the doppler shift at the sodium D line would be up to 0.015 nm. Given that the MOF filter is not tuneable in wavelength, this would imply a significant orbital change in sensitivity. However, this can be overcome by selecting an orbit at L1 (e.g., SoHO), L2, L4, or L5. Most of the optical components considered for this layout are commercial and, although these are not space qualified, they have been selected in order to prove the concept. This research will continue building the proposed magnetograph and measuring its performance in the lab to complete the feasibility study of miniaturised solar magnetographs and enable this technology for future space missions.

Author Contributions: Conceptualization, A.C.R.; Formal analysis, A.C.R. and A.S.; Investigation, A.C.R., L.M.G., A.S. and D.M.L.; Writing—original draft, A.C.R.; Writing—review & editing, L.M.G., A.S. and D.M.L.; Funding acquisition, L.M.G. All authors have read and agreed to the published version of the manuscript.

Funding: This research was funded by a Royal Society Enhancement Award. LMG also received funding under the Royal Society University Research Fellowship scheme. DML is grateful to the Science Technology and Facilities Council for the award of an Ernest Rutherford Fellowship (ST/R003246/1).

Conflicts of Interest: The authors declare no conflict of interest.

References

1. Brun, A.S.; Browning, M.K. Magnetism, dynamo action and the solar-stellar connection. *Living Rev. Sol. Phys.* **2017**, *14*, 4. [CrossRef] [PubMed]
2. Temmer, M. Space weather: The solar perspective: An update to Schwenn (2006). *Living Rev. Sol. Phys.* **2021**, *18*, 4. [CrossRef]
3. Forbes, T.G. A review on the genesis of coronal mass ejections. *J. Geophys. Res.* **2000**, *105*, 23153. [CrossRef]
4. Green, L.M.; Török, T.; Vršnak, B.; Manchester, W.; Veronig, A. The origin, early evolution and predictability of solar eruptions. *Space Sci. Rev.* **2018**, *214*, 46. [CrossRef]
5. Odstrcil, D.; Pizzo, V.J. Distortion of the interplanetary magnetic field by three-dimensional propagation of coronal mass ejections in a structured solar wind. *J. Geophys. Res.* **1999**, *104*, 28225. [CrossRef]
6. Pomoell, J.; Poedts, S. EUHFORIA: European heliospheric forecasting information asset. *J. Space Weather Space Clim.* **2018**, *8*, A35. [CrossRef]
7. de Wijn, A.G.; Casini, R.; Carlile, A.; Lecinski, A.R.; Sewell, S.; Zmarzly, P.; Eigenbrot, A.D.; Beck, C.; Wöger, F.; Knölker, M. The visible spectro-polarimeter of the Daniel K. Inouye Solar Telescope. *Sol. Phys.* **2022**, *297*, 22. [CrossRef]
8. Scherrer, P.H.; Schou, J.; Bush, R.I.; Kosovichev, A.G.; Bogart, R.S.; Hoeksema, J.T.; Liu, Y.; Duvall, T.L., Jr.; Zhao, J.; Title, A.M.; et al. The Helioseismic and Magnetic Imager (HMI) Investigation for the Solar Dynamics Observatory (SDO). *Sol. Phys.* **2012**, *275*, 207. [CrossRef]
9. Tsuneta, S.; Ichimoto, K.; Katsukawa, Y.; Nagata, S.; Otsubo, M.; Shimizu, T.; Suematsu, Y.; Nakagiri, M.; Noguchi, M.; Tarbell, T.; et al. The solar optical telescope for the Hinode mission: An overview. *Sol. Phys.* **2008**, *249*, 167. [CrossRef]
10. Scherrer, P.H.; Hoeksema, J.T.; Bush, R.I. The solar oscillations investigation—Michelson Doppler imager for SOHO. *Adv. Space Res.* **1991**, *11*, 113. [CrossRef]
11. Solanki, S.K.; del Toro Iniesta, J.C.; Woch, J.; Gandorfer, A.; Hirzberger, J.; Alvarez-Herrero, A.; Appourchaux, T.; Pillet, V.M.; Pérez-Grande, I.; Kilders, E.S.; et al. The polarimetric and helioseismic imager on solar orbiter. *Astron. Astrophys.* **2020**, *642*, A11. [CrossRef]
12. Agnelli, G.; Cacciani, A.; Fofi, M. The magneto-optical filter: I: Preliminary observations in Na D lines. *Sol. Phys.* **1975**, *44*, 509. [CrossRef]
13. Forte, R.; Jefferies, S.M.; Berrilli, F.; Moro, D.D.; Fleck, B.; Giovannelli, L.; Murphy, N.; Pietropaolo, E.; Rodgers, W. The MOTH II Doppler-Magnetographs and Data Calibration Pipeline. *Proc. Int. Astron. Union* **2018**, *335*, 335–339. [CrossRef]
14. Giovannelli, L.; Berrilli, F.; Calchetti, D.; Del Moro, D.; Viavattene, G.; Pietropaolo, E.; Iarlori, M.; Rizi, V.; Jefferies, S.M.; Oliviero, M.; et al. The Tor Vergata Synoptic Solar Telescope (TSST): A robotic, compact facility for solar full disk imaging. *J. Space Weather Space Clim.* **2020**, *10*, 58. [CrossRef]
15. Calchetti, D.; Viavattene, G.; Terranegra, L.; Pietropaolo, E.; Oliviero, M.; Murphy, N.; Jefferies, S.M.; Giovannelli, L.; Moro, D.D.; Berrilli, F. Tor Vergata Synoptic Solar Telescope: Spectral characterization of potassium KI D1 MOFs. In Proceedings of the Ground-based and Airborne Telescopes VIII, Online, 14–18 December 2020; SPIE: Bellingham, WA, USA, 2020; Volume 11445, p. 114452T.

16. Berrilli, F.; Bigazzi, A.; Roselli, L.; Sabatini, P.; Velli, M.; Alimenti, F.; Cavallini, F.; Greco, V.; Moretti, P.F.; Orsini, S.; et al. The ADAHELI solar mission: Investigating the structure of Sun's lower atmosphere. *Adv. Space Res.* **2010**, *45*, 1191. [CrossRef]
17. Moretti, P.F.; Berrilli, F.; Bigazzi, A.; Jefferies, S.M.; Murphy, N.; Roselli, L.; Mauro, M.P.D. Future instrumentation for solar physics: A double channel MOF imager on board ASI Space Mission ADAHELI. *Astrophys. Space Sci.* **2010**, *328*, 313. [CrossRef]
18. Ruzmaikin, A.; Moynihan, P.I.; Vaughan, A.H.; Cacciani, A. Compact Doppler magnetograph. In Proceedings of the Missions to the Sun II, San Diego, CA, USA, 22–23 July 1998; Volume 3442, p. 115. [CrossRef]

aerospace

MDPI

Article

The Solar Particle Acceleration Radiation and Kinetics (SPARK) Mission Concept

Hamish A. S. Reid [1,*], Sophie Musset [2], Daniel F. Ryan [3], Vincenzo Andretta [4], Frédéric Auchère [5], Deborah Baker [1], Federico Benvenuto [6], Philippa Browning [7], Éric Buchlin [5], Ariadna Calcines Rosario [8], Steven D. Christe [9], Alain Jody Corso [10], Joel Dahlin [11], Silvia Dalla [12], Giulio Del Zanna [13], Carsten Denker [14], Jaroslav Dudík [15], Robertus Erdélyi [16,17,18], Ilaria Ermolli [19], Lyndsay Fletcher [20,21], Andrzej Fludra [22], Lucie M. Green [1], Mykola Gordovskyy [23], Salvo L. Guglielmino [19], Iain Hannah [20], Richard Harrison [22], Laura A. Hayes [2], Andrew R. Inglis [9,24], Natasha L. S. Jeffrey [25], Jana Kašparová [15], Graham S. Kerr [9,24], Christian Kintziger [26], Eduard P. Kontar [20], Säm Krucker [3,27], Timo Laitinen [12], Philippe Laurent [28], Olivier Limousin [28], David M. Long [29], Shane A. Maloney [30], Paolo Massa [31], Anna Maria Massone [6], Sarah Matthews [1], Tomasz Mrozek [32], Valery M. Nakariakov [33], Susanna Parenti [5], Michele Piana [6,34], Vanessa Polito [35,36], Melissa Pesce-Rollins [37], Paolo Romano [19], Alexis P. Rouillard [38], Clementina Sasso [4], Albert Y. Shih [9], Marek Stęślicki [31], David Orozco Suárez [39,40], Luca Teriaca [41], Meetu Verma [13], Astrid M. Veronig [42], Nicole Vilmer [43], Christian Vocks [13] and Alexander Warmuth [13]

1 Mullard Space Science Laboratory, University College London, Holmbury Hill Rd, Dorking RH5 6NT, UK; sarah.matthews@ucl.ac.uk
2 European Space Research and Technology Centre, 2201 Noordwijk, The Netherlands
3 FHNW Institute for Data Science, University of Applied Sciences and Arts Northwestern Switzerland, Bahnhofstrasse 6, 5210 Windisch, Switzerland
4 INAF/Capodimonte Astronomical Observatory, 80131 Naples, Italy; vincenzo.andretta@inaf.it
5 CNRS, Institut d'Astrophysique Spatiale, Université Paris-Saclay, 91405 Orsay, France; susanna.parenti@universite-paris-saclay.fr
6 MIDA, Dipartimento di Matematica, Università di Genova, 16146 Genova, Italy; massone@dima.unige.it (A.M.M.); piana@dima.unige.it (M.P.)
7 Jodrell Bank Centre for Astrophysics, University of Manchester, Manchester M13 9PL, UK; philippa.browning@manchester.ac.uk
8 Centre for Advanced Instrumentation, Durham University, Durham DH1 3LE, UK
9 NASA Goddard Space Flight Center, Heliophysics Science Division, Code 671, Greenbelt, MD 20771, USA; kerrg@cua.edu (G.S.K.); albert.y.shih@nasa.gov (A.Y.S.)
10 National Research Council of Italy, Institute for Photonics and Nanotechnologies, Via Trasea 7, 35131 Padova, Italy; alain.corso@pd.ifn.cnr.it
11 Astronomy Department, University of Maryland, College Park, MD 20740, USA
12 Jeremiah Horrocks Institute, University of Central Lancashire, Preston PR1 2HE, UK
13 DAMTP, Centre for Mathematical Sciences, University of Cambridge, Wilberforce Road, Cambridge CB3 0WA, UK; g.del-zanna@damtp.cam.ac.uk (G.D.Z.); cvocks@aip.de (C.V.); awarmuth@aip.de (A.W.)
14 Leibniz-Institut für Astrophysik Potsdam (AIP), An der Sternwarte 16, 14482 Potsdam, Germany
15 Astronomical Institute of the Czech Academy of Sciences, Fričova 298, 251 65 Ondřejov, Czech Republic; jana.kasparova@asu.cas.cz
16 Solar Physics & Space Plasma Research Center (SP2RC), School of Mathematics and Statistics, University of Sheffield, Hounsfield Road, Sheffield S3 7RH, UK; robertus@sheffield.ac.uk
17 Department of Astronomy, Eötvös Loránd University, Pázmány Péter sétány 1/A, H-1117 Budapest, Hungary
18 Gyula Bay Zoltan Solar Observatory (GSO), Hungarian Solar Physics Foundation (HSPF), Petőfi tér 3., H-5700 Gyula, Hungary
19 INAF—Catania Astrophysical Observatory, 95123 Catania, Italy; ilaria.ermolli@inaf.it (I.E.); salvatore.guglielmino@inaf.it (S.L.G.)
20 School of Physics and Astronomy, University of Glasgow, Glasgow G12 8QQ, UK; lyndsay.fletcher@glasgow.ac.uk (L.F.); iain.hannah@glasgow.ac.uk (I.H.)
21 Rosseland Centre for Solar Physics, University of Oslo, P.O. Box 1029 Blindern, NO-0315 Oslo, Norway
22 RAL Space, STFC Rutherford Appleton Laboratory, Chilton, Didcot OX11 0QX, UK; andrzej.fludra@stfc.ac.uk (A.F.); richard.harrison@stfc.ac.uk (R.H.)
23 Department of Physics, Astronomy and Mathematics, University of Hertfordshire, Hatfield AL10 9AB, UK
24 Department of Physics, Catholic University of America, Washington, DC 20064, USA
25 Department of Mathematics, Physics & Electrical Engineering, Northumbria University, Newcastle upon Tyne NE1 8ST, UK

[26] Centre Spatial de Liège, University of Liège (ULiège)—STAR Institute, 4000 Liège, Belgium; ckintziger@uliege.be
[27] Space Sciences Lab, UC Berkeley, 7 Gauss Way, Berkeley, CA 94708, USA
[28] Université Paris-Saclay, Université Paris Cité, CEA, CNRS, AIM, 75205 Paris, France; philippe.laurent@cea.fr (P.L.); olivier.limousin@cea.fr (O.L.)
[29] Astrophysics Research Centre, School of Mathematics and Physics, Queen's University Belfast, University Road, Belfast BT7 1NN, UK
[30] Dublin Institute of Advanced Studies, 31 Fitzwilliam Place, D02 XF86 Dublin, Ireland; shane.maloney@dias.ie
[31] Department of Physics & Astronomy, Western Kentucky University, Bowling Green, KY 42101, USA
[32] Centrum Badań Kosmicznych PAN, Bartycka 18A, 00-716 Warszawa, Poland; tmrozek@cbk.pan.wroc.pl
[33] Physics Department, University of Warwick, Coventry CV4 7AL, UK
[34] Osservatorio Astrofisico di Torino, Istituto Nazionale di Astrofisica, 10025 Pino Torinese, Italy
[35] Lockheed Martin Solar and Astrophysics Laboratory, Building 252, 3251 Hanover Street, Palo Alto, CA 94304, USA
[36] Department of Physics, Oregon State University, Corvallis, OR 97331, USA
[37] Istituto Nazionale di Fisica Nucleare, Sezione di Pisa, 56127 Pisa, Italy
[38] IRAP, Université Toulouse III—Paul Sabatier, CNRS, CNES, 31062 Toulouse, France
[39] Instituto de Astrofísica de Andalucía (IAA-CSIC), 18008 Granada, Spain; orozco@iaa.es
[40] Spanish Space Solar Physics Consortium (S^3PC), 18008 Granada, Spain
[41] Max Planck Institute for Solar System Research, 37077 Göttingen, Germany
[42] Institute of Physics & Kanzelhöhe Observatory for Solar and Environmental Research, University of Graz, 8010 Graz, Austria
[43] LESIA, Observatoire de Paris, Université PSL, CNRS, Sorbonne Université, Université Paris Cité, 5 place Jules Janssen, 92195 Meudon, France; nicole.vilmer@obspm.fr
* Correspondence: hamish.reid@ucl.ac.uk

Abstract: Particle acceleration is a fundamental process arising in many astrophysical objects, including active galactic nuclei, black holes, neutron stars, gamma-ray bursts, accretion disks, solar and stellar coronae, and planetary magnetospheres. Its ubiquity means energetic particles permeate the Universe and influence the conditions for the emergence and continuation of life. In our solar system, the Sun is the most energetic particle accelerator, and its proximity makes it a unique laboratory in which to explore astrophysical particle acceleration. However, despite its importance, the physics underlying solar particle acceleration remain poorly understood. The SPARK mission will reveal new discoveries about particle acceleration through a uniquely powerful and complete combination of γ-ray, X-ray, and EUV imaging and spectroscopy at high spectral, spatial, and temporal resolutions. SPARK's instruments will provide a step change in observational capability, enabling fundamental breakthroughs in our understanding of solar particle acceleration and the phenomena associated with it, such as the evolution of solar eruptive events. By providing essential diagnostics of the processes that drive the onset and evolution of solar flares and coronal mass ejections, SPARK will elucidate the underlying physics of space weather events that can damage satellites and power grids, disrupt telecommunications and GPS navigation, and endanger astronauts in space. The prediction of such events and the mitigation of their potential impacts are crucial in protecting our terrestrial and space-based infrastructure.

Keywords: particle acceleration; magnetic reconnection; instrumentation; corona; coronal mass ejections (CMEs); flares; extreme ultraviolet; X-rays; gamma rays

1. Scientific Objectives

The SPARK mission concept aims to investigate solar particle acceleration and the magnetic energy release that powers it by observing solar eruptive events, the most energetic and geo-effective drivers of space weather.

In the standard model of solar eruptive events (Figure 1; see also [1]), highly stressed magnetic fields reconnect in the low corona, thereby impulsively releasing vast amounts of energy. Depending on the magnetic configuration, plasma, magnetic field, and accelerated particles may escape into the heliosphere as coronal mass ejections (CMEs), "jets", or solar energetic particles (SEPs), which directly contribute to space weather. Accelerated particles also spiral downward around magnetic field lines ("loops") towards the chromospheric "footpoints", depositing their energy as they propagate. This heats and ionises the plasma in the chromosphere, transition region (TR), and lower corona, producing the intense broadband radiation known as a solar flare. The rapid heating creates a high-pressure region that ablates material back up along the loops in a process known as chromospheric "evaporation", which causes the loops to radiate in extreme ultraviolet (EUV) and soft X-rays. Additionally, plasma in and above the loops can be directly heated by the energy release and/or acceleration process.

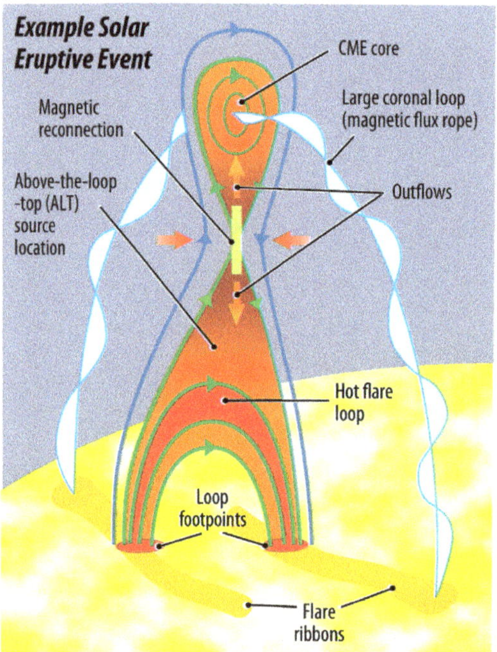

Figure 1. SPARK captures all elements of a solar eruptive event (the combination of a flare and a CME) identified in this cartoon. FOXSI images the HXR signatures of accelerated electrons and hot plasma at all locations. LISSAN captures the γ-ray signatures of accelerated ions and the most energetic electrons, and SISA reveals the lower atmospheric response and EUV structure in the corona. Figure courtesy of the FIERCE proposal team and the FOXSI SMEX proposal team.

The particles in solar eruptive events can be divided into three populations: hot plasmas, accelerated electrons, and accelerated ions. One of the most useful diagnostics for characterising thermalised and accelerated electrons is the X-ray bremsstrahlung emission they produce as they scatter in the ambient medium. The bremsstrahlung spectrum reflects the velocity distribution of the particles that produced it and can be inverted to reconstruct the spectrum of the emitting electrons [2,3]. This means hot plasma and accelerated electrons can be distinguished by their Maxwellian (thermal) and power-law (non-thermal) shaped spectra, e.g., [4–6]. Thermal emission tends to dominate in the soft X-ray (SXR; typically below 20 keV) regime, while non-thermal emission tends to dominate in the hard X-ray (HXR; typically higher than 20 keV) regime. X-rays can provide straightforward measurements of the numbers and energies of accelerated electrons not available from other wavelengths or requiring non-trivial assumptions when observed in microwaves. Hence, X-rays can provide a deeper understanding of the underlying acceleration process.

Accelerated ions are even less well understood than electrons due to the difficulties encountered in observing their γ-ray emission. Accelerated ions in the range of 1–100 MeV/nucleon can be detected via various γ-ray lines in the range of 1–10 MeV due to nuclear de-excitation, neutron capture, and positron annihilation [7–14]. Accelerated ions with energies greater than 100 MeV/nucleon can be detected via the decay of secondary pions via nuclear reactions with the ambient medium. As the pions decay, the decay products produce a broadband continuum at photon energies above 10 MeV, with a broad peak around 70 MeV from neutral pion radiation [15,16].

Although the principal points of the standard model are established, many questions remain regarding the fundamental processes of particle acceleration, impulsive energy release, and energy transport. However, the key scientific measurements required to answer these questions have not been possible with previous instruments. Solar γ-ray line emission has been imaged in one flare [17] and only localised through centroids in an additional four [14]. Consequently, the spatiotemporal evolution of accelerated ions has never been revealed. The role of ion acceleration in solar eruptive events therefore remains largely unknown, despite evidence that ions accelerated in flares may carry an energy comparable to that of accelerated electrons, e.g., [18–20]. Previous HXR spectroscopic imaging observations (e.g., RHESSI [21] and Solar Orbiter/STIX [22]) have not provided sufficient sensitivity to reliably observe accelerated electrons and direct plasma heating in the corona, where the acceleration is believed to take place. This is because the intensity of bremsstrahlung depends on the ambient density, which is typically very low in the corona, preventing observational tests of different acceleration models. Additionally, previous instruments have not provided a sufficient dynamic range ($\geq$100) to simultaneously observe the emission from the corona and the chromosphere, where the density (and, hence, emission), is much greater. This has limited our understanding of how transport effects alter the distribution of accelerated particles. Moreover, past and current X-ray images like RHESSI and STIX are limited by the use of an indirect Fourier imaging technique, which causes source areas and shapes to be only approximate. Additionally, imaging on second and subsecond time scales relevant to particle acceleration has not yet been achieved. Finally, current EUV imaging spectrographs (e.g., Hinode/EIS [23] and Solar Orbiter/SPICE [24]) have provided intriguing images of the complex structures associated with solar eruptive events, but they have not been optimised for solar eruptive events. Their typical single-slit design and operational priorities have led EUV spectra to rarely be available on the right timescales, at the right times, and in the right locations to compare with X-ray and γ-ray observations. The Solar-C Extreme-UV High-throughput Solar Telescope (EUVST; [25]) and the Multi-Slit Solar Explorer (MUSE; [26–28]) are highly complementary upcoming EUV missions that will be transformative for flare science [28]. MUSE will provide high-cadence active region-scale imaging spectroscopy, sampling select key EUV lines by rastering its 35 slits. EUVST, with its single slit, will provide much broader temperature coverage, with rich plasma diagnostics from a large number of lines. Both instruments are scheduled to launch within one year of each other and promise cospatial observations of plasma in the solar atmosphere. An instrument that would build upon their science legacy—one that co-observes with X-ray and g-ray instruments—should be the aim of a next-generation flare mission, that is, an instrument that combines rich plasma diagnostics with very high-cadence 2D spectral imaging.

SPARK will overcome all these challenges with its unique combination of high-sensitivity, fast spectroscopic imaging in γ-ray, X-ray, and EUV, optimised for solar eruptive events. It will address four specific fundamental science questions:

1. How does impulsive energy release accelerate particles in the solar atmosphere?
2. How is impulsively released energy transported and dissipated in the solar atmosphere?
3. What are the physical low-corona origins of space weather events?
4. How is the corona above active regions heated?

By addressing these questions, SPARK will elucidate fundamental physical processes that are ubiquitous throughout our Universe and drive space weather events that have direct consequences for our technologies and way of life.

1.1. *How Does Impulsive Energy Release Accelerate Particles in the Solar Atmosphere?*

Accelerated charged particles constitute a significant fraction (up to tens of percent) of the magnetic energy released in the most energetic space weather events, e.g., [19,29]. Observationally distinguishing between acceleration models requires the number, location, and evolution of multiple faint thermal and non-thermal sources near the coronal acceleration region to be characterized in the presence of much more intense chromospheric footpoint emission. SPARK's unique combination of high dynamic range and high-sensitivity imaging spectroscopy in the γ-ray, X-ray, and EUV regimes at timescales relevant to the underlying physical processes will make this possible for the first time.

1.1.1. Where and When Do Particle Acceleration and Local Plasma Heating Occur?

The two most likely models to explain the high acceleration efficiency of electrons are the Fermi acceleration process through the evolution and merging of "magnetic islands" [30,31] created by the reconnection, as well as a second-order Fermi acceleration process in the turbulent plasma of the reconnection outflow jets with or without termination shocks [32–35]. The magnetic-island model predicts that both electron acceleration and direct plasma heating occur near the reconnection site(s) in the current sheet and that direct plasma heating precedes electron acceleration [30]. Conversely, the stochastic model predicts that acceleration and direct heating occur simultaneously but significantly separated from the reconnection site in both upward and downward outflow jets (see Figure 2). Concerning ion acceleration, a detailed study of individual large events showed differences between ion and electron time evolution during the course of a flare [36]. The one flare imaged in the γ-ray line with RHESSI and the four for which emission centroid locations were calculated showed significant displacements between HXR and γ-ray line sources, indicating spatial displacements between electron and ion energy release sites [14,17]. SPARK will reveal, for the first time, where electron acceleration and direct heating occur with respect to the reconnection site, as well as under what scenarios the different acceleration models dominate, and reveal the relationship between electron and ion acceleration. This will be achieved with subsecond X-ray spectral imaging with sufficient spatial resolution to separate the various sources. SPARK will compare γ-ray and X-ray signatures of energetic electrons and ions in combination with the EUV non-Gaussian line profiles that are a signature of non-Maxwellian ion velocity distributions. Using increased X-ray sensitivity and dynamic range, SPARK will characterize the spectrum of the accelerated electrons in the corona, even in the presence of much brighter chromospheric emission. SPARK will provide complementary measurements of the coronal magnetic field from Fe X lines emitted around 1 MK. In addition, it will chart the plasma response to heating via the hot Fe XXIII and Fe XXIV (15–20 MK) spectral lines and determine the relative plasma abundances to differentiate between coronal and ablated chromospheric plasma.

1.1.2. What Are the Efficiency and Energy Content of Electron and Ion Acceleration?

The fraction of particles accelerated out of the ambient Maxwellian velocity distribution and the total energy they contain are essential constraints on acceleration models. Acceleration by magnetic islands [30] and super-Dreicer electric fields in a reconnecting current sheet [37] can accelerate a large fraction of the available electrons, while mechanisms relying on large-scale sub-Dreicer electric fields cannot [38]. SPARK will determine the number and energy of accelerated particles with an accuracy not previously possible. With a significantly enhanced X-ray dynamic range, SPARK will measure the non-thermal spectra of coronal and footpoint sources down to lower energies whilst constraining the relative number of accelerated particles of different ion species (e.g., alpha/proton ratio [36]).

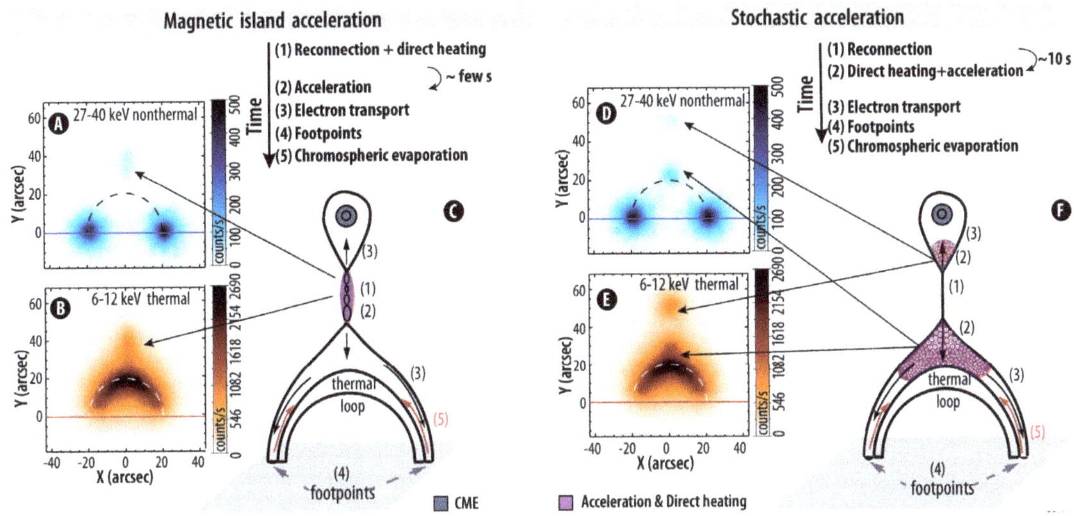

Figure 2. SPARK will distinguish between different models of flare particle acceleration. In these 2D flare cartoons, the locations and chronological order of various X-ray- and EUV-producing processes differ between the two most likely models of electron acceleration in eruptive flares: magnetic island-merging acceleration (**left**) and stochastic (second-order Fermi) acceleration (**right**). Simulated FOXSI detector response images of non-thermal electrons (blue) and thermal plasma (orange) are shown as insets. Figure courtesy of the FIERCE proposal team.

1.1.3. How Do Electron and Ion Acceleration and Transport Differ in the Flaring Atmosphere?

Theoretical studies show that differences between the acceleration and transport of electrons and ions can be used as a unique diagnostic tool for the processes in the magnetic reconnection region, as well as the geometry of the magnetic field in and around it. The lack of spatially resolved γ-ray observations of accelerated ions is therefore a significant obstacle to constructing a comprehensive solar flare model. With significantly upgraded spatial resolution in the γ-ray regime, SPARK will enable major advances in our understanding of how ions are accelerated and transported in flares and how their dynamics differ from the dynamics of energetic electrons and in using energetic ions as an important diagnostic tool for non-thermal plasma in the flaring corona.

1.1.4. Where and How Are the Most Energetic Particles Accelerated at the Sun?

Studies of small numbers of events examining γ-ray lines (1–10 MeV) and the pion continuum (>10 MeV) suggest that the accelerated ion spectrum is not a simple power law extending from non-relativistic (1–100 MeV/nucleon) to relativistic (>few hundred MeV/nucleon) regimes, e.g., [39–43]. This raises the question of whether the most energetic particles are accelerated via a different mechanism to those at lower energies. The longevity of some pion emission presents another major challenge to our understanding of how the most energetic solar particles are accelerated, e.g., [44,45]. The high-sensitivity HXR and γ-ray spectroscopy of SPARK will facilitate a comprehensive study of the timing and spectra of electron bremsstrahlung and pion decay radiation in a significant number of events for the first time. Such observations are essential to unravel the relative roles of flare and interplanetary processes in the acceleration of high-energy ions, especially in long-duration events.

1.2. How Is Impulsively Released Energy Transported and Dissipated in the Solar Atmosphere?

SPARK will probe energy-transport processes that link impulsively released magnetic energy to the resultant emission from the lower atmosphere, where the bulk of the flare energy is radiated. This will be done in two ways: by measuring hitherto poorly constrained observational inputs to the latest state-of-the-art numerical models of solar flares, e.g., [46–63] and by providing previously unachievable observations against which the model predictions will be critically interrogated. Such model inputs provided by SPARK include the non-thermal electron and ion energy distributions injected towards the lower solar atmosphere, the ribbon/footpoint source areas, and the preflare atmospheric state (e.g., coronal temperature, density, loop length, and coronal magnetic field).

1.2.1. How and Where Do Accelerated Particles Lose Their Energy in the Corona and Chromosphere?

As flare-accelerated particles exit the coronal acceleration region and propagate along loops, they lose energy through Coulomb collisions with ambient particles, wave–particle interactions, and the generation of return currents [4]. The evolution of the particle distribution as the particles propagate along flare loops depends on (and thus reveals) the relative importance of these mechanisms. The statistically significant separation between HXR and γ-ray line sources in the single resolved RHESSI γ-ray flare image and two of the four RHESSI centroid-localised γ-ray line flares [14,64] may be due to differing acceleration mechanisms. But it may also be due to different transport effects acting on the ions and electrons. We have sparse observations from EUV imaging spectrometers of the kernels of chromospheric evaporation, showing large non-thermal broadenings and upflows in the hotter lines during the impulsive phase (see, e.g., [65–67]), but a clear picture is missing. SPARK will simultaneously observe electrons throughout the flaring structure, image ion emission, and observe the spectral line response of flaring plasma at multiple temperatures. Combining X-rays and EUV imaging spectroscopy, SPARK will facilitate accurate determination of the low-energy part of the electron spectrum, as well as quantifying return current losses. SPARK will, for the first time, constrain accelerated ions transported to the chromosphere using γrays and the hottest EUV flare lines, like Fe XXIV. Moreover, the combination of X-rays and the multitemperature response of spectral lines will also provide constraints on turbulence present in the solar atmosphere [68].

1.2.2. What Are the Origins of Modulations in Solar Flare Emission?

A key observational feature in flare-associated X-ray emission is the presence of pronounced pulsations and fast-time variations. These modulations, which also appear in many stellar flares, have been observed in thermal and non-thermal emissions across all wavelength regimes, from radio to gamma rays, with characteristic time scales ranging from 0.5 to tens of seconds, e.g., [69–71]. Often, these modulations appear as regular or non-stationary oscillatory patterns, known as "quasi-periodic pulsations", e.g., [72,73]. However, despite extensive research, the origins of these short-timescale modulations in flare emissions remain debated. While some studies suggest they may be a direct signature of a repetitive impulsive energy-release process, the potential role of magnetohydrodynamic (MHD) oscillations in the flaring site or nearby, particularly in the context of longer-period pulsations, has yet to be fully determined. Similarly, a combination of these processes could be at play (see [71] for an overview of proposed mechanisms). Moreover, it is quite likely that different classes of flaring pulsations (i.e., different periods, patterns, or energies) are produced by different mechanisms. X-ray dynamic-range limitations have not yet allowed us to identify time-varying signatures from different parts of the loop, including the loop-top source. EUV imaging observations have also hindered our ability to locate the modulating emission source due to both cadence constraints and pixel saturation and bleeding during flare events. SPARK enables, for the first time, a full examination of the temporal, spatial, and spectral properties of these pulsations and their relationships across wavelengths, which are essential to determine the origins of the emission modulation. SPARK will allow us to identify the pulsations in both the thermal and non-thermal regimes in all parts of the flaring loop. Moreover, SPARK will identify whether accelerated protons have similarly associated time variability.

1.2.3. What Is the Importance of Accelerated Particles in Transporting Energy Compared with That of Other Mechanisms?

High-frequency Alfvén waves have been proposed as a means of transporting energy from a flare's magnetic reconnection site to the lower atmosphere and heating it, e.g., [74–76]. In recent years, modelling has shown that this is possible [52,60,61]. However, while Alfvén waves are undoubtedly produced during the large-scale reconfiguration of the magnetic field during flares, it is not yet known whether they play a significant role relative to accelerated particles in transporting flare energy and heating flare plasma nor whether other kinds of MHD waves, such as kink and sausage modes, contribute significantly. SPARK will, for the first time, reveal the importance of MHD waves relative to accelerated particles in transporting and dissipating energy in solar eruptive events. SPARK will examine the coronal magnetic field strength and the broadening of certain spectral lines from ions in the EUV passbands formed at different temperatures (see Section 2.3. SPARK will therefore constrain the Poynting flux as the waves propagate and dissipate their energy (see discussions in, e.g., [28]). SPARK will use variations of the chemical composition and elemental abundances to assess the role of MHD waves in transferring energy from the corona into flare kernels (cf. [77] and references therein).

1.3. What Are the Physical Low-Corona Origins of Space Weather Events?

An ESA-funded study estimated that the economic cost of a severe space weather event could be as high as EUR 15 billion (https://esamultimedia.esa.int/docs/business_with_esa/Space_Weather_Cost_Benefit_Analysis_ESA_2016.pdf, accessed on 30 October 2023). This led to the establishment of national forecasting centres across Europe and space weather as a major theme in the ESA's Space Safety programme. Despite this, many questions remain regarding the origins of space weather in the low corona, which act as an impediment to the development of timely and reliable space weather forecasts. SPARK will greatly improve our understanding of the underlying physical processes that drive these events in the low corona and inform the development of future space weather models that aim to deliver timely and accurate forecasts of flares, energetic particles, and eruptions. Knowing about the acceleration process will feed into our understanding of how active regions reach a state whereby a flare or CME is generated. Understanding of the flare initiation process will enable an improved view of the likelihood of a flare occurring in a location that is well placed to impact Earth.

1.3.1. What Are the Energy Content and Spectrum of Sun-Escaping Electrons?

Sun-escaping electrons, components of SEP space weather events, have long been studied in situ at 1 AU [78] and, more recently, closer to the Sun, e.g., [79,80]. However, such observations alone cannot be used to characterise how the electrons are accelerated because the electron distribution is modified by transport effects between the Sun and the observatory. SEP electrons can be observed remotely at the Sun as type III radio bursts (e.g., [81,82]). However, while bulk electron speeds can be inferred from the radio observations, unlike X-rays, they cannot be directly inverted to retrieve the number or energies of accelerated electrons. Therefore, the spectra and acceleration mechanism(s) of solar radio-emitting electrons remain unknown. How these accelerated electrons escape from the flare site is similarly unknown. CMEs and jets offer clear open magnetic paths for particles to escape, but confined flares do not, although interchange reconnection can pay a role [83]. Ground-based observations above 10 MHz can be used to image type IIIs [84] but may suffer from intrinsically limited spatial resolution, especially at low frequencies, on account of the radio waves scattering off density inhomogeneities between the source and observer [85,86]. SPARK will provide hitherto unachievable imaging and spectral observations of accelerated electrons as they escape the Sun [87], facilitated by unprecedented sensitivity and imaging dynamic range in the HXR regime. Such measurements will elucidate the origins of escaping electrons and how they are modified as they propagate towards Earth. SPARK will also test theories of the origins of the slow solar wind by detecting the locations in active region peripheries where particles are accelerated and escape via HXR emissions and upward flows detected in EUV-emitting plasma.

1.3.2. What Are the Dominant Initiation Mechanisms of Solar Eruptions?

Many models of the initiation of solar eruptive events involve magnetic reconnection, which results in plasma heating [88,89] and particle acceleration [90]. However, different models of CME initiation predict observationally differentiable locations of the erupting flux rope in relation to where the reconnection starts and, consequently, for the associated X-ray and EUV emissions. The internal tether-cutting model [91,92] predicts that reconnection occurs below the flux rope before the fast takeoff of the eruption. The breakout model [93] predicts that the reconnection occurs above the flux rope before fast takeoff, and the ideal MHD instability model [94] predicts that the flux rope begins to rise before reconnection occurs in either place. It is unclear if the same mechanisms driving the large-scale CMEs are also at play in these smaller events. Some models of jets involve breakout reconnection, e.g., [95], similar to the breakout model for CMEs, while others involve interchange magnetic reconnection [96–98]. SPARK will produce observations of the faint X-ray and EUV emissions linked to particle acceleration and plasma heating during the formation and initiation of solar eruptions for the first time. This will enable discrimination between the many physical processes proposed for the creation of the conditions necessary for an eruption. SPARK will also provide measurements of the plasma dynamics and of the magnetic field of the active region and filament before and during the eruption. Hence, SPARK will provide constraints on the configuration and evolution of the magnetic structure leading to solar eruptions.

1.4. How Is the Corona Above Active Regions Heated?

A long-standing enigma in solar and stellar physics is how a star's atmosphere can be orders of magnitude hotter than its surface. This temperature difference requires some form of non-radiative heating, but whether the dominant mechanism is the dissipation of Alfvén waves or impulsive heating by nanoflares has not been established [99–103]. SPARK will enable breakthroughs with respect to this fundamental problem using two approaches. First, SPARK will determine if the characteristics of energy release in the smallest detectable events are fundamentally different from those in larger flares. Secondly, SPARK will statistically determine ensemble properties of heating events too small to be detected individually.

1.4.1. Is Particle Acceleration Ubiquitous among Energy-Release Events at All Size Scales?

The number of flares as a function of their thermal energy follows a power law over several orders of magnitude [104]. This suggests that the underlying energy-release process scales similarly. If nanoflares are part of this distribution, they, too, would be expected to accelerate electrons. Indirect evidence from UV transients suggests that accelerated electrons are indeed present [105,106]. RHESSI and STIX observations have shown that in microflares, the X-ray spectral index is steeper than in larger flares [107–113], suggesting that they are less efficient at accelerating electrons. This was confirmed in a few observations of fainter microflares with NuSTAR during its limited solar campaigns [114–117]. Additional support comes from studies of the thermal–non-thermal energy partition (cf. [118]), which shows that in weaker flares there may not be a sufficient amount of energetic electrons to heat the thermal plasma. SPARK will determine how the energy-release process scales across eight orders of magnitude in energy, from the largest flaresb ($\sim 10^{33}$ ergs) down to flares at 10^{25} ergs (2 orders of magnitude smaller than those observed by RHESSI and STIX). SPARK will observe hundreds of thousands of flares below GOES C class and will provide a comprehensive investigation of events two orders of magnitude less energetic than ever before.

1.4.2. How Does Small-Scale Particle Acceleration Contribute to Coronal Heating?

The presence of temperatures exceeding 5 MK in the non-flaring active regions would provide strong evidence of impulsive, low-frequency nanoflare heating. Steady or high-frequency wave heating cannot maintain such high temperatures without violating other observational constraints [102,119]. Many studies have detected hot plasma, e.g., [120,121], but the uncertainties are large because the emission is orders of magnitude fainter than that from associated cooler plasma (cf. the review in [122]). Moreover, non-equilibrium ionization effects [123–125] and departures from a Maxwellian distribution due to the presence of accelerated particles [106,126–128] can limit the interpretation of EUV line-emission observations. SXR and HXR thermal bremsstrahlung emissions from the same plasma are not susceptible to non-equilibrium ionization effects, allowing measurements to be more clearly interpreted and the accelerated particles to be more readily detected. The FOXSI-2 sounding rocket performed X-ray measurements of high-temperature plasma in an active region [129], and SXR spectrometers flown on the SDO/EVE sounding rocket [121] and on the MinXSS CubeSat [130] have performed high-temperature measurements of spatially integrated SXR spectra. These measurements provide evidence of impulsive magnetic reconnection events contributing to active-region heating [131]. SPARK will provide important constraints on competing scenarios of coronal heating in active regions [132,133] and directly measure the predicted high-temperature X-ray signature of low-frequency nanoflare heating, improved by observations of multiple coronal emission lines from many ionisation states of Fe. Some of these lines also allow for diagnostics of accelerated electrons [126,128]. Ionization and recombination time scales will be derived through observations of the density of hot plasma, a key measurement not provided by prior EUV observations/missions. SPARK will also provide measurements of magnetic field strengths in active region loops, using the magnetically induced transition at 257.3 Å(c.f. [134] and references therein).

2. Payload

SPARK utilises three scientific instruments to provide imaging spectroscopy in the γ-ray, X-ray, and EUV regimes: LISSAN, FOXSI, and SISA, respectively. Figure 3 shows a model of the spacecraft, highlighting the accommodation of the three scientific instruments.

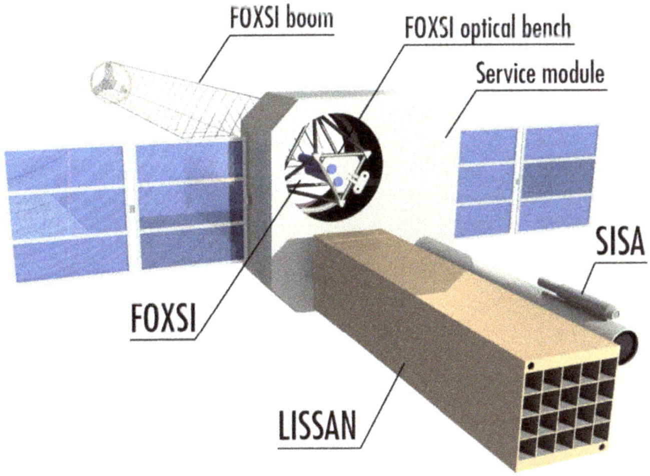

Figure 3. SPARK spacecraft model illustrating the payload accommodations.

2.1. Large Imaging Spectrometer for Solar Accelerated Nuclei (LISSAN)

LISSAN will, for the first time, reveal the dynamics of accelerated ions in solar flares via spectroscopic imaging between 40 keV and 100 MeV on time scales of less than 10 s. These capabilities will also allow it to observe high-energy X-ray emissions from energetic electrons, providing diagnostics for both types of accelerated particles. This will be achieved by using high-resolution scintillators with an energy resolution of 0.1 MeV at 6.1 MeV and an angular resolution of 8″ FWHM. At 6.1 MeV, LISSAN will achieve 40x RHESSI's sensitivity (5 photons/cm^2) and a spectral resolution of 1.5 %dE/E. LISSAN will also achieve 40x RHESSI's sensitivity (50 photons/cm^2) in the 2.2 MeV neutron capture line. A summary of the predicted performance of LISSAN is presented in Table 1.

Table 1. LISSAN instrument performance.

LISSAN Parameter	Expected Performance
Energy Range—Low	40 keV
Energy Range—High	100 MeV
Imaging Effective Area (2.2 MeV)	100 cm^2
Spectro Effective Area (2.2 MeV)	440 cm^2
Sensitivity (2.2 MeV)	50 photons/cm^2
Sensitivity (6.1 MeV)	5 photons/cm^2
Imaging Time Resolution	1 s
Angular Resolution	8″
Field of View	12.8′ diameter
Energy Resolution (6.1 MeV)	1.5% dE/E
Largest Observable Flare	>X5

LISSAN employs an indirect Fourier imaging technique [135,136]. Pairs of 1D slotted grids (bigrids) encode spatial information into moiré patterns, each of which is measured by a pixelated, spectroscopic, photon-counting γ-ray detector. Each bigrid samples an angular scale along a single direction on the plane of the sky. The resulting moiré patterns therefore represent spatial Fourier components of the field of view (visibilities), which can be combined into images via Fourier-based image reconstruction algorithms similar to those used in radio interferometry. This imaging concept has been successfully demonstrated by Solar Orbiter/STIX [22,137] in the 4–150 keV spectral range. LISSAN is composed of 20 subcollimators, 15 of which contain bigrids for imaging spectroscopy. The angular scales and directions they sample depend on the pitch (slit width) and orientation of the bigrids. The resulting visibilities can be represented as complex numbers on the (u, v) plane. One possible visibility configuration is displayed in the left panel of Figure 4, with its associated point-spread function in the central panel. The right panel shows an image of the 31 March 2022 M9.7 flare that LISSAN would have produced with this configuration in the 50–84 keV range using the CLEAN image reconstruction algorithm. Two non-thermal footpoint sources are clearly visible.

Of LISSAN's five remaining gridless subcollimators, one monitors the background, and four are used to boost sensitivity for spectroscopy. This is because the absence of bigrids increases the photon throughput by a factor four.

The detector of each subcollimator comprises 16 "fingers" of crystal. In one direction, this segmentation allows the moiré pattern to be measured. The other improves light collection and, therefore, spectral resolution and provides a redundant measurement of the moiré pattern. This guarantees the energy resolution needed to measure the Doppler profiles of the C and O lines at 4.4 and 6.1 MeV, respectively.

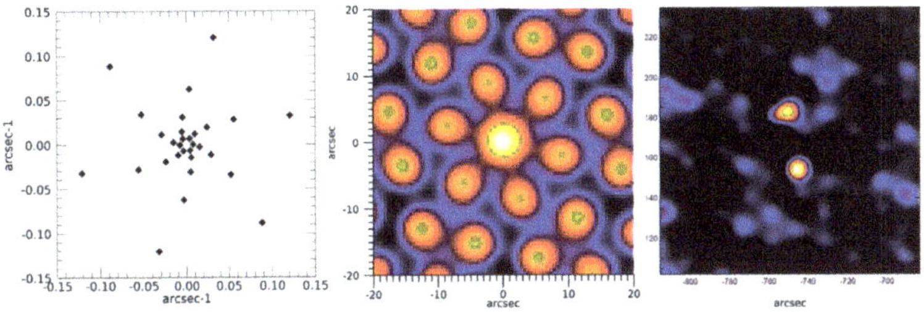

Figure 4. Left: One possible (u, v)-coverage for LISSAN; **middle**: associated point-spread function (dirty map of a point source on axis; this image contains both the X-ray source and instrumental artefacts to be removed with adequate cleaning algorithms); **right**: simulation of a LISSAN image of the 50–84 keV emission from the two hard X-ray footpoints during the M9.7 flare on 31 March 2022, which was observed by STIX on Solar Orbiter. This image was obtained by running the CLEAN algorithm on the dirty image.

2.2. Focusing Optics X-ray Solar Imager (FOXSI)

FOXSI combines grazing-incidence hard X-ray focusing optics with small, fast, pixelated detectors to produce images of the Sun at high spectral, spatial, and temporal resolution over the spectral range of 3–50 keV [138]. This strategy offers dramatic improvements in image quality, dynamic range, and sensitivity over the indirect (Fourier-based) imaging techniques of current and previous state-of-the-art solar X-ray spectroscopic imagers, e.g., RHESSI and Solar Orbiter/STIX. FOXSI will be able to reliably image faint thermal and non-thermal sources in the solar corona, even in the presence of brighter ones, for the first time. Such images will enable FOXSI to elucidate a ground-breaking new understanding of particle acceleration and the evolution of solar eruptive events. FOXSI will not intrinsically integrate images over preset time or energy intervals but instead record the energy, position, and arrival time of individual photons, allowing images and spectra to be produced ex post facto in accordance with specific science goals. FOXSI's design and measurement strategies have been proven through successful flights of several solar sounding rocket and balloon instruments [139–144]. Moreover, FOXSI will build on the success of non-solar space-based direct-focusing X-ray imagers (e.g., NuSTAR and Hitomi) and optimised for the resolution requirements and high fluxes of solar observations. This is achieved via judicious design of the optics' effective area, fast-counting detectors, and movable attenuators that can be deployed during the largest flares. This prevents corruption of the measured spectrum, e.g., via pile up, which is common in observations of even small solar flares by non-solar telescopes, e.g., NuSTAR. A summary of the predicted performance of FOXSI is presented in Table 2.

Table 2. Expected performance of SPARK/FOXSI.

FOXSI Parameter	Expected Performance
Energy Range—Low	3 keV
Energy Range—High	55 keV
Imaging Dynamic Range 1	20:1 beyond 20″ separation
Imaging Dynamic Range 2	1000:1 beyond 45″ separation
Effective Area (at 20 keV)	40 cm^2
Sensitivity	0.2 photons/cm^2
Imaging Time Resolution	0.1 s
Angular Resolution	6.3″ FWHM
Field of View	9.8′ × 9.8′
Energy Resolution	0.8 keV FWHM
Largest Observable Flare	>X10

FOXSI also includes a soft X-ray spectrometer (FOXSI-STC) that provides spatially integrated high-resolution spectra (0.2 keV FWHM at 1.5 keV) in the range of 0.8–15 keV. The combination of emission lines and thermal continuum emissions in this energy range provides additional plasma temperature and composition information averaged over all the plasma in the FOXSI FOV that FOXSI cannot access to due to its lower energy resolution (0.8 keV) and higher low-energy cutoff (3 keV). FOXSI-STC is composed of two identical spectrometers with different apertures optimised for low and high flux, respectively. This enables it to measure the X-ray fluxes of even the largest flares. FOXSI-STC will be used to control the movable attenuators of FOXSI's HXR focusing telescopes. This approach enables fewer attenuator motions compared to past and current instruments, which have to frequently remove their attenuators for short periods of time (i.e., peek) to check whether the flux has reduced to an acceptable level. A summary of the predicted performance of FOXSI-STC is presented in Table 3.

Table 3. Expected performance of SPARK/FOXSI-STC.

FOXSI-STC Parameter	Expected Performance
Energy Range—Low	0.8 keV
Energy Range—High	15 keV
Effective Area	0.01 cm^2
Energy Resolution	0.2 keV FWHM below 1.5 keV
Field of View	$9.8' \times 9.8'$
Time Resolution	0.5 s
Largest Observable Flare	>X10

2.3. *Spectral Imager of the Solar Atmosphere (SISA)*

SISA (Spectral Imager of the Solar Atmosphere) is an integral field spectrograph (IFS) [145], providing the simultaneous spectra of a bidimensional field of view of 100 arcseconds by 250 arcseconds using image slicer technology. Two spectral ranges will be covered, centred around 18.5 nm and 25 nm, with 1 arcsecond spatial resolution and a spectral resolving power of R∼3650–5160. A spectral ranges of 170–195 Å and 245–260 Å are required to measure the parameters of 1 MK plasma and the hotter 15 MK plasma, respectively. This wavelength range includes lines sensitive to coronal magnetic field strength (Fe X 25.7 nm; see, e.g., [134,146,147]) that can be deduced using a ratio of magnetically sensitive Fe X lines formed at 1 MK. The wavelength range also includes lines sensitive to electron temperature/non-Maxwellian electron distributions [148,149]. It also has a wide range of lines to measure electron densities from coronal (e.g., Fe IX, Fe XI, Fe XII, Fe XV, and Ca XV) to flare temperatures (Fe XXI) and the FIP bias. It observes He II and a wide range of lines, many at flare temperatures (e.g., Fe XVII, Fe XX, Fe XXI, Fe XXII, Fe XXIII, and Fe XXIV).

In order to achieve the temporal resolution of 1 second required to capture the rapid development of the plasma environment during flare energy release, the simultaneous observation of a 2D field of view without the traditional use of slit scanning systems is a key factor. The integral field spectroscopy technique is a novel proposal for the extreme ultraviolet (EUV) regime and benefits from a wide heritage of integral field spectrographs operating in ground-based and space-based telescopes. This strategy has significant advantages over traditional EUV scanning spectroscopy (e.g., EIS, IRIS, and Solar-C/EUVST), making 2D images over two orders of magnitude faster than before. The wavelength range of SISA is both different and wider than that of the upcoming Multi-Slit Solar Explorer (MUSE; [26]) spacecraft, offering a wide array of plasma diagnostics so that 2D maps of, for example, electron density (at multiple temperatures) and the magnetic field will be obtained. We note that several SISA diagnostic, such as the measurements of coronal magnetic fields and departures from electron Maxwellian distributions, are not available to SOLAR-C/EUVST. A summary of the predicted performance of SISA is presented in Table 4. The ability to obtain diagnostics at 1 s cadence is based on estimates of the signal in active region cores and flares and the strawman design (described in the SISA paper), consisting of a single multilayer for the three reflecting surfaces and a 20 cm aperture of the off-axis telescope.

SISA will be composed of two subsystems: the telescope, an off-axis parabolic mirror; and the integral field spectrograph, an array of curved slicer mirrors and curved gratings. The slicer mirrors are placed at the telescope focus and decompose (slice) the image of the field of view using an array of powered rectangular mirrors, each with a different tilt angle around the X and Y axes. These will produce a pupil image per slice reflected in different directions towards the curved gratings, which perform three functions: (i) dispersion of the incoming beam into its constituent wavelengths, (ii) imaging of the beams on the detector with the required magnification, and (iii) control of the location of the exit pupil. The orientations of the gratings are fixed. Each grating will produce the spectrum of each slice of the field, as shown in Figure 5. The tilt of each grating will be defined to distribute the spectra on the detectors. SISA will be the first integral field spectrograph in the EUV spectral range.

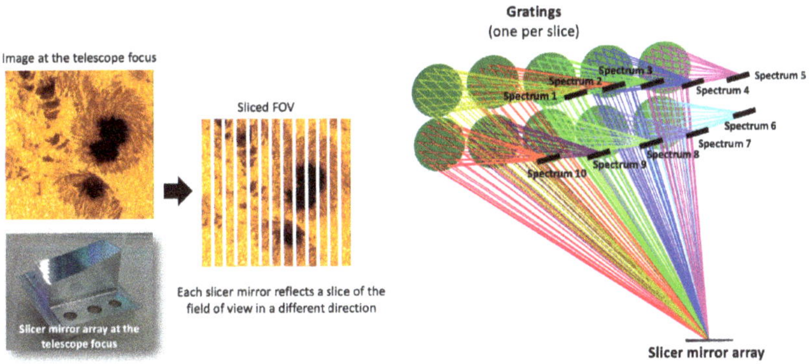

Figure 5. Left: Sketch of the functionality of an image slicer. The image slicer acts as a field reformatter, slicing the entrance field of view and generating one pupil per slice (Note that despite apparent gaps in the sliced FOV in the figure, it provides a contiguous map when combined). **Right**: SISA conceptual layout with a reduced number of slicers. Each curved grating produces the spectrum of each slice of the field. The tilt angles of the gratings offer flexibility in the geometrical distribution of the spectra on the detector.

Table 4. Instrument requirements and expected performance of SISA.

SISA Parameter	Expected Performance
Spectral Window 1	178–184 Å
Spectral Window 2	221–264 Å
Spectral Resolution	0.05 Å FWHM
Spectral Resolving Power (R)	3560–5160
Field of View	$100'' \times 250''$
Spatial Resolution	$1''$ in 2 pixels
Temporal Resolution (high signal)	1 s
Temporal Resolution (low signal)	10 s

2.4. Mass and Power

The required resources for LISSAN, FOXSI (including FOXSI-STC), and SISA are given in Table 5. The mass estimates include a 20% margin on each instrument, whilst the power requirements include a 30% margin on each instrument. The operating temperatures are the most stringent constraints for each instrument; particularly relevant are the front-end electronics (FEE) for LISSAN and the focal-plane assembly (FPA) for both FOXSI and SISA.

Table 5. Required resources for LISSAN, FOXSI (including FOXSI-STC), and SISA.

Resource	LISSAN	FOXSI	SISA
Mass	370 kg	120 kg	78 kg
Volume	1.96 m^3	(105 cm)^3 (stowed)	0.5 m^3
Power	125 W (peak)	170 W (average)	130 W (average)
Data Rate	25 Mbits/s (peak)	1 Mbits/s (peak)	50 Mbits/s (average)
Operating Temp.	0 °C (FEE)	−20–0 °C (FPA)	<-40 °C (FPA)

3. Proposed Mission Configuration and Profile

To meet its science objectives, SPARK must be launched at a time when medium-to-large solar flares can be observed. This can be achieved at any time in the solar cycle except solar minimum.

3.1. System-Level Requirements

The overlapping fields of view (FOV) of LISSAN, FOXSI and SISA are shown in Figure 6. A pointing accuracy of 10 arcsecs is required due to the FOV of all of SPARK's instruments and the need to point to a chosen active region. The performance drift error (PDE) is driven by the spatial resolution of SISA and is 0.1 arcsec within a time interval of one second. The requirement will be fulfilled by further attenuating the spacecraft PDE with a tip/tilt system. Each instrument suite will carry its own aspect system to overcome uncertainty in coalignment between the instruments, as the precise knowledge of the positions of the γ-ray, X-ray, and EUV emissions relative to one another is key to fulfilling the scientific objectives of the mission. Therefore, each instrument will provide precise knowledge of the pointing.

LISSAN and the FOXSI HXR telescopes will operate in one nominal observing mode. SISA will have two operational modes. The first mode (cadence 1) will be for observing flaring active regions when there is an abundance of EUV radiation. The second mode (cadence 2) is optimised for weaker signals when the Sun is less active, requiring slower exposure times. A safe mode will be implemented for each instrument to react to instrument or spacecraft failure.

3.2. Operations

The SPARK payload is designed to provide synchronized observations that address specific science questions. Since the instruments will always observe the same targets, the science operations will not require a large degree of flexibility. As instruments have a FOV smaller than the full Sun, target selection will be required. Targets will typically be solar active regions most likely to produce energetic flares. SPARK will allow the community to submit observing plans for targeted observations.

Science and housekeeping data recorded in the onboard mass memory will be brought down in raw format for processing on the ground into level 0 format. The nature of multiple downlink stations may require that data be aggregated and sorted before this processing. Further pipeline processing will bring data to level-2-derived products via level-1-calibrated data. Minimal data processing will happen on board, and all the data will be downlinked for processing on the ground.

3.3. Spacecraft Design

The primary drivers for the SPARK spacecraft design are the accommodation of the extendable boom for FOXSI and the large mass of LISSAN. The boom will be deployed in orbit, and alignment between the optics on the spacecraft and the detectors at the tip of the extendable structure will be performed using the FOXSI tip/tilt mechanism.

SPARK's payload includes imaging instruments and therefore requires a three-axis system to minimise spatial blurring. The combination of individual instrument stability requirements leads to an overall requirement for the PDE of 0.1″ over 1 second. The spacecraft absolute performance error (APE) is 10″ to allow a 10% error on the smaller SISA field of view pointing at the correct target. The requirement for the relative performance error (RPE) is 1″ for integration times of 1 s. The attitude sensors should include a fine Sun sensor and a star tracker in order to determine spacecraft pointing relative to the Sun. An inertial reference unit is required to determine changes in attitude over time.

SPARK's baseline L1 orbit provides a stable thermal environment such that the thermal control on the instruments can maintain any required temperature. The spacecraft will have one side constantly facing the Sun and one side facing cold space all the time. SPARK's thermal requirements can be met by a passive cooling system consisting of cold fingers and radiators.

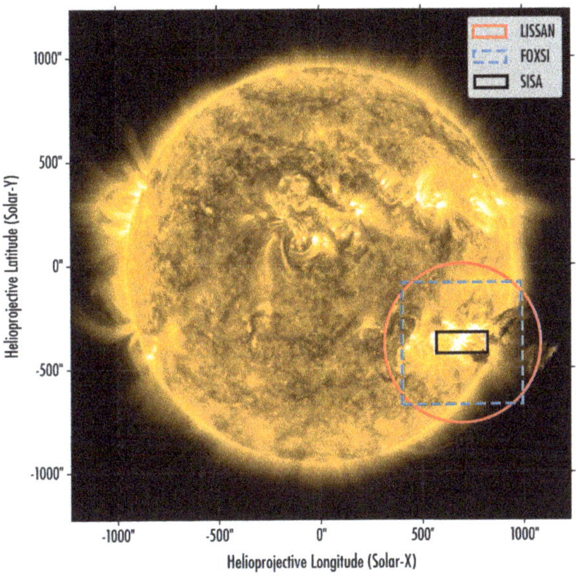

Figure 6. Fields of view of the three imaging instruments of SPARK overplotted on the EUV 171 Å Sun from AIA.

SPARK will provide science data downlink to Earth using a K-band 26 GHz antenna. Even with a reduced ground station contact of 4 hours to obtain 850 Gbits/day (similar to Euclid), this would be enough to downlink the entire maximum daily data volumes of 80 Gbit (LISSAN) and 86 Gbit (FOXSI). The SISA maximum daily data volume of 4.3 Tbits/day would be stored using onboard mass memory storage of at least 4 TB, with synoptic data being communicated to the Earth to choose a subset of events to download and/or periods to downlink with reduced cadence.

4. Current Status of SPARK

SPARK was initially submitted to the ESA in 2010 as an M-class mission proposal that included a modified version of LISSAN and FOXSI with different supporting instruments. More recently, the relevant particle acceleration and transport topical questions were presented as an ESA Voyage 2050 white paper in 2020 and subsequently published [150].

SPARK in its current form was proposed to the ESA in 2022 as an M-class mission and reached Phase-2. Development of the individual instruments proposed for SPARK continues, funded by national efforts. The United Kingdom is developing EUV image slicer technology for SISA, France is developing X-ray detectors for FOXSI, Germany is developing grids for LISSAN, and Belgium is developing improved detectors for SISA.

There is clear support for the goals and implementation of SPARK across the broad European scientific community in solar physics and beyond. The implementation of SPARK in this form presents an exciting opportunity for paradigm-shifting observations in the field of astrophysical particle acceleration and transport, using data from our local laboratory, the Sun.

Author Contributions: The writing of this paper was led by H.A.S.R. The paper was based upon the SPARK mission concept that was coordinated by H.A.S.R., S.M. (Sophie Musset), and D.F.R. All other authors contributed by providing scientific expertise or instrumentation expertise in the development of the LISSAN, FOXSI, and SISA concepts and/or the paper itself. All authors have read and agreed to the published version of the manuscript.

Funding: H.A.S.R. and S.A.M. were funded by the Science, Technology, and Facilities Council (STFC) under consolidated grant ST/W001004/1. D.M.L. is grateful to the STFC award of an Ernest Rutherford Fellowship (ST/R003246/1). Jaroslav Dudík was funded by the Czech National Science Foundation (Grant No. GACR 22-07155S). Jaroslav Dudík and J.K. were supported by project RWO:67985815. D.O.S. acknowledges financial support from through grants AEI/MCIN/10.13039/501100011033/(RTI2018-096886-C5, PID2021-125325OB-C5, PCI2022-135009-2), and ERDF "A way of making Europe" and "Center of Excellence Severo Ochoa" awards to IAA-CSIC (CEX2021-001131-S). S.M. and T.M. were supported by the National Science Centre, Poland (Grant No. 2020/39/B/ST9/01591). G.S.K. acknowledges financial support from NASA's Early Career Investigator Program (Grant No. NASA 80NSSC21K0460). G.D.Z. acknowledges support from STFC (UK) via the consolidated grants to the atomic astrophysics group at DAMTP, University of Cambridge (ST/P000665/1. and ST/T000481/1).

Data Availability Statement: No new data were created or analyzed in this study. Data sharing is not applicable to this article.

Acknowledgments: The concept for the FOXSI instrument on board SPARK was based upon the extensive work that was carried out by the FOXSI SMEX team [138] and by the FIERCE team (see https://doi.org/10.5281/zenodo.3674078). Figures 1 and 2 are from the work carried out by both these teams.

Conflicts of Interest: The authors declare no conflict of interest.

Abbreviations

The following abbreviations are used in this manuscript:

AIA	Atmospheric imaging assembly
CME	Coronal mass ejection
EIS	EUV Imaging Spectrometer on board Hinode
EUV	Extreme ultraviolet
EUVST	Extreme Ultraviolet High-Throughput Spectroscopic Telescope
EVE	EUV Variability Experiment on board SDO
FEE	Front-end electronics
FIERCE	Fundamentals of Impulsive Energy Release in the Corona Explorer
FPA	Focal-plane assembly
FOV	Field of view
FOXSI	Focusing Optics X-ray Solar Imager
FOXSI-STC	FOXSI's Spectrometer for Temperature and Composition
FWHM	Full-width half maximum
GOES	Geostationary Operational Environmental Satellite
HXR	Hard X-ray
IFS	Integral field spectrograph
IRIS	Interface Region Imaging Spectrograph
LISSAN	Large Imaging Spectrometer for Solar Accelerated Nuclei
MHD	Magnetohydrodynamic
MUSE	Multi-Slit Solar Explorer
NuSTAR	Nuclear Spectroscopic Telescope Array

RHESSI	Reuven Ramaty High-Energy Solar Spectroscopic Imager
SDO	Solar Dynamics Observatory
SEP	Solar energetic particle
SISA	Spectral Imager of the Solar Atmosphere
SPARK	Solar Particle Acceleration, Radiation, and Kinetics mission
STIX	Spectrometer/Telescope for Imaging X-rays
SXR	Soft X-ray

References

1. Shibata, K. New observational facts about solar flares from YOHKOH studies—Evidence of magnetic reconnection and a unified model of flares. *Adv. Space Res.* **1996**, *17*, 9–18. [CrossRef]
2. Brown, J.C. The Deduction of Energy Spectra of Non-Thermal Electrons in Flares from the Observed Dynamic Spectra of Hard X-Ray Bursts. *Sol. Phys.* **1971**, *18*, 489–502. [CrossRef]
3. Piana, M.; Massone, A.M.; Kontar, E.P.; Emslie, A.G.; Brown, J.C.; Schwartz, R.A. Regularized Electron Flux Spectra in the 2002 July 23 Solar Flare. *Astrophys. J. Lett.* **2003**, *595*, L127–L130. [CrossRef]
4. Holman, G.D.; Aschwanden, M.J.; Aurass, H.; Battaglia, M.; Grigis, P.C.; Kontar, E.P.; Liu, W.; Saint-Hilaire, P.; Zharkova, V.V. Implications of X-ray Observations for Electron Acceleration and Propagation in Solar Flares. *Space Sci. Rev.* **2011**, *159*, 107–166. [CrossRef]
5. Kontar, E.P.; Brown, J.C.; Emslie, A.G.; Hajdas, W.; Holman, G.D.; Hurford, G.J.; Kašparová, J.; Mallik, P.C.V.; Massone, A.M.; McConnell, M.L.; et al. Deducing Electron Properties from Hard X-ray Observations. *Space Sci. Rev.* **2011**, *159*, 301–355. [CrossRef]
6. Benz, A.O. Flare Observations. *Living Rev. Sol. Phys.* **2017**, *14*, 2. [CrossRef]
7. Chupp, E.L. High-Energy Neutral Radiations from the Sun. *Annu. Rev. Astron Astrophys.* **1984**, *22*, 359–387. [CrossRef]
8. Ramaty, R. Nuclear processes in solar flares. In *Physics of the Sun: Volume II: The Solar Atmosphere*; Springer: Dordrecht, The Netherlands, 1986; Volume 2, pp. 291–323.
9. Chupp, E.L. Evolution of our understanding of solar flare particle acceleration: (1942–1995). In *High Energy Solar Physics;* Ramaty, R., Mandzhavidze, N., Hua, X.M., Eds.; American Institute of Physics Conference Series; American Institute of Physics: College Park, MD, USA, 1996; Volume 374, pp. 3–34. [CrossRef]
10. Trottet, G.; Vilmer, N. The Production of Flare-Accelerated Particles at the Sun. In *European Meeting on Solar Physics;* Simnett, G.M., Alissandrakis, C.E., Vlahos, L., Eds.; Springer: Berlin/Heidelberg, Germany, 1997; Volume 489, p. 219. [CrossRef]
11. Share, G.H.; Murphy, R.J. Gamma Ray Spectroscopy in the Pre-HESSI Era. In *High Energy Solar Physics Workshop—Anticipating Hess!* Ramaty, R., Mandzhavidze, N., Eds.; Astronomical Society of the Pacific Conference Series; American Institute of Physics: College Park, MD, USA, 2000; Volume 206, p. 377.
12. Vilmer, N.; MacKinnon, A.L.; Trottet, G.; Barat, C. High energy particles accelerated during the large solar flare of 1990 May 24: X/γ-ray observations. *Astron. Astrophys.* **2003**, *412*, 865–874. [CrossRef]
13. Share, G.H.; Murphy, R.J. Gamma Radiation From Flare-Accelerated Particles Impacting the Sun. *Geophys. Monogr. Ser.* **2006**, *165*, 177. [CrossRef]
14. Vilmer, N.; MacKinnon, A.L.; Hurford, G.J. Properties of Energetic Ions in the Solar Atmosphere from γ-Ray and Neutron Observations. *Space Sci. Rev.* **2011**, *159*, 167–224. [CrossRef]
15. Stecker, F.W. The Cosmic γ-Ray Spectrum from Secondary Particle Production in Cosmic-Ray Interactions. *Astrophys. Space Sci.* **1970**, *6*, 377–389. [CrossRef]
16. Murphy, R.J.; Dermer, C.D.; Ramaty, R. High-Energy Processes in Solar Flares. *Astrophys. J. Suppl.* **1987**, *63*, 721. [CrossRef]
17. Hurford, G.J.; Schwartz, R.A.; Krucker, S.; Lin, R.P.; Smith, D.M.; Vilmer, N. First Gamma-Ray Images of a Solar Flare. *Astrophys. J. Lett.* **2003**, *595*, L77–L80. [CrossRef]
18. Ramaty, R.; Mandzhavidze, N. High energy processes in solar flares. In Proceedings of the Cosmic Explosions: Tenth AstroPhysics Conference, College Park, MD, USA, 11–13 October 1999; Holt, S.S., Zhang, W.W., Eds.; American Institute of Physics Conference Series; American Institute of Physics: College Park, MD, USA, 2000; Volume 522, pp. 401–410. [CrossRef]
19. Emslie, A.G.; Dennis, B.R.; Shih, A.Y.; Chamberlin, P.C.; Mewaldt, R.A.; Moore, C.S.; Share, G.H.; Vourlidas, A.; Welsch, B.T. Global Energetics of Thirty-eight Large Solar Eruptive Events. *Astrophys. J.* **2012**, *759*, 71. [CrossRef]
20. Aschwanden, M.J.; Caspi, A.; Cohen, C.M.S.; Holman, G.; Jing, J.; Kretzschmar, M.; Kontar, E.P.; McTiernan, J.M.; Mewaldt, R.A.; O'Flannagain, A.; et al. Global Energetics of Solar Flares. V. Energy Closure in Flares and Coronal Mass Ejections. *Astrophys. J.* **2017**, *836*, 17. [CrossRef]
21. Lin, R.P.; Dennis, B.R.; Hurford, G.J.; Smith, D.M.; Zehnder, A.; Harvey, P.R.; Curtis, D.W.; Pankow, D.; Turin, P.; Bester, M.; et al. The Reuven Ramaty High-Energy Solar Spectroscopic Imager (RHESSI). *Sol. Phys.* **2002**, *210*, 3–32. [CrossRef]
22. Krucker, S.; Hurford, G.J.; Grimm, O.; Kögl, S.; Gröbelbauer, H.P.; Etesi, L.; Casadei, D.; Csillaghy, A.; Benz, A.O.; Arnold, N.G.; et al. The Spectrometer/Telescope for Imaging X-rays (STIX). *Astron. Astrophys.* **2020**, *642*, A15. [CrossRef]
23. Culhane, J.L.; Harra, L.K.; James, A.M.; Al-Janabi, K.; Bradley, L.J.; Chaudry, R.A.; Rees, K.; Tandy, J.A.; Thomas, P.; Whillock, M.C.R.; et al. The EUV Imaging Spectrometer for Hinode. *Sol. Phys.* **2007**, *243*, 19–61. [CrossRef]

24. SPICE Consortium; Anderson, M.; Appourchaux, T.; Auchère, F.; Aznar Cuadrado, R.; Barbay, J.; Baudin, F.; Beardsley, S.; Bocchialini, K.; Borgo, B.; et al. The Solar Orbiter SPICE instrument. An extreme UV imaging spectrometer. *Astron. Astrophys.* **2020**, *642*, A14. [CrossRef]
25. Shimizu, T.; Imada, S.; Kawate, T.; Suematsu, Y.; Hara, H.; Tsuzuki, T.; Katsukawa, Y.; Kubo, M.; Ishikawa, R.; Watanabe, T.; et al. The Solar-C (EUVST) mission: The latest status. In *Space Telescopes and Instrumentation 2020: Ultraviolet to Gamma Ray*; den Herder, J.W.A., Nikzad, S., Nakazawa, K., Eds.; Society of Photo-Optical Instrumentation Engineers (SPIE) Conference Series; SPIE: Bellingham, WA, USA, 2020; Volume 11444, p. 114440N. [CrossRef]
26. De Pontieu, B.; Martínez-Sykora, J.; Testa, P.; Winebarger, A.R.; Daw, A.; Hansteen, V.; Cheung, M.C.M.; Antolin, P. The Multi-slit Approach to Coronal Spectroscopy with the Multi-slit Solar Explorer (MUSE). *Astrophys. J.* **2020**, *888*, 3. [CrossRef]
27. De Pontieu, B.; Testa, P.; Martínez-Sykora, J.; Antolin, P.; Karampelas, K.; Hansteen, V.; Rempel, M.; Cheung, M.C.M.; Reale, F.; Danilovic, S.; et al. Probing the Physics of the Solar Atmosphere with the Multi-slit Solar Explorer (MUSE). I. Coronal Heating. *Astrophys. J.* **2022**, *926*, 52. [CrossRef]
28. Cheung, M.C.M.; Martínez-Sykora, J.; Testa, P.; De Pontieu, B.; Chintzoglou, G.; Rempel, M.; Polito, V.; Kerr, G.S.; Reeves, K.K.; Fletcher, L.; et al. Probing the Physics of the Solar Atmosphere with the Multi-slit Solar Explorer (MUSE). II. Flares and Eruptions . *Astrophys. J.* **2022**, *926*, 53. [CrossRef]
29. Aschwanden, M.J.; Kontar, E.P.; Jeffrey, N.L.S. Global Energetics of Solar Flares. VIII. The Low-energy Cutoff. *Astrophys. J.* **2019**, *881*, 1. [CrossRef]
30. Drake, J.F.; Swisdak, M.; Che, H.; Shay, M.A. Electron acceleration from contracting magnetic islands during reconnection. *Nature* **2006**, *443*, 553–556. [CrossRef] [PubMed]
31. Arnold, H.; Drake, J.F.; Swisdak, M.; Guo, F.; Dahlin, J.T.; Chen, B.; Fleishman, G.; Glesener, L.; Kontar, E.; Phan, T.; et al. Electron Acceleration during Macroscale Magnetic Reconnection. *Phys. Rev. Lett.* **2021**, *126*, 135101. [CrossRef] [PubMed]
32. Miller, J.A.; Cargill, P.J.; Emslie, A.G.; Holman, G.D.; Dennis, B.R.; LaRosa, T.N.; Winglee, R.M.; Benka, S.G.; Tsuneta, S. Critical issues for understanding particle acceleration in impulsive solar flares. *J. Geophys. Res.* **1997**, *102*, 14631–14660. [CrossRef]
33. Petrosian, A.R.; Boulesteix, J.; Comte, G.; Kunth, D.; Lecoarer, E. An interferometric study of the blue compact dwarf galaxy IZW 18. *Astron. Astrophys.* **1997**, *318*, 390–404.
34. Vlahos, L.; Anastasiadis, A.; Papaioannou, A.; Kouloumvakos, A.; Isliker, H. Sources of solar energetic particles. *Philos. Trans. R. Soc. Lond. Ser. A* **2019**, *377*, 20180095. [CrossRef]
35. Vlahos, L.; Isliker, H. Particle acceleration and heating in a turbulent solar corona. *Plasma Phys. Control. Fusion* **2019**, *61*, 014020. [CrossRef]
36. Kiener, J.; Gros, M.; Tatischeff, V.; Weidenspointner, G. Properties of the energetic particle distributions during the October 28, 2003 solar flare from INTEGRAL/SPI observations. *Astron. Astrophys.* **2006**, *445*, 725–733. [CrossRef]
37. Litvinenko, Y.E.; Craig, I.J.D. Flare Energy Release by Flux Pile-up Magnetic Reconnection in a Turbulent Current Sheet. *Astrophys. J.* **2000**, *544*, 1101–1107. [CrossRef]
38. Holman, G.D.; Benka, S.G. A Hybrid Thermal/Nonthermal Model for the Energetic Emissions from Solar Flares. *Astrophys. J. Lett.* **1992**, *400*, L79. [CrossRef]
39. Alexander, W.M.; Tanner, W.G.; McDonald, R.A.; Schaub, G.E.; Stephenson, S.L.; McDonnell, J.A.M.; Maag, C.R. The Status of Measurement Technologies Concerning Micrometer and Submicrometer Space Articulate Matter Capture, Recovery, Velocity, and Trajectory. In Proceedings of the Particle Capture, Recovery and Velocity/Trajectory Measurement Technologies, Houston, TX, USA, 27–28 September 1993; Zolensky, M.E., Ed.; p. 11.
40. Dunphy, P.P.; Chupp, E.L.; Bertsch, D.L.; Schneid, E.J.; Gottesman, S.R.; Kanbach, G. Gamma-Rays and Neutrons as a Probe of Flare Proton Spectra: The Solar Flare of 11 June 1991. *Sol. Phys.* **1999**, *187*, 45–57. [CrossRef]
41. Kocharov, L.G.; Lee, J.W.; Zirin, H.; Kovaltsov, G.A.; Usoskin, I.G.; Pyle, K.R.; Shea, M.A.; Smart, D.F. Neutron and electromagnetic emissions during the 1990 May 24 solar flare. *Sol. Phys.* **1994**, *155*, 149–170. [CrossRef]
42. Kocharov, L.; Debrunner, H.; Kovaltsov, G.; Lockwood, J.; McConnell, M.; Nieminen, P.; Rank, G.; Ryan, J.; Schoenfelder, V. Deduced spectrum of interacting protons accelerated after the impulsive phase of the 15 June 1991 solar flare. *Astron. Astrophys.* **1998**, *340*, 257–264.
43. Vilmer, N.; Krucker, S.; Trottet, G.; Lin, R.P. Hard X-ray and metric/decimetric spatially resolved observations of the 10 April 2002 solar flare. *Adv. Space Res.* **2003**, *32*, 2509–2515. [CrossRef]
44. Kanbach, G.; Bertsch, D.L.; Fichtel, C.E.; Hartman, R.C.; Hunter, S.D.; Kniffen, D.A.; Kwok, P.W.; Lin, Y.C.; Mattox, J.R.; Mayer-Hasselwander, H.A. Detection of a long-duration solar gamma-ray flare on June 11, 1991 with EGRET on COMPTON-GRO. *Astron. Astrophys. Suppl.* **1993**, *97*, 349–353.
45. Ryan, J.M.; Lockwood, J.A.; Debrunner, H. Solar Energetic Particles. *Space Sci. Rev.* **2000**, *93*, 35–53. [CrossRef]
46. Allred, J.C.; Hawley, S.L.; Abbett, W.P.; Carlsson, M. Radiative Hydrodynamic Models of the Optical and Ultraviolet Emission from Solar Flares. *Astrophys. J.* **2005**, *630*, 573–586. [CrossRef]
47. Allred, J.C.; Kowalski, A.F.; Carlsson, M. A Unified Computational Model for Solar and Stellar Flares. *Astrophys. J.* **2015**, *809*, 104. [CrossRef]
48. Allred, J.C.; Kerr, G.S.; Gordon Emslie, A. Solar Flare Heating with Turbulent Suppression of Thermal Conduction. *Astrophys. J.* **2022**, *931*, 60. [CrossRef]

49. Kowalski, A.F.; Hawley, S.L.; Carlsson, M.; Allred, J.C.; Uitenbroek, H.; Osten, R.A.; Holman, G. New Insights into White-Light Flare Emission from Radiative-Hydrodynamic Modeling of a Chromospheric Condensation. *Sol. Phys.* **2015**, *290*, 3487–3523. [CrossRef]
50. Kowalski, A.F.; Allred, J.C.; Daw, A.; Cauzzi, G.; Carlsson, M. The Atmospheric Response to High Nonthermal Electron Beam Fluxes in Solar Flares. I. Modeling the Brightest NUV Footpoints in the X1 Solar Flare of 2014 March 29. *Astrophys. J.* **2017**, *836*, 12. [CrossRef]
51. Rubio da Costa, F.; Kleint, L.; Petrosian, V.; Liu, W.; Allred, J.C. Data-driven Radiative Hydrodynamic Modeling of the 2014 March 29 X1.0 Solar Flare. *Astrophys. J.* **2016**, *827*, 38. [CrossRef]
52. Kerr, G.S.; Fletcher, L.; Russell, A.J.B.; Allred, J.C. Simulations of the Mg II k and Ca II 8542 lines from an AlfvÉn Wave-heated Flare Chromosphere. *Astrophys. J.* **2016**, *827*, 101. [CrossRef]
53. Kerr, G.S.; Carlsson, M.; Allred, J.C.; Young, P.R.; Daw, A.N. SI IV Resonance Line Emission during Solar Flares: Non-LTE, Nonequilibrium, Radiation Transfer Simulations. *Astrophys. J.* **2019**, *871*, 23. [CrossRef]
54. Kerr, G.S.; Allred, J.C.; Polito, V. Solar Flare Arcade Modeling: Bridging the Gap from 1D to 3D Simulations of Optically Thin Radiation. *Astrophys. J.* **2020**, *900*, 18. [CrossRef]
55. Brown, S.A.; Fletcher, L.; Kerr, G.S.; Labrosse, N.; Kowalski, A.F.; De La Cruz Rodríguez, J. Modeling of the Hydrogen Lyman Lines in Solar Flares. *Astrophys. J.* **2018**, *862*, 59. [CrossRef]
56. Polito, V.; Testa, P.; Allred, J.; De Pontieu, B.; Carlsson, M.; Pereira, T.M.D.; Gošić, M.; Reale, F. Investigating the Response of Loop Plasma to Nanoflare Heating Using RADYN Simulations. *Astrophys. J.* **2018**, *856*, 178. [CrossRef]
57. Polito, V.; Testa, P.; De Pontieu, B. Can the Superposition of Evaporative Flows Explain Broad Fe XXI Profiles during Solar Flares? *Astrophys. J. Lett.* **2019**, *879*, L17. [CrossRef]
58. Zhu, Y.; Kowalski, A.F.; Tian, H.; Uitenbroek, H.; Carlsson, M.; Allred, J.C. Modeling Mg II h, k and Triplet Lines at Solar Flare Ribbons. *Astrophys. J.* **2019**, *879*, 19. [CrossRef]
59. Reep, J.W.; Polito, V.; Warren, H.P.; Crump, N.A. The Duration of Energy Deposition on Unresolved Flaring Loops in the Solar Corona. *Astrophys. J.* **2018**, *856*, 149. [CrossRef]
60. Reep, J.W.; Russell, A.J.B. Alfvénic Wave Heating of the Upper Chromosphere in Flares. *Astrophys. J. Lett.* **2016**, *818*, L20. [CrossRef]
61. Reep, J.W.; Russell, A.J.B.; Tarr, L.A.; Leake, J.E. A Hydrodynamic Model of Alfvénic Wave Heating in a Coronal Loop and Its Chromospheric Footpoints. *Astrophys. J.* **2018**, *853*, 101. [CrossRef]
62. Reep, J.W.; Bradshaw, S.J.; Crump, N.A.; Warren, H.P. Efficient Calculation of Non-local Thermodynamic Equilibrium Effects in Multithreaded Hydrodynamic Simulations of Solar Flares. *Astrophys. J.* **2019**, *871*, 18. [CrossRef]
63. Cheung, M.C.M.; Rempel, M.; Chintzoglou, G.; Chen, F.; Testa, P.; Martínez-Sykora, J.; Sainz Dalda, A.; DeRosa, M.L.; Malanushenko, A.; Hansteen, V.; et al. A comprehensive three-dimensional radiative magnetohydrodynamic simulation of a solar flare. *Nat. Astron.* **2019**, *3*, 160–166. [CrossRef]
64. Hurford, G.J.; Krucker, S.; Lin, R.P.; Schwartz, R.A.; Share, G.H.; Smith, D.M. Gamma-Ray Imaging of the 2003 October/November Solar Flares. *Astrophys. J. Lett.* **2006**, *644*, L93–L96. [CrossRef]
65. del Zanna, G.; Schmieder, B.; Mason, H.; Berlicki, A.; Bradshaw, S. The Gradual Phase of the X17 Flare on October 28, 2003. *Sol. Phys.* **2006**, *239*, 173–191. [CrossRef]
66. Milligan, R.O.; Dennis, B.R. Velocity Characteristics of Evaporated Plasma Using Hinode/EUV Imaging Spectrometer. *Astrophys. J.* **2009**, *699*, 968–975. [CrossRef]
67. Milligan, R.O. Spatially Resolved Nonthermal Line Broadening during the Impulsive Phase of a Solar Flare. *Astrophys. J.* **2011**, *740*, 70. [CrossRef]
68. Kontar, E.P.; Perez, J.E.; Harra, L.K.; Kuznetsov, A.A.; Emslie, A.G.; Jeffrey, N.L.S.; Bian, N.H.; Dennis, B.R. Turbulent Kinetic Energy in the Energy Balance of a Solar Flare. *Phys. Rev. Lett.* **2017**, *118*, 155101. [CrossRef] [PubMed]
69. Nakariakov, V.M.; Melnikov, V.F. Quasi-Periodic Pulsations in Solar Flares. *Space Sci. Rev.* **2009**, *149*, 119–151. [CrossRef]
70. McLaughlin, J.A.; Nakariakov, V.M.; Dominique, M.; Jelínek, P.; Takasao, S. Modelling Quasi-Periodic Pulsations in Solar and Stellar Flares. *Space Sci. Rev.* **2018**, *214*, 45. [CrossRef]
71. Zimovets, I.V.; McLaughlin, J.A.; Srivastava, A.K.; Kolotkov, D.Y.; Kuznetsov, A.A.; Kupriyanova, E.G.; Cho, I.H.; Inglis, A.R.; Reale, F.; Pascoe, D.J.; et al. Quasi-Periodic Pulsations in Solar and Stellar Flares: A Review of Underpinning Physical Mechanisms and Their Predicted Observational Signatures. *Space Sci. Rev.* **2021**, *217*, 66. [CrossRef]
72. Nakariakov, V.M.; Kolotkov, D.Y.; Kupriyanova, E.G.; Mehta, T.; Pugh, C.E.; Lee, D.H.; Broomhall, A.M. Non-stationary quasi-periodic pulsations in solar and stellar flares. *Plasma Phys. Control. Fusion* **2019**, *61*, 014024. [CrossRef]
73. Hayes, L.A.; Inglis, A.R.; Christe, S.; Dennis, B.; Gallagher, P.T. Statistical Study of GOES X-Ray Quasi-periodic Pulsations in Solar Flares. *Astrophys. J.* **2020**, *895*, 50. [CrossRef]
74. Emslie, A.G.; Sturrock, P.A. Temperature minimum heating in solar flares by resistive dissipation of Alfvén waves. *Sol. Phys.* **1982**, *80*, 99–112. [CrossRef]
75. Fletcher, L.; Hudson, H.S. Impulsive Phase Flare Energy Transport by Large-Scale Alfvén Waves and the Electron Acceleration Problem. *Astrophys. J.* **2008**, *675*, 1645–1655. [CrossRef]
76. Russell, A.J.B.; Fletcher, L. Propagation of Alfvénic Waves from Corona to Chromosphere and Consequences for Solar Flares. *Astrophys. J.* **2013**, *765*, 81. [CrossRef]

77. Laming, J.M. The FIP and Inverse-FIP Effects in Solar Flares. *Astrophys. J.* **2021**, *909*, 17. [CrossRef]
78. Malandraki, O.E.; Crosby, N.B. Solar Energetic Particles and Space Weather: Science and Applications. In *Solar Particle Radiation Storms Forecasting and Analysis*; Malandraki, O.E., Crosby, N.B., Eds.; Astrophysics and Space Science Library; Springer: Cham, Switzerland, 2018; Volume 444, pp. 1–26. [CrossRef]
79. McComas, D.J.; Christian, E.R.; Cohen, C.M.S.; Cummings, A.C.; Davis, A.J.; Desai, M.I.; Giacalone, J.; Hill, M.E.; Joyce, C.J.; Krimigis, S.M.; et al. Probing the energetic particle environment near the Sun. *Nature* **2019**, *576*, 223–227. [CrossRef] [PubMed]
80. Gómez-Herrero, R.; Pacheco, D.; Kollhoff, A.; Espinosa Lara, F.; Freiherr von Forstner, J.L.; Dresing, N.; Lario, D.; Balmaceda, L.; Krupar, V.; Malandraki, O.E.; et al. First near-relativistic solar electron events observed by EPD onboard Solar Orbiter. *Astron. Astrophys.* **2021**, *656*, L3. [CrossRef]
81. Pick, M.; Vilmer, N. Sixty-five years of solar radioastronomy: Flares, coronal mass ejections and Sun Earth connection. *Astron. Astrophys. Rev.* **2008**, *16*, 1–153. [CrossRef]
82. Reid, H.A.S.; Ratcliffe, H. A review of solar type III radio bursts. *Res. Astron. Astrophys.* **2014**, *14*, 773–804. [CrossRef]
83. Masson, S.; Antiochos, S.K.; DeVore, C.R. Escape of Flare-accelerated Particles in Solar Eruptive Events. *Astrophys. J.* **2019**, *884*, 143. [CrossRef]
84. Reid, H.A.S. A review of recent type III imaging spectroscopy. *Front. Astron. Space Sci.* **2020**, *7*, 56. [CrossRef]
85. Steinberg, J.L.; Aubier-Giraud, M.; Leblanc, Y.; Boischot, A. Coronal Scattering, Absorption and Refraction of Solar Radiobursts. *Astron. Astrophys.* **1971**, *10*, 362.
86. Kontar, E.P.; Chen, X.; Chrysaphi, N.; Jeffrey, N.L.S.; Emslie, A.G.; Krupar, V.; Maksimovic, M.; Gordovskyy, M.; Browning, P.K. Anisotropic Radio-wave Scattering and the Interpretation of Solar Radio Emission Observations. *Astrophys. J.* **2019**, *884*, 122. [CrossRef]
87. Saint-Hilaire, P.; Krucker, S.; Christe, S.; Lin, R.P. The X-ray Detectability of Electron Beams Escaping from the Sun. *Astrophys. J.* **2009**, *696*, 941–952. [CrossRef]
88. Longcope, D.W.; Guidoni, S.E.; Linton, M.G. Gas-dynamic Shock Heating of Post-flare Loops Due to Retraction Following Localized, Impulsive Reconnection. *Astrophys. J. Lett.* **2009**, *690*, L18–L22. [CrossRef]
89. Reeves, K.K.; McCauley, P.I.; Tian, H. Direct Observations of Magnetic Reconnection Outflow and CME Triggering in a Small Erupting Solar Prominence. *Astrophys. J.* **2015**, *807*, 7. [CrossRef]
90. Zharkova, V.V.; Arzner, K.; Benz, A.O.; Browning, P.; Dauphin, C.; Emslie, A.G.; Fletcher, L.; Kontar, E.P.; Mann, G.; Onofri, M.; et al. Recent Advances in Understanding Particle Acceleration Processes in Solar Flares. *Space Sci. Rev.* **2011**, *159*, 357–420. [CrossRef]
91. Moore, R.L.; Sterling, A.C.; Hudson, H.S.; Lemen, J.R. Onset of the Magnetic Explosion in Solar Flares and Coronal Mass Ejections. *Astrophys. J.* **2001**, *552*, 833–848. [CrossRef]
92. Lin, J.; Forbes, T.G. Effects of reconnection on the coronal mass ejection process. *J. Geophys. Res.* **2000**, *105*, 2375–2392. [CrossRef]
93. Antiochos, S.K.; DeVore, C.R.; Klimchuk, J.A. A Model for Solar Coronal Mass Ejections. *Astrophys. J.* **1999**, *510*, 485–493. [CrossRef]
94. Török, T.; Kliem, B. Confined and Ejective Eruptions of Kink-unstable Flux Ropes. *Astrophys. J. Lett.* **2005**, *630*, L97–L100. [CrossRef]
95. Wyper, P.F.; DeVore, C.R.; Antiochos, S.K. A Breakout Model for Solar Coronal Jets with Filaments. *Astrophys. J.* **2018**, *852*, 98. [CrossRef]
96. Pariat, E.; Dalmasse, K.; DeVore, C.R.; Antiochos, S.K.; Karpen, J.T. A model for straight and helical solar jets. II. Parametric study of the plasma beta. *Astron. Astrophys.* **2016**, *596*, A36. [CrossRef] [PubMed]
97. Pariat, E.; Dalmasse, K.; DeVore, C.R.; Antiochos, S.K.; Karpen, J.T. Model for straight and helical solar jets. I. Parametric studies of the magnetic field geometry. *Astron. Astrophys.* **2015**, *573*, A130. [CrossRef]
98. Wyper, P.F.; DeVore, C.R.; Karpen, J.T.; Antiochos, S.K.; Yeates, A.R. A Model for Coronal Hole Bright Points and Jets Due to Moving Magnetic Elements. *Astrophys. J.* **2018**, *864*, 165. [CrossRef]
99. Parnell, C.E.; De Moortel, I. A contemporary view of coronal heating. *Philos. Trans. R. Soc. Lond. Ser. A* **2012**, *370*, 3217–3240. [CrossRef]
100. Klimchuk, J.A. On Solving the Coronal Heating Problem. *Sol. Phys.* **2006**, *234*, 41–77. [CrossRef]
101. Klimchuk, J.A. Key aspects of coronal heating. *Philos. Trans. R. Soc. Lond. Ser. A* **2015**, *373*, 20140256. [CrossRef] [PubMed]
102. Klimchuk, J.A.; Luna, M. The Role of Asymmetries in Thermal Nonequilibrium. *Astrophys. J.* **2019**, *884*, 68. [CrossRef]
103. Viall, N.M.; Borovsky, J.E. Nine Outstanding Questions of Solar Wind Physics. *J. Geophys. Res. (Space Phys.)* **2020**, *125*, e26005. [CrossRef] [PubMed]
104. Hannah, I.G.; Hudson, H.S.; Battaglia, M.; Christe, S.; Kašparová, J.; Krucker, S.; Kundu, M.R.; Veronig, A. Microflares and the Statistics of X-ray Flares. *Space Sci. Rev.* **2011**, *159*, 263–300. [CrossRef]
105. Testa, P.; De Pontieu, B.; Allred, J.; Carlsson, M.; Reale, F.; Daw, A.; Hansteen, V.; Martinez-Sykora, J.; Liu, W.; DeLuca, E.E.; et al. Evidence of nonthermal particles in coronal loops heated impulsively by nanoflares. *Science* **2014**, *346*, 1255724. [CrossRef] [PubMed]
106. Dudík, J.; Dzifčáková, E.; Meyer-Vernet, N.; Del Zanna, G.; Young, P.R.; Giunta, A.; Sylwester, B.; Sylwester, J.; Oka, M.; Mason, H.E.; et al. Nonequilibrium Processes in the Solar Corona, Transition Region, Flares, and Solar Wind (Invited Review). *Sol. Phys.* **2017**, *292*, 100. [CrossRef]

107. Benz, A.O.; Grigis, P.C. Microflares and hot component in solar active regions. *Sol. Phys.* **2002**, *210*, 431–444. [CrossRef]
108. Krucker, S.; Christe, S.; Lin, R.P.; Hurford, G.J.; Schwartz, R.A. Hard X-ray Microflares down to 3 keV. *Sol. Phys.* **2002**, *210*, 445–456. [CrossRef]
109. Christe, S.; Hannah, I.G.; Krucker, S.; McTiernan, J.; Lin, R.P. RHESSI Microflare Statistics. I. Flare-Finding and Frequency Distributions. *Astrophys. J.* **2008**, *677*, 1385–1394. [CrossRef]
110. Warmuth, A.; Mann, G. Constraints on energy release in solar flares from RHESSI and GOES X-ray observations. II. Energetics and energy partition. *Astron. Astrophys.* **2016**, *588*, A116. [CrossRef]
111. Hannah, I.G.; Christe, S.; Krucker, S.; Hurford, G.J.; Hudson, H.S.; Lin, R.P. RHESSI Microflare Statistics. II. X-Ray Imaging, Spectroscopy, and Energy Distributions. *Astrophys. J.* **2008**, *677*, 704–718. [CrossRef]
112. Battaglia, A.F.; Saqri, J.; Massa, P.; Perracchione, E.; Dickson, E.C.M.; Xiao, H.; Veronig, A.M.; Warmuth, A.; Battaglia, M.; Hurford, G.J.; et al. STIX X-ray microflare observations during the Solar Orbiter commissioning phase. *Astron. Astrophys.* **2021**, *656*, A4. [CrossRef]
113. Saqri, J.; Veronig, A.M.; Warmuth, A.; Dickson, E.C.M.; Battaglia, A.F.; Podladchikova, T.; Xiao, H.; Battaglia, M.; Hurford, G.J.; Krucker, S. Multi-instrument STIX microflare study. *Astron. Astrophys.* **2022**, *659*, A52. [CrossRef]
114. Glesener, L.; Krucker, S.; Duncan, J.; Hannah, I.G.; Grefenstette, B.W.; Chen, B.; Smith, D.M.; White, S.M.; Hudson, H. Accelerated Electrons Observed Down to <7 keV in a NuSTAR Solar Microflare. *Astrophys. J. Lett.* **2020**, *891*, L34. [CrossRef] [PubMed]
115. Cooper, K.; Hannah, I.G.; Grefenstette, B.W.; Glesener, L.; Krucker, S.; Hudson, H.S.; White, S.M.; Smith, D.M.; Duncan, J. NuSTAR observations of a repeatedly microflaring active region. *Mon. Not. RAS* **2021**, *507*, 3936–3951. [CrossRef] [PubMed]
116. Duncan, J.; Glesener, L.; Grefenstette, B.W.; Vievering, J.; Hannah, I.G.; Smith, D.M.; Krucker, S.; White, S.M.; Hudson, H. NuSTAR Observation of Energy Release in 11 Solar Microflares. *Astrophys. J.* **2021**, *908*, 29. [CrossRef] [PubMed]
117. Polito, V.; Peterson, M.; Glesener, L.; Testa, P.; Yu, S.; Reeves, K.K.; Sun, X.; Duncan, J. Multi-wavelength observations and modeling of a microflare: Constraining non-thermal particle acceleration. *Front. Astron. Space Sci.* **2023**, *10*, 1214901. [CrossRef]
118. Warmuth, A.; Mann, G. Thermal-nonthermal energy partition in solar flares derived from X-ray, EUV, and bolometric observations. Discussion of recent studies. *Astron. Astrophys.* **2020**, *644*, A172. [CrossRef]
119. Schmelz, J.T.; Kashyap, V.L.; Saar, S.H.; Dennis, B.R.; Grigis, P.C.; Lin, L.; De Luca, E.E.; Holman, G.D.; Golub, L.; Weber, M.A. Some Like It Hot: Coronal Heating Observations from Hinode X-ray Telescope and RHESSI. *Astrophys. J.* **2009**, *704*, 863–869. [CrossRef]
120. Brosius, J.W.; Daw, A.N.; Rabin, D.M. Pervasive Faint Fe XIX Emission from a Solar Active Region Observed with EUNIS-13: Evidence for Nanoflare Heating. *Astrophys. J.* **2014**, *790*, 112. [CrossRef]
121. Caspi, A.; Woods, T.N.; Warren, H.P. New Observations of the Solar 0.5–5 keV Soft X-Ray Spectrum. *Astrophys. J. Lett.* **2015**, *802*, L2. [CrossRef]
122. Del Zanna, G.; Andretta, V.; Cargill, P.J.; Corso, A.J.; Daw, A.N.; Golub, L.; Klimchuk, J.A.; Mason, H.E. High resolution soft X-ray spectroscopy and the quest for the hot (5–10 MK) plasma in solar active regions. *Front. Astron. Space Sci.* **2021**, *8*, 33. [CrossRef]
123. Bradshaw, S.J.; Klimchuk, J.A. What Dominates the Coronal Emission Spectrum During the Cycle of Impulsive Heating and Cooling? *Astrophys. J. Suppl.* **2011**, *194*, 26. [CrossRef]
124. Golub, L.; Hartquist, T.W.; Quillen, A.C. Comments on the Observability of Coronal Variations. *Sol. Phys.* **1989**, *122*, 245–261. [CrossRef]
125. Reale, F.; Orlando, S. Nonequilibrium of Ionization and the Detection of Hot Plasma in Nanoflare-heated Coronal Loops. *Astrophys. J.* **2008**, *684*, 715–724. [CrossRef]
126. Lörinčík, J.; Dudík, J.; del Zanna, G., Dzifčáková, E.; Mason, H.E. Plasma Diagnostics from Active Region and Quiet-Sun Spectra Observed by Hinode/EIS: Quantifying the Departures from a Maxwellian Distribution. *Astrophys. J.* **2020**, *893*, 34. [CrossRef]
127. Dzifčáková, E.; Dudík, J.; Zemanová, A.; Lörinčík, J.; Karlický, M. KAPPA: A Package for the Synthesis of Optically Thin Spectra for the Non-Maxwellian κ-distributions. II. Major Update to Compatibility with CHIANTI Version 10. *Astrophys. J. Suppl.* **2021**, *257*, 62. [CrossRef]
128. Del Zanna, G.; Polito, V.; Dudík, J.; Testa, P.; Mason, H.E.; Dzifčáková, E. Diagnostics of Non-Maxwellian Electron Distributions in Solar Active Regions from Fe XII Lines Observed by the Hinode Extreme Ultraviolet Imaging Spectrometer and Interface Region Imaging Spectrograph. *Astrophys. J.* **2022**, *930*, 61. [CrossRef]
129. Ishikawa, S.n.; Glesener, L.; Krucker, S.; Christe, S.; Buitrago-Casas, J.C.; Narukage, N.; Vievering, J. Detection of nanoflare-heated plasma in the solar corona by the FOXSI-2 sounding rocket. *Nat. Astron.* **2017**, *1*, 771–774. [CrossRef]
130. Moore, C.S.; Caspi, A.; Woods, T.N.; Chamberlin, P.C.; Dennis, B.R.; Jones, A.R.; Mason, J.P.; Schwartz, R.A.; Tolbert, A.K. The Instruments and Capabilities of the Miniature X-Ray Solar Spectrometer (MinXSS) CubeSats. *Sol. Phys.* **2018**, *293*, 21. [CrossRef]
131. Marsh, A.J.; Smith, D.M.; Glesener, L.; Klimchuk, J.A.; Bradshaw, S.J.; Vievering, J.; Hannah, I.G.; Christe, S.; Ishikawa, S.n.; Krucker, S. Hard X-Ray Constraints on Small-scale Coronal Heating Events. *Astrophys. J.* **2018**, *864*, 5. [CrossRef]
132. Mandrini, C.H.; Démoulin, P.; Klimchuk, J.A. Magnetic Field and Plasma Scaling Laws: Their Implications for Coronal Heating Models. *Astrophys. J.* **2000**, *530*, 999–1015. [CrossRef]
133. Cranmer, S.R.; Winebarger, A.R. The Properties of the Solar Corona and Its Connection to the Solar Wind. *Annu. Rev. Astron Astrophys.* **2019**, *57*, 157–187. [CrossRef]
134. Brooks, D.H.; Warren, H.P.; Landi, E. Measurements of Coronal Magnetic Field Strengths in Solar Active Region Loops. *Astrophys. J. Lett.* **2021**, *915*, L24. [CrossRef]

135. Hurford, G.J. X-ray imaging with collimators, masks and grids. *Issi Sci. Rep. Ser.* **2010**, *9*, 223–234.
136. Piana, M.; Emslie, A.; Massone, A.M.; Dennis, B.R. *Hard X-ray Imaging of Solar Flares*; Springer: Berlin/Heidelberg, Germany, 2022.
137. Massa, P.; Hurford, G.J.; Volpara, A.; Kuhar, M.; Battaglia, A.F.; Xiao, H.; Casadei, D.; Perracchione, E.; Garbarino, S.; Guastavino, S.; et al. STIX imaging I—Concept. *arXiv* **2023**, arXiv:2303.02485. [CrossRef]
138. Christe, S.; Alaoui, M.; Allred, J.; Battaglia, M.; Baumgartner, W.; Buitrago-Casas, J.C.; Chen, B.; Chen, T.; Dennis, B.; Drake, J.; et al. The Focusing Optics X-ray Solar Imager (FOXSI). *Bull. Am. Astron. Soc.* **2023**, *55*, 065. [CrossRef]
139. Krucker, S.; Christe, S.; Glesener, L.; Ishikawa, S.n.; Ramsey, B.; Takahashi, T.; Watanabe, S.; Saito, S.; Gubarev, M.; Kilaru, K.; et al. First Images from the Focusing Optics X-Ray Solar Imager. *Astrophys. J. Lett.* **2014**, *793*, L32. [CrossRef]
140. Christe, S.; Glesener, L.; Buitrago-Casas, C.; Ishikawa, S.N.; Ramsey, B.; Gubarev, M.; Kilaru, K.; Kolodziejczak, J.J.; Watanabe, S.; Takahashi, T.; et al. FOXSI-2: Upgrades of the Focusing Optics X-ray Solar Imager for its Second Flight. *J. Astron. Instrum.* **2016**, *5*, 1640005–1640625. [CrossRef]
141. Glesener, L.; Krucker, S.; Christe, S.; Ishikawa, S.n.; Buitrago-Casas, J.C.; Ramsey, B.; Gubarev, M.; Takahashi, T.; Watanabe, S.; Takeda, S.; et al. The FOXSI solar sounding rocket campaigns. In *Space Telescopes and Instrumentation 2016: Ultraviolet to Gamma Ray*; den Herder, J.W.A., Takahashi, T., Bautz, M., Eds.; Society of Photo-Optical Instrumentation Engineers (SPIE) Conference Series; SPIE: Bellingham, WA, USA, 2016; Volume 9905, p. 99050E. [CrossRef]
142. Musset, S.; Buitrago-Casas, J.C.; Glesener, L.; Bongiorno, S.; Courtade, S.; Athiray, P.S.; Vievering, J.; Ishikawa, S.n.; Narukage, N.; Furukawa, K.; et al. Ghost-ray reduction and early results from the third FOXSI sounding rocket flight. In *UV, X-ray, and Gamma-Ray Space Instrumentation for Astronomy XXI*; Society of Photo-Optical Instrumentation Engineers (SPIE) Conference Series; SPIE: Bellingham, WA, USA, 2019; Volume 11118, p. 1111812. [CrossRef]
143. Christe, S.D.; Shih, A.; Rodriguez, M.; Cramer, A.; Gregory, K.; Edgerton, M.; Gaskin, J.; Wilson-Hodge, C.; Apple, J.; Stevenson Chavis, K.; et al. The high energy replicated optics to explore the sun mission: A hard x-ray balloon-borne telescope. In *Solar Physics and Space Weather Instrumentation V*; Fineschi, S., Fennelly, J., Eds.; Society of Photo-Optical Instrumentation Engineers (SPIE) Conference Series; SPIE: Bellingham, WA, USA, 2013; Volume 8862, p. 886206. [CrossRef]
144. Christe, S.; Shih, A.Y.; Rodriguez, M.; Cramer, A.; Gregory, K.; Gaskin, J.; Chavis, K.; Smith, L.; HOPE/HEROES Team. The High Energy Replicated Optics to Explore the Sun (HEROES). In Proceedings of the AAS/Solar Physics Division Abstracts #44, July 2013; AAS/Solar Physics Division Meeting; Volume 44, p. 100.76.
145. Allington-Smith, J. Basic principles of integral field spectroscopy. *New Astron. Rev.* **2006**, *50*, 244–251. [CrossRef]
146. Del Zanna, G.; Mason, H.E. Solar UV and X-ray spectral diagnostics. *Living Rev. Sol. Phys.* **2018**, *15*, 5. [CrossRef] [PubMed]
147. Landi, E.; Hutton, R.; Brage, T.; Li, W. Hinode/EIS Measurements of Active-region Magnetic Fields. *Astrophys. J.* **2020**, *904*, 87. [CrossRef]
148. Dudík, J.; Del Zanna, G.; Mason, H.E.; Dzifčáková, E. Signatures of the non-Maxwellian κ-distributions in optically thin line spectra. I. Theory and synthetic Fe IX-XIII spectra. *Astron. Astrophys.* **2014**, *570*, A124. [CrossRef]
149. Dudík, J.; Mackovjak, Š.; Dzifčáková, E.; Del Zanna, G.; Williams, D.R.; Karlický, M.; Mason, H.E.; Lörinčík, J.; Kotrč, P.; Fárník, F.; et al. Imaging and Spectroscopic Observations of a Transient Coronal Loop: Evidence for the Non-Maxwellian K Distributions. *Astrophys. J.* **2015**, *807*, 123. [CrossRef]
150. Matthews, S.A.; Reid, H.A.S.; Baker, D.; Bloomfield, D.S.; Browning, P.K.; Calcines, A.; Del Zanna, G.; Erdelyi, R.; Fletcher, L.; Hannah, I.G.; et al. The high-energy Sun—probing the origins of particle acceleration on our nearest star. *Exp. Astron.* **2022**, *54*, 335–360. [CrossRef]

aerospace MDPI

Article

Topology Optimization of a Single-Point Diamond-Turning Fixture for a Deployable Primary Mirror Telescope

Cyril Bourgenot [1,*], Valdis Krumins [2], David G. Bramall [1] and Abdul M. Haque [2]

1 NETPark Research Institute, Centre for Advanced Instrumentation, Durham University, Joseph Swan Road, Sedgefield TS21 3FB, UK; d.g.bramall@durham.ac.uk
2 Advanced Manufacturing Research Centre, University of Sheffield, Wallis Way, Catcliffe S60 5TZ, UK; v.krumins@amrc.co.uk (V.K.); a.haque@amrc.co.uk (A.M.H.)
* Correspondence: cyril.bourgenot@durham.ac.uk

Abstract: CubeSats, known for their compact size and cost effectiveness, have gained significant popularity. However, their limited size imposes restrictions on the optical aperture and, consequently, the Ground Resolution Distance in Earth Observation missions. To overcome this limitation, the concept of deployable optical payloads with segmented primary mirrors which can unfold like petals has emerged, enabling larger synthetic apertures and enhanced spatial resolution. This study explores the potential benefits of leveraging Additive Manufacturing (AM) and Topology Optimization (TO) in the realm of ultra-precision machining, specifically single-point diamond machining. The goal is to reduce fixture weight while improving stiffness to minimize deformations caused by rotational and cutting forces which compromise optical performance. Through Finite Element Analysis, this research compares conventionally machined fixtures with those produced using AM and TO techniques. The results reveal that concept designs created via TO can achieve a remarkable 68% reduction in weight. This reduction makes the assembly, including the machining fixture and 12 U deployable segments, manageable by a single operator without the need for specialized lifting equipment. Moreover, these innovative designs lead to substantial reductions of up to 86% and 51% in deformation induced by rotational and cutting forces, respectively.

Keywords: topology optimization; single-point diamond machining; deployable optics; additive manufacturing; CubeSat; lightweight; earth observation

1. Introduction

With the availability and lower cost of small satellite platforms such as CubeSat, there has been a constant effort among the instrumentation community to miniaturize all the components, including the optical payload, to fit in the confined space defined by the elementary CubeSat Unit (U) $10 \times 10 \times 10$ cm in a platform ranging from 1 U up to 16 U. However, the limited size of the platform also restricts the optical aperture of the payload, which in turn limits the achievable Ground Resolution Distance (GRD), a key parameter in Earth Observation (EO) missions. Diffraction and sensor pixels' size, when projected on the ground, contribute to the GRD. The diffraction is only defined by the aperture diameter, orbital altitude, and the wavelength. A 1 U or 3 U CubeSat with a 10 cm aperture diameter in a Low Earth Orbit (500 Km), working in the visible region at a 500 nm wavelength, will deliver at least a GRD of 3 m, while a 12 U with a 20 cm aperture diameter typically achieves a 1.5 m GRD.

Considering the launch cost scale with the spacecraft mass and volume [1], an optical payload with a deployable optics and telescope therefore becomes an attractive concept to benefit from lower launch platform costs, without compromising on the telescope aperture size. A space telescope equipped with a deployable primary mirror can, in principle, recreate, with actuated individual segments, a larger synthetic aperture than the launch vehicle size, enabling higher spatial resolution than what a non-deployable telescope would have delivered.

Historically, the James Webb Space Telescope remains the only deployable telescope currently in operation [2]. Its successful launch and operation have confirmed the viability and potential of deployable optics in space. Efforts are now being made to adapt this principle to smaller, more cost-effective platforms, such as CubeSats. Various design proposals and associated alignment concepts have been detailed in recent publications. In [3,4], concepts of a deployable secondary mirror mechanism have been described, with an extendable barrel linking the primary and secondary mirrors. In [5], a deployable primary mirror composed of four petals reconstruct a 1.5 m diameter aperture. A phasing concept for the four segments is also described, maximizing the image sharpness metric. In [6], a mechanical design of a large non-imaging deployable telescope is described. In [7], a 200 mm diameter synthetic primary mirror deployed from a 1 U CubeSat bus is described along with the holographic method to phase the three segments together using laser point sources In [8], an advanced concept of a 300 mm synthetic primary mirror fitting in a 1.5 U CubeSat space enveloped is described with the phasing principle. The paper also gives a preliminary conceptual design for a deployable secondary mirror. In [9], a similar concept is extended to a 6 U platform.

Topology Optimization (TO) and Additive Manufacturing (AM) have gained increasing prominence as tools for designing and directly 3D printing optics [10–14]. Furthermore, the application of this technology shows great potential in enhancing ultra-precision machining. The potential advantages they bring include faster prototyping, increased design flexibility, reduced material waste, and the ability to create complex geometries that were previously unattainable. When machining the optical surfaces of deployable primary mirrors, two distinct strategies can be considered:

The first strategy involves machining each segment independently and sequentially 'on-axis', with the blank centered on the polishing or diamond-turning spindle axis, and the tool describing the equivalent freeform component of the off-axis segment with three synchronized axes. The second option is to machine the four segments in their operational position, forming the synthetic aperture in its deployed configuration. This configuration offers several advantages:

- it is faster, as all segments are machined in one go rather than sequentially as in the first option;
- it requires fewer axes, thereby minimizing potential sources of error;
- it presents a primary mirror arrangement identical to the operational configuration, thus facilitating post-machining metrology when the mirrors are phased together.

However, the main drawback of this off-axis configuration is that the assembly could become large and heavy, making it challenging for a single operator to handle. Additionally, there may be specific limitations, especially with large diameters, related to possible distortion induced by rotational forces on an ultra-precise lathe.

This paper focuses on the design aspects of a machining fixture and conducts a comparative analysis of the advantages offered by AM and TO when compared to a Conventionally Machined (CM) fixture. Both types of fixtures (AM and CM) are intended for use with a single-point diamond-turning ultra-precise lathe. They have been specifically designed to accommodate four identical segments, each with dimensions of approximately 200 $\times$ 200 mm and a mass of around 1 kg. These segments, when deployed, collectively form a synthetic aperture with a diameter of ϕ 600 mm. Our findings highlight the advantages of AM and TO, offering a fresh perspective on fixture design.
The folded segments fit within a 16 U CubeSat structure, with a 12 U volume nominally allocated for the optical system and an additional 4 Units (4 U) for potential avionics and satellite bus systems.

To assess the effectiveness of the proposed improvements, both the CM and AM designs are evaluated using Finite Element Analysis (FEA) against three requirements:

- The mass of the fixture;

- The deformation induced by the rotational force during machining;
- The deformation induced by the diamond tool cutting.

In the following section, we describe the conventional design fixture case study and provide background information about the modeling of rotational and cutting forces. In Section 3, we detail the topology optimization parameters, including the starting design and the boundary constraints applied during the optimization process. Moving on to Section 4, we present the results of two AM designs and discuss their performance improvements compared to the CM design. Finally, in Section 5, we engage in a discussion of the results and the opportunities they offer for future high-precision machining of large, segmented apertures.

2. Case Study—Conventional Fixture

The objective of a machining fixture is to accurately position and securely hold the four segments, which collectively form a rotationally symmetrical master surface, in their designated locations. Additionally, the fixture must ensure a reliable interface transfer between the off-axis segments and the backing plate.

To accomplish this, the conventionally designed fixture, depicted in Figure 1, comprises three discrete components:

- Part A: The backing plate, which is affixed to the machine spindle via the central set of counterbored holes.
- Part B: Removable wedges that serve to connect the mirror segments to the backing plate. These wedges also adjust the tilt of the segments according to their off-axis values and can be positioned in the X and Z directions via shims (Axis labels differ from the conventional machine ones; instead, they align with the FEA conducted in Section 4).
- Part C: The four individual segments themselves.

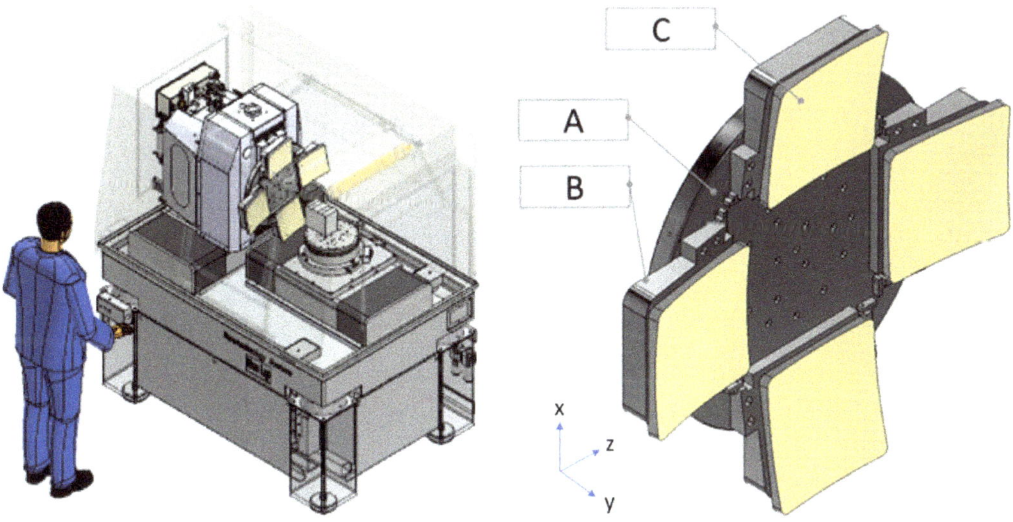

(**a**) (**b**)

Figure 1. (**a**) Example of a conventional fixture during machining on a Moore Nanotech 350FG at the Centre for Advanced Instrumentation at Durham University. (**b**) Conventional machining fixture composed of three elements. Element A is the backing plate, B are four wedges used to apply the off-axis angle on each segment and finely position each mirror on the backing plate, and C are the four segments.

As the layout of the four segments has been designed with the constraint of maintaining a thin profile to fit within the limited space of the platform, their rear faces have been aligned parallel to the tangential plane to the center of the aperture. Consequently, the use of wedges becomes necessary to compensate for this tilt. The presence of this tilt, which complicates the interface and introduces possible sources of misalignment, justifies the use of TO and its associated benefits.

2.1. Mechanical Target Requirements

The list of requirements, which defines the design objectives, is given in Table 1.

Table 1. Design objectives of the fixture.

Requirement Type	Target Requirement
Mass of assembly excluding mirrors	<15 Kg
Deformation induced by centrifugal force at 400 RPM	as low as possible [1]
Deformation induced by cutting forces of 10 N	as low as possible [1]

[1] This requirement should be in line with the residual wavefront error requirement of the optics or telescope and should be informed via a proper tolerance analysis.

The first requirement concerns the mass of the fixture, stipulating that it should not exceed 15 kg. This specification underscores the need for the fixture to be lightweight enough to be easily lifted onto the spindle of the machine by an operator, without relying on a crane. Since the part is machined in a vertical position as shown on Figure 1, a heavier mass imposes additional load (bending moment) on the spindle. This increased load can potentially limit the spindle's rotational speed, thereby extending the overall machining time. Moreover, a greater weight distributed over a larger diameter amplifies the moment and, consequently, the centrifugal force. This increase in centrifugal force induces displacements on the part during machining that give rise to form errors in the recovered mirror shape when the part is removed from the machine; hence, it is the second requirement to minimize these forces. Figure 2a provides a visual representation of the force vectors applied to each segment during the machining process. The centrifugal force acting on an object can be expressed using the following equation:

$$F_{rotation} = \omega^2 \cdot r \cdot m, \tag{1}$$

Here, ω represents the rotational speed in radians per second (rad/s), r signifies the distance from the object to the axis of rotation, and m denotes the mass of the object. While the primary force is applied radially to the segment, it is anticipated that a substantial portion of this force will translate into a component along the spindle axis due to the overhang of the wedges. This localized load will induce surface deformations on the mirror, similarly to the cutting force and shown on Figure 2b, governed by Hooke's law, with surface displacement varying linearly with the applied force:

$$F_{cutting} = \frac{\Delta L}{L} \cdot A \cdot E, \tag{2}$$

In this equation, ΔL represents the surface displacement, L stands for the thickness of the part where the force is applied, A denotes the surface area on which the force is exerted, and E signifies the modulus of elasticity. Consequently, the surface displacement along the spindle axis (ϵ) can be understood as both (i) a linear function of the tool cutting force:

$$\epsilon_{cutting} = \frac{\Delta L}{L} \propto F_{cutting}, \tag{3}$$

and (ii) a quadratic function of the rotational speed:

$$\epsilon_{rotation} = \frac{\Delta L}{L} \propto \omega^2 \cdot r \cdot m, \quad (4)$$

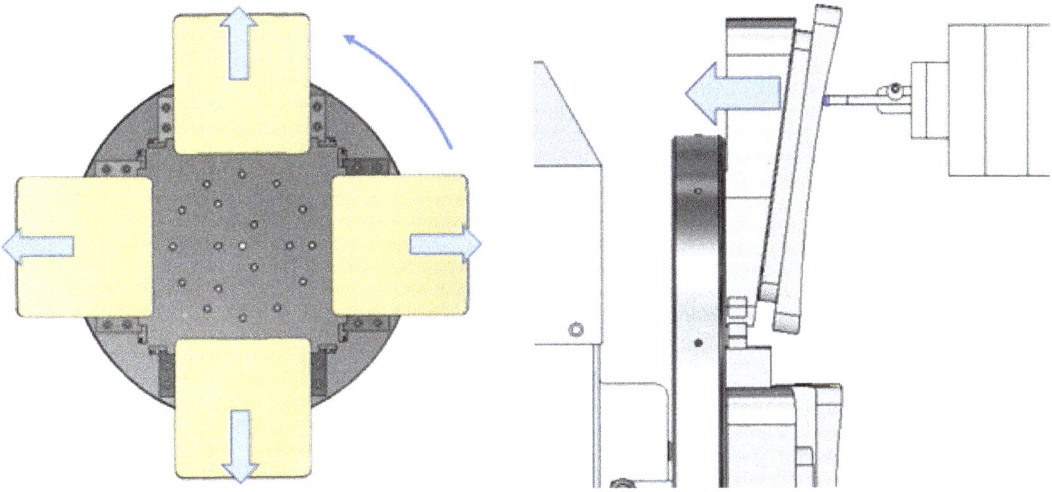

(**a**) (**b**)

Figure 2. (**a**) Direction of the force induced by centrifugal force when the fixture rotates at high speed. (**b**) Diamond tool exerting a cutting force on the segment and machining fixture.

2.2. Finite Element Analysis Parameters

FEA using Nastran [15] was used to simulate the displacement occurring on the mating faces of the fixture subjected to stress from both cutting and rotational forces. We isolated a single body part from the assembly file to simplify the transfer of deformations from the backing plate to the wedges. A mesh with element sizes of 20 mm was generated for the analysis. Fixed constraints were imposed on the bore located on the rear face of the fixture and the face in contact with the spindle, as illustrated in Figure 3a. To represent the cutting force, a 10 N load was applied along the spindle axis to the 16 raised pads situated on top of the wedges, as depicted in Figure 3b.

The cutting force, as described by [16], is directly linked to the chip area and is influenced by parameters such as the depth of the cut, tool radius, and feed rate. Further investigation has revealed that tool quality and wear also play a significant role in determining the tool force [17]. Typical cutting forces have been measured to range between 0.1 N and 1 N in the spindle's axis of rotation. We used a value of 10 N in our simulation, applying a safety factor. Equation (3) demonstrates that due to the linear relationship between displacement and force, the displacement results can be scaled proportionally when using a different cutting force.

In a similar manner to the approach followed by [18], we modeled the rotational forces applied to the fixture in Nastran using a rotation force and a velocity of 7 rev/s. It is important to note that the modeling focused exclusively on the fixture, without considering the mirror segments, in order to isolate only the fixture's contribution.

When defining the structural material parameters, a representative material with a mass density of 2.7 g/cm^3 and an Elastic Modulus of 68 GPa was selected as the material for both CM and AM cases. It is worth noting that alloys commonly used in AM, such as AlSi10Mg, or traditional aluminum alloys like 6061, exhibit very similar structural properties.

It would make sense to use a similar material for the telescope's primary mirror segment and the telescope structure to enable homogeneous thermal expansion in different temperature regimes, especially for machining, alignment, integration, and in-orbit operation.

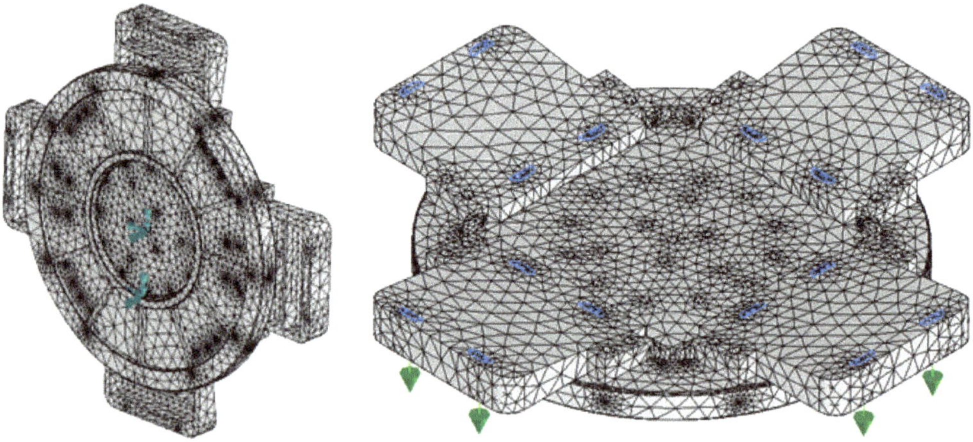

(a) (b)

Figure 3. (**a**) Fixed constraint applied to the back of the CM fixture, (**b**) 10 N force (in green) applied to the 16 interface planes (in blue) with the four deployable segments.

2.3. FEA Results—Conventional Fixture

The total mass of the assembly comprising the backing plate and the wedges is 15.5 kg.

The FEA results are presented in Figure 4 and Table 2, showcasing the deformation along the axis corresponding to the optical axis. With a cutting force of 10 N in Figure 4a, the deformation measures approximately 3.9 µm PV. Notably, the maximum deformation is observed above the overhang, highlighting the need for a significantly stiffer and more rigid support in that area. With a more realistic cutting force of 1 N, the deformation would be around 390 nm, which would become acceptable for a IR telescope, especially considering that the distortion would also be limited to the area of the pads and not the entire fixture, which is how it has been considered here. Figure 4b illustrates the displacement outcome caused by rotational forces of 400 RPM, again along the spindle axis. As significant displacement is observed at approximately 6.8 µm, it is advisable to refrain from using the fixture at rotational speeds exceeding 90 RPM, which would induce a residual deformation around 350 nm.

Table 2. FEA results obtained for the conventionally machined fixture.

Requirement Type	Target Requirement
Mass of assembly excluding mirrors	15.5 Kg
Deformation induced by centrifugal force at 400 RPM	6.8 µm
Deformation induced by cutting forces of 10 N	3.9 µm
Deformation induced by gravity	0.56 µm

Due to the lower deformation induced by gravity, our subsequent optimization will primarily focus on reducing the deformation caused by cutting and rotational forces.

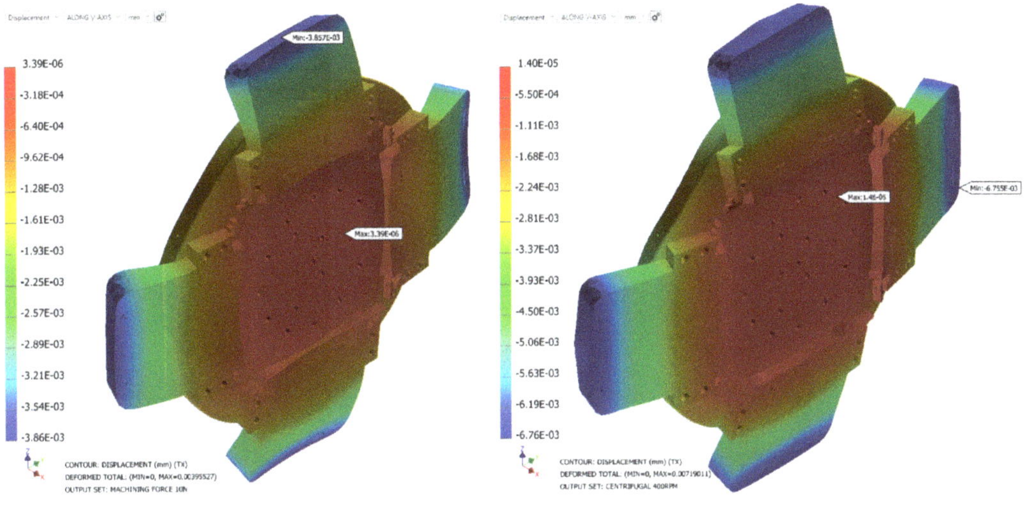

(**a**) (**b**)

Figure 4. (**a**) Displacement modeled for a cutting force equal to 10 N (**b**) and a rotational force induced by a rotation speed of 400 RPM (7 rev/s).

3. Topology Optimization

The TO was undertaken at the Advanced Manufacturing Research Centre (AMRC) and focused on a fixture design that harmonizes structural resilience with weight efficiency, suitable for deployable primary mirrors in CubeSat telescopes. For the design space, the setup was configured to optimize the fixture's design for maximum stiffness, targeting specific mass constraints. Two mass targets were explored, 5 kg and 10 kg, to drive the optimization. In addition, two design envelopes were used to explore the design variability: a short and a long design envelope. As can be seen in Figure 1a, there is free space to allow for a larger envelope along the tool axis. Table 3 lists the primary mechanical requirements used for the fixture design space configuration.

Table 3. Mechanical design requirements.

Design Requirement	Target Values
Mass of assembly excluding mirrors	<15 Kg
Deformation due to axial cutting load (10 N)	as good or better than CM ($\leq$3.9 µm)
Deformation due to acceleration at 400 RPM	as good or better than CM ($\leq$6.8 µm)

Methodology

The CAD 'Generative Design' tool within PTC Creo (version 9.0.2.0) [19] was used for the TO outputs shown in this study. Altair HyperWorks (version 2022.3) [20] was also used to run final structural analysis of the concepts. The baseline geometry, comprising the base plate, wedges, and shims, is depicted in Figure 1b. This assembly formed the reference for load direction definitions, aligned with the lathe spindle axis (Y-axis) and perpendicular axes (X and Z).

The optimization process involved delineating both a design space and a non-design space. The design space, shown in Figure 5, is where material could be added or removed, shaped by local constraints and objectives. This space was defined around the original concept, ensuring no interference with essential features like fasteners or adjacent components. Conversely, the non-design space encompassed critical interface regions and functional features that were necessary to retain. Two design space sizes were explored, including a variant that integrated the backing plate into the fixture base. The fixture's dimensions, particularly an added 200 mm length, were tailored to fit within the Single-Point Diamond Machine lathe constraints, allowing for future adjustments in design.

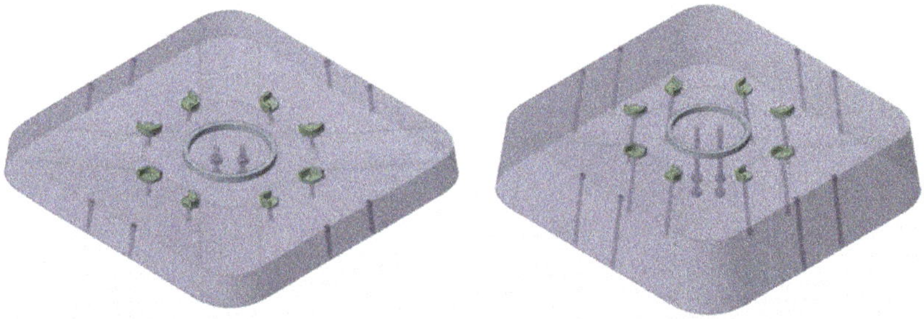

Figure 5. Baseline concept geometry for the fixture plate: (**Left**) design space envelope similar to CM; (**Right**) large envelope as machine space allows.

The fixture's theoretical fabrication was planned to use an AM process such as Laser Powder Bed Fusion (LPBF), chosen for its ability to facilitate complex and lightweight designs not possible using methods of manufacture. For this purpose, the aluminum alloy AlSi10Mg, frequently used in AM applications, was selected. The specific properties of AlSi10Mg, for guiding the design and optimization of the fixture, are detailed in Table 4.

Table 4. Material properties used for structural optimization and analysis.

Material	Young's Modulus (GPa)	Density (Kg·m^{-3})	Poisson's Ratio	Yield Stress (MPa)	Allowable Stress (MPa)
AlSi10Mg (stress relieved)	66.0	$2.68 \cdot 10^{-6}$	0.33	174	116

All degrees of freedom were fixed at the lathe spindle bolting location in four positions, as shown in Figure 6. These constraints are applied on the preserve geometries (non-design space).

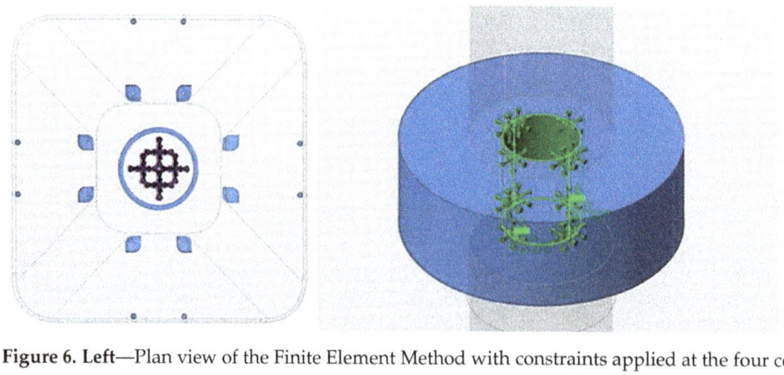

Figure 6. Left—Plan view of the Finite Element Method with constraints applied at the four central bolting locations. Dark blue indicates the fixed location constraint applied to the attachment screws, while light blue represents the interface with the mirror segments, which must not be altered during the optimization. **Right**—Close-up view of the constraint application onto the preserved geometry. Light blue denotes the area that will not be modified during the optimization, and green represents the fixed constraint applied to the geometry.

All components of the assembly, including the metal optics, were consolidated into a single entity, eliminating the need for joints or connectors. The metal optics were represented as point loads at their interface points. The load analysis comprised two categories: rotational loads at 400 RPM and an axial load of 10 N, applied at the metal optics fastener locations around the fixture's periphery. This 10 N load, an exaggerated estimate for safety, was distributed across four fastener positions as depicted in Figure 7. These forces were applied to preserved geometries (non-design spaces) such as constraints. Notably, previous analyses undertaken at Durham University (DU) included gravity loads in a vertical orientation; however, these were omitted in this study since their impact was significantly lesser compared to rotational forces. The metal optics' bodies themselves were excluded from this load evaluation to ensure direct comparability with the CM case. It is acknowledged that the inclusion of these masses in the fixture assembly will impact the study results. The primary objective of this study was to isolate and compare the effects of cutting and rotation forces on the two types of fixture.

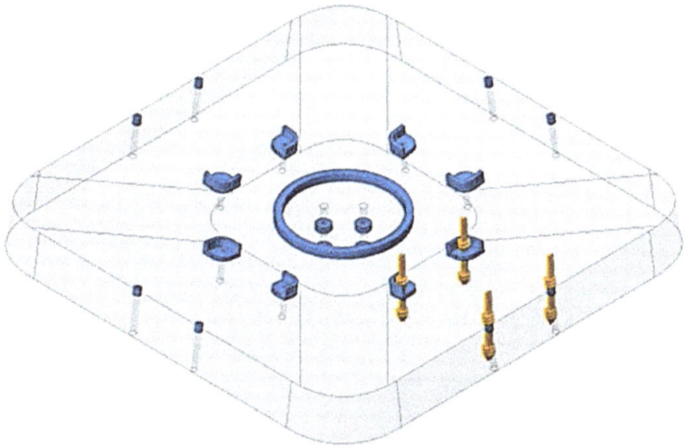

Figure 7. Application of force is shown distributed across one of the metal optics locations. Grey delineates the optimization volume, blue indicates the interface with the mirror segments, and the yellow arrows represent the force applied on the interface during machining.

PTC Creo's 'Generative Design' tool, differing from its typical p-element method with automatic polynomial adjustment, utilized an h-element-type mesh. For maximum detail, the setting generated approximately 512,000 tetrahedral elements, averaging 3.1 mm in size. Conversely, Altair HyperWorks, which also employs the h-element method, focused on mesh convergence via element sizing. The structural analysis reporting deflections used a global average mesh size of 10 mm. An automatic surface deviation method refined the mesh for curvature, reducing it to as low as 10% of the global size (1 mm minimum). An illustrative section through the utilized mesh is shown in Figure 8. To achieve manageable outputs from the TO setup, the studies were limited to 1000 iterations, with any cases reaching this limit reported as inconclusive.

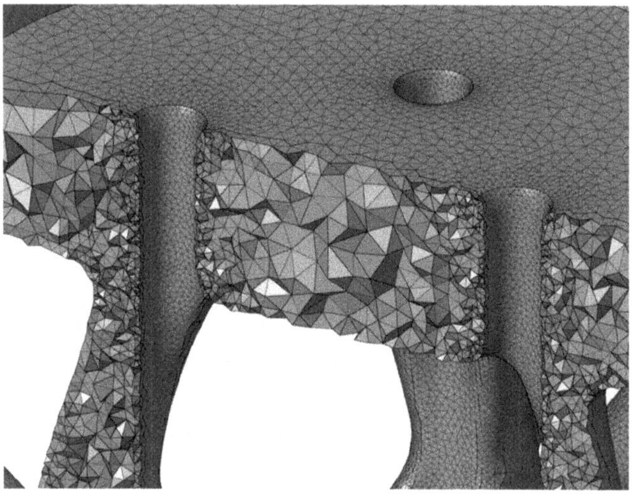

Figure 8. Example section of the mesh density used for the validation analysis of the concepts in Altair HyperWorks.

4. Results

Two TO outputs were successfully produced and met the specified design criteria, as highlighted in Table 5. An additional load case, combining both rotational and axial loads, was evaluated to better approximate the real machining setup. This combined load case resulted in the greatest deflection, with the findings presented in the subsequent figures. The TO outputs are shown in Figures 9 and 10. Detailed deflection plots for each individual load case are shown in Figures 11–13. Although peak stress levels were recorded, it is important to note that prior studies have determined that stress is not a significant concern for the materials selected for these designs.

Table 5. Results of short and long design envelopes with 5 kg and 10 kg mass targets.

Concept	Mass, kg	Peak Deflection—400 RPM, µm	Peak Deflection—10 N, µm	Peak Deflection—Combined, µm	Peak Stress, MPa
Design Criteria	<15	As good or better than CM	As good or better than CM	As good or better than CM	<116
Short Envelope	5.2	1.17	3.08	4.23	1.24
Long Envelope	10.2	1.11	2.04	3.14	1.59

The output of the TO with the shorter design envelope and 5 kg mass target, as well as the peak deflection contour plot, are shown in Figures 9 and 11.

The output of the TO with the longer design envelope and 10 kg mass target, as well as the peak deflection contour plot, are shown in Figures 10 and 12.

Comparing the results of the design envelopes, the short envelope design, with a mass of 5.2 kg, demonstrated peak deflections slightly higher than the long envelope design, which weighed 10.2 kg. Specifically, under combined load conditions, the short envelope recorded a peak deflection of 4.23 µm compared to 3.14 µm for the long envelope. This difference highlights the long envelope's slightly superior performance in minimizing deflection under load, which could be attributed to its increased mass and altered geometry. Moreover, the peak stress in the long envelope design was marginally higher (1.59 MPa) compared to the short envelope (1.24 MPa), yet both values were significantly lower than the stress limit of 116 MPa, confirming the robustness of both designs. Unlike the CM design, the TO outputs account for the stiffness of the mirror in the optimization. The resultant geometries are truss-like and have members directly connecting the mounting pads on the mirrors to the spindle. Note how the plane of the trusses acts to reduce the displacement along the spindle axis. These designs also feature reinforced support at critical points to minimize deformation under various applied loads.

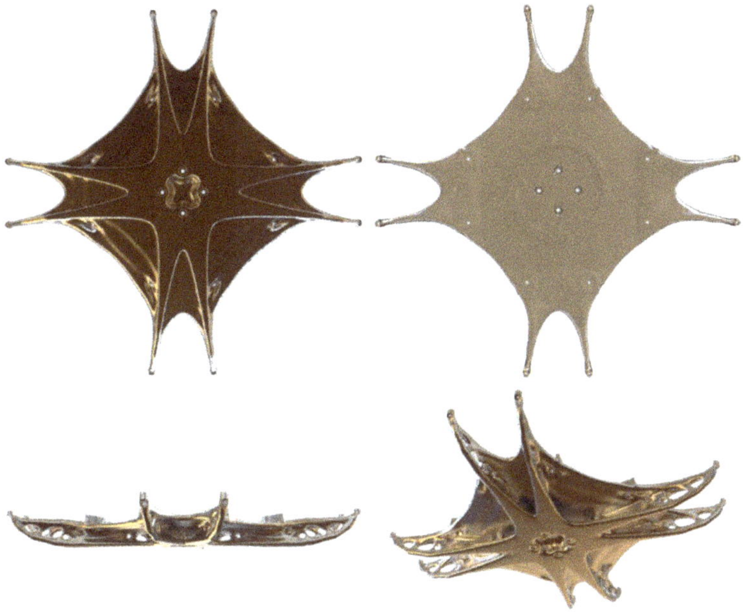

Figure 9. Topology optimized concept geometry for a 5 kg target mass and short envelope.

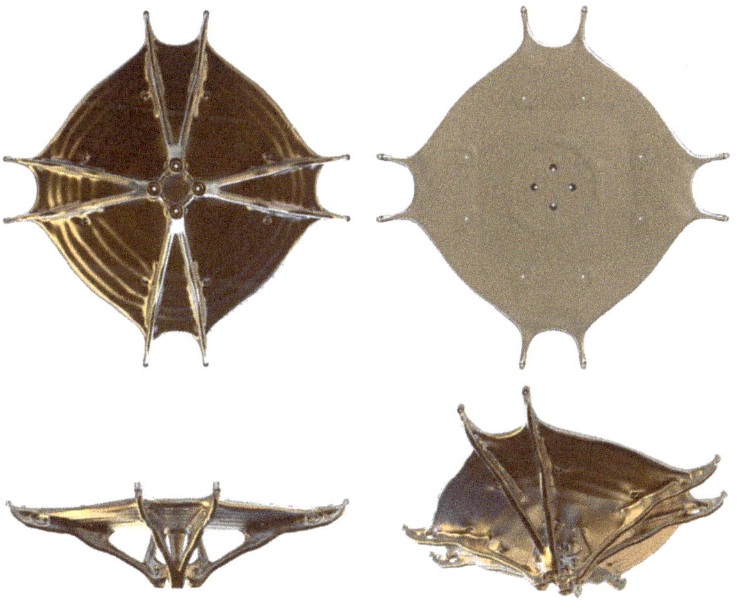

Figure 10. Topology optimized concept geometry for a 10 kg target mass and long envelope.

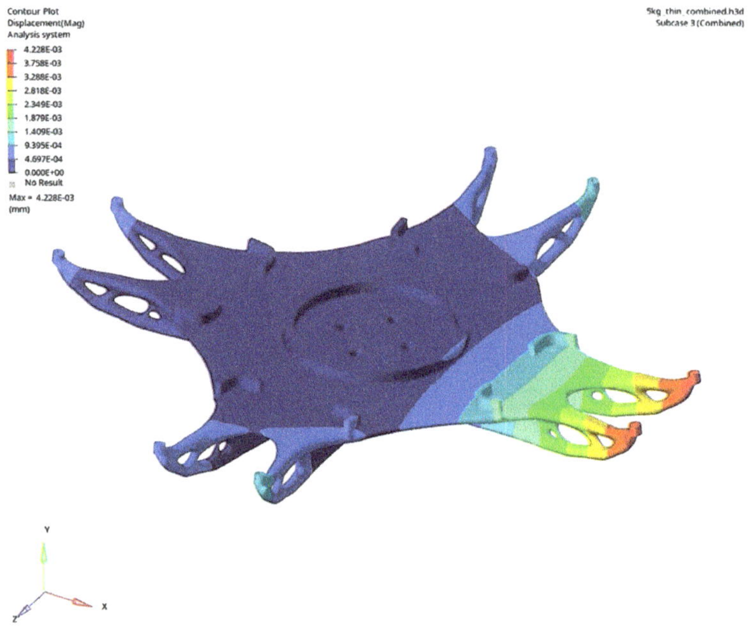

Figure 11. Peak deflection contour plot for the combined load case on the 5 kg short concept. A 4.23 µm maximum is reported.

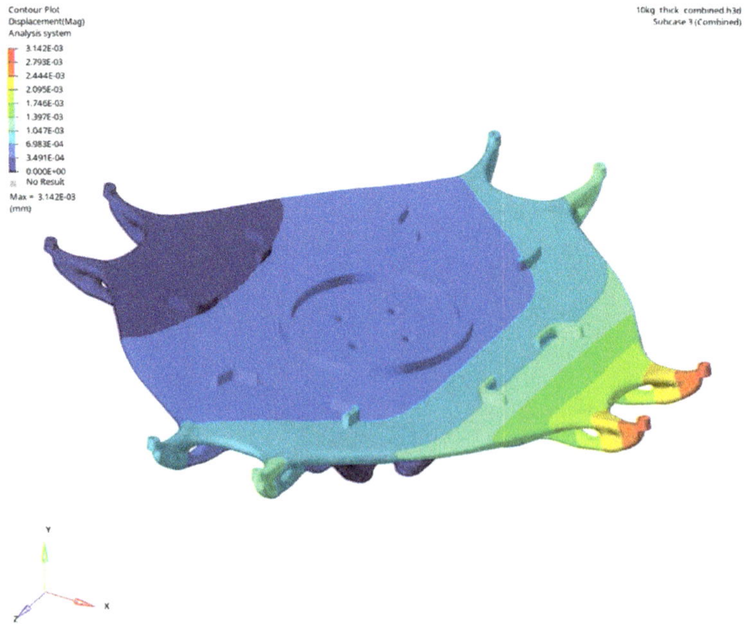

Figure 12. Peak deflection contour plot for the combined load case on the 10 kg long concept. A 3.14 µm maximum is reported.

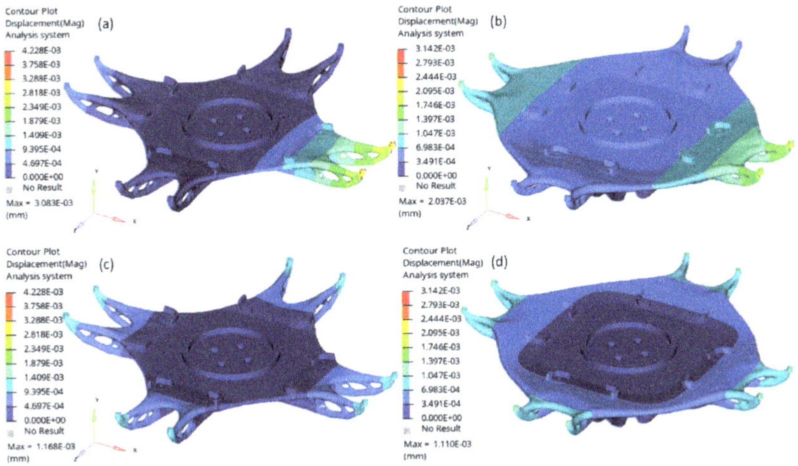

Figure 13. Peak deflection contour plot for the 10 N axial load case on the 5 kg short concept (**a**) and for the 10 kg long concept (**b**). Peak deflection contour plot for the 400 RPM load case on the 5 kg short concept (**c**) and for the 10 kg long concept (**d**).

The summarized results are presented in Table 6, where the mass and performance of the CM design are taken as a reference and compared with both TO_1 and TO_2. In addition to the actual mass and deformation values for TO_1 and TO_2, an 'improvement' factor is provided as a percentage, reflecting the reduction in mass or displacement achieved.

Table 6. Mass and FEA Results for the three Case Studies (Nominal CM Design, Design 1 AM, and Design 2 AM). The percentages represent the reduction in mass or reduction in surface displacement along the spindle axis.

	Nominal CM	TO_1 AM Short Envelope	TO_2 AM Long Envelope
Mass of assembly excluding mirrors	15.5 kg	5.2 kg 66.45%	10.2 kg 34.19%
Deformation induced by centrifugal force at 400 RPM	6.8 µm	1.17 µm 82.79%	1.11 µm 83.68%
Deformation induced by cutting forces of 10 N	3.9 µm	3.08 µm 21.03%	2.04 µm 47.69%

A significant 80% reduction in deflection was achieved when the assembly was subjected to rotational force, while improvements of 21% and 47% were observed in the deflection induced by cutting forces for the 5 Kg and 10 Kg TO, respectively. As mentioned, the recorded stress levels, though monitored, were not a primary concern, primarily due to the high stress tolerance of the AlSi10Mg material used in the designs. This material offers a favorable combination of strength and flexibility, making it well suited for applications requiring both lightness and durability.

5. Discussion

TO and AM have the potential to enhance the precision and efficiency of the Single-Point Diamond-Turning manufacturing process by optimizing fixture rigidity and resilience to the manufacturing stress. They also have the potential to further reduce the weight of large assembly and aperture, especially for cases such as deployable primary mirrors, helping with the handling of parts and the iteration between machining and "off-machine" metrology.

Although there are some clear benefits to using AM, 3D printing of large parts, such as the ones presented in this paper with a diameter of 600 mm and a thickness of 200 mm, can be challenging and the process may be long and costly. The volume capacity of current LPBF machines can also be limited to smaller dimensions. One solution could be to split the whole assembly into four rotationally symmetrical sections (one section is shown in Figure 14), which could therefore be printed on conventional LPBF machines. This segmentation not only facilitates printing but could also significantly reduce the stress and potential thermal distortions typically encountered in large-scale LPBF printing. Each section could be designed to be self-supporting at a 45-degree critical angle, optimizing the printing process by minimizing the need for additional support structures and streamlining post-processing, despite a possible increase in Z height and print time. The four subassemblies can then be reassembled on a backing plate and positioned on the machine spindle as one single unit. The flat interface with the segment can be prepared and rectified with a high-speed ball end mill [21] so that all planes are post-processed in the same setup, minimizing potential errors induced by removing and reattaching the fixture to the machine's spindle.

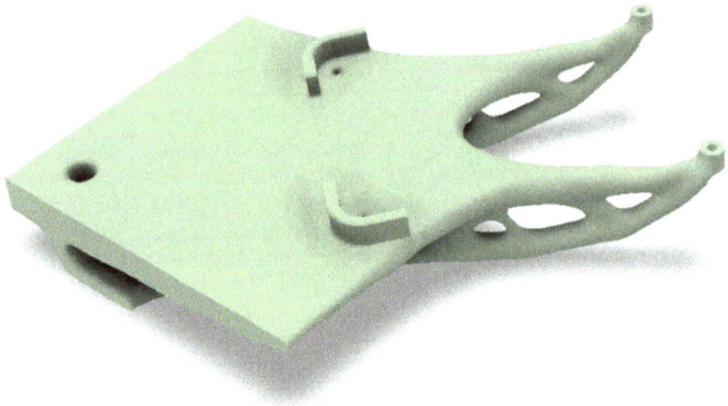

Figure 14. Split concept for a better compatibility with the production.

An alternative method of manufacture could be explored using investment wax casting [22], a process renowned for its accuracy. It involves creating a wax model of the entire TO fixture (or a sand mold). This model is then encased in a refractory material to form a mold. Once the mold is heated, the wax melts away, leaving a cavity in the shape of the fixture. Molten metal is then poured into this cavity, creating a single, unified piece that mirrors the intricate design of the TO fixture. This method could be beneficial for producing complex geometries and large-scale components, which are often challenging or impractical with LPBF due to size limitations and the need for support structures. Additionally, casting the fixture as one piece significantly reduces the risks associated with segmenting and reassembling multiple parts. It eliminates potential misalignments or inconsistencies that might arise from assembling several printed sections. This ensures a higher degree of structural integrity and uniformity in the final product.

6. Conclusions

The topology optimization process at AMRC yielded two distinct outputs, primarily focused on reducing mass while maximizing stiffness. These designs feature reinforced supports at crucial points, balancing lightweight construction with enhanced structural integrity. FEA revealed that both designs significantly outperformed the reference CM model. Specifically, TO_1 demonstrated a 66.45% mass reduction and substantial decreases in deformation (82.79% under centrifugal forces at 400 RPM and 21.03% under 10 N cutting forces). Similarly, TO_2 achieved a 34.19% mass reduction, an 83.68% reduction in deformation from centrifugal forces, and a 47.69% reduction in deformation from cutting forces. In-depth examination of the topology optimization for the metal optics fixture presented an innovative design that not only minimized mass but also met or surpassed deflection standards. Both design variants shared key structural features, such as truss-like arms extending from the spindle interface to the metal optics interfaces, enhancing bending stiffness. A distinctive plate-like feature on the design envelope's upper surface, formed in response to rotational loads, was also observed. Both designs exceeded mechanical requirements, suggesting that a more aggressive optimization approach could further reduce weight. However, our conservative strategy incorporated additional factors like rotational symmetry and dynamic acceleration criteria, ensuring compatibility with PTC Creo Version 9.0.2.0. This approach, combined with additive manufacturing considerations, underscores the transformative potential of TO and AM in refining Ultra-Precision manufacturing processes.

Author Contributions: Conceptualization, C.B.; methodology, C.B. and A.M.H.; validation, all; formal analysis, V.K. and C.B.; investigation V.K., A.M.H., D.G.B. and C.B.; resources, A.M.H. and D.G.B.; data curation, V.K., D.G.B. and C.B.; writing—original draft preparation, C.B.; writing—review and editing, all; visualization, V.K. and C.B.; supervision, A.M.H. and C.B.; project administration, A.M.H. and C.B.; funding acquisition, A.M.H. and C.B. All authors have read and agreed to the published version of the manuscript.

Funding: This research was funded by the UK Space Agency, Enabling Technology Program round 1, Grant reference number ETP01-040.

Data Availability Statement: Available on request.

Conflicts of Interest: The authors declare no conflicts of interest. The funders had no role in the design of the study; in the collection, analyses, or interpretation of the data; in the writing of the manuscript; or in the decision to publish the results.

References

1. Bearden, D.A. Small-Satellite Costs. *Crosslink* **2001**, *2*, 33–44.
2. Lightsey, P.A. James Webb Space Telescope: Large deployable cryogenic telescope in space. *Opt. Eng.* **2012**, *51*, 011003. [CrossRef]
3. Gooding, D.; Richardson, G.; Haslehurst, A.; Smith, D.; Saunders, C.; Aglietti, G.; Blows, R.; Shore, J.; Hampson, K.; Booth, M. A novel deployable telescope to facilitate a low-cost <1 m GSD video rapid-revisit small satellite constellation. In Proceedings of the International Conference on Space Optics—ICSO 2018, Chania, Greece, 9–12 October 2018; Sodnik, Z., Karafolas, N., Cugny, B., Eds.; International Society for Optics and Photonics, SPIE: Bellingham, WA, USA, 2019; Volume 11180, p. 1118009. [CrossRef]
4. Aglietti, G.S.; Honeth, M.; Gensemer, S.; Diegel, O. Deployable Optics for CubeSats. In Proceedings of the 34th Annual AIAA/USU Conference on Small Satellites, Vilamoura, Portugal, 1–6 August 2020; Volume SSC-VI-06, pp. 1–7.
5. Dolkens, D.; Kuiper, H.; Corbacho, V.V. The deployable telescope: A cutting-edge solution for high spatial and temporal resolved Earth observation. *Adv. Opt. Technol.* **2018**, *7*, 365–376. [CrossRef]
6. Lake, M.S.; Phelps, J.E.; Dyer, J.E.; Caudle, D.A.; Tam, A.; Escobedo-Torres, J.; Kasl, E.P. Deployable primary mirror for space telescopes. In Proceedings of the Advanced Telescope Design, Fabrication, and Control, Denver, CO, USA, 19–21 July 1999; Roybal, W., Ed.; International Society for Optics and Photonics, SPIE: Bellingham, WA, USA, 1999; Volume 3785, pp. 14–25. [CrossRef]
7. Muslimov, E.R.; Sakhabutdinov, A.Z.; Morozov, O.G.; Pavlycheva, N.K.; Akhmetov, D.M.; Kharitonov, D.Y. Digital Holographic Positioning Sensor for a Small Deployable Space Telescope. *Appl. Sci.* **2022**, *12*, 4427. [CrossRef]
8. Schwartz, N.; Milanova, M.; Brzozowski, W.; Todd, S.; Ali, Z.; Buron, L.; Jean-François-Sauvage; Bond, C.; Bruce, H.; Rees, P.; et al. Active deployable primary mirrors on CubeSat. In Proceedings of the 13th IAA Symposium on Small Satellites for Earth Observation (SSSEO 2021), Berlin, Germany, 27–29 April 2021. [CrossRef]
9. Schwartz, N.; Brzozowski, W.; Ali, Z.; Milanova, M.; Morris, K.; Bond, C.; Keogh, J.; Harvey, D.; Bissell, L.; Sauvage, J.F.; et al. 6 U CubeSat deployable telescope for optical Earth observation and astronomical optical imaging. In Proceedings of the Space Telescopes and Instrumentation 2022: Optical, Infrared, and Millimeter Wave, Montréal, QC, Canada, 17–22 July 2022; Coyle, L.E., Matsuura, S., Perrin, M.D., Eds.; International Society for Optics and Photonics, SPIE: Bellingham, WA, USA, 2022; Volume 12180, p. 1218031. [CrossRef]
10. Atkins, C.; Brzozowski, W.; Dobson, N.; Milanova, M.; Todd, S.; Pearson, D.; Bourgenot, C.; Brooks, D.; Snell, R.; Sun, W.; et al. Additively manufactured mirrors for CubeSats. In Proceedings of the Astronomical Optics: Design, Manufacture, and Test of Space and Ground Systems II, San Diego, CA, USA, 12–15 August 2019; Hull, T.B., Kim, D.W., Hallibert, P., Eds.; International Society for Optics and Photonics, SPIE: Bellingham, WA, USA, 2019; Volume 11116, p. 1111616. [CrossRef]

11. Atkins, C.; Brzozowski, W.; Dobson, N.; Milanova, M.; Todd, S.; Pearson, D.; Bourgenot, C.; Brooks, D.; Snell, R.; Sun, W.; et al. Lightweighting design optimisation for additively manufactured mirrors. In Proceedings of the Astronomical Optics: Design, Manufacture, and Test of Space and Ground Systems II, San Diego, CA, USA, 12–15 August 2019; Hull, T.B., Kim, D.W., Hallibert, P., Eds.; International Society for Optics and Photonics, SPIE: Bellingham, WA, USA, 2019; Volume 11116, pp. 353–371. [CrossRef]
12. Tan, S.; Ding, Y.; Xu, Y.; Shi, L. Design and fabrication of additively manufactured aluminum mirrors. *Opt. Eng.* **2020**, *59*, 013103. [CrossRef]
13. Zhang, K.; Qu, H.; Guan, H.; Zhang, J.; Zhang, X.; Xie, X.; Yan, L.; Wang, C. Design and fabrication technology of metal mirrors based on additive manufacturing: A review. *Appl. Sci.* **2021**, *11*, 10630. [CrossRef]
14. Hilpert, E.; Hartung, J.; von Lukowicz, H.; Herffurth, T.; Heidler, N. Design, additive manufacturing, processing, and characterization of metal mirror made of aluminum silicon alloy for space applications. *Opt. Eng.* **2019**, *58*, 092613. [CrossRef]
15. Autodesk Nastran 2022, Reference Manual. Available online: https://www.autodesk.com/sites/default/files/file_downloads/Autodesk-Nastran-Solver-2022-Reference-Manual.pdf (accessed on 1 January 2024).
16. Drescher, J.D.; Dow, T.A. Tool force model development for diamond turning. *Precis. Eng.* **1990**, *12*, 29–35. [CrossRef]
17. Huang, P.; Lee, W.B. Cutting force prediction for ultra-precision diamond turning by considering the effect of tool edge radius. *Int. J. Mach. Tools Manuf.* **2016**, *109*, 1–7. [CrossRef]
18. Zhang, J.; Zhang, X.; Tan, S.; Xie, X. Design and manufacture of an off-axis aluminum mirror for visible-light imaging. *Curr. Opt. Photonics* **2017**, *1*, 364–371. [CrossRef]
19. TC Creo (Version 9.0.2.0). Generative Design Interactive Ebook. Available online: https://pdsvision.com/wp-content/uploads/2022/12/Generative-Design-Interactive-Ebook.pdf (accessed on 1 January 2024).
20. Altair HyperWorks (Version 2022.3). Available online: https://help.altair.com/inspire/whatsnew/Inspire_2022.3_ReleaseNotes_English.pdf (accessed on 1 January 2024).
21. Scheiding, S.; Damm, C.; Holota, W.; Peschel, T.; Gebhardt, A.; Risse, S.; Tünnermann, A. Ultra-precisely manufactured mirror assemblies with well-defined reference structures. In Proceedings of the Modern Technologies in Space- and Ground-Based Telescopes and Instrumentation, San Diego, CA, USA, 27 June–2 July 2010; Atad-Ettedgui, E., Lemke, D., Eds.; International Society for Optics and Photonics, SPIE: Bellingham, WA, USA, 2010; Volume 7739, p. 773908. [CrossRef]
22. Viswanathan, S.; Apelian, D.; Donahue, R.J.; DasGupta, B.; Gywn, M.; Jorstad, J.L.; Monroe, R.W.; Sahoo, M.; Prucha, T.E.; Twarog, D. *Casting*; ASM International: Almere, The Netherlands, 2008. [CrossRef]

aerospace MDPI

Article

Structural Flexibility Effect on Spaceborne Solar Observation System's Micro-Vibration Response

Lin Yang [1,2], Yansong Wang [1,2,*], Lei Wei [1,2] and Yao Chen [1,2]

1 Institute of Frontiers and Interdisciplinary Sciences, Shandong University, Binhai Road 72, Qingdao 266237, China; yanglincas@sdu.edu.cn (L.Y.); weilei0906@sdu.edu.cn (L.W.); yaochen@sdu.edu.cn (Y.C.)
2 Institute of Space Sciences, Shandong University, Wenhua West Road 180, Weihai 264209, China
* Correspondence: wangyansong@sdu.edu.cn

Abstract: The spaceborne solar observation system is crucial for the study of space phenomena such as solar flares, which requires high tracking accuracy. This study presents a coupling model that integrates mechanical, electrical, and control models to investigate the structural flexibility effect on the micro-vibration response. We established a rigid–flexible model using mechanical parts. We considered the influence of flexible features while studying the dynamic responses in its operation. The state-space equations of the system showed that modal frequency, damping, and modal participation factors played significant roles. We derived transfer functions using the Laplace transform of the coupling models to better understand this mechanism, and Simulink models were thereby established. We simulated the acceleration responses of the rigid–flexible and rigid models under angle tracking modes, and the results showed significant differences. We also simulated the acceleration responses of the models under various control frequencies, and the optimal control frequency was thus obtained. Finally, we performed experiments, and the results indicated that the rigid–flexible model could better predict the motion and acceleration responses for the spaceborne solar observation system. This study provides valuable information for understanding the role of flexible features in space performance high-tracking accuracy instruments and for micro-vibration suppression research.

Keywords: satellites; 2D turntable; structural flexibility; micro-vibration; mechanical–electrical–control coupling; Simulink model

1. Introduction

The Earth is influenced by powerful electromagnetic radiation and high-energy particle radiation released from solar activity. Solar flares are among the most violent explosions in the solar atmosphere and are a major focus of space science research [1,2]. Unfortunately, high-energy particles and electromagnetic solid radiation pose a fatal threat to space facilities and deep space exploration safety. High-energy particles can easily break through the surfaces of satellites, causing charging and discharging in satellites and even destroying spaceborne equipment [3]. A spaceborne solar observation system (SSOS) is an important tool for achieving the real-time tracking, observation, and analysis of the Sun. The SSOS must have high tracking accuracy to achieve real-time and accurate observation data.

The SSOS two-dimensional (2D) turntable presented in this study is a typical mechanical–electrical–control coupled system that generates disturbing forces and torques during the tracking rotation. These forces and torques are caused by manufacturing, assembly and control errors, posing a significant threat to spacecraft pointing and tracking stability, commonly known as the micro-vibration effect [4–7]. When these forces are coupled with structural modes, resonance will occur, which may cause catastrophic impacts in many fields, such as earthquakes [8,9]. In aerospace, resonance may cause a decrease in satellite pointing accuracy and stability accuracy, further affecting the performance of satellite payloads [10]. Therefore, the flexible vibrations of the structure's turntable cannot be ignored and must be considered when designing SSOS systems.

When studying the published literature on the modeling of spaceborne 2D turntables, we discovered that the published literature is primarily focused on control system design [11,12], structural design, and structural dynamic modeling [13–17]. Wang [11] used non-magnetic technology to study turntable pointing accuracy and motion control, analyzed the influence of the servo control parameters on the dynamic performance and steady-state error, and tested the turntable angular positioning accuracy. Zou [15] investigated the micro-vibration characteristics of a spaceborne 2D turntable and conducted a modal and frequency response analysis of the structure, which were later verified by experiments. Yao [18] established the governing equations of the semi-rigid joints and integrated them into the dynamic equation of the deployable structure. The research mentioned above did not combine a control system model with a structural dynamics model. Moreover, the rigid–flexible coupling effect has not been considered, and the influence of structural high-frequency vibrations on the rigid body motion has been ignored. The challenge is that satellites and their payloads are extremely sensitive to high-frequency micro-vibrations, which can be particularly problematic for deep space exploration. For example, the pixel offset for a high-resolution camera is limited by the frequency [19]. The higher the observation frequency is, the lower the amplitude of the pixel offset that can be tolerated becomes. Several satellite images are blurred under the influence of micro-vibrations [20]. Thus, the rigid–flexible coupling cannot be ignored and it should be considered in the model of the SOSS.

To fully consider the impact of the resonance on the SOSS system, this study develops a coupling model integrating mechanical, electrical, and control models. We also investigate the impact of rigid–flexible coupling on this model by establishing a rigid–flexible model (R-F-model) and a rigid model (R-model). This study can provide a reference for the design and optimization of the SOSS system.

The following sections of the paper detail our research on the mechanical–electrical–control coupling model. In Section 2, we explain how we derived the transfer function model using the Laplace transform. In Section 3, we present simulations of the R-model and R-F-model acceleration responses under angle tracking mode. We also simulate acceleration responses under different control frequencies to determine the optimal control frequency. Section 4 outlines the experiments we performed and discusses the results. Finally, in Section 5, we summarize our findings and conclude this study.

2. Mechanical–Electrical–Control Coupling Model of the Spaceborne Solar Observation System (SSOS)

The SSOS is an exciting system that comprises a 2D turntable and an observation payload. The 2D turntable contributes majorly to the dynamic response, and it is composed of various components: a motor, a measurement feedback device, a drive controller, and a power supply. Figure 1 depicts the schematic diagram of the system, illustrating that the SSOS is a typical mechanical–electrical–control coupled system.

This section establishes the mechanical–electrical–control coupling model, which consists of mechanical, electrical, and control models. We established the R-F-model and R-model to investigate the influence of the structural flexibility on the model's accuracy.

2.1. Mechanical Model

Based on Newton's Law, the SSOS mechanical dynamics model based on its rotational degrees of freedom is expressed as follows [21]:

$$J_0\frac{d^2\theta}{dt^2} + B\frac{d\theta}{dt} = T_e - T_l, \quad (1)$$

where θ is the rotation angle, J_0 is the moment of inertia of the motor rotor, B is the damping torque coefficient, T_e is the motor torque, and T_l is the inertial torque. Due to the approximately unconstrained rotation of the rotor, the stiffness part is ignored.

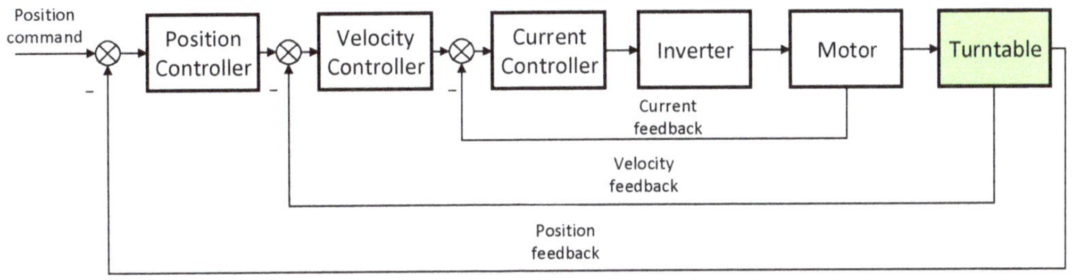

Figure 1. Schematic diagram of the spaceborne solar observation system (SSOS). (⊗ is a summation block and "−" is a minus sign).

The inertial motion of the moving components generates an inertial torque. Therefore, when the structure is assumed to be a rigid body, the angular acceleration of the rotating component generates an inertial torque. The turntable's motion includes its rigid body motion and elastic vibrations when the structure is assumed to be an elastic body. Therefore, the rigid–flexible coupling characteristics should be considered. The inertial torques of the R-model and R-F-model are established to analyze the impact of the structural flexibility, as follows.

(a) Inertial torque of R-model

Assumptions:

For the R-model, the rotating component is assumed to be a rigid body.

The inertial torque of the rigid body can be expressed as:

$$T_l = J_1 \frac{d^2\theta}{dt^2}, \tag{2}$$

where J_1 is the moment of inertia of the SSOS rotating component.

(b) Inertial torque of R-F-model

Assumptions:

For the R-F-model, the rotating component is assumed to be a flexible body, so the flexible vibration and the rotational motion of the rotating component are coupled.

The rigid–flexible coupled dynamics equations are expressed as follows:

$$\begin{bmatrix} J_1 & \boldsymbol{m}_p \\ \boldsymbol{m}_p{}^{\mathrm{T}} & \boldsymbol{I}_n \end{bmatrix} \begin{bmatrix} \frac{d^2\theta}{dt^2} \\ \frac{d^2\boldsymbol{\eta}}{dt^2} \end{bmatrix} = \begin{bmatrix} T_l \\ -2\xi\boldsymbol{\Lambda}\frac{d\boldsymbol{\eta}}{dt} - \boldsymbol{\Lambda}^2\boldsymbol{\eta} \end{bmatrix}, \tag{3}$$

where $\boldsymbol{\eta}$ is the modal coordinate vector, $\boldsymbol{m}_p$ is the modal participation factor matrix, $\boldsymbol{\Lambda}$ is the modal frequency matrix, $\boldsymbol{I}_n$ is the n-order identity matrix, n is the modal order, and ξ is the modal damping ratio. Finite element analysis can be used to determine the modal participation factors and modal frequencies, and engineering experience suggests that the modal damping ratio is between 0.03 and 0.05 [22]. The inertial torque can be obtained by solving Equation (3).

2.2. Electrical Model

This section presents the motor torque, and the voltage equation of the motor circuit is expressed as follows:

$$Ri_q + L_q \frac{di_q}{dt} = u_q - u_c, \tag{4}$$

where R is the circuit resistance, i_q is the current, L_q is the inductor, and u_q is the voltage. Moreover, $u_c = K_e d\theta/dt$ is the counter electromotive force generated by the rotation of the motor, and K_e is the counter electromotive force coefficient. The motor torque is expressed as follows:

$$T_e = K_t i_q, \tag{5}$$

where K_t is the motor torque coefficient. The motor voltage and torque equation derivations are presented in Appendix A [23].

2.3. *Servo Control Model*

The SSOS adopts a three-loop control system, as shown in Figure 1. The speed loop and the current loop are proportional–integral (PI) controllers, with proportional control of the position loop. The control instruction of the position loop is expressed as follows:

$$\omega_r = K_{p\theta} e_\theta, \tag{6}$$

where $e_\theta = \theta_0 - \theta$ is the angle error, θ_0 is the angle instruction, θ is the feedback angle, and $K_{p\theta}$ is the parameter of the position loop controller.

The speed loop control instruction is expressed as follows:

$$i_r = K_{p\omega} e_\omega + K_{i\omega} \int_0^{\Delta t} e_\omega dt, \tag{7}$$

where $e_\omega = \omega_r - \beta\omega$ is the angular velocity error, β is the angular velocity feedback coefficient, ω is the mechanical angular velocity of the motor, $\omega = d\theta/dt$, and $K_{p\omega}$ and $K_{i\omega}$ are the parameters of the speed loop controller.

The control instruction of the current loop is expressed as follows:

$$u_r = K_{pi} e_i + K_{ii} \int_0^{\Delta t} e_i dt, \tag{8}$$

where $e_i = i_r - \alpha i_q$ is the current instruction, α is the current feedback coefficient, and K_{pi} and K_{ii} are the parameters of the current loop controller.

The inverter is simplified as an inertial system, and the relationship between the output voltage and the input voltage is as follows:

$$\frac{du_q}{dt} + \frac{1}{T_{PWM}} u_q = \frac{K_{PWM}}{T_{PWM}} u_r, \tag{9}$$

where K_{PWM} and T_{PWM} are the inverter gain and the time constant, respectively.

2.4. *State-Space Equation of the System*

This section establishes the state-space equation of the system based on the discussions in Sections 2.1–2.3. The mechanical–electrical–control coupling model is obtained based on Equations (1)–(9):

$$\begin{array}{l} L_q \frac{di_q}{dt} + (K_{pi} K_{p\omega} \beta + K_e) \frac{d\theta}{dt} = G - (R + \alpha) i_q \\ J_0 \frac{d^2\theta}{dt^2} + B \frac{d\theta}{dt} = K_t i_q - T_l \end{array}, \tag{10}$$

where

$$G = K_{pi} K_{p\omega} K_{p\theta} e_\theta + K_{pi} K_{i\omega} \int_0^{\Delta t} e_\omega dt + K_{ii} \int_0^{\Delta t} e_i dt. \tag{11}$$

The state-space Equation (10) is expressed as follows:

$$\begin{bmatrix} L_q & & \\ & 1 & \\ & & J_0 \end{bmatrix} \frac{d\mathbf{x}}{dt} = \begin{bmatrix} -(R+\alpha) & 0 & -(K_{pi} K_{p\omega} \beta + K_e) \\ 0 & 0 & 1 \\ K_t & 0 & -B \end{bmatrix} \mathbf{x} + \begin{bmatrix} G \\ 0 \\ -T_l \end{bmatrix}. \tag{12}$$

(a) State-space equation of the R-model

We define $i_q = x_1, \theta = x_2, d\theta/dt = x_3$ and combine Equations (2) and (12). The state-space equation of the R-model can be expressed as follows:

$$\begin{bmatrix} L_q & & \\ & 1 & \\ & & J_0 + J_1 \end{bmatrix} \frac{d\mathbf{x}}{dt} = \begin{bmatrix} -(R+\alpha) & 0 & -(K_{pi}K_{p\omega}\beta + K_e) \\ 0 & 0 & 1 \\ K_t & 0 & -B \end{bmatrix} \mathbf{x} + \begin{bmatrix} G \\ 0 \\ 0 \end{bmatrix}. \quad (13)$$

(b) State-space equation of the R-F-model

We define $i_q = x_1, \theta = x_2, \frac{d\theta}{dt} = x_3, d\boldsymbol{\eta}/dt = [x_4, x_5, \ldots, x_{n+3}]^T$, and $\boldsymbol{\eta} = [x_{n+4}, x_{n+5}, \ldots, x_{2n+3}]^T$, and combine Equations (3) and (12). Then, the state-space equation of the R-F-model can be expressed as follows:

$$\begin{bmatrix} L_q & & & & \\ & 1 & & & \\ & & J_0 + J_1 & m_p & \\ & & {m_p}^T & I_n & \\ & & & & I_n \end{bmatrix} \frac{d\mathbf{x}}{dt} = \begin{bmatrix} -(R+\alpha) & 0 & -K_{pi}K_{p\omega}\beta - K_e & 0 & 0 \\ 0 & 0 & 1 & 0 & 0 \\ K_t & 0 & -B & 0 & 0 \\ 0 & 0 & 0 & -2\xi\boldsymbol{\Lambda} & -\boldsymbol{\Lambda}^2 \\ 0 & 0 & 0 & I_n & 0 \end{bmatrix} \mathbf{x} + \begin{bmatrix} G \\ 0 \\ 0 \\ 0 \\ 0 \end{bmatrix}. \quad (14)$$

Equation (13) lacks some critical parameters, such as the modal frequency, damping, and participation factor, which are essential in actual structures. In contrast, these parameters significantly contribute to the description of the rigid–flexible coupling characteristics in the state-space equation of the model, as shown in Equation (14). Based on the above information, the importance of the R-F-model is indicated by comparing Equations (13) and (14).

2.5. Transfer Function and Simulink Model of SSOS

It is challenging to quickly determine the impact of the relevant factors on the SSOS performance using the state-space model obtained in Section 2.4. Therefore, we established a Simulink model to simplify the process, which required the use of transfer functions obtained by performing the Laplace transform of each module in Sections 2.1–2.3. The mechanical model's transfer function is expressed as follows:

$$\frac{\theta(s)}{T_e(s) - T_l(s)} = \frac{1}{J_0 s^2 + Bs}. \quad (15)$$

The motor transfer function is expressed as follows:

$$\frac{i_q(s)}{u_q(s)} = \frac{1}{R + L_q s} - K_e s\theta(s). \quad (16)$$

The position controller transfer function is expressed as follows:

$$\frac{\omega_r(s)}{e_\theta(s)} = K_{p\theta}. \quad (17)$$

The speed controller transfer function is expressed as follows:

$$\frac{i_r(s)}{e_\omega(s)} = K_{p\omega} + \frac{K_{i\omega}}{s}. \quad (18)$$

The current controller transfer function is expressed as follows:

$$\frac{u_r(s)}{e_i(s)} = K_{pi} + \frac{K_{ii}}{s}. \quad (19)$$

The transfer function of the inverter is expressed as follows:

$$\frac{u_q(s)}{u_r(s)} = \frac{K_{PWM}}{1 + T_{PWM}s}. \tag{20}$$

The transfer function of the inertial torque of the R-model is expressed as follows:

$$\frac{T_l(s)}{\theta(s)} = J_1 s^2. \tag{21}$$

The transfer function of the inertial torque of the model is expressed as follows:

$$\frac{T_l(s)}{\theta(s)} = J_1 s^2 - \boldsymbol{m}_p\left(\boldsymbol{I}_n s^2 + \boldsymbol{\Lambda}^2 + 2\xi\boldsymbol{\Lambda}s\right)^{-1}\boldsymbol{m}_p^T s^4. \tag{22}$$

Note that $T_l(s) = T_{l0}(s) + T_f(s)$, where $T_{l0}(s)$ is the rigid torque, and $T_f(s)$ is the flexible torque. Then, Equation (22) is divided into two parts, where the transfer function of the rigid torque is expressed as follows:

$$\frac{T_{l0}(s)}{\theta(s)} = J_1 s^2. \tag{23}$$

The transfer function of the flexible torque is expressed as follows:

$$\frac{T_f(s)}{\theta(s)} = \boldsymbol{m}_p\left(\boldsymbol{I}_n s^2 + \boldsymbol{\Lambda}^2 + 2\xi\boldsymbol{\Lambda}s\right)^{-1}\boldsymbol{m}_p{}^T s^4. \tag{24}$$

As illustrated in Figure 2, we combined the transfer functions of the modules mentioned above to create two schematic diagrams, where (a) shows the schematic diagram of the R-model, and (b) shows the schematic diagram of the R-F-model.

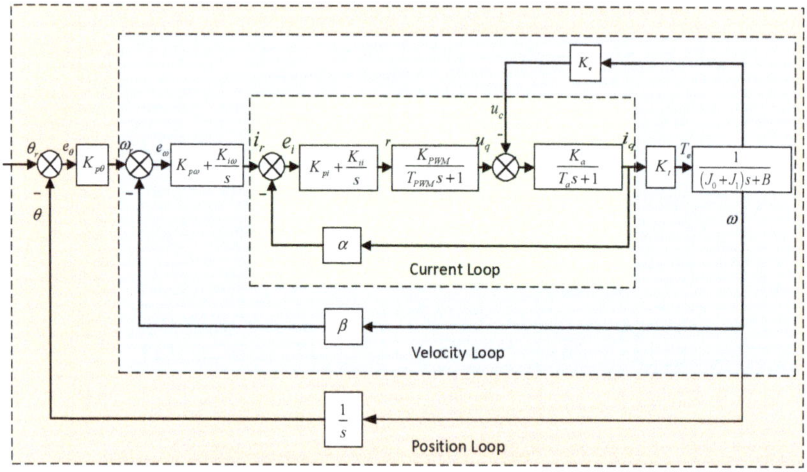

(a)

Figure 2. *Cont.*

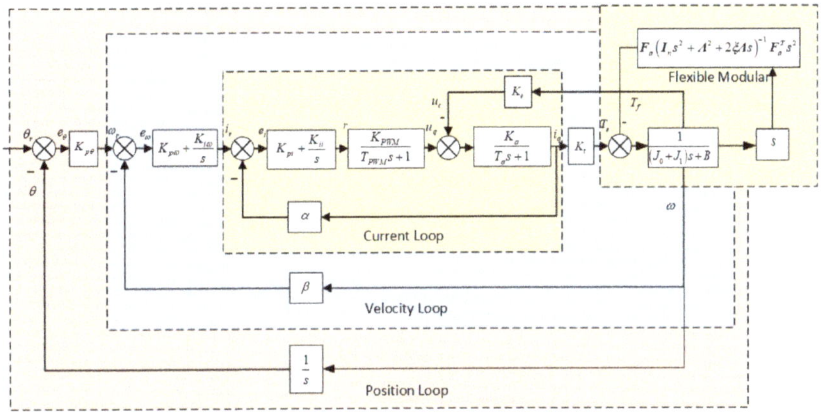

(**b**)

Figure 2. Schematic diagrams of transfer functions of the system (⊗ is a summation block and "−" is a minus sign): (**a**) schematic diagram of the rigid model (R-model), and (**b**) schematic diagram of the rigid–flexible model (R-F-model).

3. Simulations of SSOS

We established a Simulink model to analyze the SSOS dynamic characteristics based on the schematic diagram in Section 2. The natural frequency of the turntable may couple with the excitation frequency, causing resonance. Therefore, it is necessary to calculate the natural frequency of the turntable structure. These external excitations may arise from the control frequency of the motor, motor noise, and nonlinear factors. Thus, we built a finite element method (FEM) model of the 2D turntable to investigate the effect of the rigid–flexible coupling characteristics. Moreover, we obtained the modal frequencies and participation factors [8,9] of the SSOS through FEM analysis and additional calculations.

3.1. Finite Element Method (FEM) Modeling and Simulations of Two-Dimensional (2D) Turntable

3.1.1. FEM Model and Modal Analysis

The 2D turntable has dimensions of 463 mm × 336 mm × 550 mm and weighs 23 kg. The 2D turntable FEM model was established. A mesh convergence study was conducted and mesh size 10 was adopted. The first 10 modes were calculated through modal analysis. Figure 3 depicts the first three modal deformation cloud diagrams, and Table 1 presents the first 10 modal frequencies. The natural frequencies of the 2D turntable are all higher than 100 Hz, and 0.03 is the modal damping ratio.

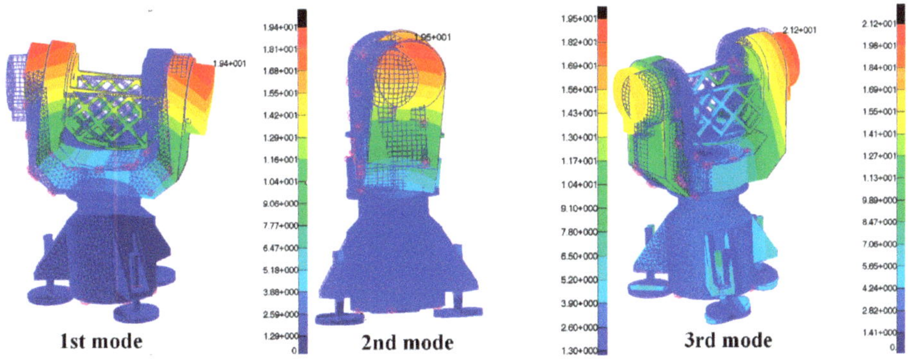

Figure 3. First three modal deformation cloud diagrams.

Table 1. Modal frequency and participation factors.

	1	2	3	4	5	6	7	8	9	10
Modal frequency	119.6	133.0	180.2	314.1	332.6	465.0	472.1	635.3	647.9	667.2
X	−0.026	−1.116	−0.111	0.004	−0.280	0.012	−0.020	−0.003	−0.003	5×10^{-5}
Y	−0.006	0.056	−0.390	-1×10^{-5}	−0.005	0.063	−0.088	−0.005	−0.006	0.001
Z	−1.124	0.024	0.016	0.389	0.003	0.002	−0.001	−0.002	−0.004	0.016

3.1.2. Modal Participation Factors

Since the turntable only moves via rotation, the rotational modal participation factor is expressed as follows:

$$\boldsymbol{m}_p = \sum_{i=1}^{N} m_i \boldsymbol{r}_i \times \boldsymbol{\Phi}_i, \tag{25}$$

where m_i represents the mass of node i, N is the total number of nodes, $\boldsymbol{r}_i \in R_{1\times 3}$ is the vector of node i relative to the reference point, and $\boldsymbol{\Phi}_i \in C_{3\times n}$ represents the modal vector matrix of node i, where n is the modal truncation order. We calculated the generalized modal effective mass and its proportional factor relative to the moment of inertia to determine the modal truncation order. The modal effective mass of the first n modes is expressed as follows:

$$\boldsymbol{M}_{eff}(n) = \boldsymbol{m}_p(n)\boldsymbol{m}_p{}^T(n). \tag{26}$$

The proportional factor of the first n modal effective masses is expressed as follows:

$$factor(n) = \frac{1}{3}\left(\frac{\boldsymbol{M}_{eff11}(n)}{J_{xx}} + \frac{\boldsymbol{M}_{eff22}(n)}{J_{yy}} + \frac{\boldsymbol{M}_{eff33}(n)}{J_{zz}}\right). \tag{27}$$

We computed the first 10 modes' participation factors (unit: $\mathrm{m} \cdot \sqrt{\mathrm{kg}}$) based on Equation (25), and the results are presented in Table 1. Moreover, we calculated the first five modes' proportional factors, based on Equations (26) and (27), as shown in Figure 4, where each column represents the proportional factor of the first n modes. The larger the proportional factor is, the closer it is to the actual vibration.

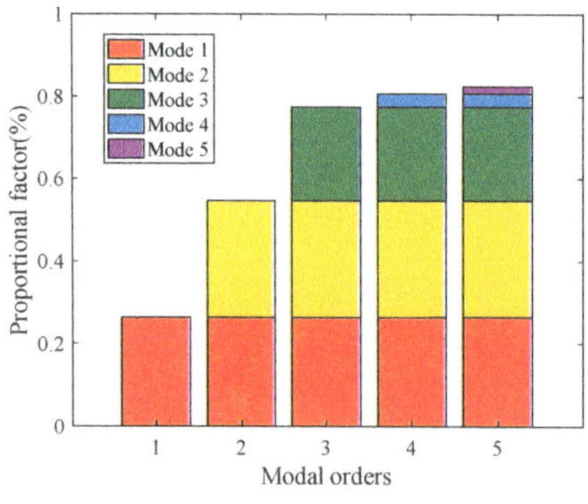

Figure 4. Proportional factors of the first five modal effective masses.

Figure 4 shows that the modal proportional factors of the first three modes reached 80%. As the modal order increased, the proportional factor increased slightly. As a result, the modal truncation order was selected as 3 in the simulation, considering both the accuracy and the calculation complexity.

3.2. Simulation of SSOS

In this section, we established a Simulink model, which was used to perform simulations with the R-model and R-F-model to analyze the impact of rigid–flexible coupling on the turntable's rotation angles and acceleration responses. Thereafter, we performed simulations with the R-model and R-F-model for different control frequencies to obtain the optimal value of the control frequency. The control frequency represents the number of instruction signals per second. In order to avoid resonance caused by coupling between the control frequency and the natural frequency of the turntable, we selected control frequencies that are lower than 1/10 of the turntable's natural frequency.

3.2.1. Simulink Model

The Simulink simulation model of the R-F-model was built based on the schematic diagram in Figure 2. Some parameters of the turntable's pitch axis are presented in Table 2, and the flexible module was disabled when R-model simulations were conducted.

Table 2. System parameters.

Parameter	Value	Parameter	Value
$K_{p\theta}$	38.18	K_e	0.19
$K_{p\omega}$	18.84	K_t	0.21
$K_{i\omega}$	6.27	L_q	9.6×10^{-4} (H)
K_{pi}	2.02	R	1.56 (Ω)
K_{ii}	315.13	T_a	6.17×10^{-4}
K_{PWM}	13.86	K_a	0.64
T_{PWM}	0.001	J_1	0.036 (kg·m^2)
B	0.0033	J_0	7.21×10^{-5} (kg·m^2)

3.2.2. Acceleration Response Simulation under Angle Tracking Mode

The SSOS mainly operates in an angle tracking mode, and its axis always points to the Sun. The solar altitude from 6:00 to 6:01 on March 24 was selected based on the simulation input data, and the control frequency was 1 Hz. Figure 5a shows the tracking angle of the R-F-model and R-model, where the red solid line is the curve of the R-F-model. The green dotted line is the curve of the R-model, and the blue dashed line is the input angle. Figure 5b depicts the R-F-model and R-model angle tracking errors. Figure 5 shows that the angles of the R-F-model and R-model are similar, and the angle tracking errors are less than 5″. Based on this figure, theoretically, the turntable's structural flexibility with high natural frequencies has little influence on the angle tracking.

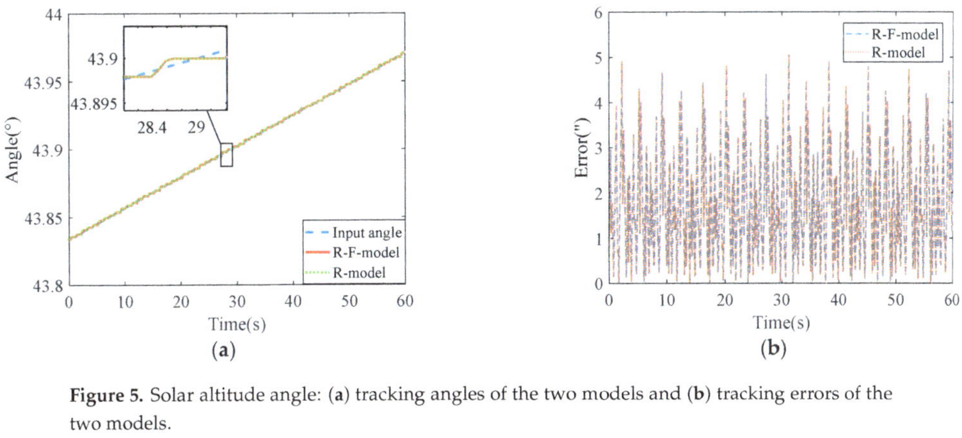

Figure 5. Solar altitude angle: (**a**) tracking angles of the two models and (**b**) tracking errors of the two models.

Figure 6 shows the acceleration spectral curves, indicating that the acceleration spectra of the R-F-model generated peak responses in high frequencies caused by structural flexibility compared with the R-model. Eventually, the extra responses caused by the structural flexibility would affect the SSOS. The simulation results indicate that the rigid–flexible coupling characteristic has little influence on the angle tracking of the simulation model. However, this rigid–flexible coupling affects the high-frequency distribution of the acceleration spectrum. This mechanism is essential for a spaceborne SSOS, as micro-vibrations at a high frequency may negatively affect the satellites' stable precision and the image definition of optical cameras. Therefore, this issue requires attention, which is verified in Section 4. Overall, the proposed coupling model can describe the difference between the R-model and the R-F-model.

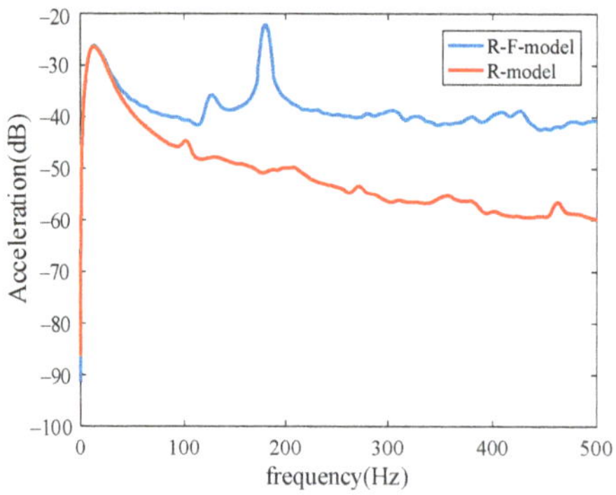

Figure 6. Acceleration spectra.

3.2.3. Acceleration Response Simulations under Various Control Frequencies

We input the angle commands at different frequencies to conduct dynamic simulations, selecting the solar altitude from 6:00 to 6:01 on March 24 as the simulation input data, where the control frequencies were 0.2, 1, 2, 5, and 10 Hz. The maximum angle tracking error curves under different control frequencies were obtained through simulations, as shown in Figure 7, where the solid blue line is the R-F-model's simulation result, and the dashed red line is the R-model's simulation result. These results indicate that the angle tracking error decreases as the control frequency increases. The maximum angle tracking error is less than 5″, and it satisfies the angle tracking requirement when the control frequency is higher than 1 Hz.

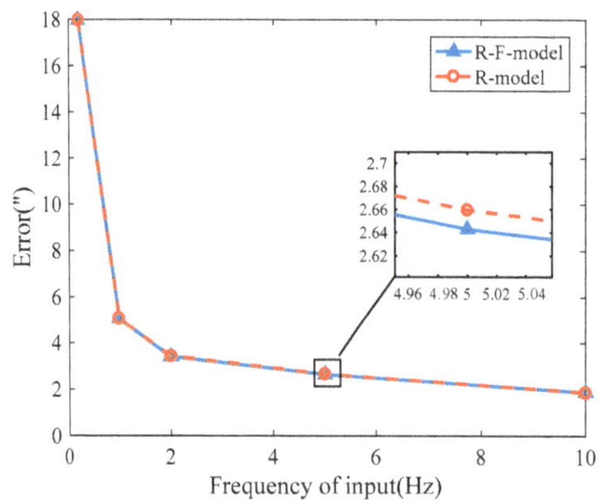

Figure 7. Maximum angle tracking errors under different control frequencies.

Figure 8 shows the acceleration spectral curves under different control frequencies, where (a) shows the acceleration spectral curve of the R-F-model, and (b) shows the acceleration spectral curve of the R-model. The third modal frequency of the turntable is 180.2 Hz (Table 1), which amplifies the vibration of the turntable, resulting in a peak acceleration at about 180 Hz in Figure 8. Moreover, the acceleration peak value increases as the control frequency increases. We calculated the acceleration response's root mean square (RMS) values under different control frequencies, as shown in Figure 9a. Interestingly, the RMS of the acceleration response of the R-F-model is two times larger than that of the R-model. The RMS of the acceleration response shows an upward trend as the control frequency increases. To obtain the optimal control frequency of the SSOS, the acceleration RMS and maximum angle tracking error were plotted, as shown in Figure 9b. The acceleration RMS and the angle tracking error shows opposite trends. When the input frequency was 2 Hz, the angle tracking error is less than $5''$, lower than the precision SSOS requirement. The angle tracking error tends to be convergent, and the acceleration RMS is relatively small at the same time. As a result, 2 Hz was selected as the control frequency when developing the SSOS.

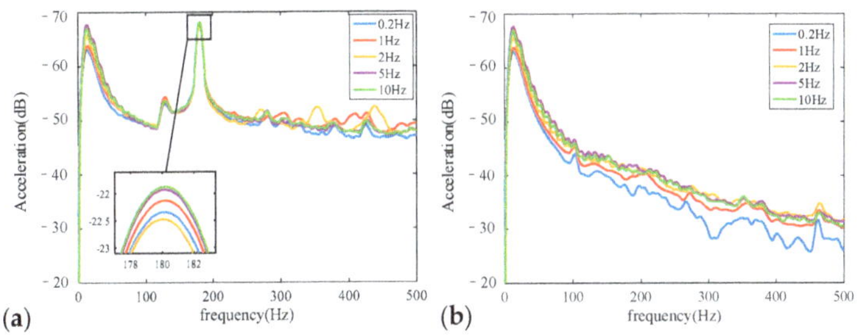

Figure 8. Acceleration spectral curves: (**a**) acceleration spectra of the R-F-model and (**b**) acceleration spectra of the R-model.

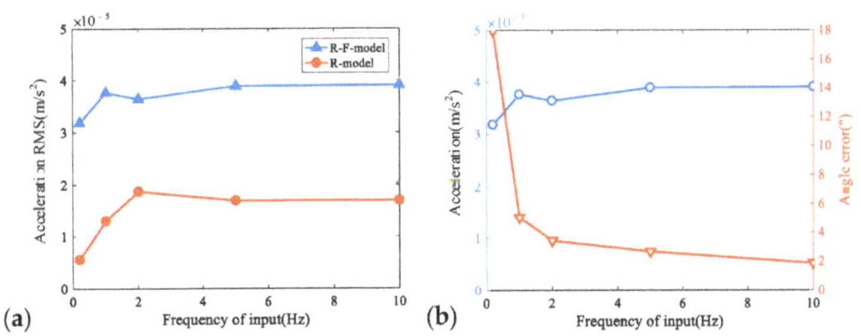

Figure 9. (**a**). Acceleration root mean square (RMS) under different control frequencies. (**b**) Acceleration RMS and maximum angle tracking error of R-F-model (the blue circles represent the acceleration RMS, and the red triangles represent the angle tracking error).

4. Experiments

We set up an SSOS experimental system to further prove the effectiveness of the coupling model and the research value of flexible features in high-accuracy spaceborne tracking instruments.

4.1. Experiment Setup

Figure 10a shows that the experimental system comprised a 2D turntable, three accelerometers, a signal acquisition system, a controller, a power supply, and a computer with control software. The type of the controller is Elmo's BMTWID20SE. The accelerometers (P1,

P2, and P3) were mounted on the rotating component of the 2D turntable. The experimental setup is shown in Figure 10b. The angle command curve was input to the control software to realize a continuous observation of the Sun, and the control frequencies were set to 1, 2, 5, and 10 Hz.

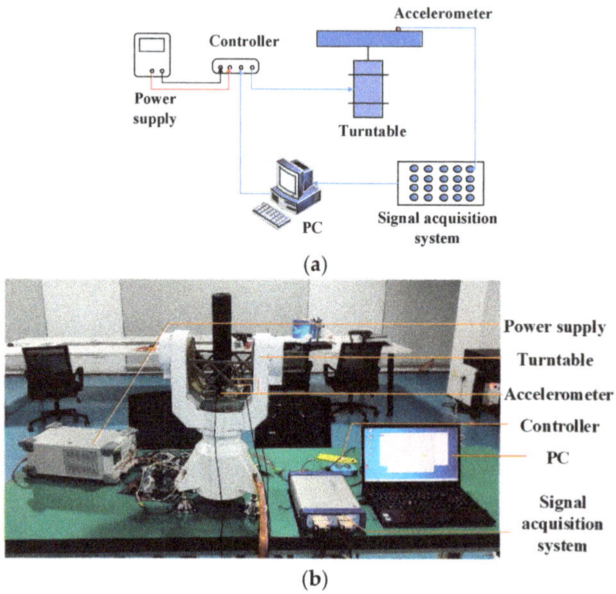

Figure 10. Experimental system: (**a**) schematic diagram and (**b**) experimental site.

4.2. Processing and Analysis of Experimental Data

Figure 11 shows the acceleration spectral curves, which are measured using accelerometers. The control frequency is 2 Hz, where the blue solid, red dashed, and yellow dotted-dashed curves represent the acceleration results measured using P1, P2, and P3, respectively. The purple dotted curve shows the R-F-model simulation result, while the green curve shows the R-model simulation result. The measured acceleration results show that there are peaks between 100 and 200 Hz, indicating that resonances are generated by the coupling of the turntable's structural modes and the excitation frequency. The R-F-model simulation result replicates the 2D turntable's vibration characteristics in the high-frequency regions, and its predictions were closer to the experimental results than those of the R-model. The results showed that the mechanical–electrical–control coupling model could better reflect the SSOS dynamic characteristics than the traditional rigid-body model when considering rigid–flexible coupling characteristics.

We conducted tests to measure the acceleration responses at different control frequencies (1, 2, 5, and 10 Hz). In Table 3, the experimental results of the acceleration peak values and the corresponding frequencies are compared with those obtained with the R-F-model. The average error of the frequencies is approximately 10.3%, while the average error of the peak value is approximately 29.9%. The frequency errors of the R-F-model are smaller than the ones of the R-model. However, the peak value errors are relatively large because it is difficult to simulate all system characteristics, such as the structural damping ratio and the nonlinear factors. However, the simulation model still simulates the high-frequency vibrations of the system. The simulation results can help to predict the micro-vibration frequencies and provide suggestions for vibration isolation design.

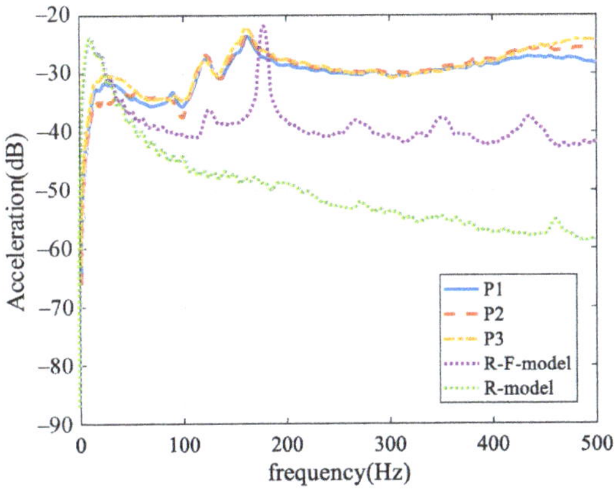

Figure 11. Comparison of experimental and simulation results of acceleration.

Table 3. Frequencies and peak values of the simulations and experiments.

	Frequency of Simulation (Hz)	Frequency of Experiment (Hz)	Frequency Error	Peak Value of Simulation (m/s^2)	Peak Value of Experiment (m/s^2)	Peak Value Error
1 Hz	177.5	161.3	10.0%	0.0077	0.0055	44.0%
2 Hz	177.1	160.8	10.1%	0.0081	0.0063	28.6%
5 Hz	177.3	160.8	10.3%	0.0092	0.0073	26.0%
10 Hz	177.6	160.5	10.7%	0.0093	0.0077	20.8%

5. Conclusions

The purpose of this study is to investigate how the elastic characteristics and control frequency impact micro-vibration acceleration responses in a high-precision SSOS. To achieve this goal, we developed the R-F-model and R-model of the SSOS, and analyzed the effect of rigid–flexible coupling by establishing the state-space model. We then built a Simulink model using the Laplace transform to derive a transfer function model. By simulating the R-F and R-model acceleration responses under the angle tracking mode, we observed differences when flexible features were considered. Specifically, the RMS of the model's acceleration response was found to be two times larger than that of the R-model. We also conducted simulations with various control frequencies to determine the optimal value. Finally, we conducted experiments and validated the theoretical model and simulation results using measured data under different control frequencies (1, 2, 5, and 10 Hz). Our findings suggest that the flexibility of the structure affects the high-frequency distribution of the acceleration spectra and poses threats to the high tracking accuracy. This study highlights the significance of elastic structures in dynamic response and provides a valuable reference for the design and micro-vibrations of spacecraft structure systems.

Author Contributions: Conceptualization, L.Y.; methodology, Y.W.; software, Y.W.; validation, L.W.; investigation, Y.C.; writing—original draft preparation, Yan-song Wang and Lin Yang.; writing—review and editing, L.W. and Y.C.; visualization, Y.W.; supervision, Y.C.; funding acquisition, L.Y. All authors have read and agreed to the published version of the manuscript.

Funding: This work was supported by the National Natural Science Foundation of China (No. 12203032) and the Natural Science Foundation of Shandong Province (ZR2022QA030).

Data Availability Statement: The data presented in this study are available on request from the corresponding author.

Conflicts of Interest: The authors declare no conflicts of interest.

Appendix A

In the d- and q-axis coordinate systems [23], the motor magnetic linkage equations are expressed as follows:

$$\begin{array}{l} \Psi_d = L_d i_d + \Psi_m \\ \Psi_q = L_q i_q \end{array}, \tag{A1}$$

where Ψ_d and Ψ_q are the magnetic linkages along the d-axis and q-axis, respectively; L_d and L_q are the inductances along the d-axis and q-axis, respectively; i_d and i_q are the currents along the d-axis and q-axis, respectively; and Ψ_m is the magnetic linkage of the permanent magnet.

The voltage equations of the motor are expressed as follows:

$$\begin{array}{l} u_d = Ri_d + \frac{d\Psi_d}{dt} - \omega_e \Psi_q = Ri_d + L_d \frac{di_d}{dt} + \frac{d\Psi_m}{dt} - \omega_e \Psi_q \\ u_q = Ri_q + \frac{d\Psi_q}{dt} + \omega_e \Psi_d = Ri_q + L_q \frac{di_q}{dt} + \omega_e \Psi_d \end{array}, \tag{A2}$$

where u_d is the voltage along the d-axis, u_q is the voltage along the q-axis, ω_e is the electrical angle, $\omega_e = p_n \frac{d\theta}{dt}$, p_n represents the pole pairs of the motor, θ is the motor angle, and R is the resistance.

The motor torque equation is expressed as follows:

$$T_e = p_n \left[\Psi_m i_q + (L_d - L_q) i_d i_q \right], \tag{A3}$$

where the first term is the magnetic excitation force, and the second term is the magnet restraining torque.

Using the i_d= 0 control strategy, the required current is the minimum, and the maximum torque can be obtained per unit current. In this case, the magnetic linkage is expressed as

$$\begin{array}{l} \Psi_d = \Psi_m \\ \Psi_q = L_q i_q \end{array}. \tag{A4}$$

and the voltage equations are expressed as

$$\begin{array}{l} u_d = \frac{d\Psi_m}{dt} - \omega_e L_q i_q \\ u_q = Ri_q + L_q \frac{di_q}{dt} + \omega_e \Psi_m \end{array} \tag{A5}$$

References

1. Li, C.; Chen, Y.; Ni, S.; Tan, B.; Zhang, Z. PIC Simulation of Double Plasma Resonance and Zebra Pattern of Solar Radio Bursts. *Astrophys. J. Lett.* **2021**, *909*, L5. [CrossRef]
2. Ni, S.; Chen, Y.; Li, C.; Zhang, Z.; Hosseinpour, M. Plasma Emission Induced by Electron Cyclotron Maser Instability in Solar Plasmas with a Large Ratio of Plasma Frequency to Gyrofrequency. *Astrophys. J. Lett.* **2020**, *891*, L25. [CrossRef]
3. Anashin, V.; Chubunov, P.; Koziukov, A.; Konyukhov, A.; Protopopov, G. Challenges and Approaches to Radiation Hardness Control of Electronic Components to In-Space High-Energy Particles Exposure. In Proceedings of the 2018 20th International Symposium on High-Current Electronics (ISHCE), Tomsk, Russia, 16–22 September 2018.
4. Nagabhushan, V.; Fitz-Coy, N.G. On-orbit jitter control in momentum actuators using a three-flywheel system. *Acta Astronaut.* **2014**, *95*, 61–81. [CrossRef]
5. Yang, L.; Wang, Y.; Wei, L. A novel identification method for micro-vibration analysis of reaction wheel assembly. *Acta Astronaut.* **2022**, *196*, 94–106. [CrossRef]
6. Lin, L.; Tan, L.; Lin, K.; Dong, W.; Yang, H. The influence of flywheel micro vibration on space camera and vibration suppression. *Mech. Syst. Signal Process.* **2018**, *100*, 360–370.
7. Sattar, M.; Wei, C. Analysis of coupled torsional disturbance behavior of micro-stepped solar array drives. *J. Sound. Vib.* **2019**, *442*, 572–597. [CrossRef]
8. Kouritem, S.A.; Elshabasy, M.M.Y.B. Tailoring the panel inertial and elastic forces for the flutter and stability characteristics enhancement using copper patches. *Compos. Struct.* **2021**, *274*, 114311. [CrossRef]
9. Bani-Hani, M.A.; Almomani, A.M.; Aljanaideh, K.F.; Kouritem, S.A. Mechanical Modeling and Numerical Investigation of Earthquake-Induced Structural Vibration Self-Powered Sensing Device. *IEEE Sens. J.* **2022**, *22*, 19237–19248. [CrossRef]
10. Yang, L.; Wang, Y.; Wei, L.; Chen, Y. Identification of reaction wheel assembly disturbances based on variable modal frequencies. *Mech. Syst. Signal Process.* **2023**, *205*, 110868. [CrossRef]

11. Wang, Z.; Wang, X.; Wang, T.; Zhang, B.; Ma, H. Accuracy Analysis and Motion Control of Two-axis Nonmagnetic Turntable Based on Ultrasonic Motor. *Mechanika* **2020**, *26*, 221–230. [CrossRef]
12. Sun, G.; Wang, M.; Yao, X.; Wu, L. Fault detection of switched linear systems with its application to turntable systems. *J. Syst. Eng. Electron.* **2011**, *22*, 120–126. [CrossRef]
13. Liu, Y.G.; Zhang, H.Z.; Hu, H.; Li, Y.L. Design and Finite Element Analysis of Large Target Tracking Turntable. In Proceedings of the Chinese Guidance, Navigation and Control Conference, Nanjing, China, 12–14 August 2016; IEEE: Piscataway, NJ, USA, 2016; pp. 2043–2047.
14. Zhang, Y.; Liu, Z.; Li, Z. Modal analysis of collimation frame for Space two-dimensional turntable. In Proceedings of the 5th International Conference on Material Science and Engineering, Guangzhou, China, 24–26 June 2016.
15. Zou, D.; Li, Z.; Liu, Z.; Cui, K.; Zhang, Y.; Zhou, L. Modal simulation and experimental verification of Space-borne two dimensional turntable. In Proceedings of the Applied Optics and Photonics China (AOPC2017), Beijing, China, 4–6 June 2017; SPIE: Bellingham, WA, USA, 2017.
16. Sun, L.; Ma, H.; Zhao, Z. Optimization Design of Aerial Work Vehicle Turntable Based on Topology Optimization. In Proceedings of the 2nd IYSF Academic Symposium on Artificial Intelligence and Computer Engineering, Xi'an, China, 8–10 October 2021; Qin, W., Ed.; SPIE: Bellingham, WA, USA, 2021.
17. Li, J. Optimization Design of the Precision Optoelectronic Tracking Turntable Frame. In Proceedings of the 5th International Symposium on Advanced Optical Manufacturing and Testing Technologies, Dalian, China, 26–29 April 2010; Luo, X., VanFreymann, G., Eds.; SPIE: Bellingham, WA, USA, 2010.
18. Yao, H.; Huang, Y.; Ma, W.; Liang, L.; Zhao, Y. Dynamic Analysis of a Large Deployable Space Truss Structure Considering Semi-Rigid Joints. *Aerospace* **2023**, *10*, 821. [CrossRef]
19. Chen, S.B.; Xuan, M.; Xin, J.; Liu, Y.; Gu, S.; Li, J.; Zhang, L. Design and experiment of dual micro-vibration isolation system for optical satellite flywheel. *Int. J. Mech. Sci.* **2020**, *179*, 105592. [CrossRef]
20. Yang, L.; Wang, Y.S. Study of the multi-excitation decoupling optimization method for the integrated structure of a high-resolution optical satellite. *J. Vib. Control* **2023**, *29*, 2086–2101. [CrossRef]
21. Taylor, J.E. *Theoretical Mechanics*; Legare Street Press: Charleston, SC, USA, 2022.
22. Heylen, W. *Modal Analysis Theory and Testing*, 2nd ed.; Katholieke Universiteit Leuven, Departement Werktuigkunde Press: Leuven, Belgium, 1998.
23. Zhu, Q.; Xiong, L.; Liu, H. A robust speed controller with smith predictor for a PMSM drive system with time delay. *Int. J. Control Autom. Syst.* **2017**, *15*, 2448–2454. [CrossRef]

aerospace

MDPI

Article

Basic Orbit Design and Maneuvers for Satellite Constellations Deployed Using Momentum Exchange Tethers

Ben Campbell [1,*] and Lawrence Dale Thomas [2]

1 Department of Mechanical and Aerospace Engineering, University of Alabama in Huntsville, Huntsville, AL 35899, USA
2 Department of Industrial & Systems Engineering and Engineering Management, University of Alabama in Huntsville, Huntsville, AL 35899, USA; ldt0001@uah.edu
* Correspondence: bac0038@uah.edu

Abstract: This article describes a new alternative approach to satellite constellation deployment by incorporating momentum exchange tethers (METs). Traditional methods of deploying satellite constellations have limitations, typically involving costly propulsion systems and extended dispersion times. METs offer a novel solution by efficiently transferring momentum between tethered objects, reducing the need for onboard propellants and streamlining the deployment process. This article discusses orbit design and maneuvers for different mission architectures using asymmetrical and symmetrical tether release techniques to deploy satellites into designated orbits. In addition, a short walkthrough of designing one possible constellation is given, showing how quickly a MET-deployed constellation can be established in low Earth orbit (LEO). This work contributes to ongoing research investigating the applicability of METs in satellite constellation deployments, which could potentially be a new opportunity for MET technology to start seeing routine usage in the space environment, and also enable new constellation architectures that have not yet been realized.

Keywords: satellite constellation; momentum exchange tether; satellite deployment; orbit design; mission design; small satellite

1. Introduction

The space industry is experiencing a surge in new satellite constellations, some of which are much larger than those originally conceived in the 1990s, creating a new category commonly referred to as "mega-constellations" [1]. Despite advancements in satellite technology through the decades, the process of deploying these constellations still relies on traditional methods, such as orbital transfer vehicles or significant propulsion systems onboard each satellite, which still pose the same challenges for constellation operators, largely revolving around dispersing their satellites in orbit [2,3]. Momentum exchange tethers (METs), a non-traditional propulsion method, could be a solution to alleviate some of the difficulties associated with satellite deployment and dispersion. MET-equipped deployers or orbital transfer vehicles (OTVs) could be made to rotate payloads on tethers and release them, providing rapid ΔV changes without necessarily requiring propellant-based systems. Early research insights have determined that they could simplify and expedite constellation deployment, reducing deployment times from years to weeks.

For this new concept blending the practices of satellite constellations and METs, this article will explain some of the basic concepts and considerations involved, which could be used to support potential future constellations. While not an exhaustive explanation of all factors to consider, it nonetheless provides a collection of building blocks to help start designing such a constellation. This article includes insights specifically on tether-deployed constellation orbit design and a conceptual deployment walkthrough. After introducing the existing technology, challenges, and significance of this research, five different questions for orbit design are discussed, the answers to which help narrow down options that can be selected for different constellation architectures. One of these options is then selected for a conceptual constellation of 12 satellites, deployed using one of the simplest approaches described here.

2. Existing Technology and Challenges

2.1. Satellite Constellations

A satellite constellation is a group of coordinated satellites designed for specific tasks, like telecommunications or Earth observation. Recent interest in constellations for applications like these has led to plans by companies like SpaceX, OneWeb, and Amazon to develop and launch large constellations, with others in development [4]. Additionally, several other constellations are currently being considered and developed by government-backed research and defense organizations. These systems are typically placed in orbit either one by one on separate launch vehicles, or in batches on a single launch with additional propulsion for spreading them out later. For missions involving multiple satellites per launch, the usage of OTVs is an additional option to consider. Regardless, these options are costly in terms of finances, materials, and time, especially for smaller organizations and academia [2,4,5]. Each approach presents challenges for constellation operators with factors that can be complex to consider and settle on a solution for, and further become even more difficult to handle when dealing with logistics and execution. The ongoing research described in this article is primarily focused on constellations constructed from batches of smaller ride-sharing satellites as opposed to larger single satellites with dedicated launchers. Some challenges associated with ride-sharing satellite delivery for satellite operators include the following:

- Flying with an OTV spacecraft offers much more flexibility, but also relies on interaction with the OTV provider, and can again be costly due to the inherent specialization of the orbit delivery process of multiple onboard spacecraft. If the propulsion system included onboard the OTV is electric based or some other low-thrust system, the time to fully deploy each constellation spacecraft can be lengthy due to the low acceleration and resultant time to achieve ΔV to spread properly.
- Including onboard propulsion systems in a single-launch constellation offers more flexibility, but also comes with design and operation challenges. Some teams may not be able to handle the complexity, mass, and cost of these systems. Ensuring their reliability and on-orbit operation is another challenge on top of this.
    - While chemical propulsion is impulsive and can be used for rapid deployments, their relatively inefficient usage of mass can make them incompatible with some small satellite designs.
    - More efficient propulsion solutions like electric propulsion are available, and are more commonly used on constellations, but require more extensive dispersion times, typically in the order of months.
    - For even smaller satellites where chemical and electric propulsion systems are too large, differential drag techniques can be used for dispersion, but this method's performance is irregular, limits application to low altitudes with enough atmospheric interaction, and contributes to orbit decay, reducing satellite lifetime.

2.2. Momentum Exchange Tethers

Tether technology in space is not a very common practice, but it has been seen on a variety of missions. The first demonstration of this technology dates back to NASA's Gemini XI mission in 1967, which incorporated a 30 m tether connecting the crewed Gemini spacecraft with the Agena target vehicle. This system rotated and induced a light artificial gravitational acceleration for roughly 3 h, then disconnected and sent Gemini and Agena in separate opposing directions as part of a tether dynamics experiment [6]. Since this time, most missions incorporating tethers have been focused on tether dynamics similar to Gemini XI, electrodynamic tethers which rely on interactions with Earth's magnetic field for propulsion or power generation purposes, or "space webs" where payloads are connected by tethers for formation flying purposes [7,8]. Momentum exchange tethers have seen some, but ultimately little usage in the space environment, and have also been used for individual deployments rather than multiple, as is being investigated by the research this article introduces.

METs are a space propulsion method that rely on the exchange of momentum between two tether-connected objects, typically a payload and a deployer or counterweight. They are not to be confused with electrodynamic tethers or space webs, which operate in different ways. The MET concept involves the mechanical rotation and separation of a spinning system, converting angular momentum into linear momentum upon release, which can alter the objects' velocities and positions [7,9,10]. Rotation in MET systems could be induced using either propulsion systems like cold gas thrusters (as used on Gemini XI), or electric motorized systems [11,12]. Operating this system is similar to a shepherd's sling spinning and releasing a stone, like in the Biblical story of David and Goliath. Generalized visualizations of such systems' operations are included later in this article. A key advantage of METs is their minimal fuel consumption (or complete avoidance thereof), making them attractive for missions with fuel supply and storage constraints. Space tethers can serve various purposes beyond propulsion, including station-keeping, attitude control, power generation, and passive deorbiting, although this article focuses solely on the tether-based deployment of constellations, particularly orbit and spacecraft design. Despite their potential benefits, METs require precise spacecraft-tether coordination and can be affected by environmental factors like solar wind, atmospheric drag, thermal effects, and micrometeoroid impacts over extended periods. Their payload capacity is constrained by mass and geometry, limiting their suitability for certain high-load missions or missions requiring higher spin rates or torques than can be achieved by feasible drive systems [13].

METs have been tested in a very limited capacity in past space missions, and traditional propulsion systems have remained dominant due to their broader applicability [7,9]. As a result, the technology readiness level (TRL) of MET technology can be estimated to be around level 6 or 7, where the technology has been demonstrated in the operational environment of space, but has not yet been applied outside of experimental research and development. METs that have flown specifically for momentum exchange purposes were designed as exceedingly long tethers, such as the SEDS-1 (1993) and SEDS-2 (1994) missions which utilized 20 km tethers [14], and the YES-2 (2007) mission which utilized a 30 km tether [15]. These missions involved the usage of tether systems to deploy only individual payloads. Many other efforts have conceptualized tether systems of similar or even larger scale for applications such as skyhooks or even space elevators, but most of these are not quite yet feasible with present resources.

This article introduces one potential application that could benefit from using tethers which has not been previously researched: satellite constellations deployed by tethers. Tether usage here occurs early in the mission lifetime, with a rotation and release of multiple payloads into desired destination orbits. This research work aims to keep the design of any applied tether systems within immediate feasibility by utilizing conventional materials and significantly shorter tether lengths on the order of meters instead of kilometers, and also put theory into practice by demonstrating this technology in the space environment.

Currently, the TRL of a MET-based constellation deployment system can be estimated to be near a level 3 or 4 at the time of this article's writing (late 2023/early 2024), but the overall research effort aims to raise this to the level 6–7 range. This TRL-raising work will culminate into a technology demonstration mission called ADRASTEA that will demonstrate symmetric and asymmetric tether release operations in the space environment following a launch partnered with NASA scheduled for August 2024. Post-flight results from this mission, when they are available, will be discussed in future publications.

3. Significance of Rapid Tether-Based Constellation Deployment

Tether-based deployment can offer two key advantages: quicker commissioning timelines and, as a result, financial benefits as well. Faster timelines result from shorter post-launch dispersion, and potentially simpler technology development depending on the constellation producer compared to traditional propulsion system development. Shortened timelines are produced with the utilization of pseudo-impulsive release maneuvers that deliver velocity change quicker than low-thrust electric propulsion or passive drag, like that used on multiple constellations such as those by Starlink/SpaceX and Planet Labs. This shortened time from launch to operation in space can allow for faster profit realization by getting technologies working in space sooner. Cost savings can potentially come from eliminating the need for more complex propulsion systems and onboard propellant for initial post-launch ΔV boosts (although smaller systems for later station-keeping may be recommended). Tether systems can also be stowed compactly in form factors comparable with the form factors of traditional propulsion systems, but also do not require expendable propellant, making them suitable for small spacecraft like CubeSats.

This potential for faster and more affordable constellation establishment could see major uses both around and beyond Earth. If current constellations are ever damaged at large or made unusable, such as by large space weather events or human-caused coordinated strikes, these assets are costly to bring back online, and can suffer periods of irregular or unavailable coverage while in recovery. Using a tether-based approach could minimize the internal cost and timelines of constellation recovery to rapidly respond to network damage. These satellites could also be used for rapid deployment of private networks, where coverage could be quickly needed over specific regions. Expectedly, constellation establishment by traditional means could take even longer if trying to work around another celestial body such as the Moon or Mars. If a tethered framework were to be applied to future constellations, it could possibly be a significant accelerator in developing the infrastructure needed in establishing manned settlements beyond Earth.

4. Materials and Methods for Designing Tether-Deployed Constellations

4.1. Conceptual Design

4.1.1. Conceptual Example "SwingSat"

Current research is exploring the potential of METs to enhance satellite constellation deployment. The scope of this work primarily revolves around three major task areas given below.

1. What are some best practices for performing satellite constellation deployments with tethers?
2. How should constellation spacecraft made for deployment by tethers be designed?
3. How does the performance of tethered constellation deployment compare against traditional deployment methods?

Of these, task areas 1 and 2 have received the most effort to date, and work is ongoing relating to task area 3. A summary of results from investigating task areas 1 and 2 make up the majority of the rest of this article, focusing on orbital maneuvers and spacecraft design. Future work will be published upon further progress in all three task areas.

For this article, a conceptual satellite constellation deployed by tethers will be discussed, and referred to by the name "SwingSat". This constellation is not a system currently being produced, but is solely for referring to a MET-based system when discussed in parallel with existing systems. It is primarily to help give a name and a visual representation to the reader for a system utilizing the concepts discussed throughout this article. An example of this system has been produced in the form factor of a 3U CubeSat, which is composed of two 1.5U halves, one being a "driver" hosting a motor drive system that rotates the "deployer" half, which hosts one or more sets of symmetrical tether deployment and release mechanisms for a set of small payloads. In operation, this system could spin up both halves in opposing directions (conserving angular momentum), extend the tethers (in this case, tape measure booms), then upon proper alignment, speed, and timing, release the tethers and payloads. A representation of this system is shown in Figure 1. Note that this design is purely conceptual and as a visual aid to assist in explaining the concept.

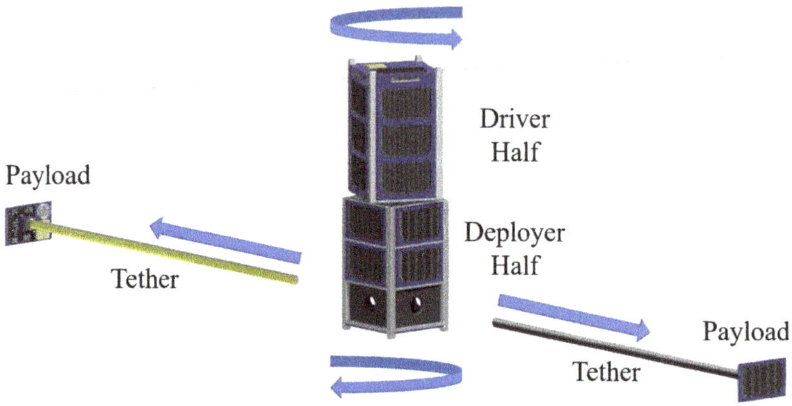

Figure 1. SwingSat concept as a 3U CubeSat.

4.1.2. Symmetric versus Asymmetric Release

The system uses paired tethers for stability upon release, as symmetrical release simplifies the return to a non-spinning state after deployment. Symmetrical release in this context involves releasing two opposing tethered objects simultaneously, which, if needed, can allow for payloads and potential counterweights of different masses and geometries as long as the center of mass remains near the central axis of rotation. The term symmetric is used to describe the geometry of the overall system, with a deployer in the center and two identical tethered payloads on either side of the deployer. Asymmetrical release involves letting go of just one tethered payload rather than two, which introduces complex disturbances due to a rapid change in the system's center of mass on one side, necessitating corrections with attitude control systems, which can be non-ideal if needing to restabilize between multiple sequenced deployments. An asymmetrical release can also alter the deployer's orbit after releasing a payload, which could require propulsive corrections depending on the nature of the mission. Symmetrical release, however, preserves the deployer's position, as the releases of opposing masses cancel out orbit-changing forces, assuming the released payloads' masses and tether lengths are identical.

With symmetrical releases, one aspect that must be considered is that releasing payloads into opposing directions means releasing payloads into different orbits. These two orbital planes of prograde and retrograde released payloads can be initially approximated as coplanar, but over significantly long periods, the effects of nodal precession can become visible due to the minute differences in altitude and velocities of the satellites in the two different planes. If these satellites include onboard propulsion for station-keeping; however, nodal precession effects can be easily mitigated.

4.1.3. Tether-Payload Release vs. Tether-Deployer Release

When considering tether release methods for the payloads, the tether can utilize a disconnection between either the tether and the payload, or between the tether and the deployer. Visual depictions of the pre-separation, separation, and post-separation states of such systems, for asymmetric and symmetric configurations, are given in Figures 2–13. It should be noted that these figures are generalizations of each release type, and the true exact locations of centers of mass are dependent on spacecraft design and may shift.

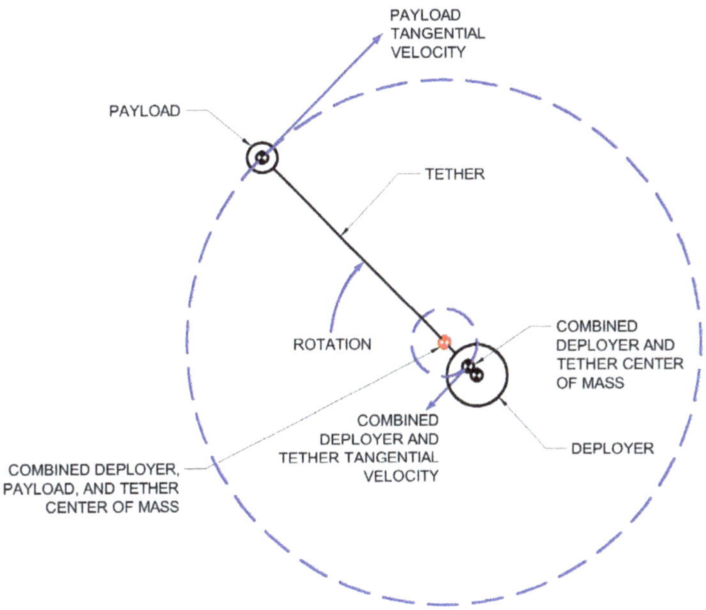

Figure 2. Asymmetric Tether-Payload Release Pre-Separation State.

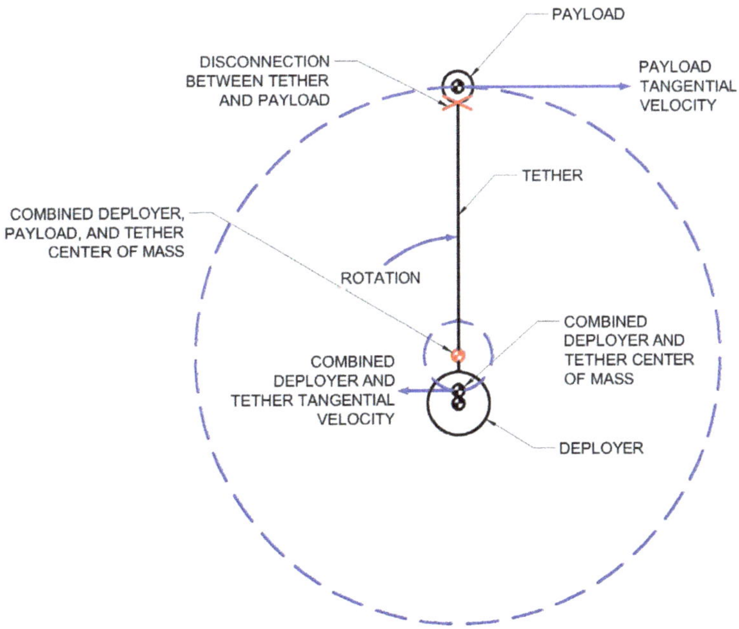

Figure 3. Asymmetric Tether-Payload Release Separation State.

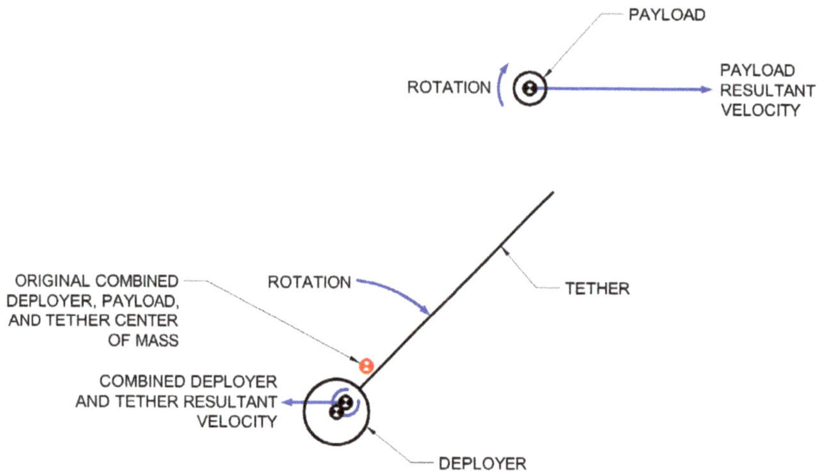

Figure 4. Asymmetric Tether-Payload Release Post-Separation State.

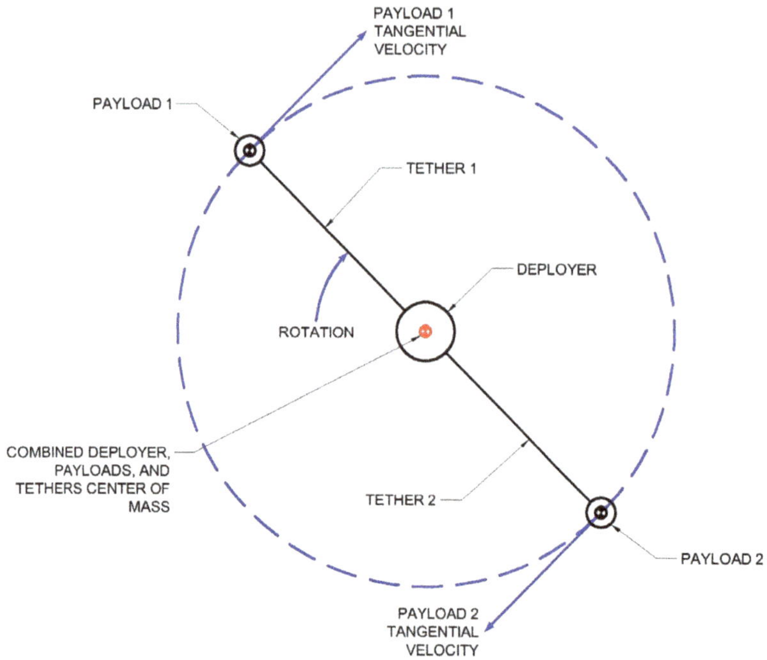

Figure 5. Symmetric Tether-Payload Release Pre-Separation State.

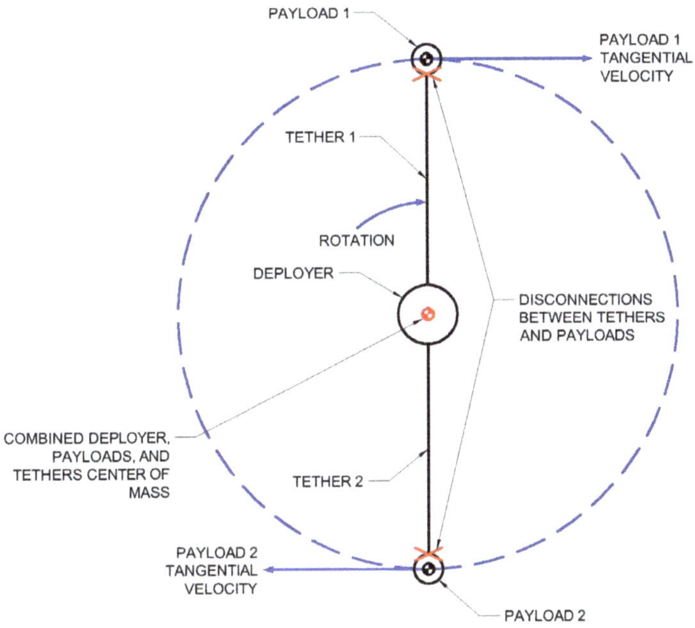

Figure 6. Symmetric Tether-Payload Release Separation State.

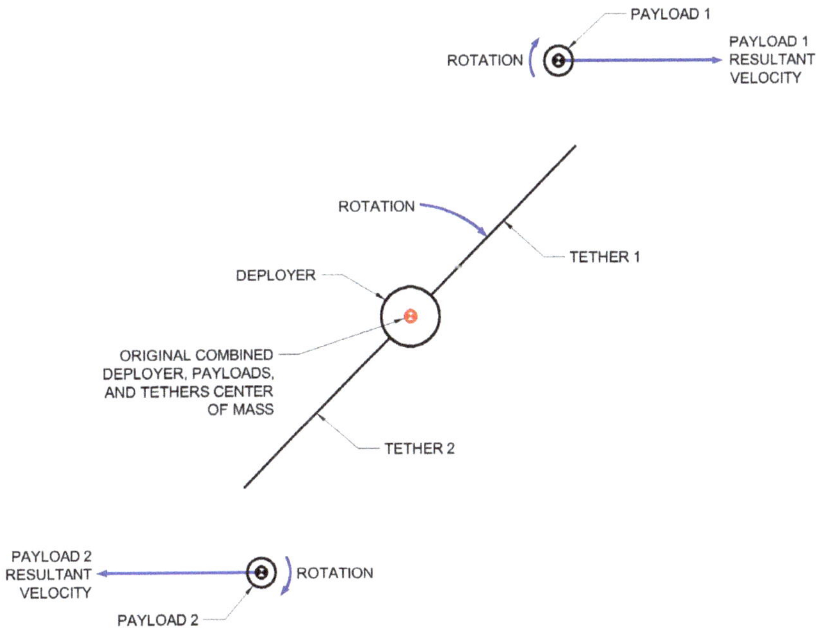

Figure 7. Symmetric Tether-Payload Release Post-Separation State.

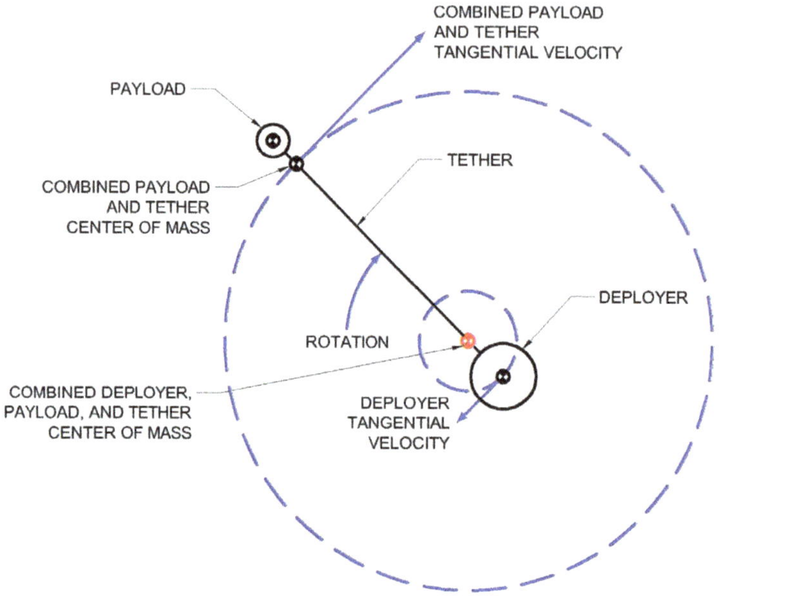

Figure 8. Asymmetric Tether-Deployer Release Pre-Separation State.

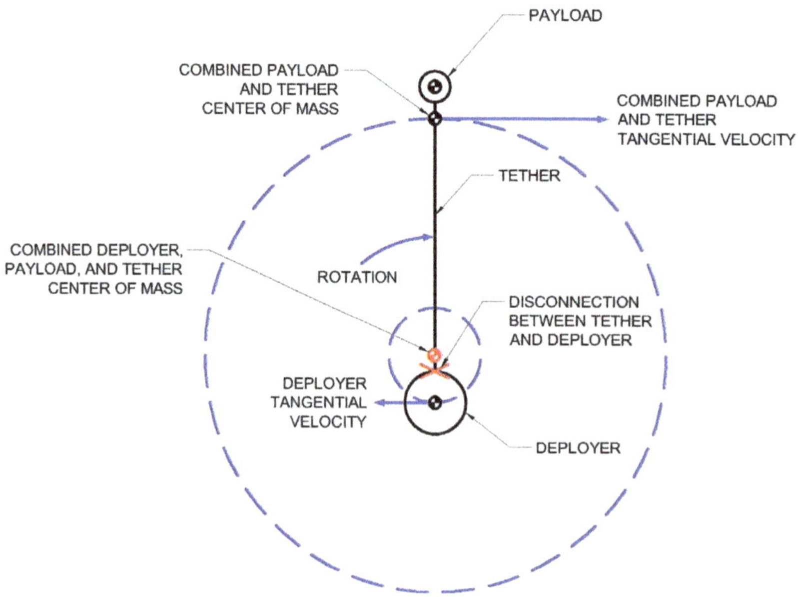

Figure 9. Asymmetric Tether-Deployer Release Separation State.

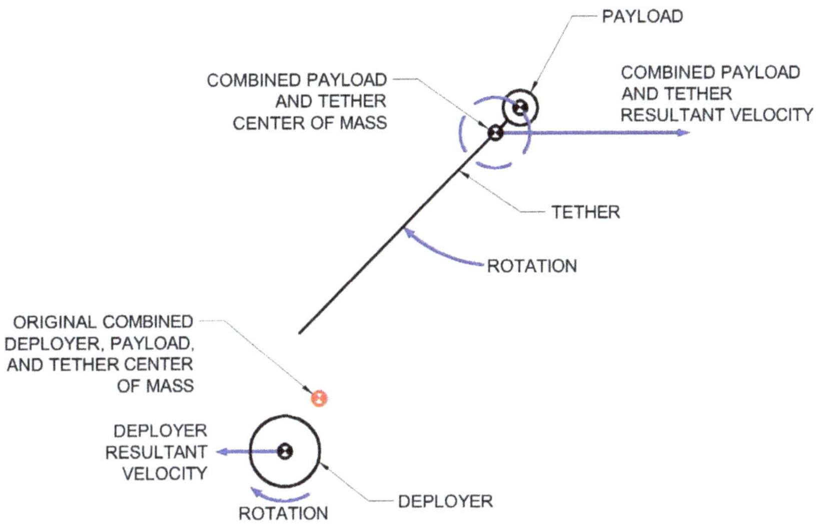

Figure 10. Asymmetric Tether-Deployer Release Post-Separation State.

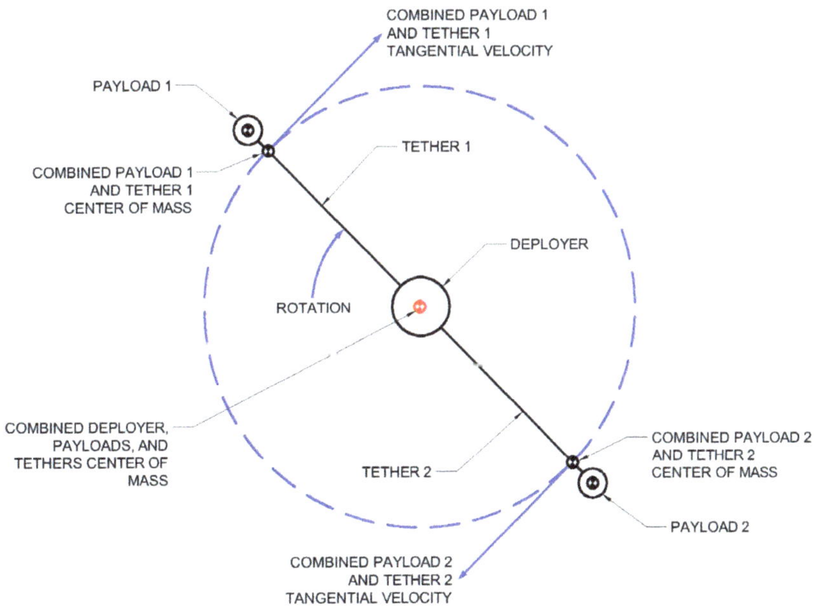

Figure 11. Symmetric Tether-Deployer Release Pre-Separation State.

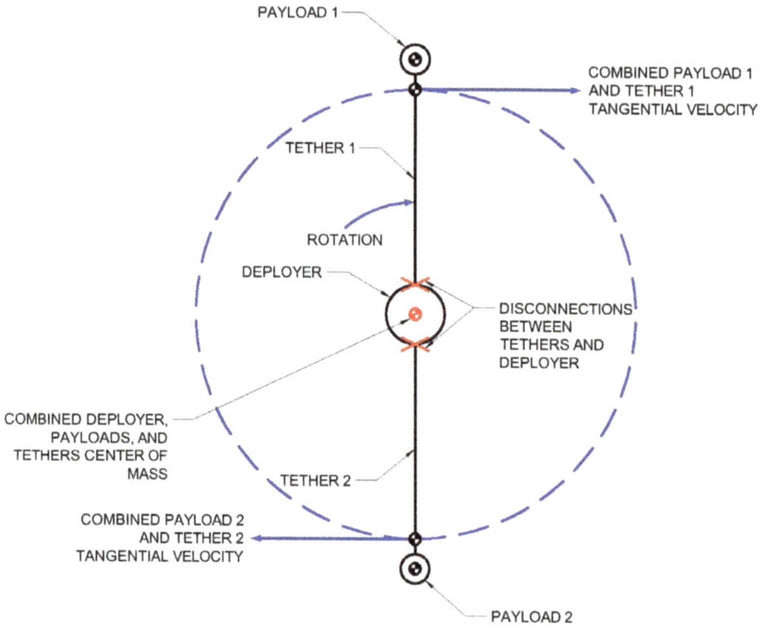

Figure 12. Symmetric Tether-Deployer Release Separation State.

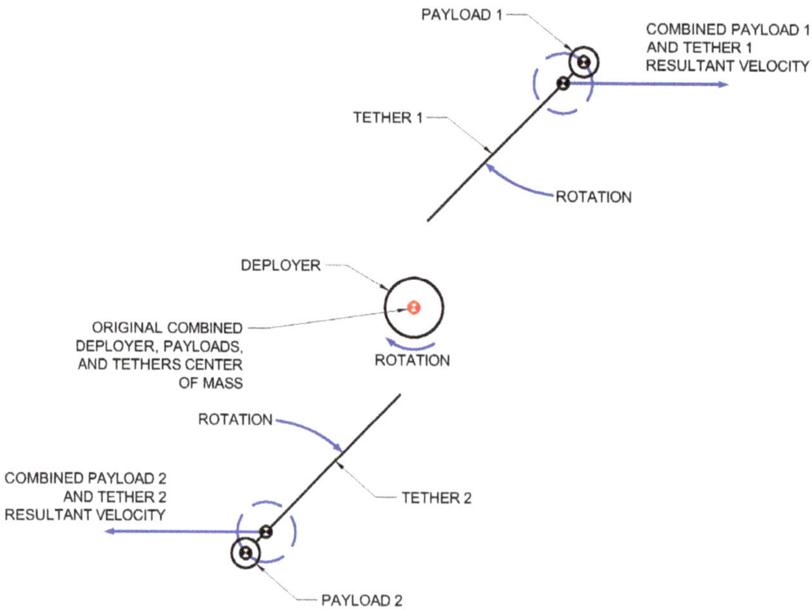

Figure 13. Symmetric Tether-Deployer Release Post-Separation State.

In designs separating the tether from the payload, the tether can be left with the deployer and leave no impact on the lifetime of the payload, aside from requiring a tether release mechanism onboard. Such systems are visualized in Figures 2–7. If this is pursued in a system utilizing symmetric releases, however, it is important that the timing of the prograde and retrograde payloads' release mechanisms are synchronized properly to avoid potentially unstable behavior that would result from delayed release sequencing. For example, if a prograde-released payload were one second before a retrograde payload releasing later off-nominally in the opposite direction, the one second period between releases would have the deployer and retrograde payload spinning in an unbalanced fashion like an asymmetric release system, and also result in the retrograde payload being released into an off-nominal orbital plane because of its angular displacement around the rotating system, thus also sending the deployer into a new off-nominal orbit as well.

To help combat the risks of non-synchronized symmetric tether-payload releases, it is possible to utilize tether-deployer releases instead. Such systems are visualized in Figures 8–13. With this method, the control of release between the deployer and tether can be controlled by a single system within the deployer, helping to ensure that both the prograde and retrograde payloads are released synchronously. An effect of this, however, is that the tethers would then remain attached to the payloads, which can be detrimental or beneficial to the payloads' operation and lifetime, depending on design and application. Negative effects of leaving the tether attached to a payload include an increase in cross-sectional area that increases atmospheric drag while in LEO (gradually decaying its orbit and reducing its lifetime), electrostatic charging while passing through regions of space with high magnetic field and radiation conditions, and additional mass and moment of inertia to handle while doing propulsive or attitude control maneuvers. While at first consideration each of these factors can be considered negative, some of them can also be reconsidered as opportunities for new functions of the payload. If the payload can actuate the tether's length, extending or retracting it on command, it can be used as a variable drag system for maneuvering and de-orbit operations. If the tether is conductive and able to be connected to other payload electrical systems, it can be repurposed into an electrodynamic tether as a power generation system and/or a low-thrust propulsion system.

4.2. Orbit Design

General constellation design entails choosing orbit parameters and satellite quantities specific to each application. For the sake of brevity, the basics of constellation design is not described in this article, but they are well-described in references [16–20]. In general, however, key factors to consider when designing a constellation include, but are not limited to, the following:

- Speed of communication between satellites and ground stations (affecting how fast information can be communicated, such as internet speed).
- Line-of-sight ranges (affecting how long ground links can last, such as length of a connection between one or more internet satellites before needing to connect to others to maintain connection).
- Revisit rate or frequency of repeated positioning of spacecraft over key locations (affecting how frequently location-specific data can be updated, such as monitoring the actively changing boundary of a wildfire).
- Range of coverage (such as limits in latitudes, affecting the limits of what regions, ground stations, and people can be covered).

Typically, constellations are constructed from one or more "rings" of satellites in orbital planes (such as a Walker constellation), but this is not necessarily always the case (such as with a Molniya-based constellation). These orbital rings are constructed in a manner where two or more spacecraft share an orbit and follow each other through the progression of their orbit path. These rings can be populated sparsely (such as with GPS), or densely (such as with Starlink).

When describing the orbits throughout this article, only simple elliptical orbits and in-plane maneuvers will be considered. While plane change maneuvers enabled by tethers are possible, they are not the primary focus of this article and will not be discussed in depth. The general geometry and parameters referenced from this point on are shown in Figure 14. Note that figures throughout this article depicting orbits have had their features exaggerated to assist in explanation, particularly regarding orbit radii and eccentricity. Some of the key orbit parameters to consider are orbit radius R, apoapsis R_a, periapsis R_p, semimajor axis a, true anomaly θ, orbital period T, gravitational parameter of the central planetary body μ, and the change in velocity needed to execute orbital maneuvers ΔV.

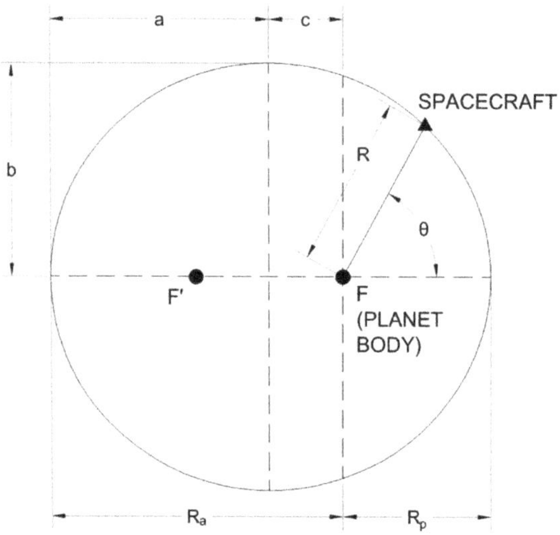

Figure 14. General orbit geometry and parameters, where a is the semimajor axis, b is the semiminor axis, c is the displacement of the planet/focus from the center of the ellipse, F and F′ are the focal points of the ellipse (F being the planetary body), R is the orbital radius, Ra is the orbit's apoapsis, Rp is the orbit's periapsis, and θ is the true anomaly.

As the tether concept allows for near-instantaneous imparting of velocity change to deployed payloads, the largest orbit changes can be made by directing this ΔV in a direction tangent to the current trajectory of the payload-deployer system. It is also possible to direct velocity changes by tethers into directions other than directly tangent with the direction of orbital motion, but it is not as energy efficient when looking mainly to spread satellites immediately after launch. If, however, it desired to execute maneuvers such as plane changes, these maneuvers are still possible if the involved systems are designed and operated properly. These kinds of maneuvers are beyond the scope of this article, but will be addressed in future work.

This ΔV tangent to the direction of orbital motion can be directed prograde or retrograde to increase or decrease a payload's orbit apoapsis R_a or periapsis R_p, thereby also affecting its semimajor axis a, eccentricity e, and orbital period T. A basic visualization of the orbits produced by prograde or retrograde release from a deployer are shown in Figure 15.

The process of deploying payloads into particular orbits forces some decisions to be made regarding the final orbit configurations of each payload and the deployer. For this system, there are currently five primary options to consider, each yielding their own benefits and drawbacks that can be selected for specific constellation design depending on application and the resources available to produce it. A flowchart describing these different options and the decisions leading to them is shown in Figure 16. Each orbit design question in the flowchart is described in the following five subsections. Future research may discover additional constellation design options, which would be discussed in future publications.

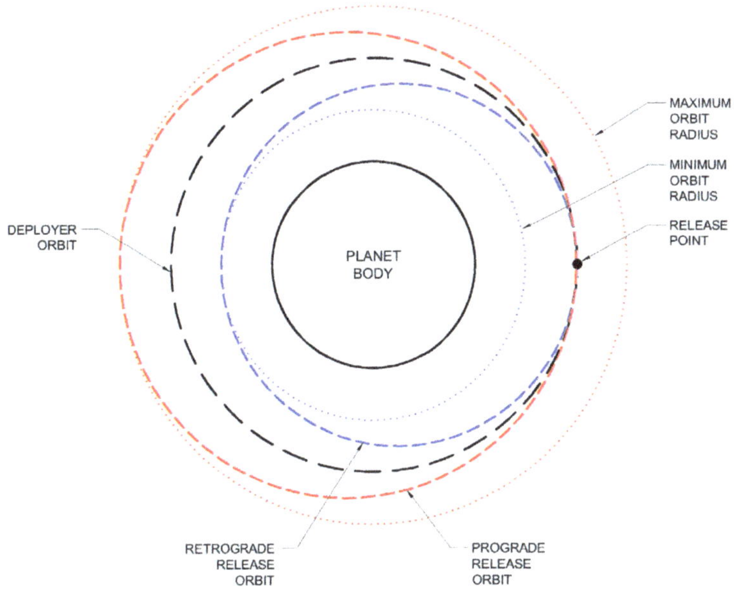

Figure 15. Basic SwingSat release orbits geometry.

4.2.1. Orbit Design Question 1: Single or Multiple Orbit Paths for Released Payloads

When considering single or multiple orbit paths, this asks whether payloads are required to be deployed solely into identical orbits with their only variation being in true anomaly. If only a single payload orbit path is permitted, this requires the system to utilize asymmetrical payload deployment, which would require either an onboard propulsion system to repeatedly boost the deployer's orbit back to its initial configuration to recover from payload momentum exchanges, or utilize symmetrical deployment with matched payloads and counterweights. The benefits of this asymmetric method entail very particular orbits with identical geometries, but the drawbacks come with devoting less resources to payloads and more to propulsion or counterweights, which reduce payload capacity and increase the number of launches needed to complete the constellation.

4.2.2. Orbit Design Question 2: Inclusion/Exclusion of Traditional Propulsion on Deployer for Asymmetric Release

When considering asymmetric release deployment, it is necessary to determine if the deployer spacecraft will include traditional propulsion, such as chemical or electric propulsion systems. If traditional propulsion is not included on the deployer, it will be necessary to include some other form of propulsion, such as deploying a tethered counterweight. Each of these options are feasible, but not advantageous for the overall intentions of tethered constellation deployments. Traditional propulsion utilizes established technology and does not contribute to space debris, but requires more design effort and the usage of propellant. Using tethered counterweights is technologically simpler, but is a poor use of spacecraft mass and volume and also contributes to space debris. If possible, these deployment methods should be avoided, but they are technically feasible options.

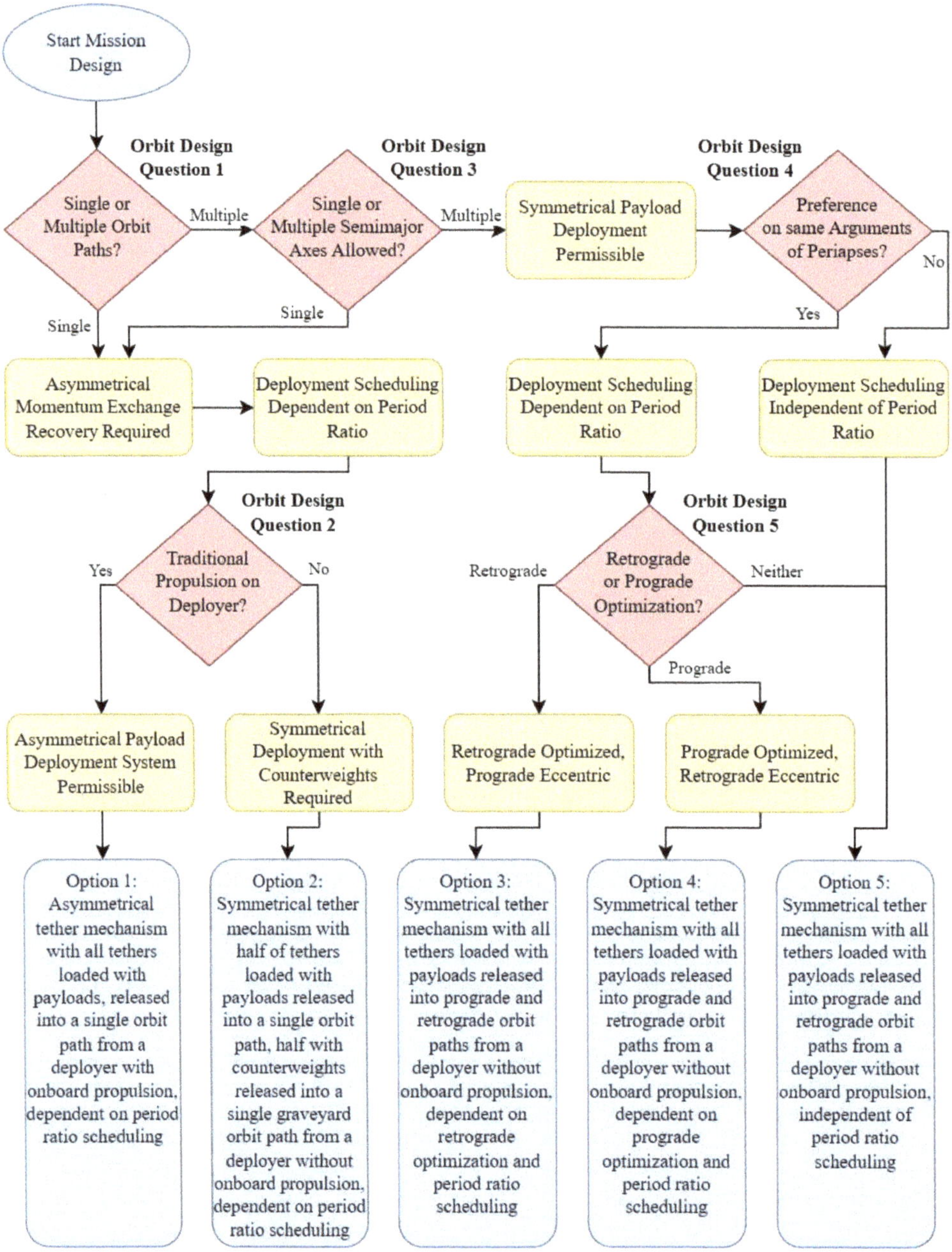

Figure 16. MET-deployed constellation design flowchart.

If the counterweights option were to be used with symmetrical releases, payloads would be released into either a prograde or retrograde release orbit, and counterweights released into the other. This particular approach is unfavorable, as it is not an efficient usage of overall mass and volume, which is critical for small satellites, and contributes to space debris. If this approach is required, it would be preferred to command the system to send payloads into prograde release orbits with the counterweights released into a retrograde orbit small enough to put the weights on a re-entry trajectory for disposal to minimize space debris' contribution, potentially with a trajectory directly intersecting with the central planetary body. This approach with prograde-released payloads is shown in Figures 17 and 18, with options for circular or elliptical orbits, and similar graphics displaying retrograde-released payloads in Figures 19 and 20.

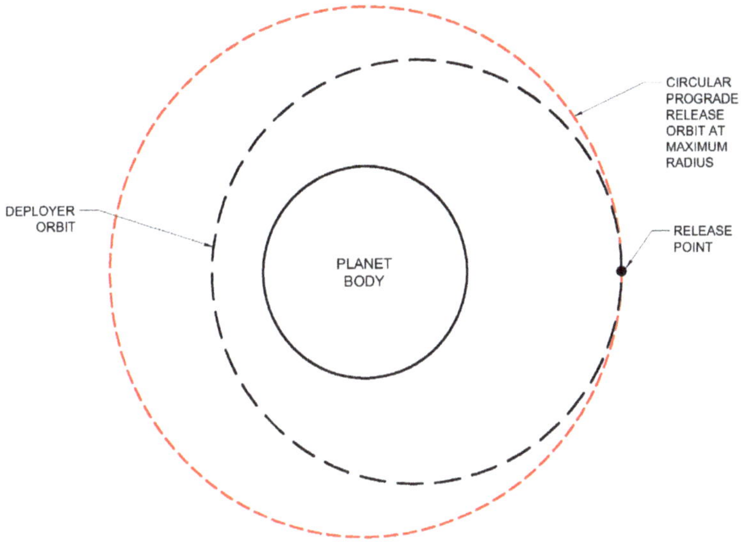

Figure 17. Prograde concentric circular release trajectory (retrograde orbit hidden).

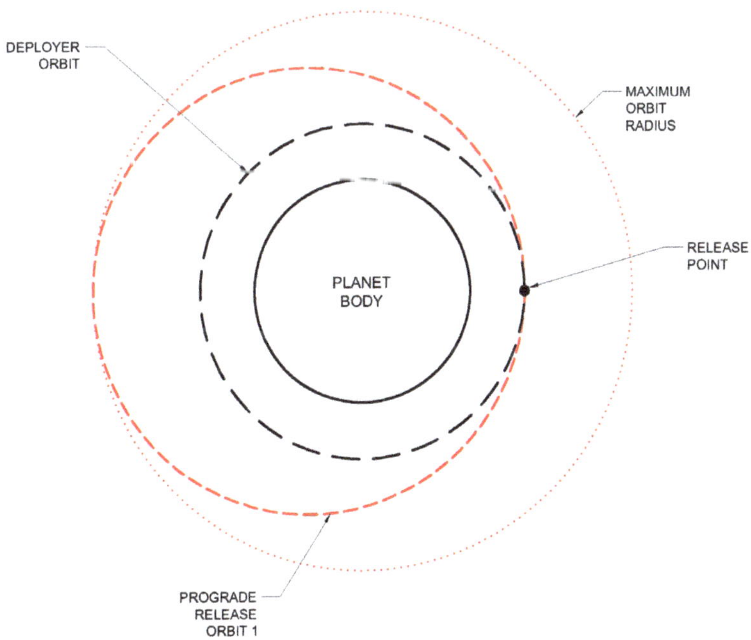

Figure 18. Prograde concentric elliptical release trajectory (retrograde orbit hidden).

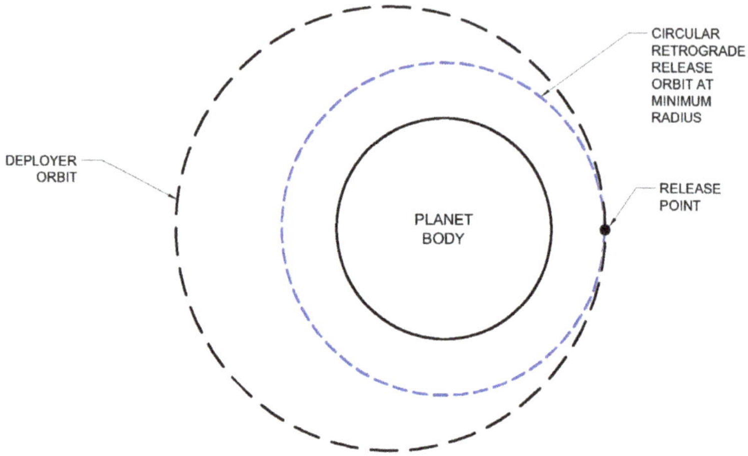

Figure 19. Retrograde concentric circular release trajectory (prograde orbit hidden).

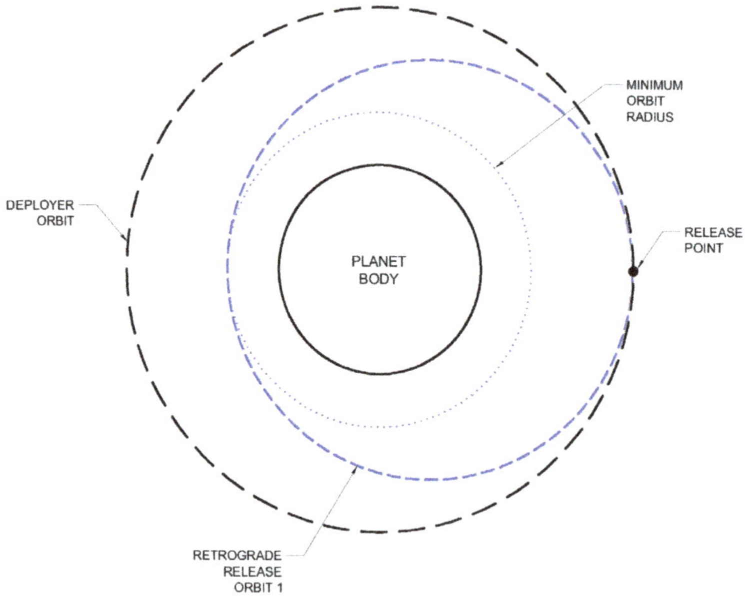

Figure 20. Retrograde concentric elliptical release trajectory (prograde orbit hidden).

4.2.3. Orbit Design Question 3: Single or Multiple Semimajor Axes for Released Payloads

When considering single or multiple semimajor axes being allowed, the key question is whether or not all payload orbits need to be dedicated to very similarly sized orbits. If this requirement is needed, this again results in requiring one of the asymmetrical payload release options. If there is allowance for variation in payload altitude however, symmetrical payload release becomes viable, and is a much more preferable option.

4.2.4. Orbit Design Question 4: Preference on Arguments of Periapses for a Prograde or Retrograde Payload Release Group

With the utilization of a symmetrical release, more options are available, which revolve around potentially optimizing the distribution of deployed payloads in either the prograde or retrograde release orbits. This is not always necessary, but if it is, it would entail the deployer and either prograde or retrograde release orbits being configured in such a way that their orbital periods would allow for a uniform distribution of payloads along an optimized orbital ring, all with the same argument of periapses. The optimized payload group's orbits could be considered concentric, as they would be identical trajectories overlapping each other. It is possible that this optimized release payload orbit be made circular, but that would eliminate the need for a preference for the arguments of periapses of that payload group. Maintaining the same arguments of periapses would require the deployer to release payloads in the same direction at the same position in its orbit for each payload in the selected optimized group. Examples of these particular orbits are depicted in Figures 17–20.

The basic progression behavior of the released payloads naturally results in the prograde payload "lagging behind" the deployer's position after one deployer orbit due to the increased orbital period of the prograde payload. The reverse is also true for the retrograde payload "leading ahead" of the deployer after one orbit due to the shorter period for the retrograde payload. A visualization of this behavior is shown in Figure 21. As more payloads get released, this will naturally produce a chain of payloads that will appear to either "lag behind" or "lead ahead" of the deployer over time as the prograde and retrograde orbits become populated.

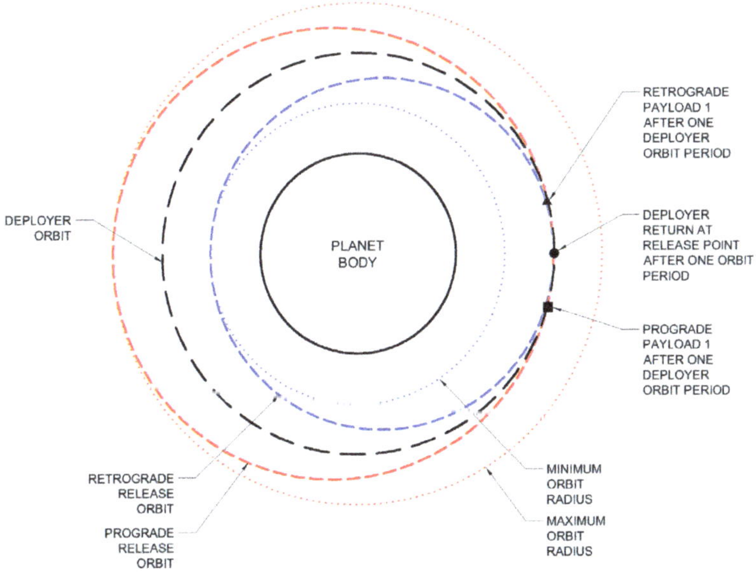

Figure 21. Basic payload orbit progression behavior post-release.

The primary benefit of the approach of optimizing one payload group is that it guarantees at least one payload group is released into a consistent pattern, uniformly distributed along a single release orbit path, but this comes at the expense of possibly producing gaps or overlaps in the distribution of payloads in the other release orbit and increasing its eccentricity, and can potentially take more time to execute when waiting for the deployer to intersect with the release orbits at the proper time to release each payload. A visualization of the different optimization options for prograde and retrograde orbit payload distributions and their resultant gaps or overlaps are shown in Figures 22 and 23. If this approach is used for optimizing specifically the prograde release orbit, the resultant geometry required for the deployer and retrograde release orbits should be carefully considered so as to not inadvertently release the retrograde payloads on a trajectory leading to a collision with the central planetary body, unless the retrograde payloads are replaced with counterweights for an asymmetrical release approach as previously described.

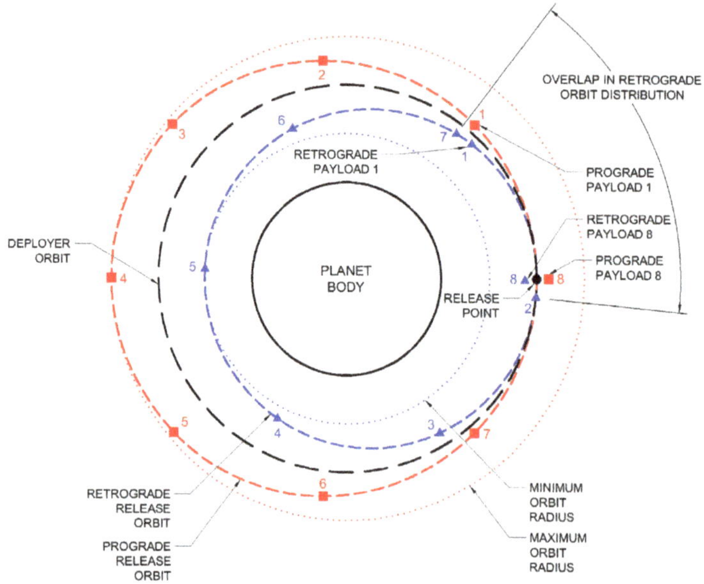

Figure 22. Basic payload orbit progression behavior with prograde release orbit distribution optimization for an example set of 8 prograde-released payloads and 8 retrograde-released payloads.

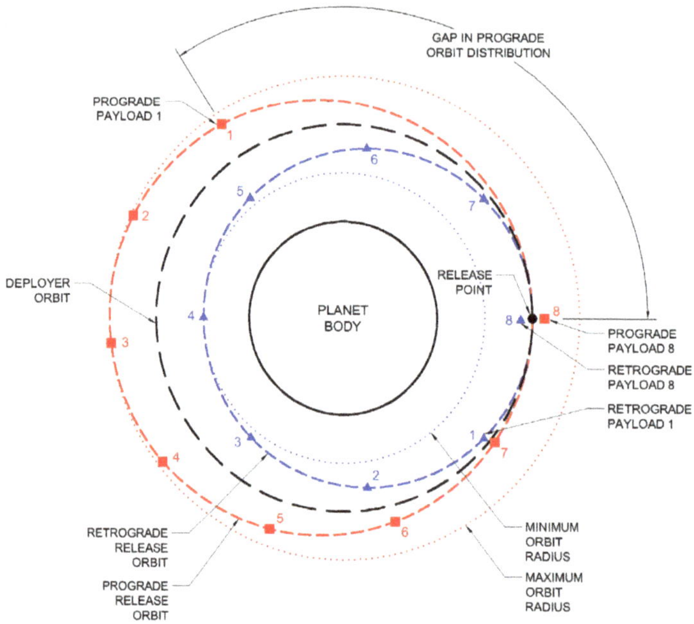

Figure 23. Basic payload orbit progression behavior with retrograde release orbit distribution optimization for an example set of 8 prograde-released payloads and 8 retrograde-released payloads.

4.2.5. Orbit Design Question 5: Prograde or Retrograde Release Orbit Optimization Dependent on Period Ratio

If choosing to optimize a prograde or retrograde release orbit payload group, orbit geometry and timing of the deployer's trajectory and release characteristics must be closely considered. This relationship between the deployer and optimized prograde/retrograde release orbit periods for proper release scheduling will here be referred to as the period ratio, T_i/T_o, where T_i is the period of the innermost orbit for optimization, and T_o the period of the outermost orbit for optimization. For prograde release orbit optimization, the innermost orbit would be the deployer orbit with the outermost orbit being the prograde release orbit. For retrograde release orbit optimization, the innermost orbit would be the retrograde release orbit with the outermost orbit being the deployer orbit. This can be visualized in Figure 24.

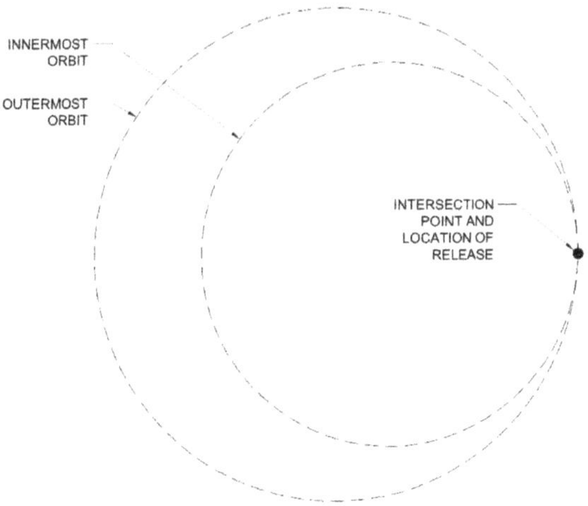

Figure 24. Basic inner and outer orbits.

The period of a single orbit T can be defined by Equation (1), where and a is the orbit's semimajor axis, μ is the gravitational parameter specific to the body being orbited, and R_a and R_p being the orbit's apoapsis and periapsis.

$$T = 2\pi\sqrt{\frac{a^3}{\mu}} = 2\pi\sqrt{\frac{\left(R_a + R_p\right)^3}{8\mu}} \tag{1}$$

The necessary period ratio T_i/T_o for a selected pair of orbits containing a deployer orbit and an evenly distributed collection of payloads in a prograde or retrograde orbit is defined in Equation (2).

$$\frac{T_i}{T_o} = \frac{kn}{kn+1} = \left(\frac{a_i}{a_o}\right)^{\frac{3}{2}} = \left(\frac{\left(R_a + R_p\right)_i}{\left(R_a + R_p\right)_o}\right)^{\frac{3}{2}} \tag{2}$$

In this equation, n is defined as an integer number of payloads to be uniformly distributed in the optimized orbit, and k is defined as the integer number of instances where the deployer intersects the optimized prograde or retrograde orbit between releases. The lowest value for k in this case is 1, which would result in the deployer releasing a payload every orbit. A value of k equal to 2 would result in the deployer releasing a payload every two orbits, and so on.

With this in mind, it can be seen that there is a balance between speed of deployment and the geometry of the orbits (apoapsis, periapsis, and eccentricity) of that must be decided on when given a set number of payloads to be dispersed. As T_i/T_o approaches 1, the deployer and optimized release orbits approach similarity, but this requires a high value for k orbit intersections between releases, which would require a longer time to execute. Conversely, as T_i/T_o approaches 0 (until k is minimized to 1), the fastest time for dispersal is achieved with one release per orbit, but this then involves more dissimilar geometries between the deployer and optimized release orbits as a consequence of the largely different orbital periods, potentially resulting in high eccentricity for one or both of them. These factors must also be considered with the resulting ΔV requirements during deployment, where similar orbit geometries with slower dispersion times require much less ΔV at deployment than dissimilar orbits with more rapid dispersion times. More discussion on the determination of ΔV is included later in this article.

4.2.6. Neutral Release Orbit Optimization Independent of Arguments of Periapses

If the optimization of distribution of either release orbits is not required, these orbits can be produced in a manner allowing the payload orbits to have varying arguments of periapses, potentially leading to either circular or flower orbit configurations with one or more release points along the deployer's orbit. The existence of gaps when optimizing the retrograde release orbit and overlaps when optimizing the prograde release orbit implies that there are one or more intermediary "neutral" options that lead to complete coverage of both release orbits, but expectedly in a slightly non-uniform distribution pattern. The benefit of not optimizing one release orbit over the other is this more equal distribution quality between both orbits, and potentially quicker dispersion times thanks to not needing to wait for specific orbit syncing between the deployer and either prograde or retrograde orbits. These configurations are shown in Figures 25 and 26.

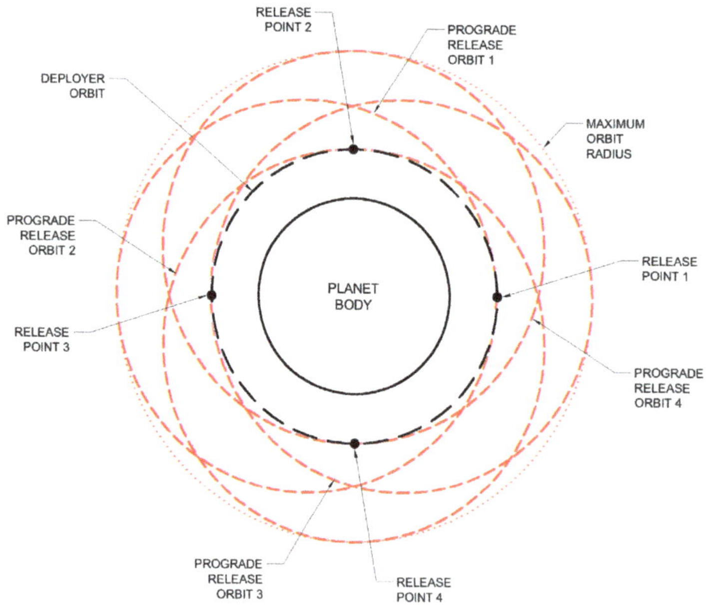

Figure 25. Prograde non-concentric elliptical release trajectories with 4 release points (retrograde orbits hidden).

There are several different approaches by which this neutral optimization can be pursued, but they are dependent on mission-specific orbit parameters and would need to be determined on a case-by-case basis. Generally though, this can be done in a way that alternates priorities between prograde and retrograde releases to populate both release orbits. It is also likely advantageous to set the deployer orbit to be circular to allow for the most balanced distribution of minimum and maximum altitudes by each payload in their respective release orbits. A graphic depicting different regions for payload distribution with neutral optimization is given in Figure 27, and is primarily composed of geometries presented in their simplified forms in Figures 25 and 26.

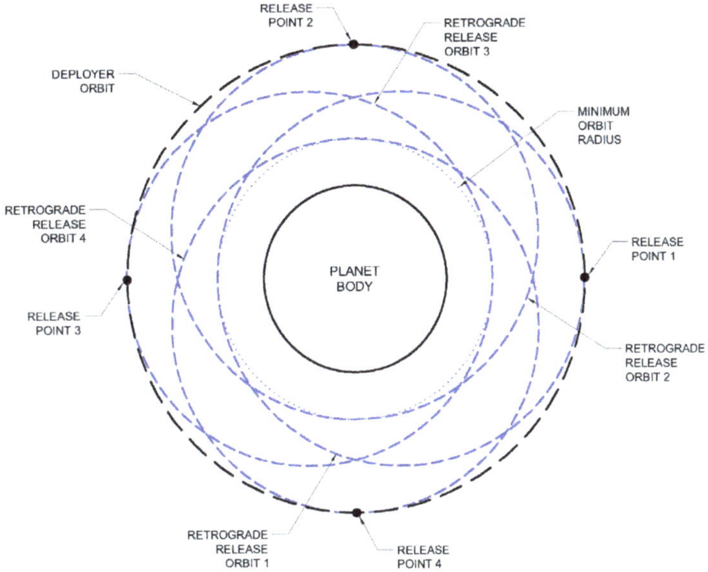

Figure 26. Retrograde non-concentric elliptical release trajectories with 4 release points (prograde orbits hidden).

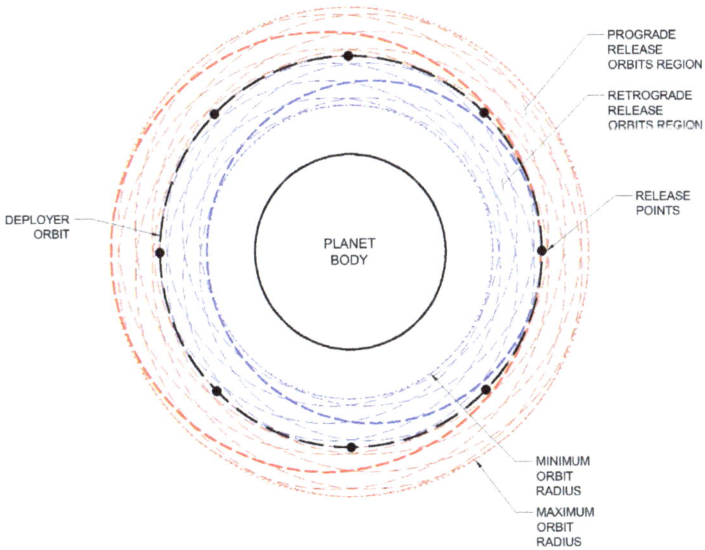

Figure 27. Basic payload orbit regions for neutral priority distribution optimization with 8 release points.

4.3. Conceptual Design Execution

Revisiting the SwingSat concept previously mentioned, the previously mentioned orbit design practices can be applied to a small theoretical constellation of 12 member satellites dispersed by a deployer spacecraft. This is just a small-scale example for the purposes of assisting in demonstrating an application of this work, and is not representative of a real constellation being developed. It is also focused solely on the generalized deployment process soon after launch, not later long-duration operations after initial dispersion. In this example, let there be 12 satellites that need to be distributed around a plane at a median altitude of 500 km. This constellation will be allowed to be deployed in a manner like design option #5 described in Figure 16. The tether deployment system will be designed to operate with the maximum ΔV imparted onto each deployed satellite being 20 m/s. The deployment speed is a fairly low and easily achievable using a tether deployment system for small satellite constellation deployment. Further details on deployer and tether design will be described in future published material by the authors. This future documentation is currently focused on tether deployment systems for satellites with individual masses of up to 500 kg, and necessary tape tether systems with masses on the order of a few kilograms made from conventional materials such as carbon fiber and spring steel.

Executing the maneuvers for satellite dispersion, the deployer will utilize neutral priority distribution optimization as depicted in Figure 27. This will be done by executing 6 symmetrical releases, each of which releasing two payloads at a time, adding up to the 12 total satellites to be deployed. For scheduling these releases, they will alternate priority for evenly spacing out satellites in the prograde and retrograde release orbits. For distribution of the satellites in each release orbit, the objective will be a nominal distribution of $360^\circ/(12/2)$ satellites = 60° between each satellite. The releases will be scheduled for an initial symmetrical release, then monitoring a difference in true anomaly $\Delta\theta$ between the deployer and first prograde-released payload until reaching $\Delta\theta = 60^\circ$ to then execute the next release. This would be followed by monitoring $\Delta\theta$ between the deployer and first retrograde-released payload until reaching $\Delta\theta = 120^\circ$ to then execute the next release. This process of alternating focus between the prograde and retrograde orbits for scheduling releases would continue for the full set of payloads to be deployed. A table describing the criteria for each release is shown in Table 1.

Table 1. Tethered payload release scheduling. Note that the first release is not preceded by any releases, hence not having an applicable true anomaly difference condition to trigger the release.

Release Number	Bodies Considered for Scheduling	True Anomaly Difference $\Delta\theta$	Date and Time of Release (UTC)
1	End of Deployer's first orbit after insertion	N/A	1 October 2023 01:34:31.657
2	Deployer and Prograde Payload 1	$\left\lvert\Delta\theta_{Deployer} - \Delta\theta_{Prograde\ 1}\right\rvert \approx 60^\circ$	2 October 2023 10:37:48.280
3	Deployer and Retrograde Payload 1	$\left\lvert\Delta\theta_{Deployer} - \Delta\theta_{Retrograde\ 1}\right\rvert \approx 120^\circ$	3 October 2023 21:15:34.810
4	Deployer and Prograde Payload 1	$\left\lvert\Delta\theta_{Deployer} - \Delta\theta_{Prograde\ 1}\right\rvert \approx 180^\circ$	5 October 2023 03:10:18.854
5	Deployer and Retrograde Payload 1	$\left\lvert\Delta\theta_{Deployer} - \Delta\theta_{Retrograde\ 1}\right\rvert \approx 240^\circ$	6 October 2023 16:56:14.554
6	Deployer and Prograde Payload 1	$\left\lvert\Delta\theta_{Deployer} - \Delta\theta_{Prograde\ 1}\right\rvert \approx 300^\circ$	7 October 2023 19:42:22.354

Following these criteria, these releases can be simulated in a software environment. In this simulation, the deployer was inserted into a simple 500 km altitude, 45° inclination circular orbit at 1 October 2023 00:00:00 UTC. The software that was used was STK Version 12.5.0 by AGI/Ansys, and idealized impulsive maneuvers were done using the Astrogator plugin. Orbit propagation was done using STK's Earth HPOP Default v10 propagator, and factors of spacecraft mass, atmospheric drag, solar radiation pressure, and radiation pressure were neglected, as those factors are highly dependent on spacecraft design, and are assumed to have little effect on the results of this basic conceptual demonstration. More detailed analysis incorporating these kinds of effects on spacecraft with masses up to 500 kg each, along with varying tether designs, will be published in future work. The deployment system was assumed to operate in idealized conditions where the payload release vectors are directly tangent with the deployer's trajectory, and releases are treated as impulsive velocity change maneuvers independent of mass and size. While allowing the deployer and onboard payloads to propagate forward in time, and releasing payloads based on the release scheduling criteria previously described, the dates and times of potential releases could be determined, and are also included in Table 1. The release maneuvers deployed payloads away at exactly 20 m/s directly prograde or retrograde. Visuals of the state of the constellation at each of the release times are given in Appendix A.

The resulting constellation produces two very similar but separate orbital planes, each with six satellites roughly evenly distributed around them, and is shown in Figure 28. Payloads that were released into prograde directions are shown in red, retrograde payloads in blue, and the deployer in white. The prograde payloads were placed into orbits with an apoapsis altitude of ~559 km, and a periapsis altitude of 500 km, and the retrograde payloads were placed into orbits with an apoapsis altitude of 500 km, and a periapsis altitude of ~416 km. The altitude bounds of 416–559 km can be narrowed closer to the 500 km median altitude with a lower release velocity than the 20 m/s used in this analysis, but this comes with the consequence of longer time periods between releases, increasing the overall dispersion time. Conversely, more rapid dispersion periods can be achieved with higher release velocities beyond 20 m/s, but this comes with the consequence of a wider altitude range, where retrograde-released payloads can be sent to precariously low altitudes and experience increased effects of atmospheric drag.

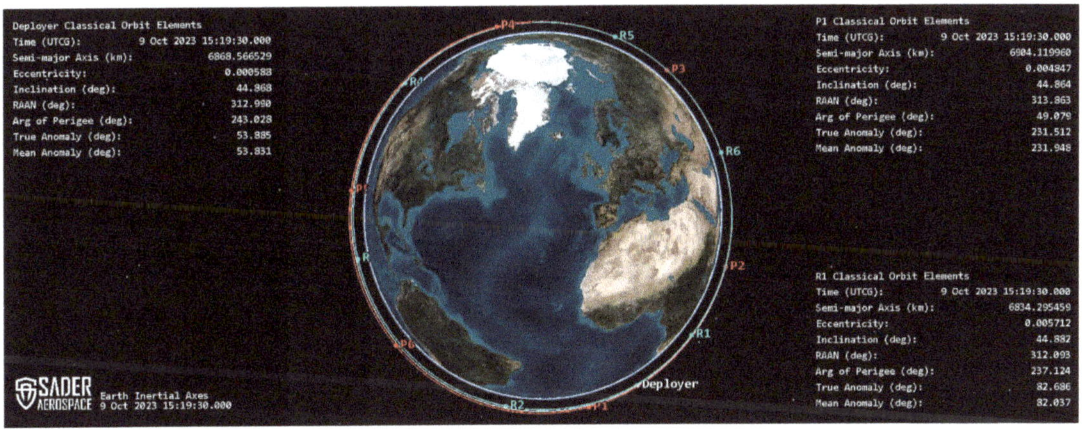

Figure 28. A 12-member constellation distributed in approximately 9 days post-launch (9 October 2023 15:19:30.000), simulated in STK software. Objects P1–P6 (red) indicate the first through sixth prograde-released payloads, and objects R1–R6 (blue) indicate the first through sixth retrograde-released payloads.

With each plane having different orbital periods because of their differing semimajor axes, there are occasional brief instances where satellites in two different planes can be in close proximity to each other in terms of true anomaly θ, but this is soon followed by their separation as one passes by another. Despite this occasional "overlapping" behavior, due to the nature of the distribution, there will never be more than approximately 60° of separation between any two adjacent satellites in the combined planes. While not (yet) optimized for precise positioning of these satellites, this deployment process can effectively distribute its satellites throughout the full spread of an orbital plane in the early phases of this particular conceptual constellation.

5. Discussion

This summary of different orbit configurations and release maneuver options is not exhaustive, but presents some key concepts to consider if a MET-deployed constellation were to be developed. Answering the five discussed orbit design questions can help in determining the nature of maneuvers needed to establish a constellation. These questions essentially all revolve around the similar theme of how similar or dissimilar the orbits for each member satellite are allowed to be, where missions allowing more dissimilar orbits can be deployed quicker and/or easier, while missions requiring more similar orbits can take longer and/or be more challenging to achieve. The deciding factors on whether these orbits can be similar or dissimilar are sourced from major requirements of the overall mission architecture itself, which is specific to the teams developing them and the proposed constellation's intended application(s). The selection of orbit design options can also inform the generation of requirements needed for associated spacecraft design.

The methods of deploying payloads by tether described here are not exhaustive, but give some initial insight to the thought processes that would likely be followed for designing such a system. The example given in this article is just one method (and the most chaotic or unoptimized of the five mission design options in Figure 16) but is still one viable means of dispersing satellites around an orbital plane for a constellation. A more optimized case would require some additional assistance in terms of attitude control and propulsion capabilities on the deployer, which would be easily achievable for moderately sized deployer spacecraft (such as ESPA class equivalent and larger).

6. Conclusions

In conclusion, this work has shed light on various orbit configuration and release maneuver options for the deployment of satellite constellations using METs. For newer satellite teams wanting to join in the constellation ecosystem, more feasible methods of establishing constellations can be helpful in this effort, with METs being one potential solution to consider. By implementing METs into constellation deployments, satellite teams can more easily handle design, resource, time, orbit delivery, and cost constraints that have been difficult in the past. The material discussed in this article gives a brief insight and demonstration of how METs and satellite constellations can be combined for rapid deployments to produce new mission architectures that could be flown. Future research and publications will include applications of the concepts described in this article to existing real-world constellations and comparing them against current industry norms, and also development and results from a small technology demonstration mission called ADRASTEA to be flown in the space environment. Current and future work in these areas will continue with the intent of improving the way constellations of any scale are created at key locations in accessible regions of space.

Author Contributions: Conceptualization, B.C.; methodology, B.C.; software, B.C.; validation, B.C.; formal analysis, B.C.; investigation, B.C.; resources, B.C. and L.D.T.; data curation, B.C.; writing—original draft preparation, B.C.; writing—review and editing, B.C. and L.D.T.; visualization, B.C.; supervision, L.D.T.; project administration, B.C. and L.D.T.; funding acquisition, B.C. and L.D.T. All authors have read and agreed to the published version of the manuscript.

Funding: This research is jointly funded by the NASA Alabama Space Grant Consortium (NASA Grant #80NSSC20M0044) and the UAH College of Engineering.

Data Availability Statement: No new data was created or analyzed in this study. Data sharing is not applicable to this article. Future supporting information (such as further publications and demonstration flight data) will be available at space.uah.edu and saderaerospace.com.

Acknowledgments: The author thanks his research advisor L. Dale Thomas for supporting this research, along with generous support from the NASA Alabama Space Grant Consortium and the UAH College of Engineering. The author also thanks the student team members from the UAH Terminus Spaceflight Research Group involved in the ADRASTEA mission for assisting in developing new technologies relating to the concepts described in this article.

Conflicts of Interest: The authors declare no conflicts of interest. The funders had no role in the design of the study; in the collection, analyses, or interpretation of data; in the writing of the manuscript; or in the decision to publish the results.

Appendix A

This appendix includes visual representations at the times of different release maneuvers while dispersing payloads around an orbital region for a conceptual 12-member constellation. The states are sequentially included below.

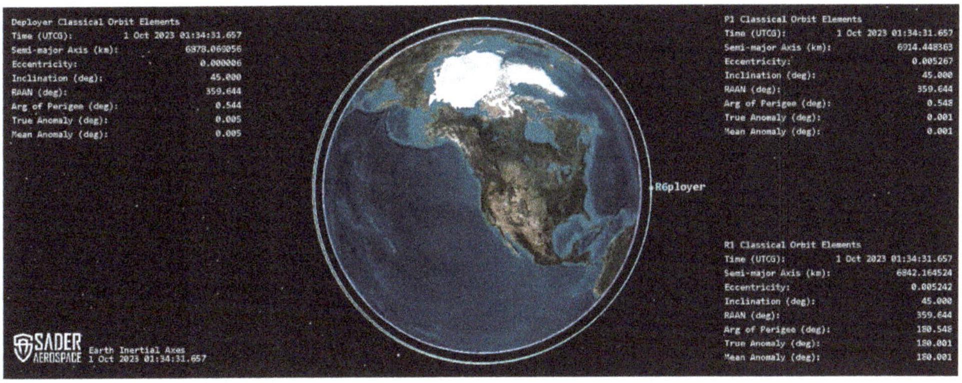

Figure A1. Constellation at the time of release maneuver 1 (1 October 2023 01:34:31.657).

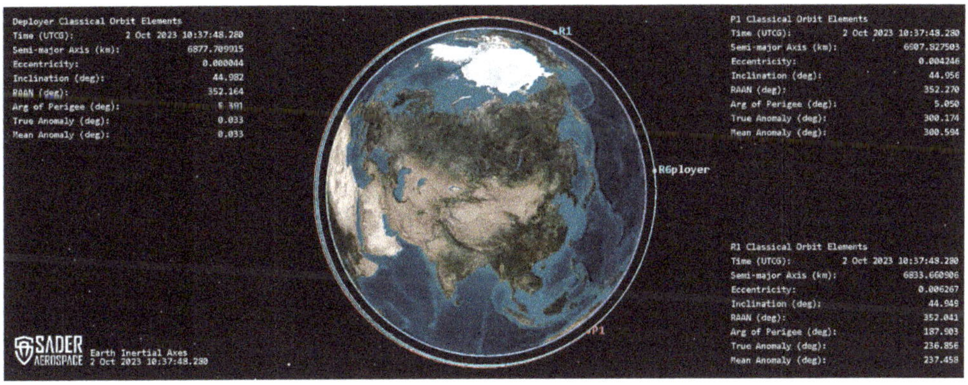

Figure A2. Constellation at the time of release maneuver 2 (2 October 2023 10:37:48.280). Object P1 (red) indicates the first prograde-released payload, and object R1 (blue) indicates the first retrograde-released payload.

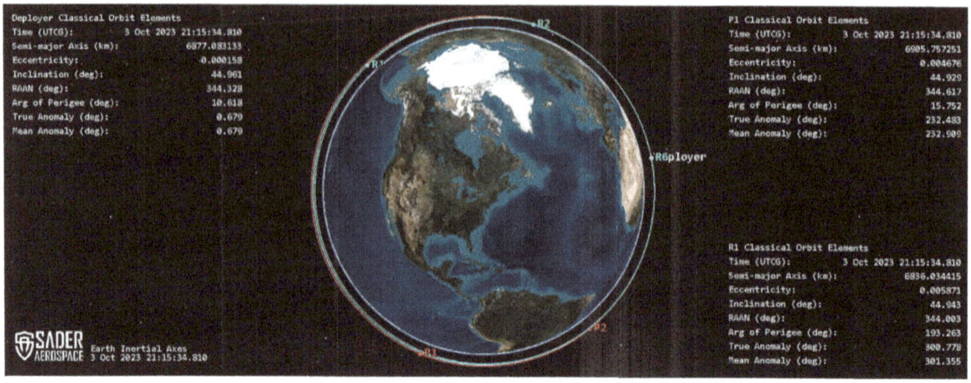

Figure A3. Constellation at the time of release maneuver 3 (3 October 2023 21:15:34.810). Objects P1 and P2 (red) indicate the first and second prograde-released payloads, and objects R1 and R2 (blue) indicate the first and second retrograde-released payloads.

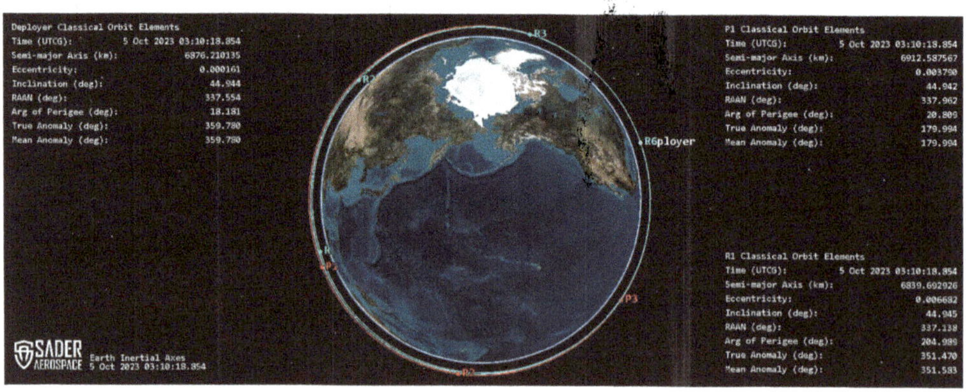

Figure A4. Constellation at the time of release maneuver 4 (5 October 2023 03:10:18.854). Objects P1–P3 (red) indicate the first through third prograde-released payloads, and objects R1–R3 (blue) indicate the first through third retrograde-released payloads.

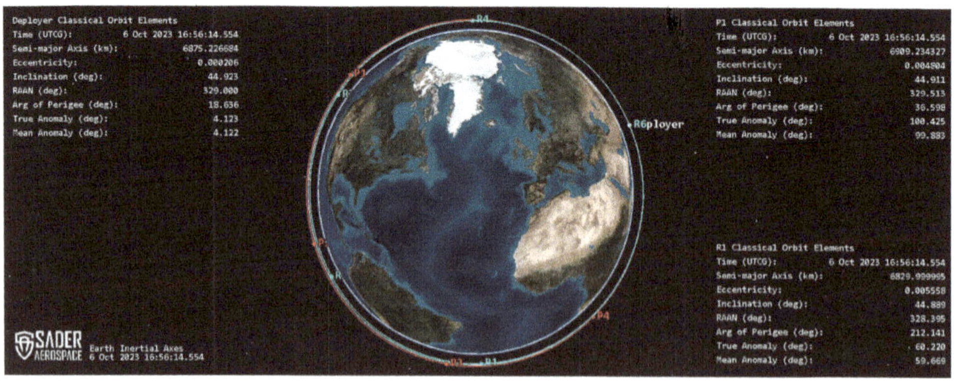

Figure A5. Constellation at the time of release maneuver 5 (6 October 2023 16:56:14.554). Objects P1–P4 (red) indicate the first through fourth prograde-released payloads, and objects R1–R4 (blue) indicate the first through fourth retrograde-released payloads.

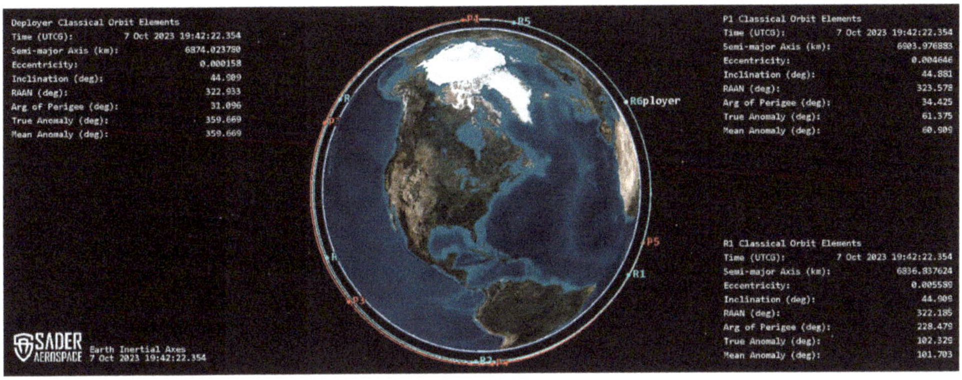

Figure A6. Constellation at the time of release maneuver 6 (7 October 2023 19:42:22.354). Objects P1–P5 (red) indicate the first through fifth prograde-released payloads, and objects R1–R5 (blue) indicate the first through fifth retrograde-released payloads.

References

1. Kulu, E. Satellite Constellations—2021 Industry Survey and Trends, Paper SSC21-XII-10. In Proceedings of the 35th Annual Small Satellite Conference, Virtual, 7–12 August 2021.
2. Crisp, N.; Smith, K.; Hollingsworth, P. Launch and deployment of distributed small satellite systems. *Acta Astronaut.* **2015**, *114*, 65–78. [CrossRef]
3. DePasquale, D.; Charania, A.C.; Kanayama, H.; Matsuda, S. Analysis of the Earth-to-Orbit Launch Market for Nano and Microsatellites. In Proceedings of the AIAA SPACE 2010 Conference and Exposition, Anaheim, CA, USA, 30 August–2 September 2010.
4. Curzi, G.; Modenini, D.; Tortora, P. Large constellations of small satellites: A survey of near future challenges and missions. *Aerospace* **2020**, *7*, 133. [CrossRef]
5. Cornara, S.; Beech, T.; Belló-Mora, M.; Martinez de Aragon, A. Satellite constellation launch, deployment, replacement and end-of-life strategies, Paper SSC99-X-1. In Proceedings of the Small Satellite Conference 1999, Logan, UT, USA, 23–26 August 1999.
6. Lang, D.D.; Nolting, R.K. Operations with Tethered Space Vehicles. In Proceedings of the Gemini Summary Conference, NASA, SP-138, Houston, TX, USA, 1–2 February 1967.
7. Hoyt, R. *Space Tethers: Technology Status and The Way Forward*; National Academy of Engineering: Washington, DC, USA, 2010.
8. Chen, Y.; Huang, R.; Ren, X.; He, L.; He, Y. History of the Tether Concept and Tether Missions: A Review. *Int. Sch. Res. Not.* **2013**, *2013*, 502973. [CrossRef]
9. Johnson, L.; Khazanov, G.; Gilchrist, B.; Hoyt, R.; Stone, N.; Lee, D.; Elder, C. Space Tethers. *J. Space Technol. Sci.* **2012**, *26*, 1_2–1_13. [CrossRef]
10. Johnson, L.; Gilchrist, B.; Estes, R.; Lorenzini, E. Overview of future NASA tether applications. *Adv. Space Res.* **1999**, *24*, 1055–1063. [CrossRef]
11. Cartmell, M.P.; McKenzie, D.J. A Review of Space Tether Research. *Prog. Aerosp. Sci.* **2008**, *44*, 1–21. [CrossRef]
12. Ziegler, S.W.; Cartmell, M.P. Using Motorized Tethers for Payload Orbital Transfer. *AIAA J. Spacecr. Rocket.* **2001**, *38*, 904–913. [CrossRef]
13. Tomlin, D.D.; Faile, G.C.; Hayashida, K.B.; Frost, C.L.; Wagner, C.Y.; Mitchell, M.L.; Vaughn, J.A.; Galuska, M.J. *Space Tethers: Design Criteria (No. NAS 1.15: 108537)*; NASA Marshall Space Flight Center: Huntsville, AL, USA, 1997.
14. Carroll, J.A. SEDS Deployer Design and Flight Performance. In Proceedings of the American Institute of Aeronautics and Astronautics, Space Programs and Technologies Conference and Exhibit, Paper 93-4764, Huntsville, AL, USA, 21–23 September 1993.
15. Kruijff, M. Tethers in Space, a Propellantless Propulsion In-Orbit Demonstration. Ph.D. Dissertation, Delft University of Technology, Delft, The Netherlands, 2011.
16. Larson, W.J.; Wertz, J.R. *Space Mission Analysis and Design*, 3rd ed.; Microcosm Press: El Segundo, CA, USA; Kluwer Academic Publishers: Norwell, MA, USA, 1999.
17. Wertz, J.R. *Orbit & Constellation Design & Management*; Microcosm Press: Hawthorne, CA, USA; Springer: Berlin/Heidelberg, Germany, 2001.
18. Marina, R.; Sanctis, M.; Rossi, T.; Lucente, M.; Mortari, D.; Bruccoleri, C.; Salvini, P.; Nicolai, V. The flower constellation set and its possible applications. *ACT Final. Rep. Arid.* **2006**, *5*, 4108.

19. Bate, R.R.; Mueller, D.D.; White, J.E. *Fundamentals of Astrodynamics*; Dover Publications Inc.: Mineola, NY, USA, 1971.
20. Campbell, B. SwingSat: Development of A Nanosatellite Constellation Deployment System by Means of Centrifugal Dispersion. In Proceedings of the American Astronautical Society 15th Wernher von Braun Memorial Symposium, Huntsville, AL, USA, 26–28 October 2022.

aerospace MDPI

Article

Spectral Imager of the Solar Atmosphere: The First Extreme-Ultraviolet Solar Integral Field Spectrograph Using Slicers

Ariadna Calcines Rosario [1,*], Frederic Auchère [2], Alain Jody Corso [3], Giulio Del Zanna [4], Jaroslav Dudík [5], Samuel Gissot [6], Laura A. Hayes [7], Graham S. Kerr [8,9], Christian Kintziger [10], Sarah A. Matthews [11], Sophie Musset [7], David Orozco Suárez [12,13], Vanessa Polito [14,15], Hamish A. S. Reid [11] and Daniel F. Ryan [16]

1 Department of Physics, Centre for Advanced Instrumentation, Durham University, Durham DH1 3LE, UK
2 Institut d'Astrophysique Spatiale, CNRS, Université Paris-Saclay, 91405 Orsay, France; frederic.auchere@universite-paris-saclay.fr
3 Institute for Photonics and Nanotechnologies, National Research Council of Italy, via Trasea 7, 35131 Padova, Italy; alain.corso@pd.ifn.cnr.it
4 Department of Applied Mathematics and Theoretical Physics, University of Cambridge, Wilberforce Road, Cambridge CB3 0WA, UK; g.del-zanna@damtp.cam.ac.uk
5 Astronomical Institute of the Czech Academy of Sciences, Fričova 298, 251 65 Ondřejov, Czech Republic
6 Royal Observatory of Belgium, Ringlaan-3-Av. Circulaire, 1180 Brussels, Belgium; samuel.gissot@oma.be
7 European Space Agency, ESTEC, Keplerlaan 1, 2201 AZ Noordwijk, The Netherlands; laura.hayes@esa.int (L.A.H.); sophie.musset@esa.int (S.M.)
8 NASA Goddard Space Flight Center, Heliophysics Science Division, Code 671, Greenbelt, MD 20771, USA; kerrg@cua.edu
9 Department of Physics, Catholic University of America, Washington, DC 20064, USA
10 Centre Spatial de Liège, STAR Institute, University of Liège (ULiège), 4031 Liège, Belgium; ckintziger@uliege.be
11 Mullard Space Science Laboratory, University College London, Holmbury St. Mary, Dorking RH5 6NT, UK; sarah.matthews@ucl.ac.uk (S.A.M.); hamish.reid@ucl.ac.uk (H.A.S.R.)
12 Spanish Space Solar Physics Consortium (S3PC), 18008 Granada, Spain; orozco@iaa.es
13 Instituto de Astrofísica de Andalucía (IAA-CSIC), 18080 Granada, Spain
14 Lockheed Martin Solar and Astrophysics Laboratory, Building 203, 3251 Hanover Street, Palo Alto, CA 94304, USA; polito@lmsal.com
15 Department of Physics, Oregon State University, Corvallis, OR 97333, USA
16 FHNW Institute for Data Science, University of Applied Sciences and Arts Northwestern Switzerland, Bahnhofstrasse 6, 5210 Windisch, Switzerland; daniel.ryan@fhnw.ch
* Correspondence: ariadna.calcines@durham.ac.uk

Abstract: Particle acceleration, and the thermalisation of energetic particles, are fundamental processes across the universe. Whilst the Sun is an excellent object to study this phenomenon, since it is the most energetic particle accelerator in the Solar System, this phenomenon arises in many other astrophysical objects, such as active galactic nuclei, black holes, neutron stars, gamma ray bursts, solar and stellar coronae, accretion disks and planetary magnetospheres. Observations in the Extreme Ultraviolet (EUV) are essential for these studies but can only be made from space. Current spectrographs operating in the EUV use an entrance slit and cover the required field of view using a scanning mechanism. This results in a relatively slow image cadence in the order of minutes to capture inherently rapid and transient processes, and/or in the spectrograph slit 'missing the action'. The application of image slicers for EUV integral field spectrographs is therefore revolutionary. The development of this technology will enable the observations of EUV spectra from an entire 2D field of view in seconds, over two orders of magnitude faster than what is currently possible. The Spectral Imaging of the Solar Atmosphere (SISA) instrument is the first integral field spectrograph proposed for observations at ~180 Å combining the image slicer technology and curved diffraction gratings in a highly efficient and compact layout, while providing important spectroscopic diagnostics for the characterisation of solar coronal and flare plasmas. SISA's characteristics, main challenges, and the on-going activities to enable the image slicer technology for EUV applications are presented in this paper.

Keywords: EUV spectroscopy; solar IFS; EUV slicers; image slicers; solar space mission; particle acceleration

1. The Need for Solar Integral Field Spectrographs in the Extreme Ultraviolet

Particle acceleration is a fundamental process arising in many astrophysical objects including active galactic nuclei, black holes, neutron stars, γ ray bursts, accretion disks, solar and stellar coronae, and planetary magnetospheres.

In our solar system, the Sun is the most energetic particle accelerator. Its proximity makes it a unique laboratory with which to explore the physics of energetic particles, which has wider importance in understanding astrophysical particle acceleration more generally. However, despite its importance and past observations, the physics underlying solar particle acceleration remains poorly understood. For the study of solar particle acceleration and the evolution of solar eruptive events (solar flares and coronal mass ejections, CMEs), it is useful to observe in the extreme ultraviolet (EUV) regime, since this wavelength range contains a number of emission lines that allow the corona and transition region to be probed and whose profiles allow the presence of non-Maxwellian particle distributions to be diagnosed. Additionally, the plasma response following the thermalisation of these energetic particles can be dramatic, with resulting heating, ionisation, mass flows, and turbulence, all of which can be diagnosed with EUV spectroscopic observations, e.g., Refs. [1–4]. The plasma in the solar corona is typically fully ionised because of the megaKelvin plasma temperatures. Such hot plasma contains ions that emit brightly in EUV wavelengths ($\lambda \sim$100–1200 Å).

Measurements in the EUV can be used to characterise the temperature, density, elemental composition, and the turbulent state of the plasma in the upper solar atmosphere, for both the quiet and active Sun. Plasma density can be diagnosed through the observation of density sensitive line ratios in all coronal features, from the quiet Sun and coronal holes to active regions, flares, jets, and coronal mass ejections. The measurements of densities in the corona are fundamental for understanding how the corona is heated. For example, the commonly observed over-density of coronal loops indicates that the corona is heated impulsively, by multiple small energy release events along a given magnetic flux tube [5–8], possibly recurring on short timescales. However, the coronal response to individual heating events, including possible transient density enhancements, is so far almost impossible to observe due to the lack of spatial coverage and resolution.

The EUV measurements are also key for understanding how the heated solar corona transitions into the solar wind. Measuring electron densities during transient processes like solar flares also allows us to understand the timescales of ionisation processes that occur during the conversion of magnetic energy to the kinetic energy of accelerated particles, which, in the standard solar flare model, occurs following magnetic reconnection, sometimes with impulsive, bursty nature as observed in, e.g., the review [1].

The traditional method for performing solar imaging spectroscopy in the EUV, e.g., Refs. [9,10] has been to use a slit spectrograph and a scanning mechanism to cover a larger field of view. Whilst 'sit-and-stare' observations can be produced within seconds for a strongly emitting EUV line from one location, the traditional slit scanning spectrograph takes minutes to create a spectrum across a 2D spatial region of the Sun depending on the desired field of view. Using integral field spectroscopy (IFS), all the spectra of a 2D field of view will be obtained simultaneously, reducing the integration time by over two orders of magnitude and without the requirement for any rastering mechanism. We aim to achieve this with SISA applying the image slicer technology. An image slicer is an integral field unit, also known as IFU, which decomposes a 2D field of view by using an array of narrow rectangular mirrors (slicer mirrors) at the telescope focal plane using different orientations to reflect slices of the image in different directions. The slicer mirrors can be flat or powered and their width defines the sampling and impacts both spatial and spectral resolutions.

Additional arrays of mirrors are used to focus these beams generating the spectrograph entrance slit, producing the desired magnification and controlling the location of the exit pupil. For SISA, in order to optimise efficiency, only the slicer mirror array is used. Thus, the concept presented in this paper combines the surfaces of the IFU with those of the spectrograph. We also note that a novel solar ultraviolet IFS concept will be used for the first time during the Solar Eruption Integral Field Spectrograph (SNIFS) NASA sounding rocket launch in Spring 2024 [11,12].

SISA is a proposal for a novel integral field spectrograph using slicers for the first time ever in the extreme ultraviolet spectral range. This manuscript presents the science cases that would benefit from this instrument, as well as the conceptual layout, whose main components are currently at a research and development stage to increase the technology readiness level (TRL) to enable this technology for space applications in the EUV.

2. SISA Science in the SPARK Framework

SISA is envisioned as one component of a suite of complementary instruments that together form The Solar Particle Acceleration Radiation and Kinetics (SPARK) mission concept, along with the Large Imaging Spectrometer for Solar Accelerated Nuclei (LISSAN) and the Focusing Optics X-ray Solar Imager (FOXSI) instruments, both of which target high-energy X-ray and γ-ray emission. SPARK is designed to address the open questions regarding particle acceleration on the Sun.

SPARK will attack the problem of particle acceleration through a powerful combination of γ-ray, X-ray, and EUV imaging and spectroscopy at high spectral, spatial, and temporal resolutions. In addition to helping to address the physics of particle acceleration, SISA shall provide the critical measurements of plasma conditions (magnetic fields, temperatures, densities, and elemental composition) in the various coronal and flare structures, both pre- and post-reconnection. As these measurements are integral to understanding how the plasma responds to particle acceleration, and critical for understanding how and when flares occur, the SISA EUV integral field spectrometer is an crucial addition to the SPARK mission. The significant advantage of SISA with respect to previous EUV spectrographs is that it will enable these crucial measurements in an integral field regime, i.e., spectroscopic measurements simultaneously over a wide field of view. Below, we briefly discuss how this instrument contributes to the key science goals to be addressed by the SPARK mission:

- **How does impulsive energy release events lead to accelerate particles in the solar atmosphere?** SISA shall provide several measurements of accelerated particles in EUV, both electrons through line ratios (see Section 3.2) and ions through line profiles (see Section 3.3). The integral field spectroscopy of SISA will offer insights as to where and when the particles are accelerated, while the fast cadence will reveal how long they persist at a given location. Measuring the spatial distribution of the accelerated particles, and their relationship to magnetic field (see Section 3.1) and field-aligned emission structures, will also offer insights as to the conditions required for particle acceleration, both in solar flares and active-region corona. The fast-cadence SISA observations with multiple hot lines (Table 1) shall also clarify the relationship of particle acceleration to plasma heating. Finally, since the emission line profiles reflect the line-of-sight distribution of ion velocities, from zero to very high velocities, they provide key information about both the high-energy particles simultaneously with the low-energy end of the distribution, which is not accessible with either LISSAN or FOXSI instruments.
- **How is impulsively released energy transported and dissipated in the solar atmosphere?** Once heated, the hot flare plasma evolves rapidly on short timescales depending on the conditions. The hot flare lines observed by SISA cover some of the largest available temperatures via EUV line spectroscopy (Table 1) and are thus favourable for characterising the hot flare plasma and its evolution. Typically, the hot plasma is observed first above the chromospheric footpoints in the form of localised bright kernels, e.g., Refs. [13,14], from which the flare loops are filled. Indications exist, however, that these kernels located within bright flare ribbons can already be pre-heated by elec-

tron beams [15,16]. These kernels have long been observed to move along ribbons, e.g., Refs. [17–19], which have been identified as a consequence of 3D slipping reconnection, e.g., Refs. [4,14,20–24], where the field lines do not reconnect in an X-point, but slip (slide) past each other as they mutually exchange their connectivities see, e.g., Refs. [25–27]. The existence of this process implies that the location of the energy deposition into the lower solar atmosphere changes with time as the slipping reconnection proceeds. In the past, it has been extremely difficult to identify the spectroscopic signatures of this process due to the slit not being able (or designed) to track a particular moving (slipping) kernel. Using sit-and-stare observations Li and Zhang [22] showed that the slipping reconnection is likely be related to periodic changes in the Si IV spectral line intensities accompanied by enhanced redshifts, as well as increased nonthermal widths, as individual kernels moved through the location of the slit. Recently, Lörinčík et al. [4] detected extremely short-lived blueshifts (upflows), lasting only seconds, and reaching about 50 km s^{-1} in chromospheric and transition-region lines, at the leading edges of the slipping kernels. The authors argued that such detection can be a matter of luck with slit spectrographs, as the slit has to be in the right place at the right time. The IFS provided by SISA shall be enormously helpful in this regard, as it can image the entire flare region and allow us to identify how the flaring atmosphere undergoing slipping reconnection responds at short temporal cadences, as well as enable the tracking of the spectral evolution of individual moving kernels as they slip along flare ribbons. In addition, the SISA thermal coverage would help in establishing whether the blueshifts at the leading edges of ribbons or individual kernels are related to the evaporation of hot flare plasma.

The fastest upflow velocities of the evaporating hot flaring plasma filling coronal loops are detected in the hottest flare lines [13], while the 'cooler' (T $< \sim$1 MK) lines show downflows at the same location, implying multi-directional flows that could occur due to rapid plasma evolution. The fast SISA cadence and the integral field spectroscopy, coupled with the available plasma diagnostics (Section 3), will provide key information on the thermal evolution of the heated plasma. For example, the diagnostics of electron density coupled with the measured timescales of plasma cooling will allow one to discern whether the plasma is in ionisation equilibrium, and at what times. Meanwhile, the emission line profiles will provide information on the presence and role of turbulence. Mass flows and the thermodynamic evolution of the atmosphere determined via spectroscopy can reveal much about the energy input into the lower atmosphere during flares, especially when combined with state-of-the-art modelling. For example, flare observations from the Interface Region Imaging Spectrograph IRIS, as seen in Refs. [10,28], have been used in tandem with flare loop models to interrogate our understanding of flare processes [29,30]. There is a major model–data discrepancy in the duration of the flare gradual phase, with models under-predicting the cooling time by an order of magnitude. SISA observations that provide plasma diagnostics over the full field of view, with high cadence, will help illuminate the source of the continued heating or energy input. Measurements of nonthermal line widths from lines formed at different temperatures will inform us about the roles of turbulence in suppressing thermal conduction, e.g., Ref. [31], and the potential role of Alfvén waves in flares, e.g., Refs. [32,33].

SISA measurements will provide key diagnostics of processes that occur within flares on short timescales. For example, a key open question in the solar flare energy release is what drives "bursty" pulsations and the oscillatory signatures observed in flare emission, known as quasi-periodic pulsations (QPPs), reviews of which can be found in Refs. [34–36]. QPPs and other oscillatory behaviour observed in flares have timescales ranging from sub-seconds to minutes, and are identified across the entire electromagnetic spectrum from radio, EUV [37], X-rays [38], and even γ-rays [39], essentially encompassing all aspects of the flaring process. The exact nature and underlying physical mechanism for the generation of these pulsations remains highly debated. It is suggested that they may be related to magnetohydrodynamic oscillations in/near the flare site, or possibly connected to the intermittent or time-dependent magnetic reconnection itself. Observational limitations of temporal cadences, spatial resolution, and saturation issues with EUV imagers have to date limited our ability to observationally identify the locations of the emission modulations and constrain the suggested models—both of which are directly linked to energy release and transport in solar flares. Some work has aimed to identify the spatial locations of the modulations, e.g., Refs. [22,37,40–42], although there are limits to the temporal cadences and spatial diagnostics, for example, there is only spatial observation along the slit position. In order to correctly identify the mechanism producing QPPs, the characteristics of the temporal, spatial, and spectral properties of pulsations and their relationships across energy ranges and temperatures are required. The SISA EUV measurements, with its high temporal cadence and its ability to perform the imaging spectroscopy of the flaring region (rather than just over the slit), will allow us to observe rapid changes in the flaring regions such as these pulsations, with information regarding where and at what temperature they originate in the flare structure.

Finally, although there is unambiguous evidence for the presence of nonthermal particles in flares, other mechanisms may also act to transport liberated magnetic energy. High-frequency Alfvén waves have been proposed as a means of transporting energy the magnetic reconnection site to the lower atmosphere and heating it, e.g., as shown in Refs. [43–45]. Modelling has revealed that those waves do indeed heat the chromosphere, and drive explosive evaporation into the corona [46,47]. While it is likely that MHD waves are produced during flares, which are fundamentally a large-scale change in the corona's magnetic field, the proportion of energy that manifests itself in the form of waves compared to energetic particles is not known. SISA's capability to measure the coronal magnetic field before, during, and after a flare will provide crucial information regarding field perturbations, density diagnostics, and the Alfvén speed, which will help to provide estimates of the Poynting flux carried by MHD waves. Furthermore, the nonthermal broadening of ions will also help constrain the Poynting flux see discussion, e.g., in Ref. [32].

- **What are the physical origins of space weather events in the low-corona?** Both coronal mass ejections (CMEs) and jets inject magnetic field and plasma into the heliosphere, where they disturb the solar wind flow. Although previous studies of CME source regions have provided details about the physical processes taking place once a CME has initiated, many questions remain regarding their initiation and similarities with smaller scale flux rope eruptions within solar jets. Observing the initiation of both CMEs and jets spectroscopically is quite difficult with current or planned instrumentation for several reasons, including the use of a single slit, long exposure times, or limited diagnostics and temperature coverage provided by the available spectral lines. SISA will capture these processes at cadences down to a few seconds from every pixel within its entire field of view. This will allow us to identify the locations, spectral properties, plasma conditions, and thus the mechanisms behind the processes of CME and jet initiations. In larger eruptive events, SISA will be able to capture the entirety of the precursor phase of the associated flare. The spatial localisation of the precursors with respect to the subsequent flare and eruption allows for the identification of the CME initiation mechanism, whether by tether-cutting, ideal MHD instability, or breakout, as can be seen in Ref. [48] for every flare observed. The high-temperature lines observed by SISA (Table 1) will provide information on the plasma properties during the onset of eruptive events, including the possibly constant, isothermal 10–15 MK onset temperatures detected by broad-band X-ray instrumentation [49]. These lines will also allow the quantification of plasma heating (via temperature and density measurements) as well as turbulence (via line broadening) in the precursor phase.
- **How is the corona above active regions heated?** It is currently thought that the solar corona is heated by individual impulsive "nanoflare" events that release small

amounts of energy at either high or low frequencies, e.g., in Refs. [6–8,50,51]. A key prediction of such impulsive energy release is the existence of small amounts of hot 5–10 MK plasma, which is difficult to detect spectroscopically with currently available instrumentation [52–56]. The spectral range of SISA contains several hot lines (Table 1) that will provide stringent constraints on the amount of plasma reaching 10 MK temperatures. The Fe XVII and Ni XVII lines (Table 1) shall provide additional constraints. SISA will also work in tandem with the HXR and SXR observations to constrain the high-T component of the impulsive energy release by nanoflares. Another key observable of impulsive energy release is that the plasma should at least temporarily be out of thermal equilibrium, showing either the presence of accelerated particles, out-of-ionisation equilibrium plasma, e.g., Refs. [57–60], or both. The coronal Fe XI lines formed in both quiet Sun and active regions offer such diagnostics for electrons [61,62] while the line profiles of multiple ionisation stages shall provide information on ion velocity distribution [63–66]. Furthermore, the temporal evolution of the line intensities of multiple ionisation stages obtained in high cadences in combination with electron density diagnostics (Section 3.7) shall provide information on both the presence of energy release events and constraints on the presence of non-equilibrium ionisation for multiple coronal structures at the same time, a feat not possible with current or planned instrumentation. Finally, it will become possible for the first time to tie all these measurements to the measurements of the underlying magnetic field (Section 3.1), thus allowing for discerning whether there are different heating mechanisms for different magnetic structures within the active and quiescent solar corona.

Table 1. A selection of SISA diagnostic spectral lines. The approximate formation temperature (log T[K]) is shown, as well as the radiances for an AR core and an M2-class flare (in erg cm^{-2} s^{-1} sr^{-1}). Numbers in brackets are DN s^{-1} per 1″ SISA pixel. bl indicates a blended line. The number of * indicates the relative diagnostic importance.

Ion	λ (Å)	log T	AR Core	M2 Flare	Notes
Ni XV	178.89	6.5	15	1.2×10^2	
Fe XXI	178.90	7.1	-	4.8×10^2 (145)	** Flare Ne
Fe XI	179.76	6.1	130 (40)	9.0×10^2 (276)	** Ne
Fe XXIII	180.04	7.2	-	8.8×10^2 (270)	* Flare
Fe XI	180.40	6.1	1.4×10^3 (429)	3.1×10^3 (950)	** (bl Fe X)
Ca XV	181.90	6.5	-	4.2×10^2 (156)	*** Ne
Fe XI	182.17	6.1	120 (47)	1.1×10^3 (429)	*** Ne
Ca XV	182.86	6.5	-	6.4×10^2 (310)	*** Ne
Fe X	184.54	6.0	220 (175)	5.5×10^2 (439)	*** Coronal B
Fe XI	184.79	6.1	76 (64)	5.0×10^2 (424)	*** Ne
Fe XII	186.89	6.2	600 (395)	2790 (1836)	(bl) ** Ne
Fe XXI	187.93	7.1	-	3.8×10^3 (1172)	*** Flare (bl Ar XIV)
Ar XIV	187.96	6.5	21	3.4×10^2	** Ne, FIP
Fe XI	188.22	6.1	710 (160)	1.6×10^3 (362)	***
Ar XV	221.15	6.5	73 (13)	3.4×10^3 (584)	*** FIP
Fe XXIII	221.34	7.2	-	5.6×10^2 (100)	Flare
S XII	221.43	6.4	90 (16)	4.1×10^2 (75)	** FIP
Fe XV	233.87	6.5	260 (247)	3.0×10^3 (2850)	** Ne
Fe IX	241.74	5.9	120 (189)	120 (189)	*** Ne
Fe XXI	242.05	7.1	-	4.1×10^3 (6540)	*** Flare Ne
He II	243.03				
Ar XIV	243.75	6.4	43	5.3×10^2	FIP (bl)
Fe XV	243.79	6.5	880 (1502)	6.5×10^3 (11095)	(bl Ar XIV)
Fe IX	244.91	5.9	80 (141)	2.3×10^2 (406)	*** Ne
Fe XXI	246.95	7.1	-	6.5×10^2 (1205)	*** Flare
Fe XXII	247.19	7.1	-	8.0×10^3 (14887)	*** Flare
Ar XIII	248.68	6.2	7 (13)	1.1×10^2 (209)	** FIP
Ni XVII	249.19	6.6	840 (1598)	9.3×10^3 (17690)	*** Flare

Table 1. *Cont.*

Ion	λ (Å)	log T	AR Core	M2 Flare	Notes
Fe XII	249.39	6.2	65 (124)	2.4×10^2 (457)	
Fe XVI	251.06	6.5	650 (1234)	1.0×10^4 (18979)	
Fe XIII	251.95	6.2	460 (864)	1.7×10^3 (3194)	
Fe XIV	252.20	6.4	280 (524)	1.8×10^3 (3367)	
Fe XXII	253.17	7.1	-	3.9×10^3 (7177)	*** Flare
Fe XVII	254.88	6.6	-	3.5×10^3 (6140)	*** Flare
Fe XXIV	255.11	7.3	-	4.8×10^4 (83656)	*** Flare
He II	256.3				(bl)
S XIII	256.68	6.5	1.1×10^3 (1797)	8.6×10^3 (14055)	** FIP
Fe X	257.26	6.0	140 (222)	150 (238)	*** coronal B
Fe XIV	257.39	6.4	420 (663)	2.1×10^3 (3317)	
Fe XI	257.55	6.1	80 (125)	2.3×10^2 (360)	** Te,NMED
Si X	258.37	6.2	420 (628)	1.5×10^3 (2243)	Ne
S X	259.50	6.2	53 (74)	1.8×10^2 (251)	*** FIP
Si X	261.06	6.2	140 (175)	4.2×10^2 (525)	Ne
Fe XVI	262.98	6.5	1.1×10^3 (1148)	1.7×10^4 (17754)	
Fe XXIII	263.77	7.2	-	2.1×10^4 (20192)	*** Flare
S X	264.2	6.2	77 (71)	257 (237)	*** FIP

3. SISA EUV Measurements and Diagnostics

Based on these scientific cases, two spectral windows have been defined for SISA: $\lambda = [178\text{–}184]$ Å and $\lambda = 246\text{–}258$ Å. There are numerous EUV spectral lines originating from a range of species that together allow a wide temperature coverage from the transition region through to coronal and flare plasma. Many density-sensitive line pairs are included within these windows, including at $T > 10$ MK, as are a pair of Fe X transitions whose ratio is proportional to the coronal magnetic field strength. Via the inclusion of Fe XI, the electron temperature and presence of non-Maxwellian electron distributions can be probed. There are also transitions of S and Ar to measure the FIP bias, relative to Fe lines, providing vital composition information.

Table 1 lists the spectral lines of interest, along with notes regarding the diagnostics they provide. There is some overlap with the EUV spectral lines observed by the Hinode/EUV Imaging Spectrograph EIS, as shown in Ref. [9], but with higher sensitivity, cadence, and simultaneous spatial coverage; for more details, see, e.g., Ref. [67]. Whilst EIS requires an estimated 50 min to scan a 100″ region with its 2″ slit and 60 s exposures to obtain enough signal for coronal magnetic field measurements, SISA would obtain the same observations within a few seconds.

Once they thermalise, the energetic particles that carry flare energy from the corona to the transition region and chromosphere heats the local plasma. The intense heating subsequently leads to ionisation and drives mass flows: so-called chromospheric evaporation and condensation. Spectral lines from species forming at different temperatures are very useful in diagnosing the conditions of the plasma before, during, and after energy deposition. They can reveal gradients in density, bulk velocity and temperature. We have typically been restricted to observations in single locations, whether that be at the flare footpoints, along the loops, or at looptops. However, to understand the full story of where flare energy is deposited and how the plasma responds, simultaneous observations, at high-cadence, of the full flaring structure, would be transformative.

3.1. Magnetic Field Measurements

The solar coronal magnetic field can be measured via the recently identified Fe X 184.54/257.26 Å line intensity ratio see, e.g., Refs. [68–70]. While the former line is an allowed transition, the latter, Fe X 257.26 Å, is a blend of several intercombination transitions. In the presence of the external magnetic field, a new decay channel to the ground configuration $3s^2\,3p^5\,{}^2P_{2/2}$ occurs due to the mixing of the upper metastable levels $3s^2\,3p^4\,3d$

$^4D_{7/2}$ and $3s^2\,3p^4\,3d\ ^4D_{5/2}$ [71]. This magnetically induced transition (MIT) increases the intensity of the 257.26 Å line, and allows for measurements of coronal magnetic field. This has already been performed using the solar EUV spectra measured by Hinode/EIS, revealing the presence of magnetic fields of up to several hundred Gauss in a C2-class flare [71] as well as in the solar coronal loops [72].

However, making such measurements routine using slit spectroscopy (as in the case of Hinode/EIS) is difficult, as the slit only samples a small portion of the solar atmosphere. In addition, the Fe X line at 256.26 Å is relatively weak, requiring long exposure times with previous instruments such as EIS. Finally, the diagnostic ratio is also sensitive to a range of parameters, such as the electron density and temperature, so additional diagnostic lines are needed. They are provided by SISA, as described below.

The SISA instrument shall allow for routine measurements of both Fe X lines in a wide field of view and at orders of magnitude faster cadence, providing revolutionary potential for magnetic field measurements both before, during, and after the reconnection process; that is, from the active region corona pre-reconnection to the post-reconnection flare loops. Such measurements will allow for the first time to provide a direct measurement of the magnetic energy released during the flare reconnection.

3.2. *Electron Temperature and Nonthermal Diagnostics*

To measure the result of transport effects on accelerated electrons, the nonthermal electron spectrum must be characterised at multiple locations in the flare structure, namely near the acceleration region, along the flare loops, and near the chromospheric footpoints.

SISA's wavelength range contains the lines of multiple Fe ions (see Table 1) from Fe IX formed at upper transition region, Fe X, Fe XI, Fe XIV, and Fe XVI formed in the corona of quiet Sun and active regions, to multiple flare lines of Fe XXI, Fe XXII, and Fe XXIV. Such wide temperature coverage allows for the characterisation of the ionisation temperature of the plasma, and in particular, the coronal structures participating in the reconnection process, from the pre-reconnection corona to the post-reconnection flare loops.

In addition, the EUV lines observed by SISA allow one to assess whether nonthermal electrons are present in certain cases. The ratios of Fe XI line intensities at 182.17/257.55 Å, 182.17/257.7 Å are sensitive to high-energy electrons at temperatures above 1 MK [61]. The Fe XI lines at 257.55 Å and 257.77 Å are located in the direct vicinity of the magnetically sensitive Fe X line at 257.26 Å. We note that, since the temperature and power-law tail parameters are the free parameters of any electron distribution, diagnostics of the high-energy electrons always need to be performed simultaneously with the diagnostics of temperature [61], and the references therein.

3.3. *Ion Temperature and Nonthermal Diagnostics*

Profiles of any emission line changes shape if energetic ions are present. This effect is detectable through the presence of enhanced line wings, which correspond to the high-energy tails in ion velocity distribution via the Doppler effect. Such high-energy ions were detected in the last decade in a limited amount of cases, mostly in flares [63–66]. The most significant enhancements of the wings were found to occur at about twice the line FWHM, see Figure 1 of Ref. [65]. This means that, in principle, any emission line with a strong enough signal-to-noise in the far wings can be used for the diagnostics of accelerated ions, independently of the temperature at which the line is formed. Thus, the wide temperature coverage of the solar corona provided by SISA fortuitously allows for the wide diagnosis of energetic ions. However, the diagnostics are complicated by the fact that the emission lines are usually broadened during the impulsive phase of flares—a well-known fact for decades, as can be seen in Section 3.1 of Ref. [73] as well as Refs. [64,66,74]. Since the broadening is indicative of the presence of turbulence, the wide spatial coverage of SISA shall enable the study of spatial correlations between the occurrence of turbulence and ion acceleration. We note that the two mechanisms can in principle be linked, with turbulence directly leading to particle acceleration in some cases, as can be seen in Ref. [75].

3.4. Electron Density Diagnostics

Measuring the electron density is a fundamental requirement for any spectrometer. Knowing the density is important for a wide range of issues, for example, to estimate the radiative losses of the plasma and thus the energy input required not only during flares, but also for the active region corona. In addition, electron density is critical for determining the ionisation and recombination timescales in the plasma, and thus whether the plasma is in ionisation equilibrium. The EUV offers a wide range of diagnostic possibilities, as can be seen in the review by Ref. [70].

The SISA spectral range and sensitivity was also chosen to enable measurements of the electron densities from a wide range of lines, formed from coronal (e.g., Fe IX, Fe XI, Fe XV, Ca XV) to flare temperatures (Fe XXI). Measurements of coronal densities have been carried out with previous spectrometers, most notably Hinode EIS, but measurements at high temperatures are nearly non-existent in the literature. The best diagnostics are in the soft X-rays around 140 Å, as discussed in Ref. [76], mainly using lines from Fe XXI and Fe XX. Such measurements have been carried out for a limited number of large X-class flares by Ref. [77] using SDO/EVE, an instrument without any spatial resolution. To estimate the ionisation and recombination times of the flaring plasma, it is essential to measure the electron density at high temperatures with good spatial resolution. SISA has two excellent line ratios from Fe XXI. One is 242.05 Å /246.95 Å, which is sensitive to high densities above 10^{12} cm^{-3}. The lines are strong and close. The other one is a ratio involving the weaker 178.9 Å line, which is already sensitive to active region densities of 10^{10} cm^{-3} or more. One issue with the 178.9 Å line is that it could be blended with a Ni XV line, depending on the relative amount of plasma at the formation temperatures of these ions. To correct for the Ni XV blend, two other density-sensitive Ni XV lines need to be observed.

Figure 1 provides a summary of the main diagnostic ratios for SISA, although other ones are available. CHIANTI version 10 [78] was used. The ratios cover a wide range of temperatures and are sensitive to a wide range of densities. Note that the Fe IX ratio is the best density diagnostic for the solar corona, but has never been observed routinely. Also note that the Fe XXI ratio is one of the best ones across all wavelengths in terms of measuring the low electron densities which are expected to be present during the heating phase. Most other density diagnostics of flare lines, from the X-rays to the UV, are sensitive above 10^{12} cm^{-3}.

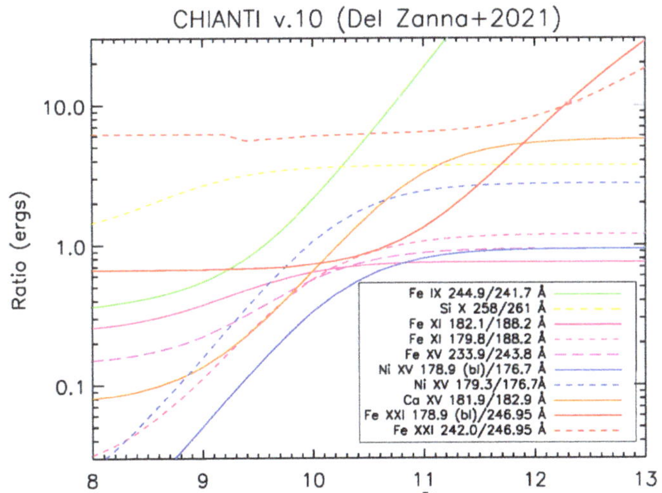

Figure 1. A selection of some important diagnostic spectral lines to measure electron densities. The line intensity ratios of Fe IX, Fe XI, Fe XV, Si X, Ni XV, and Ca XV are useful for the diagnostics of the solar corona and active regions, while the ratios of Fe XXI (red) are applicable to solar flares.

3.5. Elemental Composition Diagnostics

Within the solar corona, there is a wide range of elemental abundances. As is the case for in situ measurements, the variations are mainly associated with the first ionisation potential (FIP) of the elements, indicating that the fractionation process that creates these variations occurs in the low chromosphere, where neutrals become ionised. The ratio of a low-FIP to a high-FIP element tends to be enhanced compared to its photospheric value (the so-called FIP bias), although interesting cases of the so-called inverse FIP effect have been observed. There is ample literature, but many earlier studies turned out to be incorrect, leading to a very confusing picture. A recent review of observational aspects is given in Ref. [70].

In a nutshell, the quiescent corona has nearly photospheric abundances, at least up to 1 MK; the hot 3 MK cores of active regions have an FIP bias of about 3–4; the cool active region loops have a wide range of FIP biases. Surges and flaring loops have been known since the Skylab time (1970s) to have photospheric abundances, interpreted as the indication that the plasma filling the loops evaporated from the chromosphere with a fast timescale such that the fractionation process did not occur. Interestingly, the recent X-ray observations of small flares of the GOES B-class have shown for the first time a rapid decrease in FIP bias, followed by a gradual increase [79]. While the rapid decrease is expected to be caused by chromospheric evaporation, the gradual increase could be a signature of Alfvén waves.

There is general consensus that the FIP effect is related to the strength and topology of the magnetic field. A relatively simple hydrostatic model of the chromosphere put forward by M. Laming is able to explain the main features of the FIP effect and its inverse, as can be seen in the review and references of Ref. [80]. The FIP effect would be caused by the ponderomotive force of trapped Alfven waves in magnetically closed loops. Therefore, measuring the FIP effect and its variations during flares is important, as it could provide a way of indirectly confirming the presence of Alfven waves or at least indicate the origin of the flaring plasma: if the flare loops have an FIP bias, then the plasma originated from closed loops already present in active regions.

The diagnostics to measure the relative abundances are limited, however, as there are only a few ions formed at high temperatures from high-FIP elements. There are plenty of spectral lines from low-FIP elements such as Si, Fe, Mg, Ni, and Ca, instead. The best high-FIP ones, exploited in the 1970s and 1980s, are the H-like and He-like ions from oxygen, neon, argon, and sulphur in the X-rays. Note that sulphur has an FIP of 10 eV, but in remote-sensing observations, it varies like the other high-FIP elements. Very few weak lines from high-FIP elements are present in the EUV and UV. A few very weak lines from Ar XIV and Ar XIII have been observed with Hinode EIS, which also has stronger S X, S XII and S XIII lines.

Across the whole EUV/UV, the best diagnostic lines are the strong Li-like doublets from Ar XVI at 353.8, 389.0 Å and from S XIV at 417.7, 445.7 Å. After them, the best diagnostic is the strongest (resonance) line from Ar XV at 221.15 Å, which is within the SISA spectral range. There are also interesting Ar XIV lines as the strong 243.75 Å line, blended with Fe XV, or the 187.9 Å line, which becomes strong at high densities (but can also be blended), meaning that one would have to measure the density independently. Other good options within the SISA range are the strong S XIII at 256.6 Å and the lower-temperature S X lines, often used in previous Hinode EIS studies. Weak lines from Ar XIII are also available. There are plenty of lines from iron to be able to estimate the FIP bias measuring Ar/Fe and S/Fe in the 1–3 MK range.

3.6. Flare Lines

There are several hot flare lines in the SISA spectral range. Close to Fe X 257.26 Å there are the strong Fe XVII 254.9, Fe XXI 246.95 Fe XXII 247.19, 253.17, Ti XX 259.26, and Fe XXIV 255.1 Å lines. They span a temperature range of $\log T = [6.6–7.3]$. Nearby, there is a strong Fe XXI line at 242.05Å which could be used in combination with the 246.95 (or the 187.9 Å) to measure densities above 10^{11} cm^{-3} (Section 3.4). This is an important diagnostic.

The strong Fe XXIII intercombination line is located at longer wavelengths, at 263.76 Å. It would be an ideal electron temperature diagnostic in combination with the resonance line at 132 Å, if a second order could be observed.

Close to 184.5 Å, there are several weaker hot lines, which become observable for large flares. Fe XXIII produces a line around 180. Å. There is a relatively strong Fe XXI line around 187.93 Å; however, this line for weak flares is blended with an Ar XIV 187.96 Å line which is strongly density-sensitive, so careful deblending is needed. A weak Fe XXI around 178.9 Å is particularly important as it becomes strong at high electron densities, so it is one of the best density diagnostics for flare plasma, one that is able to measure densities above 10^{10} cm^{-3}.

3.7. Departures from Ionisation Equilibrium

The rapid heating or cooling of plasma can lead to departures from ionisation equilibrium; a state when the plasma ionic composition does not reflect its electron temperature, e.g., Refs. [57,81–86]. This transient ionisation in turn can affect other diagnostics, e.g., Refs. [87–89]. For a review of transient ionisation conditions, see Ref. [60].

Conditions favourable for the occurrence of transient ionisation are expected to arise in solar flares, where the energetic particles travel at fractions of the speed of light, and thus swiftly deposit large amounts of energy when encountering the dense solar chromosphere. This is so despite the electron densities ion flares are relatively high, leading to decrease in the timescales for collisional ionisation and recombination [84], which are in the flare plasma of the order of seconds or tens of seconds. Since the plasma evolves rapidly following the energy deposit, detecting transient ionisation conditions requires fast cadence. The previous diagnostics of transient ionisation, e.g., Refs. [81,82] focused on X-ray spectra, which were obtained with little or no spatial resolution and required scans across the wavelength range, while the detection of transient ionisation from EUV was hampered by the twin problems of slit and throughput/cadence. For the first time, SISA will offer the spatial coverage, resolution, cadence, and lines from multiple ionisation stages of iron (see Table 1), all required to detect the signatures of transient ionisation in rapidly evolving plasma along newly created flare loops.

3.8. Predicted Signal in the Main SISA Lines

Given the previous requirements, namely providing spectral line diagnostics to measure magnetic fields, electron temperatures and densities, departures from ionisation equilibrium in flaring plasma, non-Maxwellian electron distributions, variations in the FIP bias, and transient ionisation, we designed a few possible multilayers to cover the key lines, as described below. Having selected one, we then calculated the signal in the SISA lines.

As the target is to observe active regions and flares, two simulations have been carried out, one for the quiescent AR core, and one for an M2-class flare. For the AR core, the predictions are based on an AR core loop observed by Hinode, as can be seen in Ref. [67]. The Del Zanna (2013) coronal abundances are used, with an electron density of 2×10^9 cm^{-3}. For the M2 flare, the predictions are based on an M2-class flare observed by Hinode EIS. The same set of abundances is used, but with an electron density of 2×10^{10} cm^{-3}. In this case, a full DEM analysis was carried out (Del Zanna et al. 2023, in prep). CHIANTI version 10 [78] was used. Table 1 shows the radiances of a few selected SISA lines and the corresponding data numbers per second per 1″ pixel. It is clear that most lines are observable with exposures of 1 s.

4. Image Slicer Technology in the EUV Regime

Among the integral field unit (IFU) alternatives available to perform IFS, image slicers are the optimum for space: they are very compact with no moving mechanisms, lightweight, highly efficient, and all parts are made of the same material.

As shown in Figure 2, an image slicer is an optical system typically composed of two or three arrays of mirrors, placed at the telescope's image focal plane. The first array (slicer mirror array) is composed of multiple narrow mirrors with a rectangular shape distributed such that the short dimension (width) corresponds to the spectral direction and the long dimension (length) is associated with the spatial direction. Each one of the slicer mirrors have a different tilt angle around the X and Y axes (axes perpendicular to the direction of light propagation). Thus, each mirror only reflects a slice of the field of view, defined by its effective surface, in a different direction. The other arrays of mirrors are used to control the intermediate and exit pupils, re-image the slices of the field and reorganise them to generate the spectrograph entrance slit. They can also produce a magnification if required, enabling perfect coupling between the telescope and instrument, and adjusting the slit width to meet the desired resolution. Other advantages include no focal-ratio degradation and no polarisation effects.

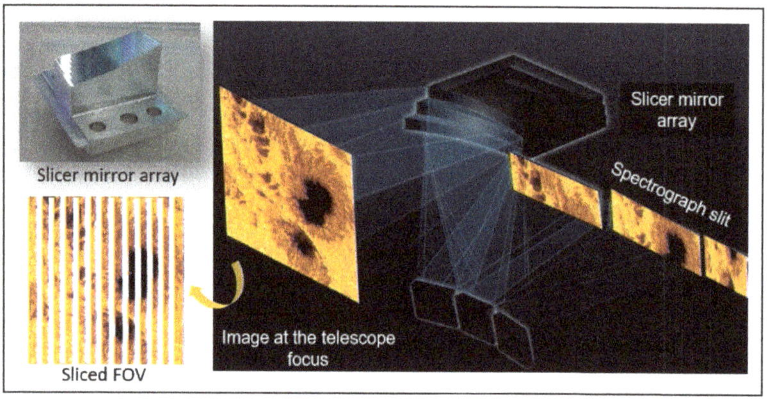

Figure 2. Conceptual sketch of an image slicer. The slicer mirror array divides the image generated by the telescope into slices that are re-imaged and reorganised by other arrays of mirrors, reformatting the 2D field of view into the spectrograph entrance slit.

Image slicers have been used for ground-based telescopes in recent decades, operating at different spectral ranges from the infrared to UV/visible, for a wide variety of science cases. Some examples are shown in Figure 3 and include the following: VLT CUBES [90], BlueMUSE [91] and MUSE [92]; GTC FRIDA [93]; GNIRS at Gemini North Telescope or ELT HARMONI [94] and METIS [95] for night-time observation. For solar physics, image slicers were first proposed for the integral field spectrograph [96,97] of the European Solar Telescope, which provides optical observations. A prototype of this image slicer is operative at GREGOR solar telescope [98], which upgraded GRIS from a long-slit spectrograph to an integral field spectrograph, and provides infrared observations. No EUV solar IFS instrument has flown, although the SNIFS rocket mission, scheduled for a 2024 launch, will observe solar Ly α using IFS [11].

There are also spaceborne image slicers onboard the James Webb Space Telescope in NIRSpec [99] and MIRI [100] operating at infrared wavelengths. At much shorter wavelengths, INFUSE [101] was the first IFS in the far ultraviolet (FUV), while SISA [102] will be the first IFS in the extreme ultraviolet, applying the image slicer technology at the shortest wavelengths ever proposed.

Finally, a different approach to the 2D coverage problem has been adopted by the multi-slit solar explorer MUSE; Refs. [3,103], a NASA Heliophysics Medium-class Explorer with a scheduled launch in 2027. Thanks to its 35-slit design, MUSE will provide high-cadence (<20 s or faster for flares), sub-arc-second resolution spectroscopic rasters over an active region size, targeting four isolated, strong spectral lines (Fe IX 171 Å at 0.7 MK, Fe XV 284 Å at 2 MK, Fe XIX at 108 Å at 10 MK, and Fe XXI 108 Å at 12 MK). With its unique capabilities, MUSE will provide new crucial discoveries for flare science. SISA would build upon the science legacy of MUSE by providing high-cadence 2D imaging from a larger set of spectral lines that offer rich diagnostic potential.

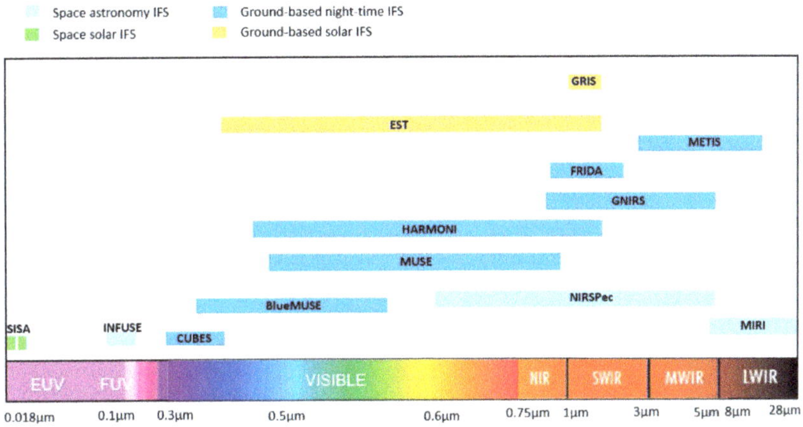

Figure 3. SISA will break new ground, both technologically, by observing using image slicers in the EUV, and scientifically, by providing the high-cadence 2D spectra of various solar phenomena. Currently operating image slicer-based IFS at different spectral ranges for solar and night-time ground-based telescopes and space missions observe at longer wavelengths than SISA.

At the shorter wavelengths of the EUV regime that SISA will explore, the application of image slicers presents additional challenges [102], like narrower slicer mirrors for high resolution solar observations, surface roughness, and coating. A slicer width of 15 μm has been defined for SISA, which is narrower than the slicer mirror width that is currently achievable both for glass and metallic slicer mirrors. This width is even more difficult to achieve on curved substrates (powered slicer mirrors), whose use would enable the reduction in the number of optical components to maximise throughput. At these short wavelengths the surface roughness is a key factor to minimise stray light that could potentially compromise the instrument sensitivity.

These technological challenges of the image slicer technology for the EUV are being addressed and developed in the UK by a collaboration between Durham University and University College London (UCL) within two projects: MINOS and LUCES. MINOS (Manufacturing of Image slicer Novel technology for Space), funded by Durham University, produced in 2023 a prototype that represents the state-of-the-art of glass slicers, achieving a surface roughness of 0.2 nm RMS in Fused Silica spherical slicer mirrors with a width of 70 μm. This is the thinnest width currently achievable on glass spherical slicers and the best surface roughness for slicers to date. This is a significant improvement towards the specifications required for SISA and the next generation of solar space missions. This demonstrator was manufactured by Bertin Winlight, in France.

LUCES (Looking Up Image Slicers Optimum Capabilities in the EUV for Space) is funded by the Enabling Technology Programme of the UK Space Agency and will produce nine slicer mirror demonstrators in metal using different widths (350 μm, 70 μm, 50 μm, 15 μm) to determine the narrowest achievable width with diamond machining techniques at Durham University, and on different substrates to compare the surface roughness. The widths for the slicer demonstrators of LUCES were selected based on the demand of the solar community both for ground-based and space observations. The widest width (350 μm) was selected as a first goal to reduce the current minimum width manufactured in metal; 70 μm and 50 μm were chosen based on the widths required for the GRIS spectrograph at GREGOR and for the integral field spectrograph of the European Solar Telescope. The thinnest width, 15 μm, represents a challenge and is the current specification for the slicer mirrors for SISA. LUCES started in April 2023 and will end in October 2024.

SISA, along with LISSAN and FOXSI, were proposed as part of the SPARK mission to an ESA's M-class solicitation in 2022. Though selected for Phase 1, SPARK was not selected for continued development in Phase 2, in part due to concerns over the relatively low TRL of aspects of the instruments. The SISA team is actively engaged in on-going research to raise the TRL of SISA, developing the image slicer technology for EUV applications. Further developments, such as the deposition of a multi-layer coating on the slicers, the design and manufacturing of a SISA prototype, and its tests in a relevant environment, are planned to achieve TRL 6 within the next couple of years, subject to funding.

5. SISA Instrument Proposal

5.1. Specifications

The Spectral Imager of the Solar Atmosphere (SISA) is the first EUV imaging spectrometer using image slicer technology in the $\lambda = [180–260]$ Å range. SISA will observe a threshold field of view (FoV) of $100'' \times 100''$ (with a goal of $100'' \times 250''$), obtaining spectra of the whole 2D field of view in less than 3 s for high-signal objects (e.g., flares) and less than 30 s for low-signal ones. This field of view will be observed in two spectral windows simultaneously, Window 1 $\lambda = [178–184]$ Å, and Window 2 $\lambda = [246–258]$ Å, with a spectral resolving power (R) of $R = [3560–5160]$, a spatial resolution of $< 3''$, and a spectral resolution of 0.05 Å full width at half maximum (FWHM).

The defined spatial resolution is required to satisfactorily resolve flare loops and ribbon/footpoint sources, including newly reconnected ones, whilst the spectral resolution is required to resolve the main diagnostic lines and measure nonthermal line profiles. The FoV of $100'' \times 100''$ can capture the core of large X-class flares, the entirety of smaller flares, or a portion of an active region. A high signal temporal resolution (Cadence 1) of 3 s will capture much of the dynamics of many flares, whilst being two orders of magnitude faster than previously observed for 2D EUV imaging spectroscopy. A low signal temporal resolution (Cadence 2) of 30 s will capture the slower dynamics present in the active regions or quiet Sun areas.

The two spectral ranges are required to measure the parameters of both 1 MK plasma and the much hotter 15 MK plasma that appears during flares, and include lines sensitive to the coronal magnetic field strength, electron temperature/non-Maxwellian electron distributions, and a wide range of lines to measure electron densities from coronal (e.g., Fe IX, Fe XI, Fe XII, Fe XV, Ca XV) to flare temperatures (Fe XXI), and the FIP bias. The specifications for SISA are presented in Table 2.

Table 2. Specifications for the instrument SISA.

Field of view	$100'' \times 100''$
Spectral window 1	178–184 Å
Spectral window 2	246–258 Å
Spectral resolution	0.05 Å FWHM
Spatial resolution	$<3''$
Spectral resolving power (R)	3560–5160
Temporal resolution (high signal)	<3 s
Temporal resolution (low signal)	<30 s

5.2. Layout

The solar EUV photon flux is very low, demanding a sensitive instrument. SISA combines the surfaces of the integral field unit (IFU) with those of the spectrograph in a compact and highly efficient layout that uses only an array of powered slicer mirrors and an array of curved diffraction gratings as shown in Figure 4. This layout is based on a design presented by Calcines et al., 2018 [104] for applications for ultra-compact integral field spectrograph and is inspired in IGIS [105], which first proposed the combination of the IFU and the spectrograph components. SISA does not have an entrance slit, it covers a

2D field of view without any scanning mechanisms, obtaining all the spectra of a 2D region of the Sun within one exposure.

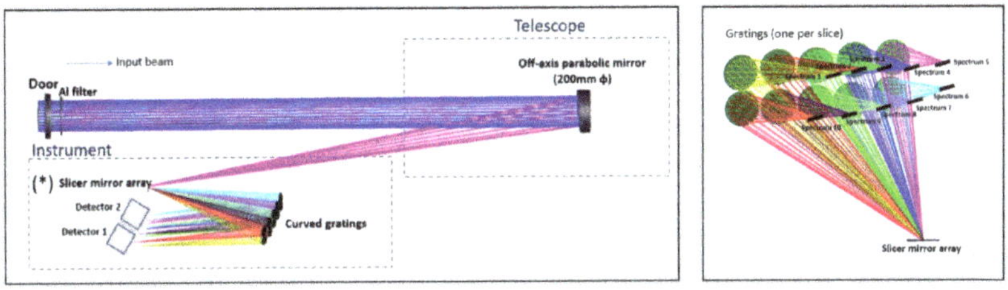

Figure 4. SISA's conceptual layout (**left**) and view of the spectra per slice generated by each curved grating (**right**).

Two subsystems are identified: the telescope and the integral field spectrograph, with a total of three surfaces. To optimise efficiency, the telescope is composed of one off-axis parabolic mirror, with an aperture of 200 mm diameter and an effective focal length of 3 m. At its focal plane, the array of slicer mirrors, each one of them with different orientations, divides the image of the field into thin slices of 15 µm. To cover a field of view of $100'' \times 100''$, a hundred slicer mirrors are needed, with a rectangular shape and whose dimensions are 3.64 mm length by 15 µm width. The slicer mirror array decomposes the incoming beam into as many sub-beams as the number of slicer mirrors, generating a pupil for each one of them. At the pupil position, an array of curved gratings will disperse the light and focus the beams at their focal length. Each grating will produce the spectrum associated with a slice of the field. The different orientations for the gratings are defined to distribute the spectra on the detectors. These orientations are fixed, with no moving mechanisms.

The design of SISA includes different trade-off analyses, such as the definition of the telescope effective focal length versus resolution and slicer mirror width. For the dispersive element, two options were considered: a larger diffraction grating versus an array of gratings, one per slice. Whilst a larger grating would provide higher spectral resolving power, curved gratings present aberrations for off-axis fields. The use of an array facilitates the organisation of the spectra on the detectors. This is currently the selected option.

On-going studies analyse the optimum distribution of the gratings in order to minimise stray light and optimise the optical quality. The use of a grating per slicer mirror enables some flexibility in the geometrical distribution, which is a key factor in the optimisation of these two important points.

5.3. Components

As mentioned in the previous section, SISA is composed of two subsystems: (1) the telescope and (2) the instrument. A door protection mechanism will be installed at the front of the instrument. This door will be used to provide protection against acoustic vibration and venting during the launch. It will also protect the instrument from contamination during assembly, integration and verification (AIV), launch and in-orbit, and control the venting and reduce air flow during on-ground purging and launch. CSL is considered as the future supplier for the door based on their previous experience designing the door for the extreme ultraviolet imager (EUI) onboard Solar Orbiter.

The aperture of the telescope (200 mm) was selected to optimise throughput with respect to existing EUV instruments. The same supplier of the METIS/Solar Orbiter mirrors is considered, TOPTEC, Institute of Plasma Physics of the Czech Academy of Sciences with TRL 9. The manufacturing feasibility of mirrors with apertures up to 300 mm has been confirmed.

The slicer mirror technology presents the technology readiness level (TRL) 9 in the infrared. For the EUV, this technology is currently under development. MINOS and LUCES projects will raise the TRL in the EUV to TRL 4. Although the slicers are the component with the lowest TRL for this spectral range, its application is a game-changer for EUV solar spectroscopy. The on-going developments within the UK in this field, together with a plan in place to achieve TRL 6 within the next several years, will enable this technology to be considered for the next generation of solar space missions.

Curved holographic gratings have already been used for other instruments like EIS and the upcoming EUV High-Throughput Spectroscopic Telescope, EUVST [106], with TRL 9. For the gratings for SISA, IAS in France is currently considered as the future lead institution. The possibility of using metallic curved gratings made of the same material as the slicer mirrors will also be investigated.

The SISA instrument requires a large-size image sensor with high EUV sensitivity. The baseline will be to use the back-illuminated CMOS radiation-hardened active-pixel image sensor (APS), developed by CSL/ROB (Belgium). Compared to CCD technology, CMOS-APS detectors do not require a mechanical shutter, have a greater radiation tolerance (low charge transfer), low power consumption, high speed, and dynamic range, with a non-destructive readout for low readout noise. Additional features include random addressing, windowing and anti-blooming. SISA's detectors must be cooled to a temperature lower than -40 °C (ideally around -60 °C), with a stability of ±5 °C.

5.4. *Efficiency*

The efficiency of the instrument will be limited by two important factors: (1) the possibility of using only two optical components for the integral field spectrograph, and only one for the telescope based on the performance of optical quality, and (2) the coating.

For the simultaneous observation of the two spectral windows, a dual-band multilayer coating is being designed for SISA at the National Research Council of Italy, Padova. Considering the required operational spectral bands for SISA, the most efficient coatings in terms of efficiency were identified to be the three-material multilayer Al/Mo/B_4C [107] or, alternatively, the Al/Mo/SiC [108]. Both coatings show very good long-term stability, and in the case of Al/Mo/SiC, its reliability has been proven through its successful use in the Solar Orbiter/EUI instrument. The multiband response required for SISA was obtained using a design consisting of a superposition of two periodic multilayer stacks separated by two or three buffer layers. The coating optimization was performed using a genetic algorithm, following the procedure previously described in Ref. [109]. Figure 5 shows the three options obtained with an Al/Mo/B_4C multilayer, among which Option 3 is the preferred alternative due to it offering the broadest response, with a signal near 220 Å. Similar results can be obtained with the Al/Mo/SiC multilayer.

A preliminary estimation of the throughput has been performed. The effective area for SISA is represented in Figure 6 (in black), compared to Hinode/EIS (in blue). To calculate the effective area, we have considered the following components: (1) a front Al filter, the same used for SDO/AIA, but with an improved mesh transmission of 95%, compared to the AIA one (85%). Such an improved mesh has already flown on the Hi-C sounding rockets. The AIA filters have shown little degradation at the SISA wavelengths, probably thanks to their carbon layer, but alternative coatings will be considered. (2) Two mirror reflectivities (one for the primary and one for the slicer mirrors) using the multilayer option 3; (3) A third reflectivity using the same multilayer for the grating, with an efficiency of 35%, the same measured for the Hinode EIS grating in the long wavelength channel; and (4) The same quantum efficiency of the EIS CCD, 0.85. A secondary focal plane filter has not been included to maximise throughput, although a rotating wheel of redundant focal plane filters should be included.

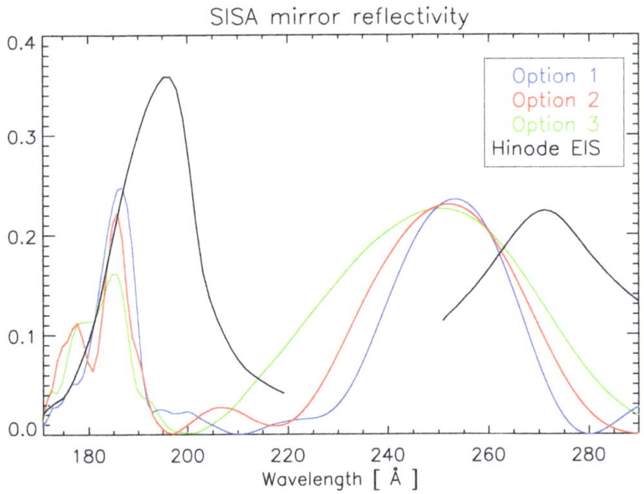

Figure 5. Reflectivities for the three multi-layer coatings under design for SISA.

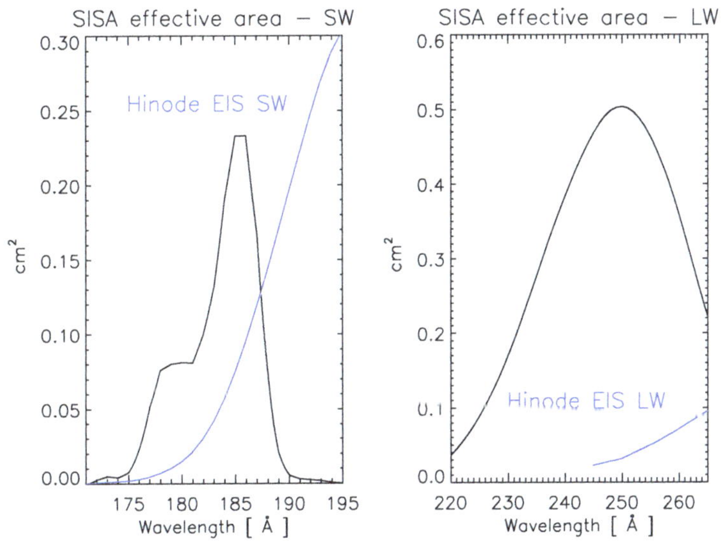

Figure 6. Preliminary estimates of SISA's effective area (black), compared to Hinode/EIS short-wavelength (SW) and long-wavelength (LW) channels (blue).

The significant improvement over the EIS effective areas is due to the slightly larger aperture (20 cm instead of 15), and the fact that a single multilayer is used (the EIS aperture is sectioned with two different multilayers). Also, the multilayer has been designed to have maximum reflectance close to the main diagnostic lines around 250 Å.

6. Conclusions

Integral field spectroscopy is one of the most powerful techniques applied in astronomy. The ability to obtain the spectra of a 2D field of view simultaneously reduces the integration time by over two orders of magnitude. Despite this great advantage, this technique has not been used in the extreme ultraviolet regime, which can only be observed from space due to the limitations and challenges for the existing integral field unit alternatives.

The image slicer technology seems to be the optimum choice, offering a compact, robust, and highly efficient solution with no moving mechanisms. The SPARK (Solar Particle Acceleration Radiation and Kinetics) mission proposal led by the UK (UCL) includes the first ever integral field spectrograph in the EUV: SISA. SISA consists of a three-mirror system including the telescope and the instrument.

A preliminary study has identified optimal spectral ranges to allow many plasma diagnostics, some of which are not available to current (Hinode EIS) or future EUV instruments (EUVST, MUSE). With a relatively small aperture of 20 cm diameter, a single multilayer and the slicers, we expect to perform the types of measurements that Hinode EIS has been producing about 100 times faster.

Although the image slicer technology currently presents the low TRL at EUV wavelengths, the UK (Durham University and UCL) has on-going research and development projects that will soon enable this technology for the next generation of space missions, including solar physics and all fields of astronomy. These are focused on the most critical aspects for EUV applications: the manufacturing feasibility of slicer mirrors to reduce the current state-of-the-art to the specifications required by SISA; the optimisation of the surface roughness for the slicer mirrors to minimise stray light, and to increase the feasibility of the application of multi-layer dielectric coatings to narrow slicers in the EUV with high efficiency for the simultaneous spectral windows covered by SISA.

SISA, currently under design study as part of the SPARK payload mission proposal described in Reid et al. [110], promises to be a game changer in our understanding of our star, the Sun.

Author Contributions: The paper was based upon the SPARK mission concept that was coordinated by H.A.S.R., S.M. and D.F.R. All authors contributed by providing scientific expertise or instrumentation expertise in the development of the SISA instrument concept and contributed to the research goals. Their contributions include: conceptualisation, investigation, formal analysis, writing, and funding acquisition. Methodology: the optical design was led by A.C.R. Conceptualisation: The optical design is based on an idea for EUV image slicers proposed by S.A.M. Methodology and software: A.J.C. developed a model to obtain the optimal multilayer coatings for SISA, with the wavelength selection coordinated by G.D.Z., who also developed the models for the effective areas, calculated the expected signal (cf. Table 1), and developed/described the spectroscopic diagnostics listed in the paper. Writing: the science sections were mainly written by G.D.Z., J.D. and G.S.K. (Sections 2 and 3). The paper review and editing was performed by: G.D.Z., G.S.K., J.D., V.P., H.A.S.R. and A.C.R. Project administration: H.A.S.R. (principal investigator of SISA). All authors have read and agreed to the published version of the manuscript.

Funding: H.A.S.R. and S.A.M. were funded by the UK Science, Technology, and Facilities Council (STFC) under the consolidated grant ST/W001004/1. H.A.S.R. also acknowledges funding from STFC grant ST/X002012/1. G.D.Z. acknowledges support from STFC via the consolidated grants to the atomic astrophysics group at DAMTP, University of Cambridge (ST/P000665/1. and ST/T000481/1). J.D. acknowledges the Czech National Science Foundation, Grant No. GACR 22-07155S, as well as institutional support RWO:67985815 from the Czech Academy of Sciences. G.S.K. acknowledges financial support from NASA's Early Career Investigator Program (Grant# NASA 80NSSC21K0460) as well as from the PHaSER cooperative agreement (80NSSC21M0180). D.O.S. acknowledges financial support from the grants AEI/MCIN/10.13039/501100011033/ (RTI2018-096886-C5, PID2021-125325OB-C5, PCI2022-135009-2) and ERDF "A way of making Europe" and "Center of Excellence Severo Ochoa" award to IAA-CSIC (CEX2021-001131-S). L.A.H. and S.M. are supported by ESA Research Fellowships. A.C.R. acknowledges the ETP funding programme of the UK Space Agency and Durham University Seedcorn Funding.

Data Availability Statement: New data are contained within the article. The simulation of EUV multilayer coatings efficiency as well as the estimation of the SISA effective areas were obtained using new custom-made codes. Data sharing is not applicable to this article.

Acknowledgments: J.D. wishes to acknowledge useful discussions with E. Dzifčáková.

Conflicts of Interest: The authors declare no conflicts of interest.

Abbreviations

The following abbreviations are used in this manuscript:

AIA	Atmospheric Imaging Assembly
AIV	Alignment Integration and Verification
CCD	Charge-Coupled Device
CME	Coronal Mass Ejection
CUBES	Cassegrain U-Band Efficient Spectrograph
EIS	EUV Imaging Spectrometer Onboard Hinode
ELT	Extremely Large Telescope
EUI	Extreme Ultraviolet Imager (onboard Solar Orbiter)
EUV	Extreme Ultraviolet
EUVST	Extreme Ultraviolet High-Throughput Spectroscopic Telescope
EVE	EUV Variability Experiment Onboard SDO
FIP	First Ionisation Potential
FOXSI	Focusing Optics X-ray Solar Imager
FRIDA	Infrared Imager and Dissector for Adaptive optics
Full Width Half Maximum	
FUV	Far Ultraviolet
FWHM	Full Width Half Maximum
GOES	Geostationary Operational Environmental Satellite
GNIRS	Gemini Near-Infrared Spectrograph
GRIS	Gregor Infrared Spectrograph
GTC	Gran Telescopio Canarias
HARMONI	High-Angular-Resolution Monolithic Optical and Near-Infrared Integral field spectrograph
HXR	Hard X-ray
IFS	Integral Field Spectrograph
IFU	Integral Field Unit
INFUSE	Integral Field Ultraviolet Spectroscopic Experiment
LISSAN	Large Imaging Spectrometer for Solar Accelerated Nuclei
LUCES	Looking Up Image Slicer Optimum Capabilities in the EUV for Space
METIS	Mid-Infrared E-ELT Imager and Spectrograph
MHD	Magnetohydrodynamic
MINOS	Manufacturing of Image Slicer Novel technology for Space
MIRI	Mid-Infrared Instrument
MUSE	Multi-Slit Solar Explorer
QPP	Quasi-Periodic Pulsations
SDO	Solar Dynamics Observatory
SISA	Spectral Imaging of the Solar Atmosphere
SNIFS	Solar Eruption Integral Field Spectrograph
SPARK	Solar Particle Acceleration, Radiation, and Kinetics Mission
SXR	Soft X-ray
TRL	Technology Readiness Level

References

1. Fletcher, L.; Dennis, B.R.; Hudson, H.S.; Krucker, S.; Phillips, K.; Veronig, A.; Battaglia, M.; Bone, L.; Caspi, A.; Chen, Q.; et al. An Observational Overview of Solar Flares. *Space Sci. Rev.* **2011**, *159*, 19–106. [CrossRef]
2. Milligan, R.O. Extreme Ultra-Violet Spectroscopy of the Lower Solar Atmosphere During Solar Flares (Invited Review). *Sol. Phys.* **2015**, *290*, 3399–3423. [CrossRef]
3. De Pontieu, B.; Testa, P.; Martínez-Sykora, J.; Antolin, P.; Karampelas, K.; Hansteen, V.; Rempel, M.; Cheung, M.C.M.; Reale, F.; Danilovic, S.; et al. Probing the Physics of the Solar Atmosphere with the Multi-slit Solar Explorer (MUSE). I. Coronal Heating. *Astrophys. J.* **2022**, *926*, 52. [CrossRef]
4. Lörinčík, J.; Dudík, J.; Polito, V. Blueshifted Si IV 1402.77 Å Line Profiles in a Moving Flare Kernel Observed by IRIS. *Astrophys. J.* **2022**, *934*, 80. [CrossRef]
5. Klimchuk, J.A. On Solving the Coronal Heating Problem. *Sol. Phys.* **2006**, *234*, 41–77. [CrossRef]
6. Klimchuk, J.A. Key aspects of coronal heating. *Philos. Trans. R. Soc. Lond. Ser. A* **2015**, *373*, 20140256. [CrossRef]

7. Cargill, P.J. Active Region Emission Measure Distributions and Implications for Nanoflare Heating. *Astrophys. J.* **2014**, *784*, 49. [CrossRef]
8. Schmelz, J.T.; Winebarger, A.R. What can observations tell us about coronal heating? *Philos. Trans. R. Soc. Lond. Ser. A* **2015**, *373*, 20140257. [CrossRef]
9. Culhane, J.L.; Harra, L.K.; James, A.M.; Al-Janabi, K.; Bradley, L.J.; Chaudry, R.A.; Rees, K.; Tandy, J.A.; Thomas, P.; Whillock, M.C.R.; et al. The EUV Imaging Spectrometer for Hinode. *Sol. Phys.* **2007**, *243*, 19–61. [CrossRef]
10. De Pontieu, B.; Title, A.M.; Lemen, J.R.; Kushner, G.D.; Akin, D.J.; Allard, B.; Berger, T.; Boerner, P.; Cheung, M.; Chou, C.; et al. The Interface Region Imaging Spectrograph (IRIS). *Sol. Phys.* **2014**, *289*, 2733–2779. [CrossRef]
11. Chamberlin, P.C.; Schmit, D.J.; Daw, A.N.; Polito, V.; Gong, Q.; Milligan, R.O. The Solar eruptioN Integral Field Spectrograph (SNIFS) Sounding Rocket. In Proceedings of the AGU Fall Meeting Abstracts, Online, 1–17 December 2020; Volume 2020, p. SH056-03.
12. Knoer, V.; Chamberlin, P.; Daw, A.; Gong, Q.; Milligan, R.; Polito, V.; Schmit, D. A Novel Integral Field Spectrograph Design for taking High-Cadence Spectral Solar Images: SNIFS. In Proceedings of the AGU Fall Meeting Abstracts, Online, 13–17 December 2021; Volume 2021, p. SH55B-1837.
13. Young, P.R.; Doschek, G.A.; Warren, H.P.; Hara, H. Properties of a Solar Flare Kernel Observed by Hinode and SDO. *Astrophys. J.* **2013**, *766*, 127. [CrossRef]
14. Dudík, J.; Polito, V.; Janvier, M.; Mulay, S.M.; Karlický, M.; Aulanier, G.; Del Zanna, G.; Dzifčáková, E.; Mason, H.E.; Schmieder, B. Slipping Magnetic Reconnection, Chromospheric Evaporation, Implosion, and Precursors in the 2014 September 10 X1.6-Class Solar Flare. *Astrophys. J.* **2016**, *823*, 41. [CrossRef]
15. Panos, B.; Kleint, L.; Huwyler, C.; Krucker, S.; Melchior, M.; Ullmann, D.; Voloshynovskiy, S. Identifying Typical Mg II Flare Spectra Using Machine Learning. *Astrophys. J.* **2018**, *861*, 62. [CrossRef]
16. Polito, V.; Kerr, G.S.; Xu, Y.; Sadykov, V.M.; Lorincik, J. Solar Flare Ribbon Fronts. I. Constraining Flare Energy Deposition with IRIS Spectroscopy. *Astrophys. J.* **2023**, *944*, 104. [CrossRef]
17. Qiu, J.; Lee, J.; Gary, D.E.; Wang, H. Motion of Flare Footpoint Emission and Inferred Electric Field in Reconnecting Current Sheets. *Astrophys. J.* **2002**, *565*, 1335–1347. [CrossRef]
18. Fletcher, L.; Pollock, J.A.; Potts, H.E. Tracking of TRACE Ultraviolet Flare Footpoints. *Sol. Phys.* **2004**, *222*, 279–298. [CrossRef]
19. Sun, X.; Hoeksema, J.T.; Liu, Y.; Aulanier, G.; Su, Y.; Hannah, I.G.; Hock, R.A. Hot Spine Loops and the Nature of a Late-phase Solar Flare. *Astrophys. J.* **2013**, *778*, 139. [CrossRef]
20. Dudík, J.; Janvier, M.; Aulanier, G.; Del Zanna, G.; Karlický, M.; Mason, H.E.; Schmieder, B. Slipping Magnetic Reconnection during an X-class Solar Flare Observed by SDO/AIA. *Astrophys. J.* **2014**, *784*, 144. [CrossRef]
21. Li, T.; Zhang, J. Slipping Magnetic Reconnection Triggering a Solar Eruption of a Triangle-shaped Flag Flux Rope. *Astrophys. J. Lett.* **2014**, *791*, L13. [CrossRef]
22. Li, T.; Zhang, J. Quasi-periodic Slipping Magnetic Reconnection During an X-class Solar Flare Observed by the Solar Dynamics Observatory and Interface Region Imaging Spectrograph. *Astrophys. J. Lett.* **2015**, *804*, L8. [CrossRef]
23. Gou, T.; Liu, R.; Wang, Y.; Liu, K.; Zhuang, B.; Chen, J.; Zhang, Q.; Liu, J. Stereoscopic Observation of Slipping Reconnection in a Double Candle-flame-shaped Solar Flare. *Astrophys. J. Lett.* **2016**, *821*, L28. [CrossRef]
24. Pan, H.; Gou, T.; Liu, R. Sigmoid Formation through Slippage of a Single J-shaped Coronal Loop. *Astrophys. J.* **2022**, *937*, 77. [CrossRef]
25. Aulanier, G.; Pariat, E.; Démoulin, P.; Devore, C.R. Slip-Running Reconnection in Quasi-Separatrix Layers. *Sol. Phys.* **2006**, *238*, 347–376. [CrossRef]
26. Aulanier, G.; Janvier, M.; Schmieder, B. The standard flare model in three dimensions. I. Strong-to-weak shear transition in post-flare loops. *A&A* **2012**, *543*, A110. [CrossRef]
27. Janvier, M.; Aulanier, G.; Pariat, E.; Démoulin, P. The standard flare model in three dimensions. III. Slip-running reconnection properties. *A&A* **2013**, *555*, A77. [CrossRef]
28. De Pontieu, B.; Polito, V.; Hansteen, V.; Testa, P.; Reeves, K.K.; Antolin, P.; Nóbrega-Siverio, D.E.; Kowalski, A.F.; Martinez-Sykora, J.; Carlsson, M.; et al. A New View of the Solar Interface Region from the Interface Region Imaging Spectrograph (IRIS). *Sol. Phys.* **2021**, *296*, 84. [CrossRef]
29. Kerr, G.S. Interrogating Solar Flare Loop Models with IRIS Observations 1: Overview of the Models, and Mass flows. *Front. Astron. Space Sci.* **2022**, *9*, 1060856. [CrossRef]
30. Kerr, G.S. Interrogating Solar Flare Loop Models with IRIS Observations 2: Plasma Properties, Energy Transport, and Future Directions. *Front. Astron. Space Sci.* **2023**, *9*, 1060862. [CrossRef]
31. Allred, J.C.; Kerr, G.S.; Gordon Emslie, A. Solar Flare Heating with Turbulent Suppression of Thermal Conduction. *Astrophys. J.* **2022**, *931*, 60. [CrossRef]
32. Cheung, M.C.M.; Martínez-Sykora, J.; Testa, P.; De Pontieu, B.; Chintzoglou, G.; Rempel, M.; Polito, V.; Kerr, G.S.; Reeves, K.K.; Fletcher, L.; et al. Probing the Physics of the Solar Atmosphere with the Multi-slit Solar Explorer (MUSE). II. Flares and Eruptions. *Astrophys. J.* **2022**, *926*, 53. [CrossRef]
33. Ashfield, W.; Longcope, D. A Model for Gradual-phase Heating Driven by MHD Turbulence in Solar Flares. *Astrophys. J.* **2023**, *944*, 147. [CrossRef]
34. Nakariakov, V.M.; Melnikov, V.F. Quasi-Periodic Pulsations in Solar Flares. *Space Sci. Rev.* **2009**, *149*, 119–151. [CrossRef]

35. McLaughlin, J.A.; Nakariakov, V.M.; Dominique, M.; Jelínek, P.; Takasao, S. Modelling Quasi-Periodic Pulsations in Solar and Stellar Flares. *Space Sci. Rev.* **2018**, *214*, 45. [CrossRef]
36. Zimovets, I.V.; McLaughlin, J.A.; Srivastava, A.K.; Kolotkov, D.Y.; Kuznetsov, A.A.; Kupriyanova, E.G.; Cho, I.H.; Inglis, A.R.; Reale, F.; Pascoe, D.J.; et al. Quasi-Periodic Pulsations in Solar and Stellar Flares: A Review of Underpinning Physical Mechanisms and Their Predicted Observational Signatures. *Space Sci. Rev.* **2021**, *217*, 66. [CrossRef]
37. Clarke, B.P.; Hayes, L.A.; Gallagher, P.T.; Maloney, S.A.; Carley, E.P. Quasi-periodic Particle Acceleration in a Solar Flare. *Astrophys. J.* **2021**, *910*, 123. [CrossRef]
38. Hayes, L.A.; Inglis, A.R.; Christe, S.; Dennis, B.; Gallagher, P.T. Statistical Study of GOES X-ray Quasi-periodic Pulsations in Solar Flares. *Astrophys. J.* **2020**, *895*, 50. [CrossRef]
39. Nakariakov, V.M.; Foullon, C.; Myagkova, I.N.; Inglis, A.R. Quasi-Periodic Pulsations in the Gamma-Ray Emission of a Solar Flare. In Proceedings of the EGU General Assembly Conference Abstracts: EGU General Assembly Conference Abstracts, Vienna, Austria, 2–7 May 2010; p. 8615.
40. Brosius, J.W.; Daw, A.N.; Inglis, A.R. Quasi-periodic Fluctuations and Chromospheric Evaporation in a Solar Flare Ribbon Observed by Hinode/EIS, IRIS, and RHESSI. *Astrophys. J.* **2016**, *830*, 101. [CrossRef]
41. Mishra, S.K.; Sangal, K.; Kayshap, P.; Jelínek, P.; Srivastava, A.K.; Rajaguru, S.P. Origin of Quasi-periodic Pulsation at the Base of a Kink-unstable Jet. *Astrophys. J.* **2023**, *945*, 113. [CrossRef]
42. Xu, J.; Ning, Z.; Li, D.; Shi, F. Quasi-Periodic Pulsations in an M-Class Solar Flare. *Universe* **2023**, *9*, 215. [CrossRef]
43. Emslie, A.G.; Sturrock, P.A. Temperature minimum heating in solar flares by resistive dissipation of Alfvén waves. *Sol. Phys.* **1982**, *80*, 99–112. [CrossRef]
44. Fletcher, L.; Hudson, H.S. Impulsive Phase Flare Energy Transport by Large-Scale Alfvén Waves and the Electron Acceleration Problem. *Astrophys. J.* **2008**, *675*, 1645–1655. [CrossRef]
45. Russell, A.J.B.; Fletcher, L. Propagation of Alfvénic Waves from Corona to Chromosphere and Consequences for Solar Flares. *Astrophys. J.* **2013**, *765*, 81. [CrossRef]
46. Reep, J.W.; Russell, A.J.B. Alfvenic Wave Heating of the Upper Chromosphere in Flares. *Astrophys. J. Lett.* **2016**, *818*, L20. [CrossRef]
47. Reep, J.W.; Russell, A.J.B.; Tarr, L.A.; Leake, J.E. A Hydrodynamic Model of Alfvénic Wave Heating in a Coronal Loop and Its Chromospheric Footpoints. *Astrophys. J.* **2018**, *853*, 101. [CrossRef]
48. Chifor, C.; Tripathi, D.; Mason, H.E.; Dennis, B.R. X-ray precursors to flares and filament eruptions. *A&A* **2007**, *472*, 967–979. [CrossRef]
49. Hudson, H.; Fletcher, L.; Hannah, I.; Hayes, L.; Simoes, P. Hot onsets of solar flares. In Proceedings of the AGU Fall Meeting Abstracts, Online, 13–17 December 2021; Volume 2021, p. SH22B-02.
50. Warren, H.P.; Reep, J.W.; Crump, N.A.; Ugarte-Urra, I.; Brooks, D.H.; Winebarger, A.R.; Savage, S.; De Pontieu, B.; Peter, H.; Cirtain, J.W.; et al. Observation and Modeling of High-temperature Solar Active Region Emission during the High-resolution Coronal Imager Flight of 2018 May 29. *Astrophys. J.* **2020**, *896*, 51. [CrossRef]
51. Upendran, V.; Tripathi, D. On the Impulsive Heating of Quiet Solar Corona. *Astrophys. J.* **2021**, *916*, 59. [CrossRef]
52. Cargill, P.J. Some Implications of the Nanoflare Concept. *Astrophys. J.* **1994**, *422*, 381. [CrossRef]
53. Reale, F.; McTiernan, J.M.; Testa, P. Comparison of Hinode/XRT and RHESSI Detection of Hot Plasma in the Non-Flaring Solar Corona. *Astrophys. J. Lett.* **2009**, *704*, L58–L61. [CrossRef]
54. Schmelz, J.T.; Kashyap, V.L.; Saar, S.H.; Dennis, B.R.; Grigis, P.C.; Lin, L.; De Luca, E.E.; Holman, G.D.; Golub, L.; Weber, M.A. Some Like It Hot: Coronal Heating Observations from Hinode X-ray Telescope and RHESSI. *Astrophys. J.* **2009**, *704*, 863–869. [CrossRef]
55. Winebarger, A.R.; Warren, H.P.; Schmelz, J.T.; Cirtain, J.; Mulu-Moore, F.; Golub, L.; Kobayashi, K. Defining the "Blind Spot" of Hinode EIS and XRT Temperature Measurements. *Astrophys. J. Lett.* **2012**, *746*, L17. [CrossRef]
56. Parenti, S.; del Zanna, G.; Petralia, A.; Reale, F.; Teriaca, L.; Testa, P.; Mason, H.E. Spectroscopy of Very Hot Plasma in Non-flaring Parts of a Solar Limb Active Region: Spatial and Temporal Properties. *Astrophys. J.* **2017**, *846*, 25. [CrossRef]
57. Bradshaw, S.J.; Del Zanna, G.; Mason, H.E. On the consequences of a non-equilibrium ionisation balance for compact flare emission and dynamics. *A&A* **2004**, *425*, 287–299. [CrossRef]
58. Bradshaw, S.J.; Klimchuk, J.A. What Dominates the Coronal Emission Spectrum During the Cycle of Impulsive Heating and Cooling? *Astrophys. J. Suppl. Ser.* **2011**, *194*, 26. [CrossRef]
59. Dzifčáková, E.; Dudík, J.; Mackovjak, Š. Non-equilibrium ionization by a periodic electron beam. I. Synthetic coronal spectra and implications for interpretation of observations. *A&A* **2016**, *589*, A68. [CrossRef]
60. Dudík, J.; Dzifčáková, E.; Meyer-Vernet, N.; Del Zanna, G.; Young, P.R.; Giunta, A.; Sylwester, B.; Sylwester, J.; Oka, M.; Mason, H.E.; et al. Nonequilibrium Processes in the Solar Corona, Transition Region, Flares, and Solar Wind (Invited Review). *Sol. Phys.* **2017**, *292*, 100. [CrossRef]
61. Lörinčík, J.; Dudík, J.; del Zanna, G.; Dzifčáková, E.; Mason, H.E. Plasma Diagnostics from Active Region and Quiet-Sun Spectra Observed by Hinode/EIS: Quantifying the Departures from a Maxwellian Distribution. *Astrophys. J.* **2020**, *893*, 34. [CrossRef]
62. Del Zanna, G.; Polito, V.; Dudík, J.; Testa, P.; Mason, H.E.; Dzifčáková, E. Diagnostics of Non-Maxwellian Electron Distributions in Solar Active Regions from Fe XII Lines Observed by the Hinode Extreme Ultraviolet Imaging Spectrometer and Interface Region Imaging Spectrograph. *Astrophys. J.* **2022**, *930*, 61. [CrossRef]

63. Jeffrey, N.L.S.; Fletcher, L.; Labrosse, N. First evidence of non-Gaussian solar flare EUV spectral line profiles and accelerated non-thermal ion motion. *A&A* **2016**, *590*, A99. [CrossRef]
64. Jeffrey, N.L.S.; Fletcher, L.; Labrosse, N. Non-Gaussian Velocity Distributions in Solar Flares from Extreme Ultraviolet Lines: A Possible Diagnostic of Ion Acceleration. *Astrophys. J.* **2017**, *836*, 35. [CrossRef]
65. Dudík, J.; Polito, V.; Dzifčáková, E.; Del Zanna, G.; Testa, P. Non-Maxwellian Analysis of the Transition-region Line Profiles Observed by the Interface Region Imaging Spectrograph. *Astrophys. J.* **2017**, *842*, 19. [CrossRef]
66. Polito, V.; Dudík, J.; Kašparová, J.; Dzifčáková, E.; Reeves, K.K.; Testa, P.; Chen, B. Broad Non–Gaussian Fe XXIV Line Profiles in the Impulsive Phase of the 2017 September 10 X8.3-class Flare Observed by Hinode/EIS. *Astrophys. J.* **2018**, *864*, 63. [CrossRef]
67. Del Zanna, G. The multi-thermal emission in solar active regions. *A&A* **2013**, *558*, A73. [CrossRef]
68. Li, W.; Li, M.; Wang, K.; Brage, T.; Hutton, R.; Landi, E. A Theoretical Investigation of the Magnetic-field-induced Transition in Fe X, of Importance for Measuring Magnetic Field Strengths in the Solar Corona. *Astrophys. J.* **2021**, *913*, 135. [CrossRef]
69. Chen, Y.; Li, W.; Tian, H.; Bai, X.; Hutton, R.; Brage, T. Application of a Magnetic-field-induced Transition in Fe X to Solar and Stellar Coronal Magnetic Field Measurements. *Res. Astron. Astrophys.* **2023**, *23*, 022001. [CrossRef]
70. Del Zanna, G.; Mason, H.E. Solar UV and X-ray spectral diagnostics. *Living Rev. Sol. Phys.* **2018**, *15*, 5. [CrossRef]
71. Landi, E.; Li, W.; Brage, T.; Hutton, R. Hinode/EIS Coronal Magnetic Field Measurements at the Onset of a C2 Flare. *Astrophys. J.* **2021**, *913*, 1. [CrossRef]
72. Brooks, D.H.; Warren, H.P.; Landi, E. Measurements of Coronal Magnetic Field Strengths in Solar Active Region Loops. *Astrophys. J. Lett.* **2021**, *915*, L24. [CrossRef]
73. Antonucci, E. Solar flare spectral diagnosis: Present and future. *Sol. Phys.* **1989**, *121*, 31–60. [CrossRef]
74. Del Zanna, G.; Berlicki, A.; Schmieder, B.; Mason, H.E. A Multi-Wavelength Study of the Compact M1 Flare on October 22, 2002. *Sol. Phys.* **2006**, *234*, 95–113. [CrossRef]
75. Bian, N.H.; Emslie, A.G.; Stackhouse, D.J.; Kontar, E.P. The Formation of Kappa-distribution Accelerated Electron Populations in Solar Flares. *Astrophys. J.* **2014**, *796*, 142. [CrossRef]
76. Del Zanna, G.; Andretta, V.; Cargill, P.J.; Corso, A.J.; Daw, A.N.; Golub, L.; Klimchuk, J.A.; Mason, H.E. High resolution soft X-ray spectroscopy and the quest for the hot (5-10 MK) plasma in solar active regions. *Front. Astron. Space Sci.* **2021**, *8*, 33. [CrossRef]
77. Del Zanna, G.; Woods, T.N. Spectral diagnostics with the SDO EVE flare lines. *A&A* **2013**, *555*, A59. [CrossRef]
78. Del Zanna, G.; Dere, K.P.; Young, P.R.; Landi, E. CHIANTI—An Atomic Database for Emission Lines. XVI. Version 10, Further Extensions. *Astrophys. J.* **2021**, *909*, 38. [CrossRef]
79. Mondal, B.; Sarkar, A.; Vadawale, S.V.; Mithun, N.P.S.; Janardhan, P.; Del Zanna, G.; Mason, H.E.; Mitra-Kraev, U.; Narendranath, S. Evolution of Elemental Abundances during B-Class Solar Flares: Soft X-ray Spectral Measurements with Chandrayaan-2 XSM. *Astrophys. J.* **2021**, *920*, 4. [CrossRef]
80. Laming, J.M. The FIP and Inverse FIP Effects in Solar and Stellar Coronae. *Living Rev. Sol. Phys.* **2015**, *12*, 2. [CrossRef]
81. Doschek, G.A.; Feldman, U.; Kreplin, R.W.; Cohen, L. High-resolution X-ray spectra of solar flares. III - General spectral properties of X1-X5 type flares. *Astrophys. J.* **1980**, *239*, 725–737. [CrossRef]
82. Doschek, G.A.; Tanaka, K. Transient Ionization and Solar Flare X-ray Spectra. *Astrophys. J.* **1987**, *323*, 799. [CrossRef]
83. Bradshaw, S.J.; Testa, P. Quantifying the Influence of Key Physical Processes on the Formation of Emission Lines Observed by IRIS. I. Non-equilibrium Ionization and Density-dependent Rates. *Astrophys. J.* **2019**, *872*, 123. [CrossRef]
84. Smith, R.K.; Hughes, J.P. Ionization Equilibrium Timescales in Collisional Plasmas. *Astrophys. J.* **2010**, *718*, 583–585. [CrossRef]
85. Lee, J.Y.; Raymond, J.C.; Reeves, K.K.; Shen, C.; Moon, Y.J.; Kim, Y.H. Nonequilibrium Ionization Effects on Solar EUV and X-ray Imaging Observations. *Astrophys. J.* **2019**, *879*, 111. [CrossRef]
86. Shen, C.; Raymond, J.C.; Murphy, N.A. Nonequilibrium Ionization Modeling of Petschek-type Shocks in Reconnecting Current Sheets in Solar Eruptions. *Astrophys. J.* **2023**, *943*, 111. [CrossRef]
87. Olluri, K.; Gudiksen, B.V.; Hansteen, V.H. Non-equilibrium Ionization Effects on the Density Line Ratio Diagnostics of O IV. *Astrophys. J.* **2013**, *767*, 43. [CrossRef]
88. Olluri, K.; Gudiksen, B.V.; Hansteen, V.H.; De Pontieu, B. Synthesized Spectra of Optically Thin Emission Lines. *Astrophys. J.* **2015**, *802*, 5. [CrossRef]
89. Shi, T.; Landi, E.; Manchester, W. Nonequilibrium Ionization Effects on Coronal Plasma Diagnostics and Elemental Abundance Measurements. *Astrophys. J.* **2019**, *882*, 154. [CrossRef]
90. Calcines, A.; Wells, M.; O'Brien, K.; Morris, S.; Seifert, W.; Zanutta, A.; Evans, C.; Di Marcantonio, P. Design of the VLT-CUBES image slicers. *Exp. Astron.* **2023**, *55*, 267–280. [CrossRef]
91. Richard, J.; Bacon, R.; Jeanneau, A. BlueMUSE: Science cases and requirements. In Proceedings of the Ground-Based and Airborne Instrumentation for Astronomy VIII, Online, 14–18 December 2020; Volume 11447, p. 114470S. [CrossRef]
92. Laurent, F.; Adjali, L.; Arns, J.; Bacon, R.; Boudon, D.; Caillier, P.; Daguisé, E.; Delabre, B.; Dubois, J.P.; Godefroy, P.; et al. MUSE integral field unit: Test results on the first out of 24. In Proceedings of the Modern Technologies in Space- and Ground-Based Telescopes and Instrumentation, San Diego, CA, USA, 27 June–2 July 2010; Volume 7739, p. 77394M. [CrossRef]
93. Cuevas, S.; Eikenberry, S.S.; Sánchez, B. FRIDA integral field unit manufacturing. In Proceedings of the Advances in Optical and Mechanical Technologies for Telescopes and Instrumentation, Montréal, QC, Canada, 22–27 June 2014; Volume 9151, p. 91514N. [CrossRef]

94. Laurent, F.; Boudon, D.; Kosmalski, J.; Loupias, M.; Raffault, G.; Remillieux, A.; Thatte, N.; Bryson, I.; Schnetler, H.; Clarke, F.; et al. ELT HARMONI: Image slicer preliminary design. In Proceedings of the Ground-Based and Airborne Instrumentation for Astronomy VII, Austin, TX, USA, 10–15 June 2018; Volume 10702, p. 1070296, [CrossRef]
95. Brandl, B.R.; Bettonvil, F.; van Boekel, R.; Glauser, A.; Quanz, S.P.; Absil, O.; Feldt, M.; Garcia, P.J.V.; Glasse, A.; Guedel, M.; et al. Status update on the development of METIS, the mid-infrared ELT imager and spectrograph. In Proceedings of the Ground-Based and Airborne Instrumentation for Astronomy IX, Montréal, QC, Canada, 17–23 June 2022; Volume 12184, p. 1218421. [CrossRef]
96. Calcines, A.; López, R.L.; Collados, M. MuSICa: The Multi-Slit Image Slicer for the est Spectrograph. *J. Astron. Instrum.* **2013**, *2*, 1350009. [CrossRef]
97. Calcines, A.; López, R.L.; Collados, M. A High Resolution Integral Field Spectrograph for the European Solar Telescope. *J. Astron. Instrum.* **2013**, *2*, 1350007. [CrossRef]
98. Calcines, A.; López, R.L.; Collados, M.; Vega Reyes, N. MuSICa image slicer prototype at 1.5-m GREGOR solar telescope. In Proceedings of the Ground-Based and Airborne Instrumentation for Astronomy V, Montréal, QC, Canada, 22–27 June 2014; Volume 9147, p. 91473I. [CrossRef]
99. Dubbeldam, C.M.; Robertson, D.J.; Ryder, D.A.; Sharples, R.M. Prototyping of diamond machined optics for the KMOS and JWST NIRSpec integral field units. In Proceedings of the Society of Photo-Optical Instrumentation Engineers (SPIE) Conference Series, Orlando, FL, USA, 24–31 May 2006; Volume 6273, p. 62733F. [CrossRef]
100. Lee, D.; Wells, M.; Dickson, C.J.; Shore, P.; Morantz, P. Development of diamond machined mirror arrays for integral field spectroscopy. In Proceedings of the Society of Photo-Optical Instrumentation Engineers (SPIE) Conference Series, Orlando, FL, USA, 24–31 May 2006; Volume 6273, p. 62731Y. [CrossRef]
101. Witt, E.M.; Fleming, B.T.; Green, J.C.; France, K.; Williams, J.; Sukegawa, T.; Siegmund, O.; Chafetz, D.; Tecza, M.; Levy, A.; et al. INFUSE: Assembly and alignment of a rocket-borne FUV integral field spectrograph. In Proceedings of the UV, X-ray, and Gamma-Ray Space Instrumentation for Astronomy XXII, San Diego, CA, USA, 1–5 August 2021; Volume 11821, p. 118210F. [CrossRef]
102. Calcines Rosario, A.; Matthews, S.; Reid, H. Exploring the application of image slicers for the EUV for the next generation of solar space missions. In Proceedings of the Space Telescopes and Instrumentation 2022: Ultraviolet to Gamma Ray, Montréal, QC, Canada, 1–5 August 2022; Volume 12181, p. 121810K. [CrossRef]
103. De Pontieu, B.; Martínez-Sykora, J.; Testa, P.; Winebarger, A.R.; Daw, A.; Hansteen, V.; Cheung, M.C.M.; Antolin, P. The Multi-slit Approach to Coronal Spectroscopy with the Multi-slit Solar Explorer (MUSE). *Astrophys. J.* **2020**, *888*, 3. [CrossRef]
104. Calcines, A.; Bourgenot, C.; Sharples, R. Design of freeform diffraction gratings: Performance, limitations and potential applications. In Proceedings of the Advances in Optical and Mechanical Technologies for Telescopes and Instrumentation III, Austin, TX, USA, 10–15 June 2018; Volume 10706, p.107064Z. [CrossRef]
105. Bourgenot, C.; Robertson, D.J.; Stelter, D.; Eikenberry, S. Towards freeform curved blazed gratings using diamond machining. In Proceedings of the Advances in Optical and Mechanical Technologies for Telescopes and Instrumentation II, Edinburgh, UK, 26 June–1 July 2016; Volume 9912, p. 99123M. [CrossRef]
106. Shimizu, T.; Imada, S.; Kawate, T.; Ichimoto, K.; Suematsu, Y.; Hara, H.; Katsukawa, Y.; Kubo, M.; Toriumi, S.; Watanabe, T.; et al. The Solar-C_EUVST mission. In Proceedings of the UV, X-ray, and Gamma-Ray Space Instrumentation for Astronomy XXI, San Diego, CA, USA, 11–15 August 2019; Volume 11118, p. 1111807. [CrossRef]
107. Corso, A.J.; Pelizzo, M.G. Extreme ultraviolet multilayer nanostructures and their application to solar plasma observations: A review. *J. Nanosci. Nanotechnol.* **2019**, *19*, 532–545. [CrossRef]
108. Hu, M.H.; Le Guen, K.; André, J.M.; Jonnard, P.; Meltchakov, E.; Delmotte, F.; Galtayries, A. Structural properties of Al/Mo/SiC multilayers with high reflectivity for extreme ultraviolet light. *Opt. Express* **2010**, *18*, 20019–20028. [CrossRef] [PubMed]
109. Corso, A.J.; Del Zanna, G.; Polito, V. Future perspectives in solar hot plasma observations in the soft X-rays. *Exp. Astron.* **2021**, *51*, 453–474. [CrossRef]
110. Reid, H.A.S.; Musset, S.; Ryan, D.F.; Andretta, V.; Auchère, F.; Baker, D.; Benvenuto, F.; Browning, P.; Buchlin, É.; Calcines Rosario, A.; et al. The Solar Particle Acceleration Radiation and Kinetics (SPARK) Mission Concept. *Aerospace* **2023**, *10*, 1034. [CrossRef]

aerospace

MDPI

Article

Optimizing Space Telescopes' Thermal Performance through Uncertainty Analysis: Identification of Critical Parameters and Shaping Test Strategy Development

Uxia Garcia-Luis [1], Alejandro M. Gomez-San-Juan [1,*], Fermin Navarro-Medina [1], Carlos Ulloa-Sande [1], Alfonso Yñigo-Rivera [2] and Alba Eva Peláez-Santos [3]

1 Escola de Enxeñaría Aeronautica e do Espazo, Aerospace Technology Research Group, atlanTTic, Universidade de Vigo, Pabellón Manuel Martínez-Risco, As Lagoas, S/N, 32004 Ourense, Ourense, Spain; uxia.garcia.luis@uvigo.gal (U.G.-L.); fermin.navarro.medina@uvigo.gal (F.N.-M.); carlos.ulloa@uvigo.gal (C.U.-S.)

2 Instituto de Astrofísica de Canarias (IAC), c/ Via Láctea, S/N, 38205 La Laguna, Tenerife, Spain; alfonso.ynigo.rivera@iac.es

3 ESA-ESTEC, co ATG Europe BV, Keplerlaan 1, 2201AZ Noordwijk, The Netherlands; albaeva.pelaezsantos@esa.ext.int

* Correspondence: alejandromanuel.gomez@uvigo.gal

Abstract: The integration of uncertainty analysis methodologies allows for improving design efficiency, particularly in the context of instruments that demand precise pointing accuracy, such as space telescopes. Focusing on the VINIS Earth observation telescope developed by the Instituto de Astrofísica de Canarias (IAC), this paper reports an uncertainty analysis on a thermal model aimed at improving cost savings in the future testing phases. The primary objective was to identify critical parameters impacting thermal performance and reduce overdesign. Employing the Statistical Error Analysis (SEA) method across several operational scenarios, the research identifies key factors, including the Earth's infrared temperature and albedo, and the spacecraft's attitude and environmental conditions, as the variables with major influences on the system's thermal performance. Ultimately, the findings suggest that uncertainty-based analysis is a potent tool for guiding thermal control system design in space platforms, promoting efficiency and reliability. This methodology not only provides a framework for optimizing thermal design and testing in space missions but also ensures that instruments like the VINIS telescope maintain optimal operating temperatures in diverse space environments, thereby increasing mission robustness and enabling precise resource allocation.

Keywords: uncertainty; SEA; telescope; design; efficiency

1. Introduction

1.1. The Relevance of Thermal Testing in Spacecraft Instruments

In aerospace engineering, the tolerance for error is extremely limited, requiring rigorous analysis. This is particularly important in spacecraft development, which is constrained by weight, space availability, and high launch costs [1]. The opportunity to correct design flaws once a satellite is in orbit is minimal, highlighting the importance of avoiding errors at the design and manufacturing phases [2]. The primary strategy to mitigate these risks is error prevention, which is especially challenging in system elements which require high accuracy.

The importance of thermal testing in this context cannot be overstated. Given the harsh space conditions, spacecraft instruments must be designed to operate reliably within a range of thermal environments. These conditions vary widely, from the intense cold of shadowed space to the severe heat when exposed directly to the sun. Thermal testing, therefore, plays a pivotal role in spacecraft development. It serves as a critical check against the theoretical models used in the design phase, offering a practical assessment of how spacecraft instruments will perform under the thermal conditions they will face in space. This testing phase enables engineers to identify and address potential thermal performance issues, thereby reducing the risk of failure and the costly prospects of in-orbit adjustments.

Moreover, the evolving landscape of space exploration demands ever-increasing levels of precision in spacecraft instrumentation, especially for missions involving Earth observation, planetary exploration, and deep space scientific research. Instruments like the VINIS Earth observation telescope, developed by the Instituto de Astrofísica de Canarias (IAC), exemplify the need for such precision. These missions depend not only on the accurate collection of data but also on the reliability of instruments to operate within expected thermal parameters.

It is within this context that the integration of uncertainty analysis methodologies emerges as a key strategy for enhancing design efficiency. By conducting a thorough uncertainty analysis on the thermal model, engineers can identify critical parameters that impact thermal performance. This approach not only aids in reducing overdesign but also contributes to cost savings in future testing phases. Thus, uncertainty-based analysis is not merely a theoretical exercise; it is a practical tool that guides the design of thermal control systems, ensuring spacecraft instruments can withstand the variable conditions of space. This methodology enhances mission robustness, enables precise resource allocation, and, ultimately, secures the success of space missions.

This paper is organized as follows. Section 1 outlines the importance of thermal testing in spacecraft design due to the strict constraints in aerospace engineering and introduces uncertainty calculation as a tool for thermal analysis. Section 2 details the methods for uncertainty calculation applied to spacecraft thermal control. The results in Section 3 present insights into thermal performance and the impact of uncertainties, leading to Section 4's discussion on the key thermal drivers. The paper concludes in Section 5, emphasizing the role of uncertainty analysis in enhancing the design and reliability of spacecraft thermal control systems.

1.2. Uncertainty Calculation

In complex system design, the use of mathematical models is essential. They are crucial for initial system trade-off analyses, design refinement, and performance evaluation, aligning with the system objectives. The accuracy and efficiency of these models in representing the physical system and predicting behavior within their application domain are critical, necessitating thorough validation [3].

The fidelity of models in mirroring reality depends on several factors. These include the unknown parameters, the limitations in capturing certain phenomena, and the challenges in accounting for all possible configurations [4–7]. Parameters, typically defined under ideal or constrained test conditions, may not always represent the real system accurately. Engineers often introduce equivalent effective parameters to manage the geometric complexities and interactions between components, a task that grows more complex with increasing system scale [8].

Recognizing and integrating the inherent uncertainties in the numerical model outcomes is essential throughout all the analysis stages. This integration significantly influences decision-making in design and project management, particularly in the early phases where approximately 85% of the project's lifecycle costs are determined [9]. Addressing uncertainties related to manufacturing, material properties, positioning, mounting, environmental conditions, and modelling processes is crucial for ensuring robustness, reliability, and safety of the systems.

The framework of Uncertainty Quantification (UQ) is applicable in various fields, which ranges from deterministic models based on physical laws to models for complex socio-economic phenomena (like the efficiency of sustainability policies in the European Union [10]), groundwater modelling [11], ocean dynamics [12], remote sensing [13], and aircraft design [14]. This approach increases the likelihood of mission success and assists in efficient resource utilization and risk mitigation. Despite its benefits, UQ is not widely implemented in the design of space missions. The diversity of methods presented in the specialized literature can sometimes make this analysis challenging.

Uncertainty analysis is often conducted alongside sensitivity analysis, yet these two concepts have different focuses, and their distinction is often blurred in the existing literature. Sensitivity analysis evaluates the effects of variations in input parameters on the outputs, while uncertainty analysis encompasses the assessment of all possible outcomes and their associated probabilities.

Methods for sensitivity analysis are classified based on the characteristics of the mathematical model, such as nonlinearities or the effects of simultaneous parameter variations, and how they influence the results of uncertainty analysis. These methods fall into three primary categories: screening methods, local sensitivity analysis, and global sensitivity analysis [15]. Screening methods prioritize parameters by importance without detailed quantification. Local methods focus on uncertainties close to the nominal solution but may overlook significant effects. Global methods cover the entire range of parameter validity, capturing important model characteristics, including nonlinearities and wide parameter validity intervals.

Another way to categorize these methods is based on the type of information sought from the analysis [16]. This includes screening methods, importance measurements (which compare statistical values of parameters derived from outputs to those of input parameters), and techniques for deeper exploration of sensitivities (providing comprehensive knowledge of the model's behavior in parameter space).

Uncertainties in modeling are classified as either random, arising from the inherent unpredictability of natural phenomena, or epistemic, resulting from gaps in information during system modeling [17]. Random uncertainties are represented using probability distributions and require expert knowledge to develop representative distributions for the parameters involved [18]. Epistemic uncertainties are further divided into phenomenological uncertainties, linked to unknown information in innovative projects, and model-associated uncertainties, originating from the limitations in the precision of the mathematical model [19].

1.3. Application of Uncertainty Calculation in the Aerospace Domain

Uncertainty significantly influences aerospace designs, whose impact can be either advantageous or detrimental to the system performance. Assessing and understanding the extent and impact of this uncertainty is essential. Therefore, uncertainty analysis is a useful tool in aerospace engineering for risk management. By quantifying uncertainties in system components and processes, engineers can more effectively predict and mitigate potential failures, enhancing the safety, reliability, and effectiveness of space missions. This approach requires a comprehensive understanding of how different uncertainties interact within the overall system [20,21].

UQ in aerospace engineering is important not only for design and manufacturing improvement but also for the planning and developing of testing campaigns. By identifying the system elements most susceptible to uncertainty and which have a higher impact on performance, engineers can focus their testing strategies on these critical aspects, improving testing efficiency and the overall reliability of the system. This process represents an integration of theoretical analysis, empirical data, and practical experience, evolving continuously with new insights and contributing to the development of robust and resilient aerospace systems.

1.4. Application of Uncertainty Calculation in Spacecraft Thermal Analysis

A two-phase methodology is commonly used in the process of thermal control and design for space systems [22–24]. Initially, engineers establish scenarios that represent worst-case design conditions, combining the most challenging environmental factors (such as solar irradiance or planetary infrared radiation) with demanding operational conditions (like internal heat dissipation and the degradation of material and optical properties).

Following the analysis of these scenarios, predetermined design margins, also known as uncertainty margins, are applied to the results. These margins are designed to address uncertainties in parameters during the initial analysis.

The most common practice in thermal design is to adhere to predefined margins, determined from statistical analyses of temperature predictions versus actual measurements from several spacecraft operations once in orbit. These margins, reflecting the deviation between predicted and actual temperatures, have not seen significant updates since the last revision by NASA in 1994 [25], though more recent studies [26–30] suggest the need for mission-specific margins, considering their unique nature.

The application of uncertainty analysis in thermal control provides an alternative to the traditional fixed-margin approach. It employs the results of uncertainty assessments to establish design margins. This methodology is broadly categorized into two types of methods: One-At-a-Time (OAT) and stochastic methods. OAT approaches, exemplified by Statistical Error Analysis (SEA), offer simplicity and speed at the cost of model simplification. In contrast, stochastic methods like Monte Carlo Simulation (MCS), involving hundreds of solves of the complete thermal model, yield results closer to the actual thermal performance but also bring longer computation times and a lack of detailed insights into how each parameter individually contributes to the overall uncertainty.

In terms of data availability for uncertainty parameters in thermal systems, there is a significant lack of information. The probability distributions associated with these parameters, often assumed to be Gaussian normal, may vary significantly across different project phases. In the initial stages, distributions with higher uncertainty, such as uniform or triangular distributions, are more commonly assigned due to undefined parameter values [31,32]. However, as the project progresses and specific choices are made (e.g., selecting a particular type of paint for a radiator), the distribution may shift towards a normal distribution due to random variations between samples.

1.5. Research Objectives

The primary objective of this research was to assess the effectiveness of uncertainty analysis methodologies for identifying design drivers and allowing for a more efficient test campaign planning. The research focused on a conceptual design of a satellite platform capable of accommodating the VINIS Earth observation telescope, an initiative of the IAC. The study sought to apply the methodology to identify the thermal parameters that mostly influence the system's performance. This approach is intended to enhance the design process, ensuring the telescope operates optimally in its intended environment. A critical aspect of this objective is to mitigate the risk of overdesign by avoiding the implementation of unnecessary margins or to identify points in the design where those predefined margins could be insufficient, which might be even worse. To achieve this main objective, the following secondary objectives have been outlined:

- Development of a conceptual thermal model for a platform capable of supporting the VINIS instrument.
- Conduct an analysis of the expected thermal environment that the VINIS instrument will encounter. This analysis will be based on the platform as defined in the previous objective, ensuring a comprehensive understanding of the thermal challenges and requirements in the operational context.
- Design and propose a passive protection system for the VINIS instrument. The aim of this system is to maintain the instrument's temperature within a controlled range of −10 °C to +30 °C, thereby ensuring optimal performance and longevity under varying space conditions.

2. Materials and Methods

2.1. Uncertainty Calculation Methods Applied to the Thermal Domain

There is no standardized method for calculating uncertainties for space thermal analysis. However, as mentioned in the introduction, there are three prevalent methods generally adopted: the application of pre-established margins, OAT methods, and stochastic methods.

2.1.1. Fixed Margins

The most employed approach in space thermal design is the use of fixed margins, which are based on statistical analyses that compare outcomes from mathematical models with actual flight data [33–37]. This technique is implemented from the start of a project and continues throughout the design phase. However, these safety factors, often based on engineering judgment and applied to worst-case scenario results, lack a reliable quantification method, leading to ambiguity. Although they are intended to add conservatism, these factors can inadvertently lead to a false sense of safety. Furthermore, basing system designs on such margins can result in overdesign, incurring unnecessary expenses, especially in large missions.

2.1.2. OAT Methods

Among OAT methods, SEA is a conventional approach, originating from the methods of Kline and McKlintock in 1953 for the statistical analysis of experimental samples [38]. SEA computes uncertainties by multiplying sensitivity coefficients $\partial T_i/\partial x_k$ (temperature derivatives of parameters) by parameter uncertainties at a specified confidence level. The final uncertainty is obtained from the contributions' Root Sum Square (RSS). Given the uncertainty values w_{xk} for the x_k parameters for a defined level of confidence σ_n, the uncertainty of each of the T_i temperatures can be expressed as follows:

$$w_{Ti} = \left[\left(\frac{\partial T(x_1, \ldots, x_k)}{\partial x_1} w_{x1} \right)^2 + \ldots + \left(\frac{\partial T(x_1, \ldots, x_k)}{\partial x_k} w_{xk} \right)^2 \right]^{1/2} \tag{1}$$

Despite its simplicity and suitability for initial approximations, SEA has limitations, including its assumption of linear models, statistical independence, and normal distribution of parameters. It aligns with procedures endorsed by ESA per their ECSS, but recent standards do not provide methods for uncertainty calculation.

2.1.3. Stochastic Methods

Recent advances in computational resources have promoted the adoption of stochastic methods such as MCS for uncertainty analysis assessment in mathematical models [39]. Unlike other methods, MCS does not depend on oversimplified assumptions. Nevertheless, it requires examining probability distribution data for each parameter uncertainty. These data, which can be found in several sources [40–43], usually presume uniform or normal distributions. In scenarios where specific data are not available, uniform distributions are used to represent an equal likelihood of all values within the defined range.

2.2. Spacecraft Thermal Control

2.2.1. Thermal Environment

In adherence to the specifications laid out by the IAC for the VINIS instrument, the thermal analyses are conducted in a predefined orbital scenario. This orbit scenario, consistent across all cases, is a Sun-Synchronous Orbit (SSO) at an altitude of 550 km, with an inclination of 97.7° and a Local Time of Ascending Node (LTAN) at 11:00 AM (Right Ascension of Ascending Node (RAAN) = 30°). The primary distinction between the hot and cold orbit scenarios lies in the Earth–Sun distance: for the cold cases, it is set at Earth's aphelion (1.52×10^6 km), and for the hot cases, at the perihelion (1.47×10^6 km). In an SSO, the spacecraft experiences key factors that significantly affect its thermal environment. Notably, the spacecraft faces periodic eclipse events, leading to cyclic thermal conditions due to the lack of direct solar radiation, which requires a thermal control system capable of effectively dissipating heat when exposed to the sun and retaining it during eclipses. The orbit's polar nature also subjects the spacecraft to varying thermal conditions at different latitudes, influenced by Earth's albedo and infrared radiation. This demands a thermal control system that is both robust and adaptable, which is able to handle a range of thermal inputs.

The thermal environment must maintain a temperature range of $-20/+40$ °C at the platform level and $-10/+30$ °C for the telescope, including the baffle and optical bench. For specific components like the Back End Electronics (BEE) and Focal Plane Assembly (FPA), the maximum allowable temperatures exceed these ranges, reaching up to 65 °C.

2.2.2. Thermal Model

A comprehensive systems engineering analysis was conducted to determine the most suitable platform to allocate the VINIS telescope. An initial assessment indicated that while it is feasible to find platforms with a payload mass capacity of 20 kg, it is challenging to find one that offers a volume compatible with VINIS and the desired EELV Secondary Payload Adapter (ESPA) size. A critical aspect of the platform design is maintaining the telescope's Earth orientation to prevent freezing, which may occur if the satellite loses attitude control. This issue potentially needs a protective cover for the telescope, thus increasing the complexity of the design.

After evaluating up to three configurations with their corresponding preliminary thermal models in the initial project stage, it was concluded that the detailed thermal response of the telescope would be assessed using a platform comprising a prismatic square structure made of sandwich panels. This structure includes a service module at the rear side of the telescope frame. The entire structure would be enveloped in a Multi-Layer Insulation (MLI) blanket, providing thermal stability, and equipped with a thermal radiator. Additionally, it would feature two deployable solar panels capable of being oriented on a perpendicular axis to two platform external faces, as shown in Figure 1. While the inclusion of deployable solar panels adds technical complexity, this configuration allows for maximized energy extraction with minimum size.

The platform's structure is primarily made of sandwich panels with aluminum core and CFRP skins for their advantageous thermo-mechanical properties. The telescope assembly, instead, is designed based on the iso-thermal and iso-static mounting principles, ensuring thermal stability and uniform temperature distribution. The structure which surrounds both the Primary (M1) and Secondary (M2) metallic mirrors is made of the same aluminum material. Additionally, aluminum 6061-T6 is used in other components, such as the radiator and antenna.

In terms of thermo-optical properties, the complete platform is wrapped in MLI blankets, with Kapton on the outer layer and VDA on the inner. Optical elements, including mirrors, receive a reflective silver coating to enhance their performance, while solar panels are modeled with a gallium arsenide surface finish with absorptance depending on the panel's efficiency. The employment of black paint across most of the elements serves dual purposes: thermal control and stray light mitigation within the telescope. Conversely, white paint is applied to the radiators and antennas to reflect heat.

Figures 1–3 provide an overview of the Geometrical Mathematical Model (GMM) across three levels of simplification, setting the stage for the subsequent presentation of results. The three levels of simplification begin with the integration of the VINIS platform and telescope assembly. The focus then shifts to the telescope assembly, examining components like the baffle and mounting panels. The most detailed level centers on the VINIS telescope itself.

As for the Thermal Mathematical Model (TMM), there are several points which also deserve to be emphasized. The connection between the electronics and the radiator was modeled through a linear conductor of 0.4 W/K in both cases. After adjusting the TMM, it was concluded that the radiator should have dimensions of 0.3 m × 0.3 m. Regarding the insert-type joints on the sandwich panels, they were computed accounting for (i) the contact conductance between the insert and any external surface, (ii) the through-conductance of the potting material, and (iii) the contact between potting and the honeycomb core. These calculations led to a linear conductor in between the telescope–platform electronic box–vertical optical bench of 0.004 W/K. As for the connection of the BEE and FPA to their respective supports, these were carried out through a contact conductance of 300 W/m^{2}K. Finally, it was made a capacitance adjustment according to a VINIS CAD model.

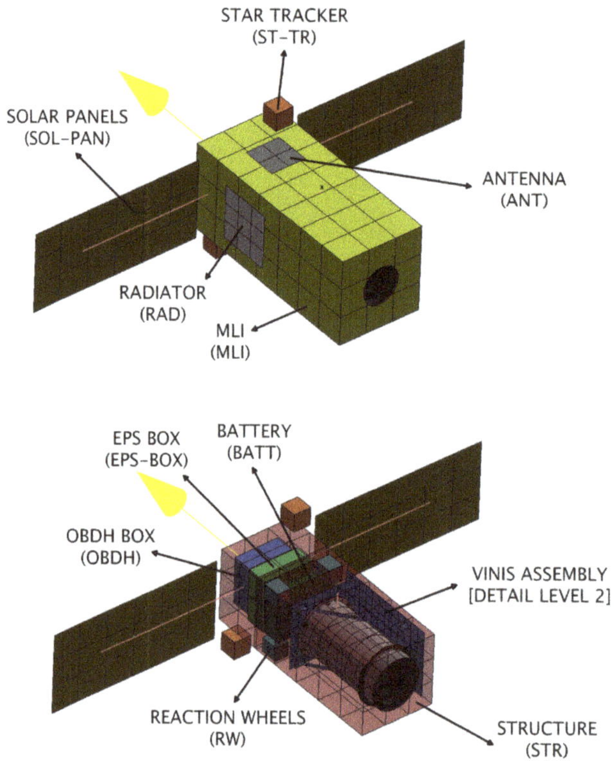

Figure 1. Elements included in the VINIS platform (first level of detail).

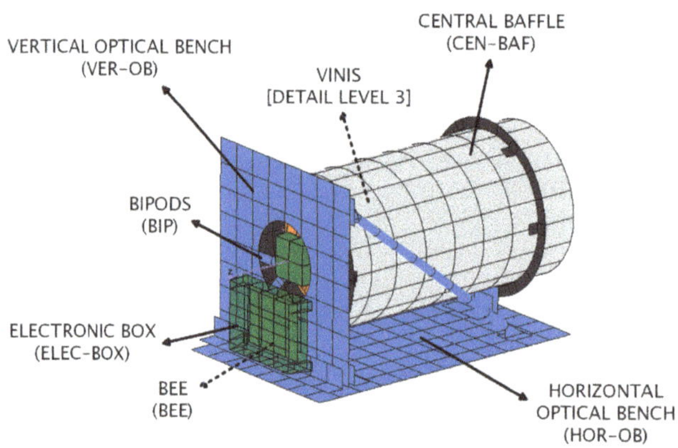

Figure 2. Elements included in the VINIS telescope assembly (second level of detail).

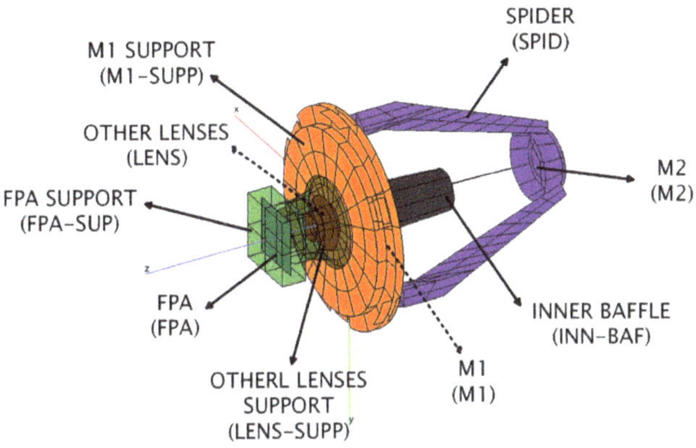

Figure 3. Elements included in VINIS telescope (third level of detail).

2.2.3. Thermal Cases

The analysis cases were derived from exploring several thermal scenarios, focusing on the effects of different operational modes and spacecraft attitudes. The spacecraft's operational modes include the following:

- Sun Pointing. Used for powering subsystems and recharging batteries. The solar panels are aligned perpendicularly to the Sun, while the telescope points towards nadir to avoid excessively cold temperatures.
- Science. The telescope is directed towards Earth for observations, with solar panels optimized for maximum sun exposure.
- Communications. Here, the spacecraft's antenna is oriented towards Earth, specifically focusing on the Canary Islands. This mode's positioning is subject to change in the early design stages due to limited information.
- Off-pointing. Activated during subsystem failure, directing the telescope away from Earth to minimize thermal impact.

The study also delves into the internal heat dissipation of the satellite, acknowledging the high design margins typical of preliminary project stages. These margins are set at +10% of the nominal value for components like the Electrical Power System (EPS), On Board Data Handling (OBDH), reaction wheels, and instrument components (BEE and FPA), aiming to avoid overdesign.

To assess the thermal effectiveness of these configurations, four thermal cases are examined: Consecutive Imaging Case (CIC), Hot Operational Case (HOC), Cold Operational Case (COC), and Cold Non-Operational Case (CNOC), each with unique operational modes and power dissipation profiles. Key aspects are presented in Tables 1–3 and include the following:

- CIC. Follows a descending Sun-Synchronous Orbit (SSO), emerging from an eclipse over the North Pole, and includes three science measurements and a communication window over the Canary Islands.
- HOC. Combines ground contact and scientific operations under continuous sunlight.
- COC. Entails multiple Sun Pointing orbits followed by scientific measurements after emerging from an eclipse.
- CNOC. Keeps an off-pointing mode throughout the orbit with constant heat dissipation from the EPS and OBDH subsystems.

Table 1. Operation modes and total power dissipation under the CIC scenario.

Operational Mode and Attitude	% Orbit Time	Total Power Dissipation [W]
Sun Pointing 1	25.0	16.5
Science Pointing	5.3	16.5
Science Imaging 1	1.0	33.5
Science Post-Processing 1	2.6	22.5
Science Imaging 2	1.1	33.5
Science Post-Processing 2	2.6	22.5
Science Imaging 3	1.1	33.5
Science Post-Processing 3	1.7	22.5
Comms Pointing	4.4	16.5
Comms Data Link	2.5	26.5.
Sun Pointing 2	52.8	16.5

Table 2. Operation modes and total power dissipation under the COC scenario.

Operational Mode and Attitude	% Orbit Time	Total Power Dissipation [W]
Sun Pointing 1	68.7	14.85
Science Pointing	5.5	14.85
Science Imaging	1.1	30.25
Science Post-Processing	4.9	20.25
Sun Pointing 2	19.8	14.85

Table 3. Operation modes and total power dissipation under the HOC scenario.

Operational Mode and Attitude	% Orbit Time	Total Power Dissipation [W]
Sun Pointing 1	10.8	18.15
Comms Pointing	4.9	18.15
Comms Data Link	6.4	18.15
Science Pointing	6.2	18.15
Science Imaging	1.0	36.75
Science Post-Processing	4.3	24.75
Sun Pointing 2	66.4	18.15

3. Results

3.1. Thermal Results (Baseline)

During the evaluation of the results, an investigation was conducted on the evolution of maximum and minimum temperatures, average temperatures, temperature slopes, and gradients within parts. It was demonstrated that the incorporation of MLI and a radiator as passive control measures generally ensured a satisfactory satellite performance.

For the computation of the CIC results, the dissipation profile was based on averages between the HOC and COC. This approach provides a comprehensive understanding of the expected orbital conditions under a representative scenario. Given the high frequency of science operations, as indicated in Table 1, this scenario resembles a variant of the HOC. In this context, a more pronounced thermal gradient within the telescope was observed, presumably due to increased mean dissipations from continuous operations. Nonetheless, all components met the operational temperature ranges within the nominal operation profile.

In the HOC, a local thermal gradient of approximately 30 °C (Figure 4) was observed in the outermost part of the central baffle, particularly towards the end of the science phase. The evolution (in green) and temporal variation (blue) of this gradient during the orbit can be seen in Figure 5. This local effect, induced by direct solar radiation during polar observations, has the potential to adversely affect optical performance by elevating the temperature of the central baffle beyond established limits. Although passive control measures may not suffice to mitigate these effects, adjusting the satellite's attitude could be an effective strategy.

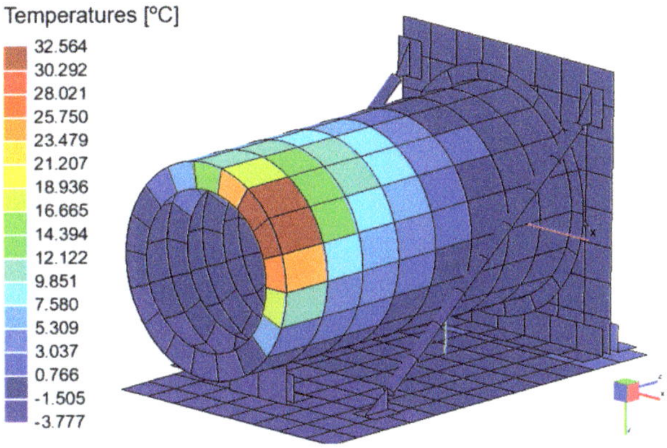

Figure 4. Baffle local thermal gradient under the HOC.

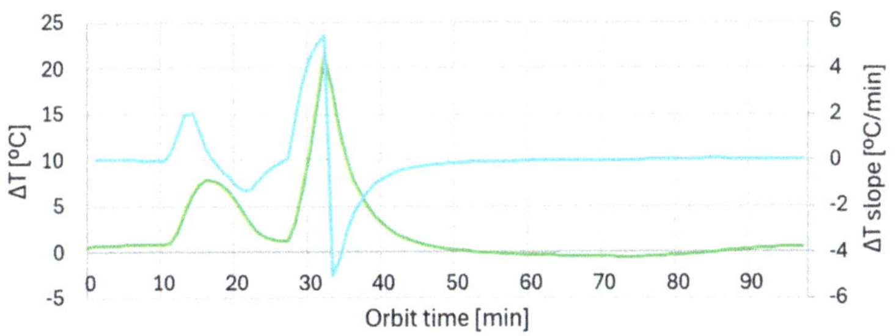

Figure 5. Variation of the temperature gradient (green curve) and temperature gradient slope (blue curve) on the main baffle under the HOC.

The COC analysis identified significant thermal gradients, particularly at the end of the science imaging phase. These gradients, predominantly orthogonal to the observation direction, were observed mainly on the main baffle and the horizontal optical bench. Additionally, a through-thickness temperature gradient was noted on the vertical optical bench, primarily arising from the temperature disparity between the cold outer space and the heat dissipation of electronic components. Nevertheless, the study also highlighted that the use of the same material across the telescope assembly, including the M1, M2, and spider, effectively minimized thermal gradients.

Finally, during the CNOC, numerous subsystems failed to meet temperature requirements. Passive control measures proved inadequate in elevating the low temperatures affecting the instrument, needing the inclusion of active thermal control mechanisms, such as heaters.

3.2. Uncertainty Analysis

The uncertainty analysis was performed by applying the SEA method. Although it presents some limitations, it is an excellent tool for a first approximation. In this way, the uncertainty associated with the parameters with a random nature was calculated. Additionally, a modeling uncertainty was considered. The typical value in uncorrelated models ranges between 2 to 5 °C. Due to the level of detail of the model, the design phase, and the type of geometry and MLI considered, a 3 °C uncertainty was added to the calculated margins. For example, if $w = 11.5$ °C was the calculated uncertainty, $w = 14.5$ °C was the considered uncertainty value. The values for the uncertainty of the different parameters were obtained from [44] and were for a 2σ confidence level (95.4%

confidence interval). Table 4 presents the different types of parameters. In total, the analysis encompassed the evaluation of up to 104 different parameters.

Table 4. Main input parameters, symbols, and uncertainty applied.

Parameter Type	Symbol	Uncertainty Applied
IR emissivity	ε_i	5%
Solar absorptance	α_i	5%
Sun temperature	T_S	21 K
Earth temperature	T_E	16 K
Earth albedo	a	0.1
Heat dissipation	Q_i	20%
Thermal capacitance	C_i	7.5%
Thermal conductivity (general)	k_i	10%
Thermal conductivity (composite)	$k_i,\ i = CFRP/Honeycomb$	30%
Contact coefficients	h_i	100%
Linear conductor	GL_i	100%
MLI efficiency	MLI_{EFF}	50%

The findings for the absolute minimum and maximum temperatures, along with their corresponding uncertainty margins across three scenarios, are presented in Tables 5–7. The maximum and minimum temperatures correspond to the ones reached by the satellite element, regardless of the orbit moment and whether they correspond to the same thermal node or not. Emphasis is placed on the results corresponding to the VINIS telescope assembly (components of the second level of detail, Figure 2). This focus is driven by the study's objective, which is to devise passive methods to maintain the temperature of the telescope's surrounding structure within the necessary operational range.

In the last part of the uncertainty analysis, a results decomposition from all three scenarios was performed. In this way, a physical understanding of the problem was attained, and all the relevant parameters to be tracked in future stages of the design were identified. Among the 104 parameters assessed, only 18 accounted for over 99% of the total uncertainty within any section of the telescope or its assembly. Furthermore, out of these 18 critical parameters (detailed in Table 8), 6 were associated with local phenomena, with the other 12 being the drivers of the overall thermal behavior.

Table 5. CIC scenario, absolute minimum and maximum temperature for each element along the orbit, for the VINIS telescope assembly (temperatures in °C) and SEA method.

	T_{min}	T_{max}	w_{min}	w_{max}	$T_{min}+w_{min}$	$T_{max}+w_{max}$	T_{min_req}	T_{max_req}
BIP	15.3	17.3	11.2	11.3	4.1	28.6	−10.0	30.0
CEN-BAF	0.5	16.9	10.1	11.2	−9.6	28.1	−10.0	30.0
ELEC-BOX	21.2	21.8	11.3	11.3	9.9	33.1	−20.0	65.0
BEE	24.0	29.9	11.2	11.6	12.8	41.5	−20.0	65.0
VER-OB	17.2	18.4	11.3	11.3	5.9	29.7	−10.0	30.0
HOR-OB	6.9	17.6	10.6	11.4	−3.7	29.0	−20.0	40.0

Table 6. HOC scenario, absolute minimum and maximum temperature for each element along the orbit, for the VINIS telescope assembly (temperatures in °C) and SEA method.

	T_{min}	T_{max}	w_{min}	w_{max}	$T_{min}+w_{min}$	$T_{max}+w_{max}$	T_{min_req}	T_{max_req}
BIP	16.2	17.7	11.6	11.6	4.6	29.3	−10.0	30.0
CEN-BAF	3.6	55.5	10.9	19.8	−7.3	75.3	−10.0	30.0
ELEC-BOX	20.2	20.6	11.6	11.6	8.6	32.2	−20.0	65.0
BEE	21.7	25.2	11.5	11.5	10.2	36.7	−20.0	65.0
VER-OB	17.1	18.6	11.6	11.7	5.5	30.3	−10.0	30.0
HOR-OB	10.2	18.7	11.3	11.8	−1.1	30.5	−20.0	40.0

Table 7. COC scenario, absolute minimum and maximum temperature for each element along the orbit, for the VINIS telescope assembly (temperatures in °C) and SEA method.

	T_{min}	T_{max}	w_{min}	w_{max}	$T_{min}+w_{min}$	$T_{max}+w_{max}$	T_{min_req}	T_{max_req}
BIP	2.4	4.0	10.5	10.6	−8.1	14.6	−10.0	30.0
CEN-BAF	−9.4	4.0	9.8	10.6	−19.2	14.6	−10.0	30.0
ELEC-BOX	6.5	6.7	10.6	10.6	−4.1	17.3	−20.0	65.0
BEE	7.8	10.7	10.5	10.5	−2.7	21.2	−20.0	65.0
VER-OB	3.3	4.9	10.6	10.6	−7.3	15.5	−10.0	30.0
HOR-OB	−3.2	5.1	10.1	10.7	−13.3	15.8	−20.0	40.0

Table 8. Principal contributors to thermal uncertainty.

Group	Symbol	Name
Optical properties	$\alpha_{BLACK-V}$	VINIS black paint solar absorptance (telescope)
Optical properties	ε_{KAPT-S}	Satellite Kapton IR emissivity (MLI)
Optical properties	α_{KAPT-S}	Satellite Kapton solar absorptance (MLI)
Optical properties	$\varepsilon_{WHITE-S}$	Satellite white paint IR emissivity (radiators)
Environment	T_E	Earth temperature
Environment	a	Earth albedo
Heat dissipation	Q_{BEE}	BEE heat dissipation
Heat dissipation	Q_{EPS}	EPS heat dissipation
Heat dissipation	Q_{FPA}	FPA heat dissipation
Heat dissipation	Q_{OBDH}	OBDH heat dissipation
Heat dissipation	Q_{RW}	RW heat dissipation
Thermal capacitance	$C_{BEE-BOX}$	BEE structure thermal capacitance
Thermal capacitance	$C_{BEE-PCB}$	BEE PCB thermal capacitance
Contact coefficient	h_{FPA}	FPA thermal contact coefficient
Linear conductor	GL_{INS}	Honeycomb insert linear conductor
Linear conductor	$GL_{RAD-EPS}$	EPS to radiator linear conductor
Linear conductor	$GL_{RAD-OBDH}$	OBDH to radiator linear conductor
MLI efficiency	MLI_{EFF}	MLI effective emittance

The analysis detailed below evaluates the specific impact of uncertainty from each parameter, quantifying it as a percentage of the overall temperature uncertainty. This evaluation covers all components comprising the VINIS telescope assembly, as depicted in Figure 2. Consequently, every segment of the pie chart illustrates the contributions to the RSS of Equation (1), with each expressed as a percentage of w_{Ti}^2.

Figure 6 provides a detailed overview of each parameter's contribution to the total thermal uncertainty, highlighting the contributions to the CIC scenario and minimum temperature (T_{min}) conditions. Generally, these contributions do not significantly differ from other scenarios and maximum temperature (T_{max}) conditions. Nevertheless, an exception occurs to the CEN-BAF in the HOC scenario, which exhibits a substantial difference; these findings are compiled in Figure 7.

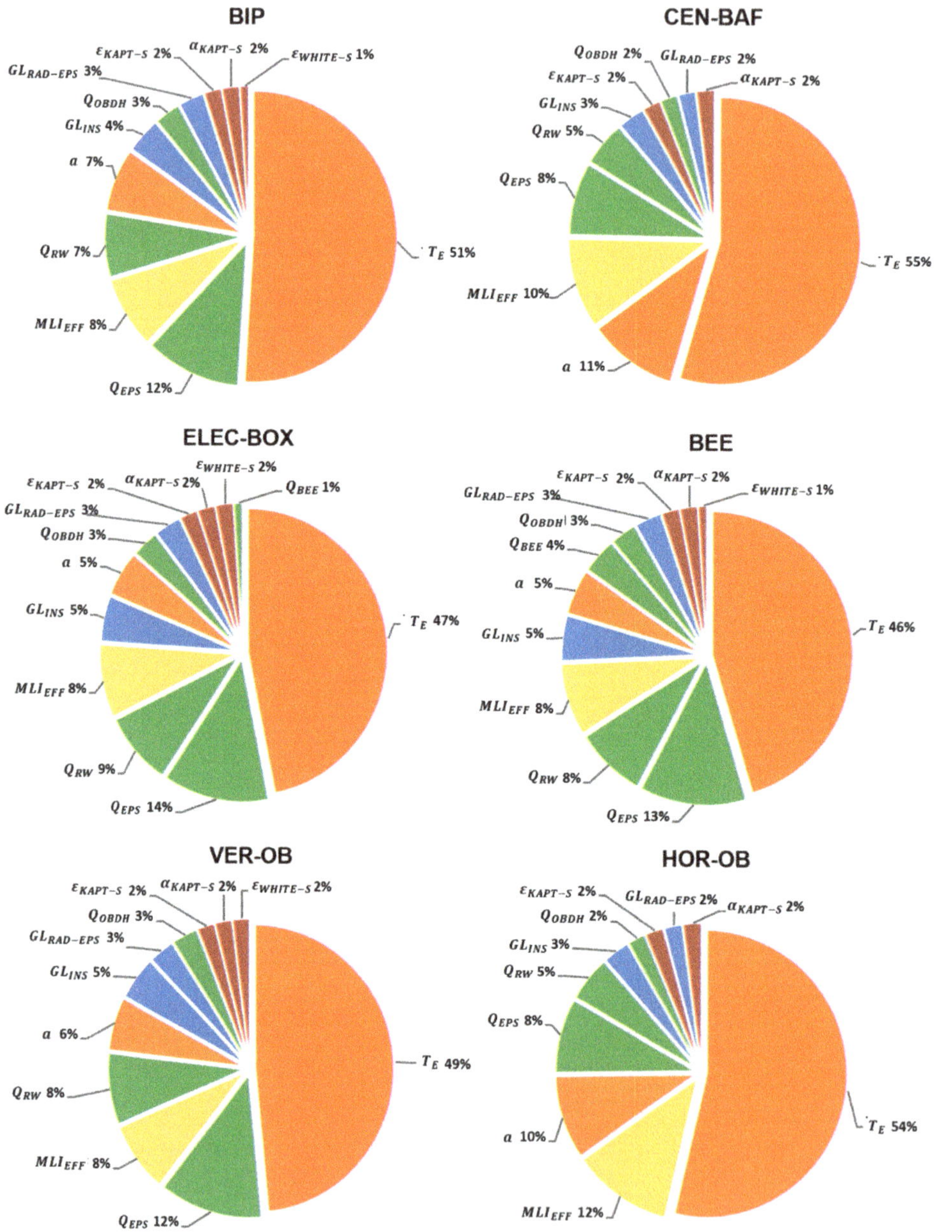

Figure 6. Breakdown of primary factors contributing to the thermal uncertainty of each of the VINIS telescope assembly components (second level of detail). CIC case and minimum temperature (T_{min}) results. Color coding is used to differentiate parameter groups: optical properties in red, environmental parameters in orange, heat dissipation in green, linear conductors in blue, and MLI efficiency in yellow.

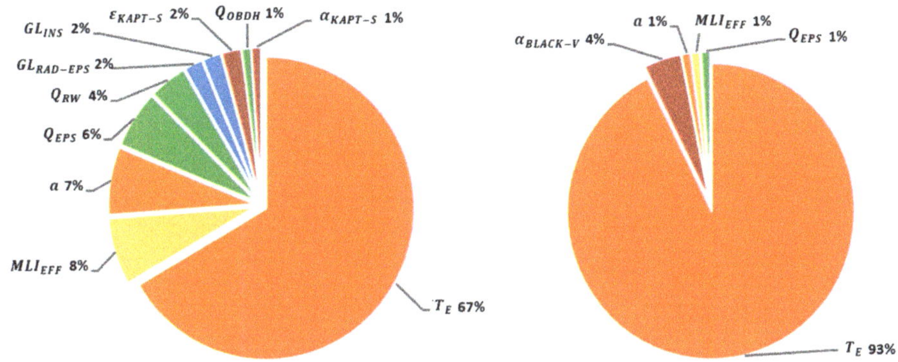

Figure 7. Breakdown of primary factors contributing to the thermal uncertainty of each of the CEN-BAF and CIC cases, for both the minimum temperature (T_{min}, **left**) and maximum temperature (T_{max}, **right**) results. Color coding is used to differentiate parameter groups: optical properties in red, environmental parameters in orange, heat dissipation in green, linear conductors in blue, and MLI efficiency in yellow.

4. Discussion

4.1. Evaluation of the Uncertainty Results

The results from Tables 5–7 show the predicted temperatures as a result of applying the margins to the calculated temperatures and the thermal requirements to compare if the results are within the allowable limits.

Concerning the telescope assembly, it is noted that the environmental temperatures generally remain within the required range for all scenarios, except for the telescope baffle (CEN-BAF). The baffle's temperature deviations are largely attributed to the extreme orientations adopted during the imaging of Earth's poles in both hot and cold cases. This indicates that, firstly, temperature control within the telescope assembly is achievable through precise pointing strategies, and secondly, a careful planning of telescope pointing operations is essential to prevent achieving temperatures outside the required range, which could negatively impact optical performance. Moreover, the highest and lowest temperatures typically occur at the most external sections of the telescope. Radiative heat transfer tends to mitigate these temperature gradients, suggesting that these variations may ultimately be within the limits.

Lastly, it is noteworthy that the uncertainty associated with the baffle's maximum temperature in the HOC is nearly twice as large as that in the other scenarios. This underscores the thermal challenges posed by the direct influx of solar radiation into the telescope, emphasizing the need for effective thermal management strategies to counteract such effects.

A comparative analysis of the uncertainty margins calculated using the SEA method with the ± 10 K margin proposed by the ECSS for a Phase B study [44] reveals that, except for the w_{max} at the central baffle (CEN-BAF), the margins are generally close to the recommended limits. This trend holds true for the telescope components (third level of detail, Figure 3). However, for the VINIS platform parts (first level of detail, Figure 1), the calculated uncertainties exceed the recommended range in some cases.

4.2. Main Thermal Drivers

Upon analyzing the uncertainty decomposition, it becomes evident that the Earth's IR temperature stands as the predominant source of uncertainty in all scenarios, with albedo also contributing significantly to most cases. Such implications could be anticipated due to the low satellite's altitude, as well as its attitude, exposing it to a wide spectrum of IR temperatures and albedo changes. These fluctuations stem from the Earth's varied surface features, including oceans, forests, and cloud-covered regions. Thus, these elements of uncertainty are inherent to the satellite's operation and cannot be acted upon, but the design can accommodate this uncertainty in the optical performance budget. There are other sources of uncertainty which are elements of the design of the satellite, and these should be the next focus of attention.

Furthermore, the platform's thermal design parameters emerge as crucial factors influencing the VINIS's thermal performance. The parameters that are the main sources of uncertainty in this regard regulate the main heat paths that drive the thermal behavior. They can be seen in Figure 8 and are the following:

1. The service module's heat dissipation represents the major source of controllable uncertainty in the problem (Q_{EPS}, Q_{RW}, Q_{OBDH})
2. The heat flux through the MLI, influenced by the MLI's efficiency and the thermo-optical properties of its outer layer (MLI_{EFF}, ε_{KAPT-S}, α_{KAPT-S}).
3. The heat path through the radiator, including the conductive couplings to it, and the IR emissivity of its white paint coating ($GL_{RAD-EPS}$, $GL_{RAD-OBH}$, $\varepsilon_{WHITE-S}$).

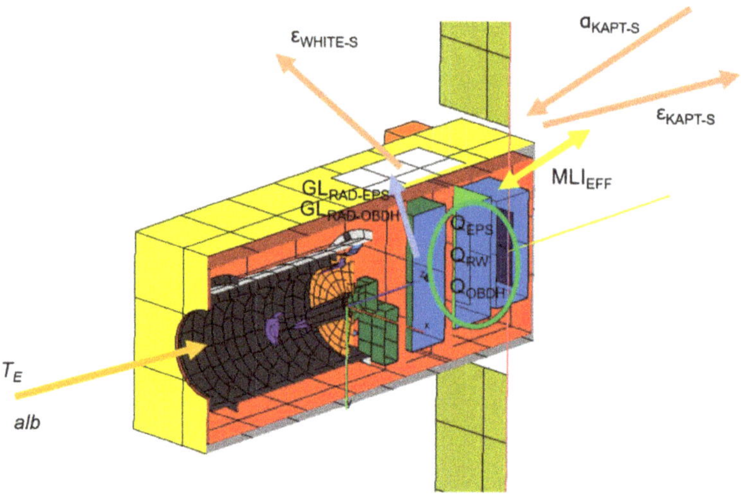

Figure 8. Main contributors to the thermal uncertainty of the VINIS thermal model.

Keeping these parameters under check and reducing their uncertainty as early in the project as possible will ensure that the thermal design is reliable and robust. This can be achieved by tailoring the engineering test program.

Other parameters, such as the inserts' conductive coupling, also contribute to the total uncertainty. Nevertheless, its implications are challenging to quantify due to their presence in most of the platform's conductive heat paths. Additionally, certain parameters may become significant in the detailed thermal analysis of specific components, such as the FPA.

4.3. Proposed Thermal Testing Model

The benefits of adopting an uncertainty-based analysis methodology will be demonstrated during the testing process. By identifying the key thermal drivers through this analysis, the test campaign can be streamlined, focusing on elements that have the greatest impact on the final design. This targeted approach leads to a more efficient and cost-effective testing strategy. It eliminates the need for the extensive testing of every component, allowing for a more focused allocation of attention and resources towards critical aspects that dictate the satellite's performance. Such an approach not only simplifies the development process but also enhances the effectiveness of thermal testing, ensuring a more reliable and robust satellite design.

Considering the uncertainty analysis findings, a proposed approach to the design process is outlined as follows:

1. Prioritize the detailed identification of heat dissipation values early in the design process, as these will significantly influence the temperature levels within the satellite, aside from the environmental and platform design factors.
2. During the development of the Structural and Thermal Model (STM), it is essential for engineers to pay special attention to the MLI. Since its performance has significant implications, the MLI employed in the STM should closely mirror that of the final design in materials, thickness, size, and openings. This alignment will help in minimizing the uncertainty associated with both its effectiveness and optical properties.
3. The STM will allow for retrieving high-resolution temperature measurements from specific points to correlate with the inserted conductive couplings, ensuring accurate thermal modeling.
4. It would be beneficial to treat the thermal straps for radiators as a degree of freedom, allowing for modifications during testing. These straps should be incorporated into the tests, yet with the flexibility to make modifications. This approach facilitates the fine-tuning of the satellite's operating temperature to achieve optimal performance, once other non-tunable parameters with large uncertainties, like the MLI effective emittance, are determined.
5. Moreover, engineers could consider a cost-effective Thermal Vacuum Chamber (TVAC) STM test that ensures the representativeness of these critical elements. Focus on the representativeness of the MLI should be the key, and considerations of the straps as per point 4 should apply. Other elements in the design may not need to be so representative, allowing for two things:
    i. Less expensive hardware/less engineering hours employed in other elements deemed to not be critical.
    ii. Leaving some elements of the design open when other subsystems still need some iterations (for example, if there are some doubts on which inserts to use in some areas or similar).

These elements do not dominate the thermal behavior and, therefore, allow for flexibility without significantly affecting the results.

This methodology diverges from traditional approaches by providing a quantitative assessment of the impact of uncertainties on thermal performance. It effectively identifies and evaluates critical thermal drivers, guiding the development of thermal management strategies. Offering a structured method for prioritizing design and testing efforts, it introduces an innovative perspective to thermal testing planning, thus improving the thermal control system's efficiency and reliability.

Furthermore, the suggested thermal testing model is a novel approach that leverages uncertainty analysis to streamline the testing process. Concentrating on key thermal drivers allows for a more targeted and cost-efficient design optimization process, ensuring crucial uncertainties are addressed promptly to enhance the overall robustness of the design.

5. Conclusions

This study on the VINIS space telescope underscores the significant advantages of incorporating an uncertainty-based analysis methodology into the satellite design process. This methodology is instrumental in identifying the key thermal drivers, streamlining the testing campaign to focus on elements that greatly influence the final design. Such a targeted approach not only makes the testing strategy more efficient and cost-effective but also eliminates the need for the exhaustive testing of every component. By concentrating efforts and resources on the critical aspects that dictate satellite performance, the development process is simplified, and the effectiveness of thermal testing is enhanced, resulting in a more reliable and robust satellite design.

Additionally, the findings from the uncertainty analysis provide valuable insights for the design process. Key recommendations include the early identification of heat dissipation values, which significantly influence internal temperature levels, and the need for a careful consideration of the MLI during the development of the STM. The performance of the MLI, closely mimicking that of the final design, is crucial, and high-resolution temperature measurements from the STM are vital for accurate thermal modeling. Flexibility in adjusting thermal straps for radiators during testing is recommended, allowing for the fine-tuning of the satellite's operating temperature. A targeted TVAC test on the STM should focus on these critical elements, ensuring the representativeness of the MLI and thermal straps, while allowing for some design elements to remain adaptable for future iterations.

As secondary conclusions drawn from the thermal behavior analysis, the study presents the following points:

- The assessment of four operational scenarios (CIC, HOC, COC, CNOC) validated the effectiveness of passive thermal control strategies in maintaining the telescope within its optimal operational temperature range, except for minor deviations under extreme pointing conditions.
- A temperature gradient in the baffle appears to be an inherent challenge that passive methods alone may not overcome. Accurate pointing is critical to ensure that the satellite's baffle temperatures remain within the required limits. Evaluating the maximum permissible thermal gradient for correct optical performance and adjusting operational profiles may be advantageous.
- Frequent scientific observations, as in the scenario, could produce significant temperature gradients, potentially needing a limit on operational frequency to prevent performance degradation.

These secondary conclusions, derived from the detailed thermal analysis, complement the primary insights gained through the uncertainty analysis, contributing to a comprehensive understanding of the design and operational challenges of the satellite.

Author Contributions: Conceptualization, U.G.-L., F.N.-M., A.Y.-R. and A.E.P.-S.; methodology, U.G.-L. and A.M.G.-S.-J.; software, U.G.-L.; validation, F.N.-M. and C.U.-S.; formal analysis, A.M.G.-S.-J.; investigation, F.N.-M. and C.U.-S.; resources, A.Y.-R. and A.E.P.-S.; data curation, U.G.-L. and A.M.G.-S.-J.; writing—original draft preparation, U.G.-L. and A.M.G.-S.-J.; writing—review and editing, F.N.-M., C.U.-S., A.Y.-R. and A.E.P.-S.; visualization, U.G.-L.; supervision, C.U.-S. and A.Y.-R.; project administration, A.M.G.-S.-J. and F.N.-M.; funding acquisition, C.U.-S. All authors have read and agreed to the published version of the manuscript.

Funding: This research was funded by MCIN/AEI/10.13039/501100011033, grant number PID2022-141669OA-I00 and by *Instituto de Astrofísica de Canarias*, grant number VINIS-EN_SE-2100. This research also was partially conducted with the support of a doctoral grant awarded to Uxia Garcia-Luis under the "*Programa de axudas á etapa predoutoral da Xunta de Galicia*" by the Galician Regional Government (*Consellería de Cultura, Educación, Formación Profesional e Universidades*).

Data Availability Statement: No data were used for the research described in this article.

Acknowledgments: The authors wish to thank the Spanish *Ministerio de Ciencia, Innovación y Universidades* and *Xunta de Galicia* for the funding received. The authors wish to extend their gratitude also to the *Cabildo de Tenerife* for the financial support that facilitated the research contracts for some members at IACTEC-Espacio, as well as allowing the use of their space telescope's preliminary design. This contribution has been instrumental in assessing the innovative uncertainty methodology, further enriching the practical application of this research.

Conflicts of Interest: The authors declare no conflicts of interest.

References

1. Massood, T.; Erkoyuncu, J.A.; Roy, R.; Harrison, A. Integrating design attributes, knowledge and uncertainty in aerospace sector. *CIRP J. Manuf. Sci. Technol.* **2014**, *7*, 83–96. [CrossRef]
2. Crespo, L.G.; Giesy, D.P.; Kenny, S.P. Robustness Analysis and Robust Design of Uncertain Systems. *AIAA J.* **2008**, *46*, 388–396. [CrossRef]

3. Blattnig, S.R.; Green, L.L.; Luckring, J.M.; Morrison, J.H.; Tripathi, R.K.; Zang, T.A. Towards a Credibility Assessment of Models and Simulations. In Proceedings of the 49th AIAA/ASME/ASCE/AHS/ASC Structures, Structural Dynamics, and Materials, Schaumburg, IL, USA, 7–10 April 2008.
4. Brynjarsdóttir, J.; Kennedy, A. Learning about physical parameters: The importance of model discrepancy. *Inverse Probl.* **2014**, *30*, 114007. [CrossRef]
5. Canavan, E.R.; Tuttle, J.G. Thermal conductivity and specific heat measurements of candidate structural materials for the JWST optical bench. *AIP Conf. Proc.* **2006**, *824*, 233–240.
6. Tuttle, J.; Di Pirro, M.; Canavan, E.; Hait, T. Thermal properties of double-aluminized Kapton at low temperatures. *AIP Conf. Proc.* **2008**, *986*, 34–41.
7. Tuttle, J.; Canavan, E. Recent NASA/GSFC cryogenic measurements of the total hemispheric emissivity of black surface preparations. *IOP Conf. Series Mater. Sci. Eng.* **2015**, *102*, 012015. [CrossRef]
8. Gilmore, D. *Spacecraft Thermal Control Handbook*, 2nd ed.; Volume 2: Cryogenics; The Aerospace Press: El Segundo, CA, USA, 2002.
9. Thunnisen, D.P. Propagating and Mitigating Uncertainty in the Design of Complex Multidisciplinary Systems. PhD Thesis, California Institute of Technology, Pasadena, CA, USA, 2005.
10. Saltelli, A.; Ratto, M.; Andres, T.; Campolongo, F.; Cariboni, J.; Gatelli, D.; Saisana, M.; Tarantola, S. *Global Sensitivity Analysis. The Primer*; John Wiley & Sons, Ltd.: London, UK, 2008.
11. Higdon, D.; Lee, H.; Holloman, C. Markov chain Monte Carlo-based approaches for inference in computationally intensive inverse problems. *Bayesian Stat.* **2003**, *7*, 181–197.
12. McKeague, I.W.; Nicholls, G.; Speer, K.; Herbei, R. Statistical inversion of South Atlantic circulation in an abyssal neutral density layer. *J. Mar. Res.* **2005**, *63*, 683–704. [CrossRef]
13. Haario, H.; Laine, M.; Lehtinen, M.; Saksman, E.; Tamminen, J. Markov Chain Monte Carlo Methods for High Dimensional Inversion in Remote Sensing. *J. R. Stat. Soc. Ser. B (Stat. Methodol.)* **2004**, *66*, 591–607. [CrossRef]
14. Roskam, J. *Airplane Design Part VIII: Airplane Cost Estimation Design Development and Manufacturing and Operating*; DAR Corporation: Lawrence, KS, USA, 1990.
15. Saltelli, A.; Chan, K.; Scott, E.M. *Sensitivity Analysis*; Wiley: New York, NY, USA, 2000.
16. Iooss, B.; Lemaître, P. A review on global sensitivity analysis methods. In *Uncertainty Management in Simulation-Optimization of Complex Systems: Algorithms and Applications*; Springer: Boston, MA, USA, 2015; pp. 101–122.
17. Ang, A.H.-S.; Tang, W.H. *Probability Concepts in Engineering: Emphasis on Applications to Civil and Environmental Engineering*, 2nd ed.; Wiley & Sons, Inc.: Hoboken, NJ, USA, 2006.
18. O'Hagan, A.; Buck, C.E.; Daneshkhah, A. *Uncertain Judgements: Eliciting Expert's Probabilities*; John Wiley & Sons: Hoboken, NJ, USA, 2006.
19. Oberkampf, W.L.; Helton, J.C.; Sentz, K. Mathematical Representation of Uncertainty. In Proceedings of the 19th AIAA Applied Aerodynamics Conference, Seattle, WA, USA, 16–19 April 2001.
20. Fu, X.; Liang, L.; Ma, W.; Cui, H.; Zhao, Y. Efficient Uncertainty Analysis of External Heat Flux of Solar Radiation with External Heat Flux Expansion for Spacecraft Thermal Design. *Aerospace* **2023**, *10*, 672. [CrossRef]
21. Liu, Z.; He, H.; Li, J.; Hao, J.; Tang, J.; Zhang, Z.; Jiang, S.; Chi, G.; Liu, R.; Wang, L.; et al. Measurement and Uncertainty Analysis of Lunar Soil Water Content via Heating Flux Method. *Aerospace* **2023**, *10*, 657. [CrossRef]
22. Avila, A. JPL Thermal Design Modeling Philosophy and NASA-STD-7009 Standard for Models and Simulations: A Case Study. In Proceedings of the 41st International Conference on Environmental Systems, Portland, OR, USA, 18–21 July 2011.
23. Ousley, W. *Requirements for Thermal Design, Analysis and Development*; Technical Report: 545-PG-8700.2.1A; NASA Thermal Engineering Branch: Greenbelt, MD, USA, 2005.
24. *ECSS-E-ST-31C*; ECSS Standards: Space Engineering—Thermal Control General Requirements. ESA-ESTEC Requirements & Standards Division: Noordwijk, The Netherlands, 2008.
25. Stout, K. Bayesian-based Simulation Model Validation for Spacecraft Thermal Systems. Ph.D. Thesis, Massachusetts Institute of Technology, Boston, MA, USA, 2015.
26. Welch, J.W. Comparison of Satellite Flight Temperatures with Thermal Predictions; SAE Technical Papers 2003-01-2460. In Proceedings of the 32nd International Conference on Environmental Systems, Vancouver, BC, Canada; 2003.
27. Welch, J.W. Comparison of Recent Satellite Flight Temperatures with Thermal Model Predictions; SAE Technical Papers 2006-01-2278. In Proceedings of the 35th International Conference on Environmental Systems, Norfolk, VA, USA; 2006.
28. Peabody, H.; Rodríguez-Ruiz, J.; Benítez, V. Thermal Margin Study for the Global Precipitation Measurement Spacecraft. In Proceedings of the 42nd International Conference on Environmental Systems, San Diego, CA, USA, 15–19 July 2012.
29. Garrison, M.B. Statistical Analysis of Thermal Analysis Margins. In Proceedings of the Thermal & Fluids Analysis Workshop, Newport News, VA, USA, 15–19 August 2011.
30. Garrison, M.B.; Peabody, H.; Peabody, S. Mission Life Thermal Analysis and Environment Correlation for the Lunar Reconnaissance Orbiter. In Proceedings of the 42nd International Conference on Environmental Systems, San Diego, CA, USA, 15–19 July 2012.
31. Thunnissen, D.P.; Tsuyuki, G.T. Margin Determination in the Design and Development of a Thermal Control System. In Proceedings of the 34th International Conference on Environmental Systems, Colorado Springs, CO, USA, 18–22 July 2004.
32. Allaire, D.L. Uncertainty Assessment of Complex Models with Application to Aviation Environmental Systems. Ph.D. Thesis, Massachussets Institute of Technology, Boston, MA, USA, 2009.

33. Gilmore, D.G. *Spacecraft Thermal Control Handbook*; Aerospace Press: El Segundo, CA, USA, 2002.
34. Haar, T.H.V.; Suomi, V.E. Satellite Observations of the Earth's Radiation Budget. *Science* **1969**, *163*, 667–669. [CrossRef] [PubMed]
35. Anderson, B.J.; Justus, C.G.; Batts, G.W. *Guidelines for the Selection of Near-Earth Thermal Environment Parameters for Spacecraft Design: NASA/TM-2001-211221*; NASA: Huntsville, AL, USA, 2001.
36. Garrison, M.B. Impacts of Bounding Worst Case Thermal Environments. In Proceedings of the 43rd International Conference on Environmental Systems, Vail, CO, USA, 14–18 July 2013.
37. Loeb, N.G.; Lyman, J.M.; Johnson, G.C.; Allan, R.P.; Doelling, D.R.; Wong, T.; Soden, B.J.; Stephens, G.L. Observed changes in top-of-the-atmosphere radiation and upper-ocean heating consistent within uncertainty. *Nat. Geosci.* **2012**, *5*, 110–113. [CrossRef]
38. Kline, S.J.; McKlintock, F.A. Describing Uncertainties in Single-Sample Experiments. *ASME Mech. Eng.* **1953**, *75*, 8.
39. Ishimoto, T.; Bevans, J.T. Method of evaluating f variance for thermal variance analysis. *AIAA J.* **1968**, *6*, 1178–1180. [CrossRef]
40. Ishimoto, T.; Bevans, J.T. Temperature variance in spacecraft thermal analysis. *J. Spacecr. Rocket.* **1968**, *5*, 1372–1376. [CrossRef]
41. Goble, R.G. Temperature Uncertainties Associated with Spacecraft Thermal Analyses. In Proceedings of the AIAA 6th Thermophysics Conference, Tullahoma, TN, USA, 26–28 April 1971.
42. Macgregor, R.K. Limitations in thermal scale modeling. *J. Spacecr. Rocket.* **1971**, *8*, 559–560. [CrossRef]
43. Zigrang, D.J. Comparison of Monte Carlo and Statistical Treatmetns of Heat-Transfer Data Uncertainties. *J. Spacecr.* **1977**, *14*, 546–549. [CrossRef]
44. *ECSS-E-HB-31-03A*; ECSS Standards: Space Engineering—Thermal Control Handbook. ESA-ESTEC Requirements & Standards Division: Noordwijk, The Netherlands, 2016.

aerospace

MDPI

Article

Steering Mirror System with Closed-Loop Feedback for Free-Space Optical Communication Terminals

Chris Graham [1], David Bramall [1], Othman Younus [2], Amna Riaz [2], Richard Binns [2], Eamon Scullion [2], Robert T. Wicks [2] and Cyril Bourgenot [1,*]

[1] Centre for Advanced Instrumentation, Durham University, NETPark Research Institute, Joseph Swan Road, Sedgefield TS21 3FB, UK; christopher.j.graham@durham.ac.uk (C.G.); d.g.bramall@durham.ac.uk (D.B.)

[2] Department of Mathematics, Physics and Electrical Engineering, Northumbria University, Newcastle upon Tyne NE1 8ST, UK; othman.i.younus@northumbria.ac.uk (O.Y.); amna.riaz@northumbria.ac.uk (A.R.); richard.binns@northumbria.ac.uk (R.B.); eamon.scullion@northumbria.ac.uk (E.S.); robert.wicks@northumbria.ac.uk (R.T.W.)

* Correspondence: cyril.bourgenot@durham.ac.uk

Abstract: Precision beam pointing plays a critical role in free-space optical communications terminals in uplink, downlink and inter-satellite link scenarios. Among the various methods of beam steering, the use of fast steering mirrors (FSM) is widely adopted, with many commercial solutions employing diverse technologies, particularly focusing on small, high-bandwidth mirrors. This paper introduces a method using lightweight, commercial off-the-shelf components to construct a custom closed-loop steering mirror platform, suitable for mirror apertures exceeding 100 mm. The approach involves integrating optical encoders into two off-the-shelf open-loop actuators. These encoders read the signal reflected on purposefully diamond-machined knurled screw knobs, providing maximum contrast between light and dark lines. The resulting steering mirror has the potential to complement or replace FSM in applications requiring a larger stroke, at the expense of motion speed. In the presented setup, the mirror tilt resolution achieved based on the encoder closed-loop signal feedback is 45 µrad, with a mean slew rate of 1.5 mrad/s. Importantly, the steering assembly is self-locking, requiring no power to maintain a steady pointing angle. Using the mirror to actively correct for a constantly moving incoming beam, a 5-fold increase in concentration of the beam spot on the center of the detector was obtained compared to a fixed position mirror, demonstrating the mirrors ability to correct for satellite platform jitter and drift.

Keywords: free-space optical communication; pointing acquisition tracking; tracking; acquisition; fast steering mirror; CubeSat; position sensing detector; low-earth orbit

1. Introduction

Free-Space Optical (FSO) communication links from Low-Earth Orbit (LEO) to the ground, as well as inter-satellite links (LEO to LEO) or LEO to Geostationary Earth Orbit (GEO), have been globally developed by major space agencies. LEO satellites, known for their decreased latency and capacity for improved bandwidth, present a promising option for delivering global internet access, especially in remote and underserved areas when compared to GEO satellites. This has resulted in increased interest from academia and industry [1]. However, implementing FSO links within LEO satellite constellations poses unique challenges not encountered by their counterparts.

The challenges of implementing FSO links within LEO satellite constellations stem from their unique orbit characteristics. LEO satellites, moving at faster speeds (7–11 km/s) than GEO satellites, face shorter communication windows, necessitating frequent handovers and increasing the risk of disruptions. Unlike GEO satellites, LEO satellites experience drag from the Earth's atmosphere, requiring frequent orbital adjustments that can disrupt communication links and add complexity.

Pointing, Acquisition, and Tracking (PAT) are critical aspects of FSO systems, ensuring the precise targeting of the optical signal from the transmitter to the receiver. PAT systems usually include a steering and tracking capability [2] to establish and maintain the link, as well as to compensate for the satellite's motion, atmospheric turbulence, vibration, and jitters.

Recent flight demonstrations have primarily focused on LEO-to-ground communication. There is an increasing demand for higher data rates for near real-time downloading of earth observation data. The first successful reported demonstration occurred in 2010 with a pioneer LEO-to-ground bidirectional link between the NFIRE satellite and a commercial laser communication terminal jointly developed by the US and Germany [3]. Subsequent achievements include the Small Optical TrAnsponder (SOTA) onboard the SOCRATES satellite in 2014 [4], showcasing stable links over an extended period. In both NFIRE and SOTA cases, the optical head was mounted on a gimbal with four transmitter channels and a single receiver.

A significant step toward miniaturization was demonstrated with the OC4 instrument on board the PIXL-1 satellite in 2021 [5]. This compact system successfully demonstrated a downlink, incorporating a laser communication terminal with a Fine Pointing Assembly (FPA) consisting of a 4-Quadrant Diode and a Fast Steering Mirror (FSM) to compensate for satellite pointing variation. NASA's AeroCube7B and 7C [6], both part of the CubeSat Optical Communications and Sensors Demonstration (OCSD) program [7], followed by the Laser Infrared CrosslinK Mission Click-A system [8] in 2017 and 2022, have also achieved LEO-to-ground links. While the AeroCube7 optical head did not include any beam steering capability or gimbal mechanism, relying on a high-performance star tracker for attitude control, and larger beam divergence for link stability, the more recent Click-A incorporated an FPA similar to the OC4 instrument for closed-loop control of the pointing system.

Another notable miniaturized system for LEO-to-ground communication is CubeCAT, featuring the optical head developed by TNO [9]. CubeCAT also incorporates an FPA similar to both CLICK-A and OC4.

Most of these groups are also developing Inter-Satellite Link (ISL) terminals for CubeSats. While for LEO-to-ground links, the telescope's aperture on the ground can be as large as needed (typically ranging from 100 mm to 600 mm) to maximize signal to noise for restricted laser power onboard the satellite, ISL telescope apertures are limited by the satellite platform dimensions. DLR is working on CubeISL, an ISL version of the OC4 [5]. With Click-B and -C, NASA plans to establish bidirectional ISL between two 6U CubeSats in the same orbit [10]. Finally, TNO is also working on an ISL version called LEOcat [11], with a size of 190 × 190 × 250 mm beyond the CubeSat platforms.

Steering mirrors are critical technologies for ensuring stability in pointing and tracking. Figure 1 illustrates the optical architectures of terminal technologies incorporating steering mirrors. When the same steering mirror is employed for both the transmitter and receiver, the steering functionality must compromise between speed (to correct for satellite jitter), size (to decrease the natural divergence induced by diffraction for the transmitter beam or maximize the beam collection capability on the receiver channel), and stroke amplitude (linked to the Field of Regard (FoR)).

In a duplex co-located PAT system, as depicted in Figure 1a, a single mirror is used to steer both the transmitted and received optical signal. While this configuration offers the advantage of a simpler design [9,10], it typically requires a separate beacon channel, along with the use of dichroic beamsplitters to separate the send and receive beams. These beam splitters necessitate the use of different data wavelengths on opposing terminals.

Alternatively, a duplex bistatic configuration, shown in Figure 1b, features separate transmitter and receiver optical systems. This approach provides greater versatility in decoupling the channels, allowing customization of the FoV and divergence for each path to meet the specific requirements of both the transmitter and receiver.

This paper describes a steering mirror design which can be used on the receiver side of the ISL. It incorporates a two-axis mirror to increase the angular FoR of the receiver satellite, enabling the use of a small and fast Avalanche Photodiode Detector (APD). Smaller APDs have a higher cutoff frequency, and current commercial technology achieves up to 4 GHz on a 40-micron photosensitive area. Commercial off-the-shelf (COTS) actuators and optical encoders are used, offering a simple and compact solution for future CubeSat laser communication terminals.

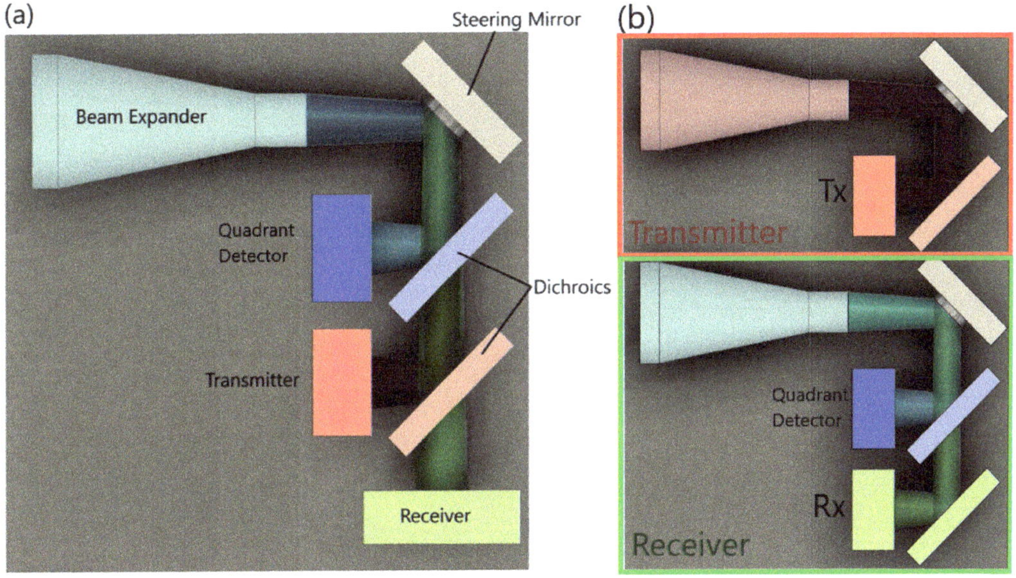

Figure 1. (**a**) Simplified transceiver terminal using a single optical train, (**b**) A simplified optical transceiver using a separated optical train design. Red corresponds to the transmitted beam data beam, green to the received data beam, and blue to the received beacon beam.

2. Steering Mirror Design and Trade-Offs

With the growing applications of steering mirrors, the market has seen the introduction of a diverse range of commercial products [12]. These solutions are available in a selection of actuator categories, including from electrostatic MEMS actuators, magnetic voice coils and piezo-stack micro-displacement amplifiers [13–15]. Each technology comes with its own advantages and trade-offs. MEMS mirrors, for instance, are light weight and have low power consumption. However, their lower actuating forces require the use of lower mass mirrors, thereby restricting aperture and optical power handling [16]. Voice coil actuators offer the advantage of higher actuation forces, enabling the use of larger mirrors with enhanced power handling and travel range. This comes at the expense of increased power consumption, heat generation, and limited resolution. Steering mirrors employing opposing pairs of piezo-stacks, which drive micro-displacement amplifiers, can move a larger-diameter mirror with high pointing resolution, at the expense of lower deflection angles. It is important to note that all of these actuator types require some power to maintain a fixed pointing angle, with the mirror returning to an uncontrolled position when powered off. A qualitative comparison of these primary actuator categories is shown in Table 1.

For the receiving channel, a large mirror aperture is desirable to improve the signal-to-noise ratio in high-speed communications. Given the successful demonstration of body pointing for the coarse steering of CubeSat laser communications in previous missions [6,17], there is flexibility in relaxing the requirement for high steering speeds on the receiving channel. This allows for a strategic trade-off, sacrificing speed for increased precision or larger scan angle.

Table 1. Qualitative comparison of fine steering mirrors [12].

Metric	Electrostatic MEMS	Voice Coil	Piezo-Stack	Stick-Slip Kinematic
Power Consumption (active)	Low	High	High	Moderate
Power Consumption (stationary)	Low	High	Moderate	Zero
Deflection Angle	Moderate	High	Low	High
Mirror Size	Low	Moderate	Moderate	High
Resolution	Moderate	Low	High	High
Optical Power Handling	Low	High	High	High

The increased size of the mirror assembly allows for a broader exploration of actuator technologies and mechanical designs. Here, a simple kinematic mount was developed, featuring a spherical ball joint to facilitate tip and tilt movement. This movement is controlled by two piezoelectric inertia stick slip actuators (PIA13VF, Thorlabs, Ely, UK). The mirror itself has a broadly rectangular aperture, measuring 130 × 90 mm. A rendering of the mirror and mount is illustrated in Figure 2, with the positions of the spherical joint and actuator tips marked in blue and red, respectively. The placement of the actuators with respect to the pivot point results in the actuators being coupled for steering the mirror about the horizontal and vertical axes, as depicted in Figure 2. To pivot around a vertical axis, both actuators must be driven in the same direction. Conversely, to pivot the mirror around its horizontal axis, the actuators must be driven differentially. The distance of the actuator from the pivot axis can be adjusted to optimize the pointing performance, mirror stroke and angular steering speed. Positioning the actuators closer to the pivot axis will increase the slew rate and angular range of the mirror, with a trade-off of reduced pointing resolution.

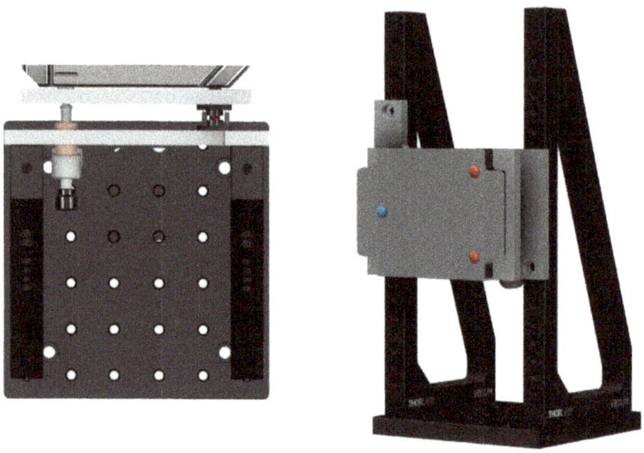

Figure 2. Rendering of a steering mirror assembly, showing a top-down and side view. The assembly uses a bespoke kinematic mount, with a spherical joint enabling motion in the desired axis, controlled by two piezoelectric inertia actuators. The position of the spherical joint is marked in blue, with the positions of the actuators marked in red. The actuators are positioned 82 mm horizontally from the pivot point, and 30 mm vertically on either side of the pivot plane. The mirror has an aperture measuring 130 × 90 mm.

Inertia stick slip actuators work by using a piezo stack to turn a lead screw. These actuators provide an extended travel range of 13 mm, coupled with a fine step size of 20 nm, making them well suited for precisely steering a mirror over a long working range. A significant advantage of these actuators is that they are self-locking when at rest, eliminating the need for constant power to maintain a set pointing angle. This characteristic is valuable in applications with tight power budgets, such as in CubeSats. The limitations of these actuators are the relatively slow travel speeds of around 3.5 mm/min, and the piezo hysteresis requiring external feedback for precise positioning.

3. Control Feedback

The stick slip piezo actuators used in this design exhibit significant hysteresis, with the size of the forward and reverse steps differing by approximately 16%, as illustrated in Figure 3a. With the actuator unloaded, the mean forward step size was measured as 24.1 nm, and the mean reverse step size as 20.3 nm. Consequently, counting the number of commanded steps does not provide an accurate measure of actuator position, as the 16% positioning error will accumulate with multiple changes in direction. An example of this is shown in Figure 3b, which depicts the position of the actuator tip over a series of forwards and backward movements, highlighting the forward drift of the 0 step position with each successive pass.

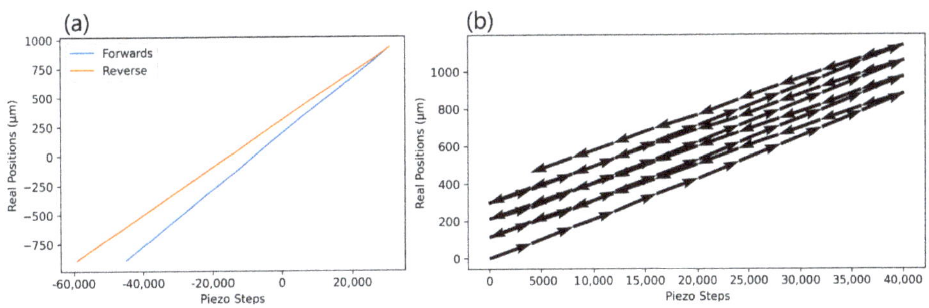

Figure 3. (**a**) Real position of piezo actuator head as a function of piezo step count. (**b**) Forward drift of 0 step position as the actuator is driven successively between 0 and 40,000 steps.

For accurate positioning of the actuators, a form of closed-loop control is essential. One option would be to use a mechanically coupled feedback sensor such as a linear variable differential transformer, with the pushrod attached to the back of the mirror. The downside of these feedback sensors is that a malfunction or jamming could mechanically lock the actuator in place. A more robust method employs non-contact sensors, such as optical encoders. The sensor receives the light directed onto a coded reflective pattern to track movement. Single-channel encoders can monitor movement speed, while quadrature encoders can be used to also track the direction of motion.

While some commercial products come with built-in optical encoders, they are generally much bulkier and heavier, weighing 200 g compared to the 55 g PIA13VFs. This increased size and weight makes these actuators less practical for use in CubeSats. Therefore, an in-house solution was developed, integrating optical encoder feedback into the COTS Thorlabs actuators. The chosen optical encoder for this setup is the AEDR-8300-1K2 quadrature encoder (Broadcom, Palo Alto, CA, USA). A critical specification for these encoders is the reflectance of the coded pattern. Rather than requiring a simple black-and-white barcode-style pattern, the brighter areas of the pattern must have a specular reflectance of over 60%. To achieve this, a fine straight knurled knob was diamond-turned, providing a highly reflective surface on the raised portions of the knurling. An image of the encoder wheel is displayed in Figure 4a. The knob features 72 grooves, corresponding to 288 pulses per revolution with the quadrature encoder. Along with the $1/4'' \times 80$ fine thread lead screw on the actuators, the encoder should provide a resolution of 1.1 μm. Since these inertial actuators require a mass on the screw thread, the only additional weight for this system is the encoder PCB, adding just 6 g to the actuator's mass. Figure 4b displays a plot of the actuator tip position against the encoder counts for 20 successive movements, using the encoder feedback for position control. This can be compared to Figure 3b, where the commanded piezo steps are used for feedback. The 12 μm spread on the actuator position originates partially from the backlash of the measurement probe, and partially from the delay in communication between the prototype encoder electronics and the motor controller.

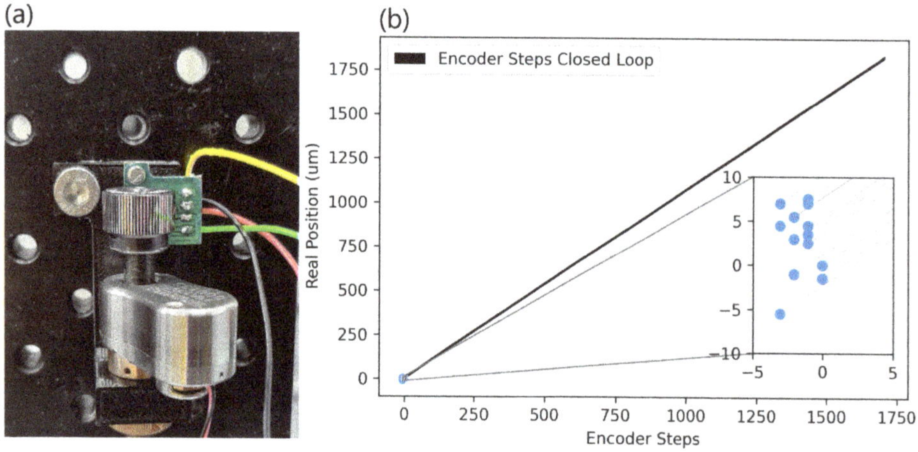

Figure 4. (**a**) Image of diamond-turned optical encoder track, (**b**) actuator position plotted against encoder step count for 15 successive movements. The inset shows a magnified view of recorded actuator positions.

4. Optical Test

To create a realistic test environment for the closed-loop actuator, a prototype slow steering mirror was constructed, with a rendering of the set-up shown in Figure 5. The mirror is flat and was machined using single-point diamond turning. A custom kinematic mount was designed to steer the mirror, employing a pair of these stick-slip actuators and a spherical ball joint to facilitate movement. The actuators are positioned 82 mm horizontally from the pivot point, and 30 mm vertically on either side of the horizontal pivot plane. The mirror is tensioned against each actuator by two extension springs.

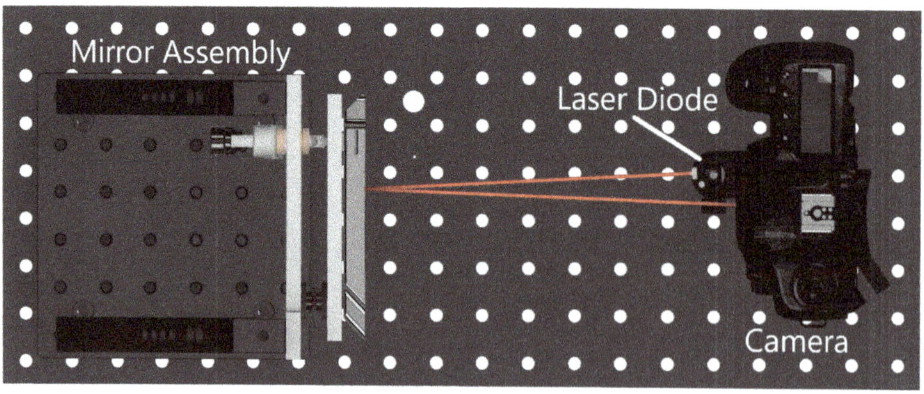

Figure 5. Rendering of the experimental optical test set-up to measure pointing repeatability and accuracy of mirror assembly.

For optical measurements of the mirror's movement, a simple set-up using a 658 nm laser diode and a DSLR camera was devised. The laser diode was positioned as close to the bore axis of the camera sensor as possible, to prevent off-axis effects. The camera was set up perpendicular to the nominal position of the mirror, and the laser was directed to land on the center of the camera sensor after reflecting off of the mirror.

The camera sensor was positioned at a distance of 280 mm from the surface of the steering mirror. The sensor has a pixel pitch of 6.41 μm, corresponding to a resolution of 11 μrad of mechanical angular movement of the mirror.

The camera and both encoders were connected to two Arduino Uno R4s (Arduino, Turin, Italy) and linked to the ThorLabs motor controller using Python (Ver 3.10.9). The mirror was then commanded to repeatedly move through a range of positions, addressing four extreme positions and a central one, with an image recorded at each point. The centroids of the laser spot were calculated for each image using a center-of-mass method and is plotted in Figure 6 [18]. Taking the mean of the five positions, the maximum spread of the spots on the camera sensor was found to be 4.2 pixels, with a standard deviation of 1.1 pixels. This corresponds to a maximum spread of 45 μrad, or 3.9 μm at the actuator tip. The standard deviation of the actuator position was calculated to be 1.0 μm. The horizontal angular slew rate of the mirror was measured as to be 1.05 mrad/s, with a vertical slew rate of 1.92 mrad/s.

The repeatability along the long axis of the mirror is limited by the delay in synchronizing both motors and actuators. Although the encoder steps are counted using hardware interrupts on the two Arduinos (one for each actuator), the step counts are read by the controlling computer via polling of the serial connection, which can result in a one-step delay at high motor speeds. The Arduinos act as converters of the raw quadrature signal into a count number read by the laptop, where the decision is taken to continue the counting or to stop the motor if the number of the desired count has been achieved. This delay could be reduced by communicating to the motor controller from the Arduinos directly.

The overall precision of the mirror mount can also be adjusted by optimizing the positions of the actuators. Moving the actuators further from the pivot point will increase angular resolution at the expense of decreasing the angular speed.

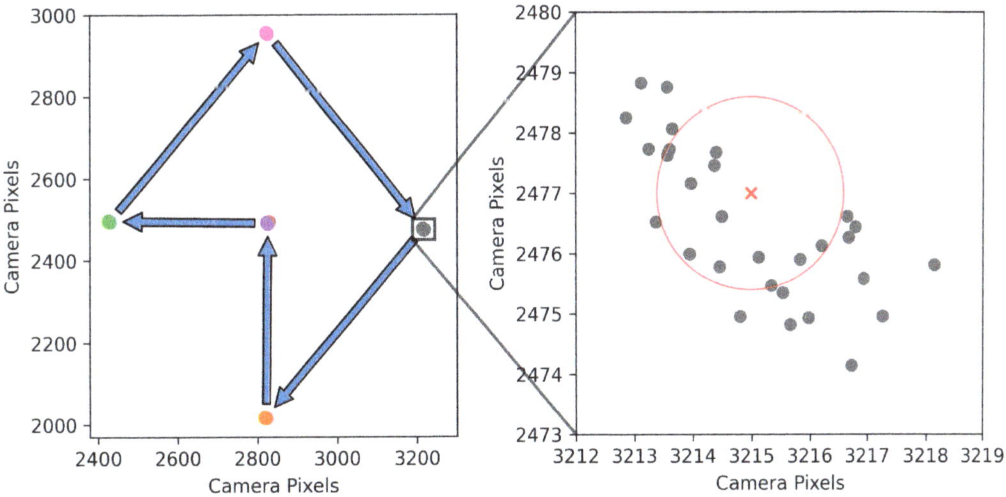

Figure 6. (**Left**) Repeatability testing of optical encoder feedback on mirror position, with the mirror being driven between 5 points (color coded) in a continuous loop. Arrows show the scanning direction. (**Right**) Close-up view of one of the rightmost point

5. Active Stabilisation Testing

Perturbations in the CubeSat's pointing direction are induced by various disturbance torques, including aerodynamic drag, solar radiation pressure, thermal radiation, Earth's gravity gradient, residual magnetism in the satellite structure, and variations in Earth's magnetic field [19]. To mitigate these disturbances during operation, the attitude determination and control system (ADCS), typically operating magneto-torquers or reaction wheels in CubeSats, dynamically adjusts the pointing direction at regular intervals. Despite the precision attainable with a star tracker for attitude measurement, the robustness of a link is ultimately contingent on the satellite's ability and accuracy in adjusting its attitude to the intended direction for laser beam transmission.

One of the key applications of a steering mirror in a communications satellite is to correct for this platform jitter, keeping the incoming beam continuously centered on the receiving detector. To simulate this scenario on an optical bench, a set-up using two steering mirrors was devised, illustrated in Figure 7. A collimated 633 nm laser beam was first directed onto a small, piezo-stack-driven fast steering mirror (silver coated P-FSM150S, Cedrat Technologies, Meylan, France), which then directed the beam onto the custom mirror under test conditions. The beam was then focused onto both a high-resolution camera sensor and a Thorlabs PSDM4 position-sensing device (PSD) using a beam-splitting cube.

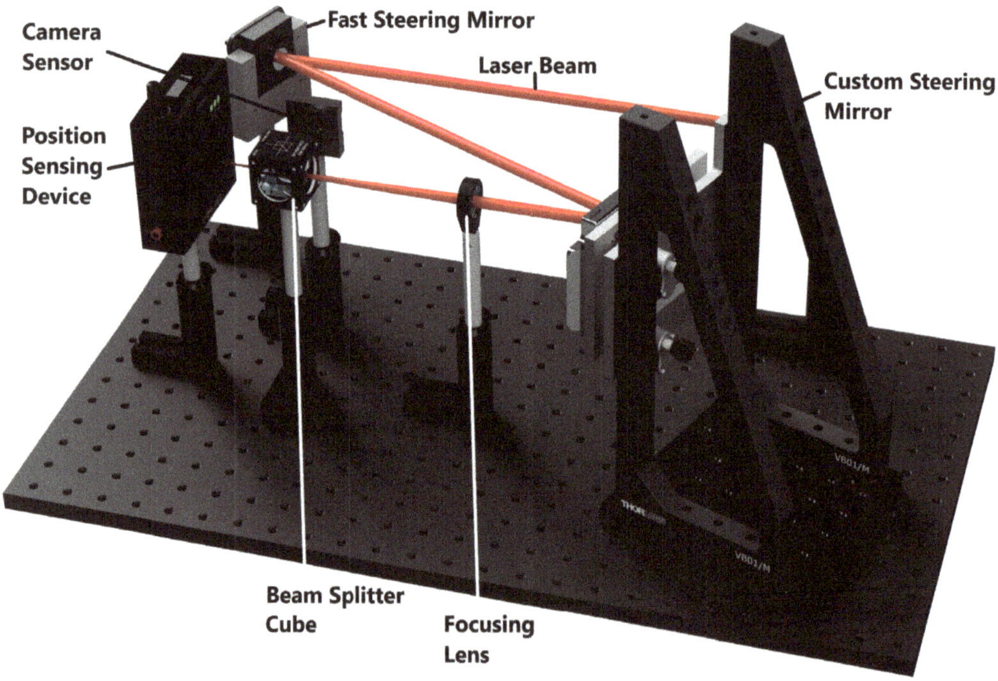

Figure 7. Rendering of the experimental optical test set-up for closed-loop control of the steering mirror.

The camera sensor was used to record the exact position of the beam spot with a high spatial resolution. Using a center-of-mass method, the centroid of the laser spot was calculated and stored in real time in order to reduce storage space needed for longer experiments.

The PSD was read out using a high-resolution data logger, with the position measurement fed into the Python control code, providing real-time feedback. The control algorithm used was a simple proportional–integral–derivative (PID) loop, taking the distance from the center position as the input. The proportional control ramps the actuator speed up as the spot moves further from the center of the sensor. The integral control acts to minimize positioning error when the spot is close to the center of the sensor, while the derivative control is optimized to reduce overshoots and control loop-induced oscillations. PID outputs are calculated separately for the X and Y displacements measured by the PSD, and summed to give the final actuator commands.

The piezo-stack-driven mirror was instructed to replicate a simulated jitter profile with a 1 Hz sampling rate, exhibiting an angular variation of ± 3.5 mrad amplitude in two Euler angles (yaw and pitch, corresponding to the x-axis and y-axis, respectively) at a maximum speed of 7 mrad/s. The simulated profile is designed to model attitude changes during the link, representing instances when the control system dynamics rectify the pointing direction to counteract disturbances. Throughout this process, the steering mirror is tasked with maintaining the laser beam centered onto a small-sized APD via the closed loop within an FPA to ensure uninterrupted communication.

The pointing angles in pitch and yaw are depicted in Figure 8a—top. The profile has been generated randomly using the available Python code (refer to Data Availability Statement). It comprises a random time series that underwent smoothing through cumulative sum and a rolling average over 60 s. To eliminate low-frequency angular drift that might exceed the FSM maximum range of $\pm 1^{\circ}$, the time series is further filtered with a high-pass filter, removing frequencies lower than 1.5×10^{-3} Hz. When the jitter profile is applied onto the FSM in the experimental setup, the mirror's movement induces a 'wandering' of the beam spot around the detector plane, represented by the blue line in Figure 8b. When active corrections are enabled on the actuated steering mirror through the closed-loop system reading from the PSD, the beam spot's position becomes significantly more focused on the camera sensor. The pointing error becomes strongly attenuated (Figure 8a—bottom) when the active mirror is engaged, and the track remains within an angular extent of 600 µrad. This signifies a more than 5-fold improvement in concentration compared to the uncorrected case, where the point track spreads over an angular extent of approximately 3400 µrad.

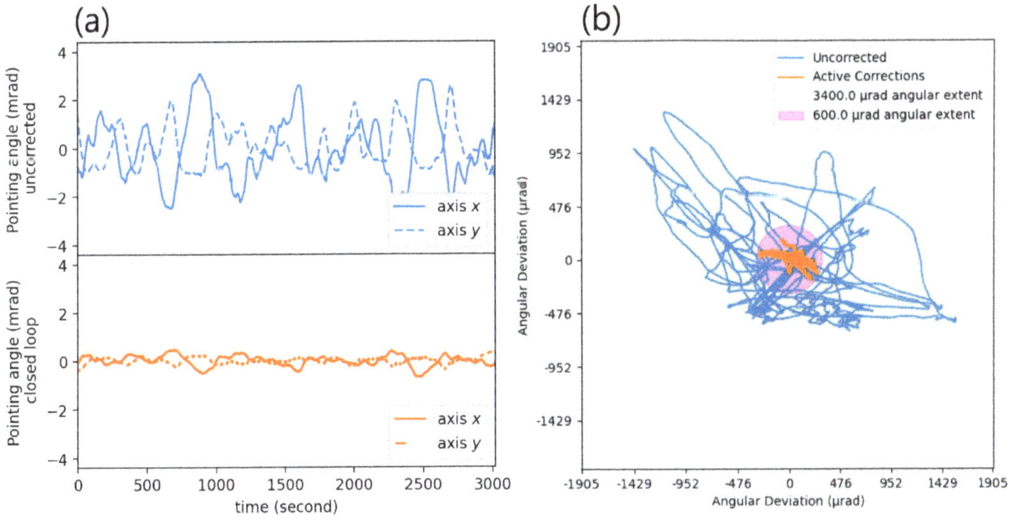

Figure 8. (**a**) Top—uncorrected pitch and yaw variation over 3000 s. Bottom—corrected pitch and yaw with closed loop enabled. (**b**) Plot of laser spot centroids measured on the camera sensor, with simulated Cubesat platform jitter being provided by a separate fast-steering mirror. The blue plot shows the track of the beam spot on the camera sensor with the piezo-steering mirror switched off, and the orange plot shows the track with real-time corrections switched on.

The outlying peaks in the actively corrected data are due to the sharp motions in the jitter profile, which correlate with satellite platform maneuvering by the ADCS. Due to the limited 8 Hz readout rate of the PSD, fast changes in the direction of the satellite platform are under sampled, resulting in a time lag before the control algorithm updates. This results in jumps of the angular displacement of the spot, seen as small tails in the orange track in Figure 8. This effect could be minimized by using faster readout electronics for the PSD, or including communication between the ADCS and mirror controllers to predict sharp movements of the ADCS. Also notable is the elliptical shape of the spot track during active corrections. This is caused by the asymmetric mounting of the actuators about the pivot plane, meaning the mirror is able to rotate more quickly about its horizontal axis than its vertical axis. This could be solved by placing the actuators equidistantly in vertical and horizontal position relative to the pivot point.

6. Thermal Tests

While the PIA13VF actuators are rated for vacuum conditions, the AEDR8300 spec sheet does not provide information on vacuum operating conditions. Additionally, neither of the AEDR or PIA13VF is rated for operation below −20 °C. A test rig consisting of an actuator, an encoder and datum switches was created, depicted in Figure 9.

The chamber was pumped down to a vacuum of 3×10^{-6} mbar, and the actuator was cooled to a temperature of 234 K (−39 °C). While the actuator moved between the two end-limit switches, the encoder pulses were recorded by an oscilloscope. The pulse train appears identical to the measurements made under room temperature and pressure. These results indicate that the system functions correctly under these extreme conditions.

Figure 9. Image of encoder/actuator test set-up inside vacuum chamber.

7. Conclusions

In summary, a novel method of adding closed-loop control to a consumer-off-the-shelf piezo actuator is described. Using low-cost electronics, the actuator can be positioned in a closed loop with a feedback error repeatability of 1.0 µm at 1 σ and an accuracy in line with the angular resolution of the encoder, defined by the period of the lines on the screw knob.

Two actuators were combined into a custom kinematic mount, producing a large-aperture steering mirror with repeatable angular positioning of less than 45 µrad. Due to the self locking nature of the actuators, the mirror does not require power to maintain a pointing direction, making it ideal for low-power applications. Using the mirror to actively center an incoming beam following a random jitter pattern, the mirror provided over a 5-fold concentration of angular extent on the detector. The actuator assembly has also been successfully tested at low temperatures under vacuum, potentially enabling the assembly to be used in space applications.

Author Contributions: Conceptualization, C.G., D.B., O.Y., A.R., R.B., E.S., R.T.W. and C.B.; methodology, C.G., D.B. and C.B.; software, C.G.; validation, C.G., D.B. and C.B.; formal analysis, C.G. and C.B.; investigation, C.G., D.B. and C.B.; resources, C.B.; data curation, C.G. and C.B.; writing—original draft preparation, C.G.; writing—review and editing, all; visualization, C.G. and C.B.; supervision, C.B.; project administration: E.S., R.T.W., R.B. and C.B.; funding acquisition, E.S., R.T.W., R.B. and C.B. All authors have read and agreed to the published version of the manuscript.

Funding: This research was funded by the UK Space Agency grant number UKSAG22_0042.

Data Availability Statement: The Python code that generates the random profile can be found at https://github.com/CyrilBourgenot/FreeSpaceOpticalCommunication.git (accessed on 17 April 2024).

Acknowledgments: The authors would like to thank e2E Services Ltd./Telespazio-UK, SMS Electronics Ltd. and Smart Sentry Ltd., partners in the NSIP Phase 3 Project "ALIGN Laser Optical Communications for CubeSats", for their input and support.

Conflicts of Interest: The authors declare no conflicts of interest.

References

1. Su, Y.; Liu, Y.; Zhou, Y.; Yuan, J.; Cao, H.; Shi, J. Broadband LEO Satellite Communications: Architectures and Key Technologies. *IEEE Wirel. Commun.* **2019**, *26*, 55–61. [CrossRef]
2. Kaymak, Y.; Rojas-Cessa, R.; Feng, J.; Ansari, N.; Zhou, M.; Zhang, T. A Survey on Acquisition, Tracking, and Pointing Mechanisms for Mobile Free-Space Optical Communications. *IEEE Commun. Surv. Tutor.* **2018**, *20*, 1104–1123. [CrossRef]
3. Fields, R.; Kozlowski, D.; Yura, H.; Wong, R.; Wicker, J.; Lunde, C.; Gregory, M.; Wandernoth, B.; Heine, F. 5.625 Gbps bidirectional laser communications measurements between the NFIRE satellite and an Optical Ground Station. In Proceedings of the 2011 International Conference on Space Optical Systems and Applications (ICSOS'11), Santa Monica, CA, USA, 11–13 May 2011; pp. 44–53. [CrossRef]
4. Kolev, D.; Takenaka, H.; Munemasa, Y.; Akioka, M.; Iwakiri, N.; Koyama, Y.; Kunimori, H.; Toyoshima, M.; Artaud, G.; Issler, J.L.; et al. Overview of international experiment campaign with small optical transponder (SOTA). In Proceedings of the 2015 IEEE International Conference on Space Optical Systems and Applications (ICSOS), New Orleans, LA, USA, 26–28 October 2015; pp. 1–6. [CrossRef]
5. Schmidt, C.; Rödiger, B.; Rosano, J.; Papadopoulos, C.; Hahn, M.T.; Moll, F.; Fuchs, C. DLR's Optical Communication Terminals for CubeSats. In Proceedings of the 2022 IEEE International Conference on Space Optical Systems and Applications (ICSOS), Virtual, 29–31 March 2022; pp. 175–180. [CrossRef]
6. Rose, T.S.; Rowen, D.W.; LaLumondiere, S.D.; Werner, N.I.; Linares, R.; Faler, A.C.; Wicker, J.M.; Coffman, C.M.; Maul, G.A.; Chien, D.H.; et al. Optical communications downlink from a low-earth orbiting 15U CubeSat. *Opt. Express* **2019**, *27*, 24382. [CrossRef] [PubMed]
7. Rose, T.S.; Rowen, D.W.; LaLumondiere, S.; Werner, N.I.; Linares, R.; Faler, A.; Wicker, J.; Coffman, C.M.; Maul, G.A.; Chien, D.H.; et al. Optical communications downlink from a 1.5U Cubesat: OCSD program. In Proceedings of the International Conference on Space Optics ICSO 2018, Chania, Greece, 9–12 October 2018; p. 111800J. [CrossRef]
8. Cahoy, K.; Grenfell, P.; Crews, A.; Long, M.; Serra, P.; Nguyen, A.; Fitzgerald, R.; Haughwout, C.; Diez, R.; Aguilar, A.; et al. The CubeSat Laser Infrared CrosslinK Mission (CLICK). In Proceedings of the International Conference on Space Optics ICSO 2018, Chania, Greece, 9–12 October 2018; p. 111800Y. [CrossRef]
9. Dresscher, M.; Human, J.D.; Witvoet, G.; Van Der Heiden, N.; Den Breeje, R.; Kuiper, S.; Fritz, E.C.; Korevaar, C.W.; Van Der Valk, N.C.; De Lange, T.J.; et al. Key Challenges and Results in the Design of Cubesat Laser Terminals, Optical Heads and Coarse Pointing Assemblies. In Proceedings of the 2019 IEEE International Conference on Space Optical Systems and Applications (ICSOS 2019), Portland, OR, USA, 14–16 October 2019. [CrossRef]
10. Tomio, H.; Grenfell, P.; Kammerer, W.; Serra, P.; Čierny, O.; Lindsay, C.; Garcia, M.; Cahoy, K.; Clark, M.; Coogan, D.; et al. Development and Testing of the Laser Transmitter and Pointing, Acquisition, and Tracking System for the CubeSat Laser Infrared CrosslinK (CLICK) B/C Mission. In Proceedings of the 2022 IEEE International Conference on Space Optical Systems and Applications (ICSOS 2022), Kyoto City, Japan, 28–31 March 2022; pp. 224–231. [CrossRef]
11. Saathof, R.; Crowcombe, W.; Kuiper, S.; van der Valk, N.; Pettazzi, F.; de Lange, D.; Kerkhof, P.; van Riel, M.; de Man, H.; Truyens, N.; et al. Optical satellite communication space terminal technology at TNO. In Proceedings of the International Conference on Space Optics ICSO 2018, Chania, Greece, 9–12 October 2018; p. 111800K. [CrossRef]
12. Milaševičius, M.; Mačiulis, L. A Review of Mechanical Fine-Pointing Actuators for Free-Space Optical Communication. *Aerospace* **2023**, *11*, 5. [CrossRef]

13. Yoo, B.W.; Park, J.H.; Park, I.H.; Lee, J.; Kim, M.; Jin, J.Y.; Jeon, J.A.; Kim, S.W.; Kim, Y.K. MEMS micromirror characterization in space environments. *Opt. Express* **2009**, *17*, 3370. [CrossRef] [PubMed]
14. Langenbach, H.; Schmid, M. Fast steering mirror for laser communication. In Proceedings of the 11th European Space Mechanisms and Tribology Symposium, ESMATS 2005, Lucerne, Switzerland, 21–23 September 2005; Volume 591, pp. 27–33.
15. Zheng, H.; Lu, S.; Zhai, Q.; Huang, B.; Long, Y.; Zhao, Y.; Qi, J. Simulation and experiment of a diamond-type micro-displacement amplifier driven by piezoelectric actuator. *J. Eng.* **2020**, *2020*, 141–147. [CrossRef]
16. do Vale Pereira, P.; Hunwardsen, M.T.; Cahoy, K. Characterization of laser thermal loading on microelectromechanical systems-based fast steering mirror in vacuum. *Opt. Eng.* **2020**, *59*, 1. [CrossRef]
17. Schieler, C.; Robinson, B.S.; Guldner, O.; Bilyeu, B.; Garg, A.S.; Riesing, K.M.; Chang, J.T.; Hakimi, F. NASA's Terabyte Infrared Delivery (TBIRD) Program: Large-Volume Data Transfer from LEO. In Proceedings of the Small Satellite Conference (SSC19-VI-02), Logan, UT, USA, 5–8 August 2019.
18. Fosu, C.; Hein, G.; Eissfeller, B. Determination of centroid of CCD star images. *Int. Arch. Photogram. Remote Sens. Spat. Inform. Sci* **2004**, *35*, 612–617.
19. Rawashdeh, S.A.; Straub, J. Attitude Analysis of Small Satellites Using Model-Based Simulation. *Int. J. Aerosp. Eng.* **2019**, *2019*, 11. [CrossRef]

aerospace

MDPI

Article

On-Board Image Enhancement on Remote Sensing Payload

Guo-Cheng Xu [1], Pei-Jun Lee [1,*], Trong-An Bui [1], Pei-Hsiang Hsu [1] and Shiuan-Hal Shiu [2]

1 Department of Electronic and Computer Engineering, National Taiwan University of Science and Technology, Taipei 106, Taiwan; d11102003@mail.ntust.edu.tw (G.-C.X.); trongan93@mail.ntust.edu.tw (T.-A.B.); m11102224@gapps.ntust.edu.tw (P.-H.H.)

2 Liscotech System Company, Ltd., Taipei 115, Taiwan; shiuanhalshiu@liscotech.com

* Correspondence: pjlee@mail.ntust.edu.tw

Abstract: CubeSats are designed to optimize applications within the strict constraints of space and power. This paper presents an On-Board Image Enhancement technique for remote sensing payloads, focusing on achieving Auto White Balance (AWB) with limited resources and enhancing the capabilities of small/microsatellites. The study introduces hardware-based techniques, including histogram adjustment, De-Bayer processing, and AWB, all tailored to minimize hardware resource consumption on CubeSats. The integrated 1U CubeSat system comprises a sensor board, an Image Data Processor (IDP) unit, and onboard computing, with a total power consumption estimated at 2.2 W. This system facilitates image capture at a resolution of 1920 × 1200 and utilizes the proposed algorithm for image enhancement on remote sensing payloads to improve the quality of images captured in low-light environments, thereby demonstrating significant advancements in satellite image processing and object-detection capabilities.

Keywords: CubeSat; satellite; FPGA; auto white balance; De-Bayer; histogram

1. Introduction

The field of small-satellite technology has witnessed rapid growth in recent years. These satellites offer high cost-effectiveness, shorter development cycles, and extensive commercial application potential in areas such as communication [1], Earth observation [2], and scientific research [3]. With continuous technological advancements, small satellites are poised to drive further innovation and applications in the space industry, opening up new markets and application domains.

Designing remote sensing payloads for small satellites presents multiple challenges. First, image resolution is of paramount importance. Due to the limited space on CubeSats [4], achieving sufficient image resolution within this confined space is crucial to ensure the accuracy and reliability of remote sensing data. Second, the use of high-dimensional lenses is restricted by the size limitations of CubeSats, which may impact the performance and functionality of the remote sensing payload. Additionally, the hardware systems on CubeSats face strict power consumption limitations, potentially restricting the performance of sensors and image-processing units. Thus, optimizing power consumption is essential to ensure a system operates effectively within a limited energy supply. Finally, it is necessary to limit the amount of incoming light to protect the sensor, which may affect the brightness and contrast of the images, consequently impacting the quality of satellite imagery. In summary, the design of remote sensing payloads is a complex task that requires multiple constraints to be overcome to ensure the performance and effectiveness of the payload.

For efficient image processing and capturing on satellites, the implementation of algorithms on embedded processors is required. Field-Programmable Gate Arrays (FPGAs) [5] are highly suitable for this purpose due to their balance between parallel processing capabilities and low power consumption [6].

This study utilizes FPGA logic for onboard object detection missions. The design of an effective imaging payload necessitates the careful consideration of power consumption, resource allocation, and image quality. The adoption of FPGA technology facilitates the design of a compact development board, enabling parallel data processing while ensuring low power consumption within the limited space of CubeSats. This research introduces FPGA firmware implementation and its technical specifications, focusing particularly on FPGA logic design. These applications have found wide usage in CubeSat. To achieve parallel-processing and low-power image capture architecture, the development of system architecture for an imaging payload on a 1U CubeSat (10 cm × 10 cm) is presented. This system efficiently coordinates the available volume and power consumption of CubeSats and is designed based on FPGA technology. The image-enhancing De-Bayer [7] and AWB [8] algorithms play a critical role in enhancing the clarity of satellite imagery. The AWB algorithm automatically corrects color deviations, ensuring the naturalness and accuracy of colors under different lighting conditions. Simultaneously, the De-Bayer algorithm reconstructs full-color images from raw image data, improving color accuracy and image detail clarity. As both algorithms utilize histograms, this study stores the parameters computed from histograms in DDR to enhance computational efficiency and leverages the parallel-processing capabilities of FPGA to accelerate the processing speed. This capability enables the leveraging of image recognition technology for the identification og objects on the Earth's surface or in maritime areas, such as detecting vehicles on roads or ships at sea.

2. Proposed Algorithm for Image Enhancement on Remote Sensing Payload

The spatial constraints of CubeSats preclude the accommodation of high-resolution lenses, leading to reduced variations in captured light [9]. This limitation results in images with a lack of lighting variations and subtle details, reducing their quality and usability. Additionally, the necessity to protect the sensor results in less incoming light being permitted, leading to darker images with diminished details. Such exposure limitations affect the dynamic range of the images, thus further diminishing their quality. Consequently, preprocessing images captured on satellites enhances their quality. It compensates for the limitations of light variations and exposure time, making the images clearer and more suitable for subsequent analysis and applications. This study proposes an algorithm for image enhancement on remote sensing payloads, including FPGA-based De-Bayer and Auto White Balance (AWB) processes to address the issues arising from light variations and exposure-time limitations. AWB automatically adjusts the color saturation of images, rendering them more true-to-life and natural, while the De-Bayer process addresses color distortion issues in images, further enhancing their quality.

2.1. FPGA-Based De-Bayer Algorithm

This study employs an innovative approach, starting with the histogram processing of RGB channel data. Subsequently, the calculated maximum and minimum values of the RGB channel parameters are stored in DDR via DMA. During the normalization process, parameter values are retrieved from DDR through DMA for normalization. Then, the image undergoes iteration over each pixel using the De-Bayer filter. Following that, color interpolation is performed on each pixel, converting the result back to RGB values to obtain the output image. Figure 1 illustrates the FPGA logic design of the De-Bayer algorithm, where RGB channel data are collected nine at a time into the De-Bayer matrix generator. Subsequently, the data are divided into groups of three using an MUX. Further, it undergoes multiplication with a kernel mask, followed by outputting channel data. Finally, the data are sent to DMA via FIFO and then written into the specified DDR location.

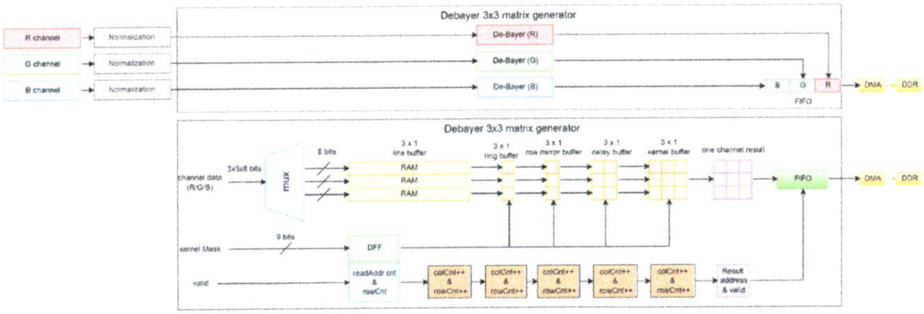

Figure 1. De-Bayer FPGA logic design.

The De-Bayer algorithm is a digital image-processing technique used to reconstruct a full-color image from incomplete color samples output by an image sensor overlaid with a Bayer filter. It assumes the reflectance of all the image channels as expressed by Equation (2), with the RGB image and De-Bayer filter iterating through the process.

$$N_c(x,y) = C(x,y) - \min(C(x,y))/\max(C(x,y)) - \min(C(x,y)) \quad (1)$$

where $N_c(x,y)$ represents the normalized pixel value by Equation (1), $C(x,y)$ represents the original channel, the pixel value at the (x,y) position, and $\min(C(x,y))$ and $\max(C(x,y))$ represent the minimum and maximum values of the channel data, respectively.

$$N_c(x,y) \times DB_C(x,y) = R_c(x,y) \quad (2)$$

where $DB_C(x,y)$ represents the De-Bayer filter and c is the channel, $N_c(x,y)$ is one channel of the image R, G, and B, and $R_c(x,y)$ is the channel result after applying the De-Bayer process.

2.2. FPGA-Based Auto White Balance (AWB) Algorithm

This FPGA-based AWB algorithm aims to preprocess and enhance low-light satellite images. First, it uses histogram preprocessing to calculate the maximum, minimum, and total values for each channel. Subsequently, these values are stored in DDR via DMA. Next, the algorithm extracts the maximum, minimum, and total parameters through DMA for image compensation. Finally, the output image is obtained. The FPGA logic design diagram, as shown in Figure 2, illustrates the process where RGB channel values enter a buffer, while addresses are calculated to read black patterns and compute black averages from ROM. Additionally, differential gain calculations are performed from ROM, followed by a right shift of 16 bits to obtain constrained minimum and maximum values. Subsequently, the results are sent to DMA via FIFO and stored in the specified location of DDR.

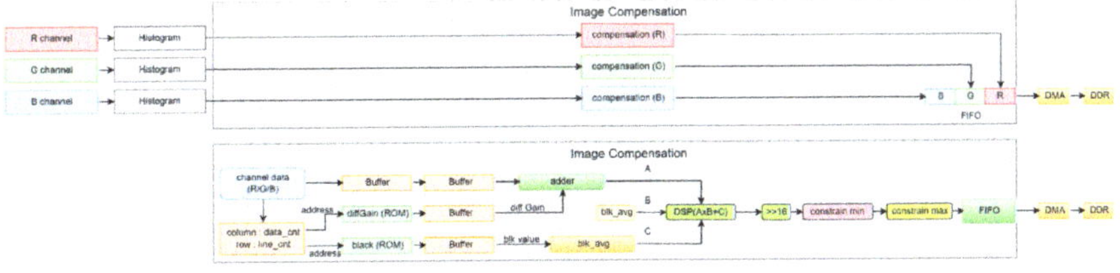

Figure 2. AWB FPGA logic design.

This assumption is articulated by Equation (4), detailing the FPGA-based AWB function.

$$H_c(v) = \sum_{x=1}^{N} \sum_{y=1}^{M} I_c(x,y) \quad (3)$$

$$I_c(x,y) = H_c(v) \times W_c \tag{4}$$

For each color channel (red, green, blue) of the RGB image, its histogram is calculated separately by Equation (3). A histogram is a statistical chart that displays the frequency of each brightness value, where $H_c(v)$ is the histogram of color channel $c \in \{R, G, B\}$ with the value $v = N \times M$, where N is the image height and M is the image width. $I_c(x,y)$ is mean the channel c place (x,y). W_c is calculated based on the assumption of the white point for channel c. The W_c parameters of white point assumption are adjusted based on the results obtained from the white reference objects in the experiment. Each color channel is adjusted using the calculated gains to achieve color balance. $I_c(x,y)$ represents the brightness value of channel c at position (x,y) after adjustment.

3. Hardware Logical On-Board Image Enhancement Design

The collection of satellite imagery faces challenges such as terrain and shadow effects, particularly in mountainous areas. Atmospheric factors like clouds and fog also impact image clarity. The proposed methodology involves histogram adjustments for contrast and brightness, De-Bayer processing for enhancing image quality, and Auto White Balance (AWB) to address lighting variations and atmospheric effects, ensuring accurate color representation. AWB adjusts white balance based on light source colors, providing realistic surface feature depiction despite changing conditions.

In the design of remote sensing payloads, the computation resource and power consumption play an important role. This paper proposes the FPGA-based algorithm and the hardware design of image enhancement with fewer computing resources and lower power consumption, which means the image enhancement can be processed in real time. The improved De-Bayer algorithm and AWB FPGA logic design is described in Figure 3. In this design, preprocessing of the histogram is performed first upon image input. These processed histograms are stored in memory through the APB Bus [10] and APB arbiter controller, along with DMA. Subsequently, the max, min, and sum parameters of the histogram are extracted via DMA and provided to the De-Bayer and AWB modules. Upon receiving the parameters, the De-Bayer module performs normalization processing, followed by De-Bayer processing, and finally saves the result to DDR via DMA. Following this, AWB adjusts white balance and color balance upon receiving parameters, eventually saving the result to DDR via DMA.

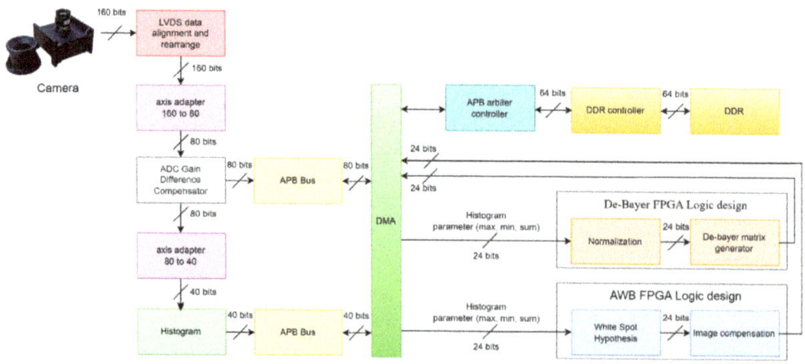

Figure 3. The main flowchart of proposed FPGA-based image enhancement on remote sensing payload.

To implement an imaging payload system on a 1U CubeSat, this study designs the image capture and image enhancement functionalities for 1U CubeSats, as illustrated in Figure 4. Figure 5 showcases the lens (Schneider Xenoplan 1.9/35 (manufacturer Schneider Kreuznach, city of Bad Kreuznach, Germany)), CMOS Sensor (Python 2000 Global Shutter (manufacturer onsemi, city of Phoenix, AZ, USA)), and the Image Data Processor (IDP) utilized. The primary task of the IDP involves receiving and preprocessing image data. The main FPGA processing unit on the Image Acquisition Board is the Microsemi SmartFusion2 M2S060 FPGA System (Aliso Viejo, CA, USA) on Chip (SoC). The FPGA facilitates basic processing of the image data.

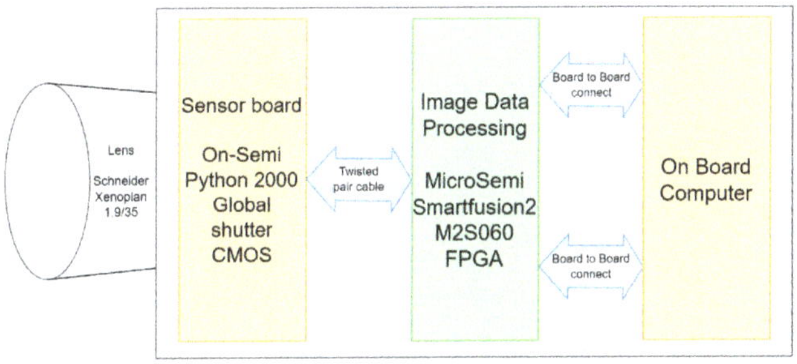

Figure 4. On-board image enhancement on remote sensing payload architecture.

(**a**)

(**b**)

Figure 5. (**a**) Module of the imaging payload on 1U CubeSat; (**b**) inside of the imaging payload on 1U CubeSat module.

As depicted in Figure 4, the sensor board captures the image through the lens and transmits the image's raw data through a twisted pair cable connected to the IDP [11]. The IDP board is responsible for image-processing functions, including histogram adjustment, De-Bayer, and AWB. The processed image result is sent to the onboard computer via a board-to-board connection.

The FPGA Logic Design of the On-Board Image Enhancement on Remote Sensing Payload

The actual 1U CubeSat architecture is illustrated in Figure 5, which depicts the design of the imaging payload system on the 1U CubeSat. The module structure of the imaging payload system is shown in part (a), and the interior of the imaging payload system is presented in part (b). This system includes the On-Semiconductor PYTHON 2000 sensor board and the Image Data Processor (IDP). Detailed specifications of the modules are provided in Table 1. Table 1 lists the sensor types as color, monochrome, and NIR-enhanced monochrome; the optical format as 2/3 inch; active pixels as 1920 $\times$ 1200; pixel size as 4.8 $\times$ 4.8 µm; and the fully functional power consumption of the sensor board [11] as 1W. Additionally, it details the SoC as Microsemi SmartFusion2 M2S060 [12], the MCU as Microsemi SmartFusion2 ARM Cortex-M3, the operating system as FreeRTOS, the on-board memory as 256 MB ECC LPDDR, and the on-board flash as 2 GB NAND and 16 MB NOR, with the fully functional power consumption of the IDP as 1.2 W. The total power consumption of the modules in this study, including the sensor board and IDP, is 2.2 W. This indicates that the design is well suited for FPGA utilization in CubeSats, highlighting an efficient power consumption profile for the IDP.

Tables 2 and 3, respectively, compare the FPGA logic design and resource utilization of the proposed de-Bayer and AWB methods with other methods.

Table 1. 1U CubeSat hardware specifications.

Sensor Board	On-Semi PYTHON 2000
Sensor types	Color, monochrome, and NIR-enhanced monochrome
Optical format	2/3 inch
Active pixels	1920 × 1200
Pixel size	4.8 × 4.8 μm
Power consumption	1 W
Image Data Processing (IDP)	
SoC	Microsemi Smartfusion2 M2S060
MCU	Microsemi Smartfusion2 ARM Cortex-M3 MCU
OS	FreeRTOS
On-board memory	256 MB ECC LPDDR
On-board flash	2 GB NAND + 16 MB NOR
Power consumption	1.2 W

Table 2. Resource usage of the De-Bayer FPGA logic design comparison.

Type	Proposed	Smart Stereo Camera [13]	Heterogeneous Systems-on-Chip [14]
Feature map	3 × 3	2560 × 720	2048 × 2048
CLB LUTs	173	32,665	5184
CLB Registers	194	None	None
LUT as memory	10	6910	48
FPGA	Microsemi Smartfusion2 M2S060	zynq xc7z030 soc	Xilinx Virtex-7

Table 3. Resource usage of the AWB FPGA logic design comparison.

Type	Proposed	Camera Systems [15]	Multi-Camera GigE Vision IP Core [16]
Feature map	3 × 3	1920 × 1080	2048 × 2048
CLB LUTs	972	5802	813
CLB Registers	2703	None	610
LUT as memory	4	None	8
FPGA	Microsemi Smartfusion2 M2S060	Zynq-7020 SoC	virtex-4 fx100-11

4. Experiment Results

Using satellite images captured in low-light environments, this study simulated the Auto White Balance (AWB) function. The input images and histogram, as shown in Figure 6, underwent AWB processing. The output images and histogram, depicted in Figure 7, demonstrate that the experimental approach effectively addresses the issue of CubeSats' inability to accommodate high-resolution lenses in space. These satellites encounter situations where lighting conditions undergo substantial changes, necessitating the restriction of incident light to safeguard the sensors and thereby generating low-dynamic-range (LDR) images. Such issues could potentially affect the brightness and contrast of the images. However, the results of this experiment successfully mitigate these concerns, resulting in clearer image rendering. The histogram images in Figures 6c,d and 7c,d, consist of lines rep-resenting the red, green, and blue channels, respectively.

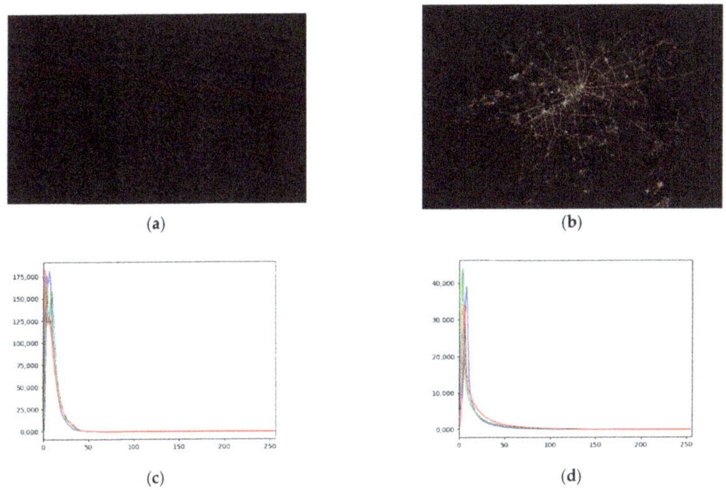

Figure 6. LDR images: airport and city satellite image test cases (1). (**a**) Airport test case (1) and histogram (**c**); (**b**) city test case (1) and histogram (**d**).

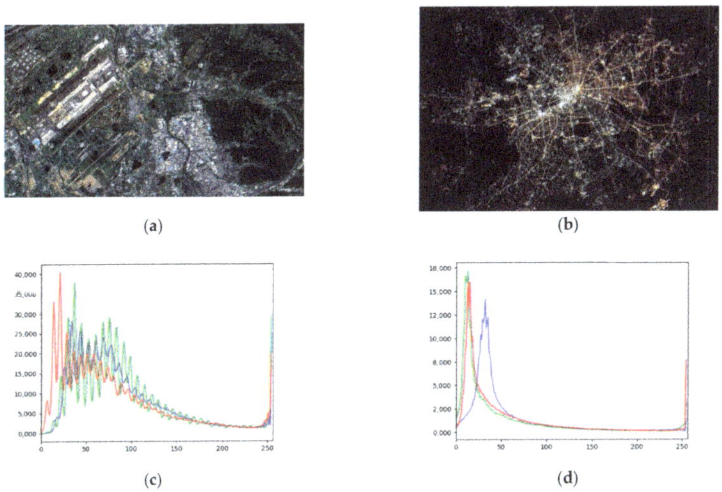

Figure 7. Simulation of proposed algorithm image enhancement on remote sensing payload of airport and city satellite image result. (**a**) Airport test case (1) and histogram (**c**); (**b**) city test case (1) and histogram (**d**).

Additionally, this paper utilizes the proposed module of the imaging payload on 1U CubeSat (Figure 5) to obtain satellite images through the sensor payload. Figure 8 displays the raw low-dynamic-range (LDR) image captured by the imaging payload. The proposed on-board image enhancement for remote sensing payload is simulated and presented in Figure 9. It is observed that the brightness of the original images significantly improves after processing by the Image Data Processor (IDP). The red rectangle highlights how the Auto White Balance (AWB) result enhances the clarity of the buildings.

(a) (b)

Figure 8. LDR images: airport and city satellite image test cases (2). (**a**) Airport test case (2); (**b**) city test case (2).

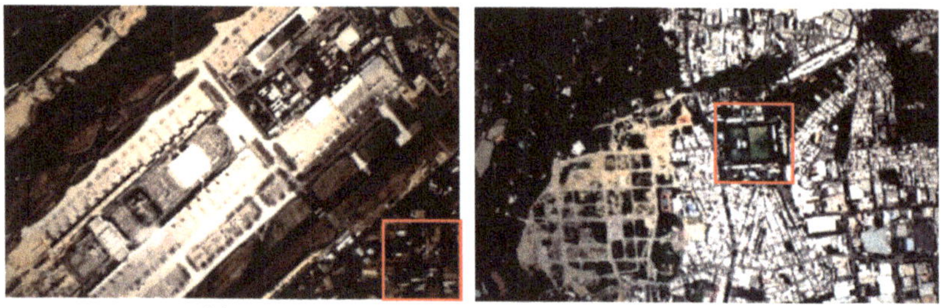

(a) (b)

Figure 9. Experiment results of proposed algorithm image enhancement on remote sensing by module of the imaging payload on 1U CubeSat. (**a**) Airport test case (2); (**b**) city test case (2).

Furthermore, the IDP captures satellite images over the sea to detect ships. These images serve the purpose of ship detection. The results of ship object detection performed pre-training with the YOLO v7 model [17] and conducting training using the AI-TOD DataSets [18] are shown in Figure 10. In the test images, two boats are present, and with YOLO v7 object detection, it is evident that the ships captured by the method proposed in this study, as shown in Figure 10b, are more accurately detected than those in the unprocessed image, as shown in Figure 10a, due to the presence of two ships in the test image.

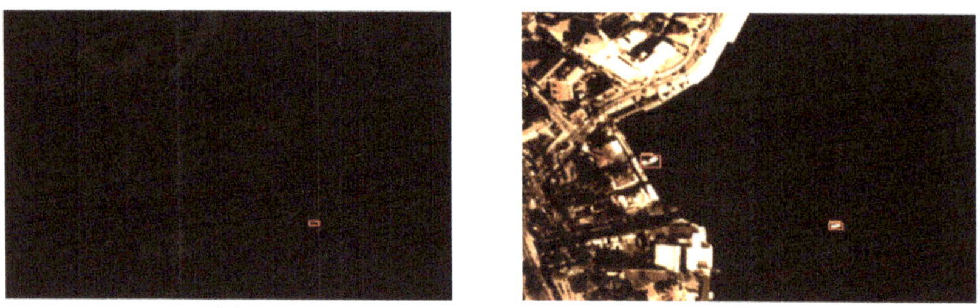

(a) (b)

Figure 10. Harbor satellite image and proposed algorithm for ship detection. (**a**) Harbor satellite image; (**b**) proposed algorithm.

5. Conclusions

This study develops an on-board image enhancement technique for remote sensing payloads aimed at enhancing image quality and facilitating object detection. Auto White Balance (AWB) is utilized for image enhancement, resulting in clearer satellite imagery. The FPGA-based design enables the use of smaller development boards and reduces power consumption, making it highly suitable for CubeSat applications. Through experiments with satellite images captured in low-light environments, this study addresses concerns related to image quality, leading to improved image rendering. Furthermore, the developed Image Data Processor (IDP) captures satellite images over the sea for ship detection. The results from YOLO v7 object detection demonstrate enhanced accuracy in identifying boats compared to traditional methods, underscoring the effectiveness of the proposed approach in satellite-based object detection.

Author Contributions: Methodology, S.-H.S.; Software, P.-H.H.; Validation, P.-H.H.; Writing—original draft, G.-C.X. and T.-A.B.; Writing—review & editing, T.-A.B.; Supervision, P.-J.L. All authors have read and agreed to the published version of the manuscript.

Funding: This work was supported in part by the National Science and Technology Council of Taiwan under Grant MOST 110-2221-E011-157-MY3, Grant NSTC 112-2622-E-011-022, and Grant NSTC 112-2218-E-006-015.

Data Availability Statement: Data is contained within the article.

Conflicts of Interest: Author Shiuan-Hal Shiu was employed by the company Liscotech System Company, Ltd. The remaining authors declare that the research was conducted in the absence of any commercial or financial relationships that could be construed as a potential conflict of interest.

References

1. Abdullah, H.H.; Elboushi, A.; Gohar, A.E.; Abdallah, E.A. An Improved S-Band CubeSat Communication Subsystem Design and Implementation. *IEEE Access* **2021**, *9*, 45123–45136. [CrossRef]
2. Wu, S.; Chen, W.; Cao, C.; Zhang, C.; Mu, Z. A multiple-CubeSat constellation for integrated earth observation and marine/air traffic monitoring. *Adv. Space Res.* **2021**, *67*, 3712–3724. [CrossRef]
3. Diwan, R.; de Kuijper, K.; Pal, P.S.; Ritter, A.; Parkinson, P.S.; Kong, A.C.; Parker, Q. 6U MeV CubeSat Mission: A low-cost approach towards gamma-ray astronomy. *arXiv* **2023**. [CrossRef]
4. Francisco, C.; Henriques, R.; Barbosa, S. A Review on CubeSat Missions for Ionospheric Science. *Aerospace* **2023**, *10*, 622. [CrossRef]
5. Xu, M.; Chen, L.; Shi, H.; Yang, Z.; Li, J.; Long, T. FPGA-Based Implementation of Ship Detection for Satellite On-Board Processing. *IEEE J. Sel. Top. Appl. Earth Obs. Remote Sens.* **2022**, *15*, 9733–9745. [CrossRef]
6. Grøtte, M.E.; Birkeland, R.; Honoré-Livermore, E.; Bakken, S.; Garrett, J.L.; Prentice, E.F.; Sigernes, F.; Orlandić, M.; Gravdahl, J.T.; Johansen, T.A. Ocean Color Hyperspectral Remote Sensing with High Resolution and Low Latency-The HYPSO-1 CubeSat Mission. *IEEE Trans. Geosci. Remote Sens.* **2022**, *60*, 1000619. [CrossRef]
7. Camera Assisted Field Strength Measurement System for Industrial Environments | VDE Conference Publication | IEEE Xplore. Available online: https://ieeexplore.ieee.org/abstract/document/6856686 (accessed on 30 March 2024).
8. Khalifat, J.; Arslan, T. A novel Dynamic Partial Reconfiguration design for automatic white balance. In Proceedings of the 2014 NASA/ESA Conference on Adaptive Hardware and Systems, AHS 2014, Leicester, UK, 14–18 July 2014; pp. 9–14. [CrossRef]
9. Bakken, S.; Henriksen, M.B.; Birkeland, R.; Langer, D.D.; Oudijk, A.E.; Berg, S.; Pursley, Y.; Garrett, J.L.; Gran-Jansen, F.; Honoré-Livermore, E.; et al. HYPSO-1 CubeSat: First Images and In-Orbit Characterization. *Remote Sens.* **2023**, *15*, 755. [CrossRef]
10. AMBA APB Protocol Specification. Available online: https://developer.arm.com/documentation/ihi0024/latest/ (accessed on 15 February 2024).
11. Liscotech. Available online: https://www.liscotech.com/ (accessed on 15 February 2024).
12. Microchip Headquarters. UG0445 User Guide SmartFusion2 SoC FPGA and IGLOO2 FPGA Fabric About Microsemi. Available online: https://www.microsemi.com (accessed on 15 February 2024).
13. Michalik, S.; Michalik, S.; Naghmouchi, J.; Berekovic, M. Real-Time smart stereo camera based on FPGA-SoC. In Proceedings of the IEEE-RAS International Conference on Humanoid Robots, Birmingham, UK, 15–17 November 2017; pp. 311–317. [CrossRef]
14. Pilato, C.; Mantovani, P.; Di Guglielmo, G.; Carloni, L.P. System-Level Optimization of Accelerator Local Memory for Heterogeneous Systems-on-Chip. *IEEE Trans. Comput.-Aided Des. Integr. Circuits Syst.* **2017**, *36*, 435–448. [CrossRef]
15. Khalifat, J.; Ebrahim, A.; Adetomi, A.; Arslan, T. A dynamic partial reconfiguration design for camera systems. In Proceedings of the 2015 NASA/ESA Conference on Adaptive Hardware and Systems, AHS 2015, Montreal, QC, Canada, 15–18 June 2015. [CrossRef]
16. Ibraheem, O.W.; Irwansyah, A.; Hagemeyer, J.; Porrmann, M.; Rueckert, U. A resource-efficient multi-camera GigE vision IP core for embedded vision processing platforms. In Proceedings of the 2015 International Conference on ReConFigurable Computing and FPGAs, ReConFig 2015, Riviera Maya, Mexico, 7–9 December 2015. [CrossRef]

17. Wang, C.-Y.; Bochkovskiy, A.; Liao, H.-Y.M. YOLOv7: Trainable bag-of-freebies sets new state-of-the-art for real-time object detectors. In Proceedings of the IEEE/CVF Conference on Computer Vision and Pattern Recognition 2023, Vancouver, BC, Canada, 17–24 June 2023; pp. 7464–7475. [CrossRef]
18. Wang, J.; Yang, W.; Guo, H.; Zhang, R.; Xia, G.S. Tiny object detection in aerial images. In Proceedings of the International Conference on Pattern Recognition, Milan, Italy, 10–15 January 2021; pp. 3791–3798. [CrossRef]

aerospace MDPI

Article

Spitzer Resurrector Mission: Advantages for Space Weather Research and Operations

Shawn M. Usman [1,*], Giovanni G. Fazio [2], Christopher A. Grasso [3,4], Ryan C. Hickox [5], Cameo Lance [1], William B. Rideout [1], Daveanand M. Singh [1], Howard A. Smith [2], Angelos Vourlidas [6], Joseph L. Hora [2], Gary J. Melnick [2], Matthew Ashby [2], Volker Tolls [2], Steven Willner [2] and Salma Benitez [1]

1 Rhea Space Activity, Inc., Washington, DC 20004, USA
2 Center for Astrophysics | Harvard & Smithsonian, Cambridge, MA 02138, USA; jhora@cfa.harvard.edu (J.L.H.)
3 Aerospace Department, College of Engineering, University of Colorado, Boulder, CO 80309, USA
4 Blue Sun Enterprises, Inc., Boulder, CO 80302, USA
5 Dartmouth College, Hanover, NH 03755, USA
6 Johns Hopkins University Applied Physics Laboratory, Laurel, MD 20723, USA
* Correspondence: shawn.usman@rheaspaceactivity.com

Abstract: In 1979, NASA established the Great Observatory program, which included four telescopes (Hubble, Compton, Chandra, and Spitzer) to explore the Universe. The Spitzer Space Telescope was launched in 2003 into solar orbit, gradually drifting away from the Earth. Spitzer was operated very successfully until 2020 when NASA terminated observations and placed the telescope in safe mode. In 2028, the U.S. Space Force has the opportunity to demonstrate satellite servicing by telerobotically reactivating Spitzer for astronomical observations, and in a separate experiment, carry out novel Space Weather research and operations capabilities by observing solar Coronal Mass Ejections. This will be accomplished by launching a small satellite, the Spitzer-Resurrector Mission (SRM), to rendezvous with Spitzer in 2030, positioning itself around it, and serving as a relay for recommissioning and science operations. A sample of science goals for Spitzer is briefly described, but the focus of this paper is on the unique opportunity offered by SRM to demonstrate novel Space Weather research and operations capabilities.

Keywords: space weather; Spitzer; astrophysics

1. Introduction

Space Weather (SWx) is a serious concern for all areas of our society, yet forecasting remains an extremely challenging problem. The size and complexity of the system (inner heliosphere) compared to the size of Earth—the 'target' of the forecast—is enormous (about >108:1 in area). Observing the system with only a handful of research missions results in sparse coverage and inconsistent data over time. Solving the sparse coverage problem is a complex and expensive task that requires a systems approach [1] and efficient collaboration among the government, military, and commercial sectors, as has been noted in several recent reports (e.g., SWx R2O2R Framework [2], NAS Workshop on the Future SWx Research & Operations Infrastructure [3]).

An obvious strategy to build up the SWx space infrastructure is to leverage the increasing number of launches via rideshares (e.g., NOAA's SWFO-L1 on the NASA IMAP mission). While this strategy is gaining popularity, with several rideshare missions (mostly CubeSats) on the books, it is not the only option. NASA's planetary missions have long hosted Heliophysics-funded payloads that leverage the cruise phase or orbit to make measurements with Heliophysics science priorities.

This inter-divisional collaboration has been successful in sampling heliospheric regions that would be programmatically difficult to reach otherwise. An example is the Interstellar Mapping and Acceleration Probe (IMAP) mission, which leverages collaboration among divisions to resolve scientific questions about the heliosphere. So, why not replicate this strategy across more government agencies including the Department of Defense (DoD) as well as the commercial sector? Dual-use instrumentation is already considered for GEO and LEO DoD missions. Can we expand this to Deep Space where our observational coverage needs are severe? Those programs, however, are expensive and thus have long development schedules preceded by cumbersome and complex acquisition procedures. Could we come up with a more agile response that lowers costs and, importantly, speeds up the observational coverage in Deep Space?

In the following, we describe a potential pathfinder opportunity to answer these questions in the affirmative. The Spitzer-Resurrector Mission (SRM) concept leverages an unusual funding opportunity, an interdisciplinary team from academia and industry, a novel mission design, and recent research and technical advances in Heliophysics to reach a cost-effective, yet innovative, solution to both research and operations gaps in SWx.

2. Mission Objectives

The Spitzer-Resurrector Mission (SRM, hereafter) is a space mission concept which recently underwent a Phase I study funded by the U.S. Space Force (USSF) under the Orbital Prime [4] call. SRM is 'two missions in one', with goals of demonstrating spacecraft recovery and control in deep space, as well as demonstrating novel SWx research and operations capabilities. The spacecraft recovery is accomplished by launching a small satellite to rendezvous with the Spitzer Space Telescope [5] (currently in hibernation) in 2030, positioning itself around it, and serving as a relay for recommissioning and science operations is discussed in Section 3.

The on-orbit servicing objectives of SRM have long-term benefits (lower risk, higher science return) for space missions in the future. A sample of science goals for Spitzer is described in Section 2.1. However, the focus of this paper is on the unique opportunity offered by SRM to demonstrate novel SWx research and operations capabilities, highlighted in Section 3.2.2.

2.1. SRM-Science Objectives

During the Phase I study we are considering several payload options (Section 3.2) to demonstrate the versatility of the mission and spacecraft design to SWx objectives. Each instrument brings a unique measurement capability that can fill important gaps in our scientific knowledge while offering novel Research-to-Operations pathways with benefits to USSF, NOAA, NASA, and various other U.S. federal agencies. Figure 1 depicts the expected imaging capabilities of SRM for tracking CMEs. The final payload mix will depend on sponsor and schedule priorities. Our list of technical and research objectives, along with the measurement approach for each, is as follows:

- Demonstrate remote sensing and reconstruction of CME magnetic fields (Faraday rotation);
- Understand the evolution of CME plasma and magnetic field in the inner heliosphere (Imager, Faraday rotation);
- Improve solar energetic particles (SEP) time-of-arrival, peak, and fluence predictions for cislunar space (SEP suite);
- Improve CME time-of-arrival predictions (Imager, Faraday rotation);
- Test real-time monitoring of solar wind impacting Earth (Imager, Faraday rotation);
- Improve forecasting of CME and SEP events (Imager, SEP suite).

The main goal of SRM is to demonstrate satellite servicing by establishing radio contact with Spitzer, verifying its health, relaying command sequences to restart its operations, and then remaining nearby to act as a communications relay so that Spitzer can resume operations. With SRM, Spitzer will be much more efficient than it was at the end of its mission because it will no longer have to pause observations for hours-long downlinks to transfer data. The potential Spitzer/Infrared Array Camera (IRAC) observations are complementary to JWST's capabilities because IRAC's wide field of view and Spitzer's rapid pointing enables it to map large areas of the sky much more quickly than JWST. In addition, it can monitor many individual objects, such as active galactic nuclei or young stellar objects, for variability. Long continuous observations are also possible, such as the continuous 21-day Spitzer observations of the Trappist-1 exoplanet system. Finally, Spitzer's unique location in the solar system, in conjunction with earth-based telescopes or JWST, EUCLID, or Roman Space Telescope observations, enables microlensing parallax observations that can determine the mass and distance of free-floating planets and exo-moons.

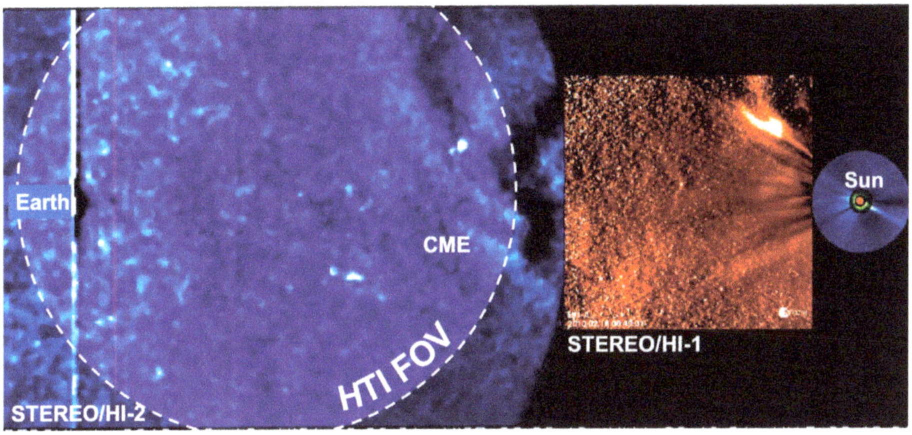

Figure 1. This snapshot of an Earth-directed CME on 16 February 2010 was captured by the imagers on STEREO-A from the Lagrangian L4 region. The SRM Imager (FoV is marked) can acquire images ~10× higher contrast and 4× higher spatial resolution than HI-2 thanks to its vantage point and larger aperture while leveraging the designs matured in Parker Solar Probe and Solar Orbiter missions.

Spitzer's location also makes possible observations of potentially hazardous asteroids that are currently not observable from near-Earth instruments. The science program proposed for the resurrected Spitzer will be described more fully in a future paper [6], in preparation.

3. Mission Design

3.1. Orbit, Trajectory Past Earth-Sun L4, and Station-Keeping near Earth-Sun L3

After a launch as a ride-along on an Evolved Expendable Launch Vehicle (EELV) Secondary Payload Adapter (ESPA) ring to lunar transfer orbit, the spacecraft uses electric propulsion to achieve the necessary Δv. The SRM trajectory, shown in Figure 2, comprises one year of thrusting to descend to 0.875 AU, coasting for roughly a year and a half, and then another year of thrusting to raise its orbit to 1.023 AU × 1.003 AU. The total transit time from the Lunar Transfer Orbit (LTO) to the final destination is approximately three and a half years.

This course takes Resurrector in front of the west side of the Sun, which is relevant for space exploration due to the magnetic connection of eruptions on this hemisphere to the Earth-Moon system. Station keeping near the Earth-Sun L3 point allows continuous beaconing to characterize solar flow and perform real-time CME detection using an Earth-based ground aperture. In addition, simultaneous helioimaging is performed to augment the radio beacon data.

The far-side location near L3 also allows Resurrector to rendezvous with and restore operations to the Spitzer Space Telescope, placed in safe mode in 2020. The spacecraft allows a two-for-one mission in space weather and spacecraft servicing/recovery.

The trajectory necessitates long-duration continuous burns of one or more ion thrusters. A multi-thruster version would allow differential thrusting to simplify the propulsion system by eliminating the need for a gimbal and tankage while simultaneously using a compact, scalable thrust array. Substantial Δv is possible using such an array, up to 6 km/s depending on the overall mass of the spacecraft.

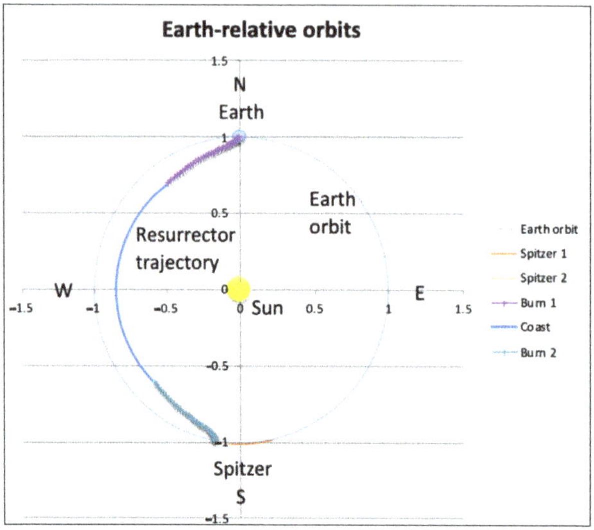

Figure 2. The trajectory of the Resurrector past the western hemisphere of the Sun before arrival near Earth-Sun L3.

3.2. *Strawman Payloads*

3.2.1. Heliospheric Transient Imager (HTI) Image

HTI is a modified version of the WISPR instrument [7] aboard the Parker Solar Probe mission, currently in operation. WISPR looks at the large-scale structure of the corona and solar wind, like CMEs. HTI retains the WISPR camera electronics box, baffle system, and Inner Telescope, repackaged into a smaller volume for imaging within 5° from the Sun (Figure 3) The current volume of 25 cm × 40 cm × 15 cm (H × L × D) will be reduced through further optimization. Based on the current on-orbit performance of WISPR, we conservatively estimate that HTI will be ~2× more sensitive than STEREO/HI-1 and >10× more sensitive than STEREO/HI-2.

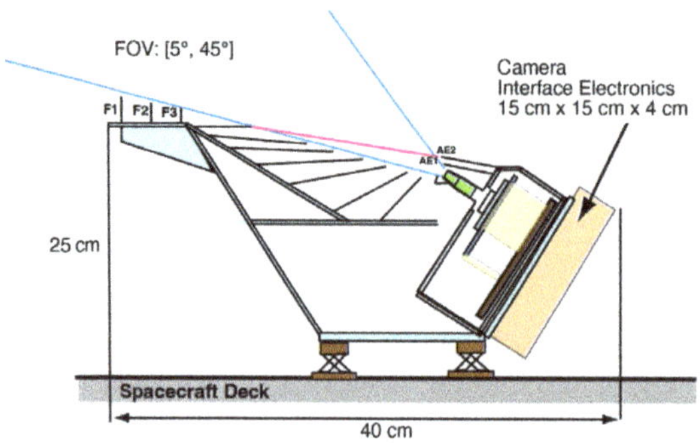

Figure 3. Side view of the WISPR instrument showing the exterior (F1–F3) and interior baffles (AE1, AE2). The HTI is a modified version of the WISPR instrument on PSP capable of observing within 5° from the Sun's center. HTI will perform >10× better than STEREO/HI-2 for the same solar elongations.

3.2.2. Faraday Rotation Experiment (FRE)

When linearly polarized radiation transverses magnetized plasmas, it undergoes rotation in its polarization plane. This effect, dubbed Faraday Rotation (FR) has been used for decades to remotely probe coronal and heliospheric structures (e.g., Levy et al. 1969 [8]; Kooi et al. 2017 [9]; and Figures 4 and 5). FR is a particularly promising remote sensing technique to derive not only the strength but also the 3D configuration of the magnetic field in CMEs while they are still in the corona. This is a valuable capability for SWx operations, as the CME magnetic field is the dominant quantity in assessing a CME's geo-effective potential. The rising interest in SWx impacts across the U.S. civilian and military sectors has renewed interest in FR studies and space applications [10,11].

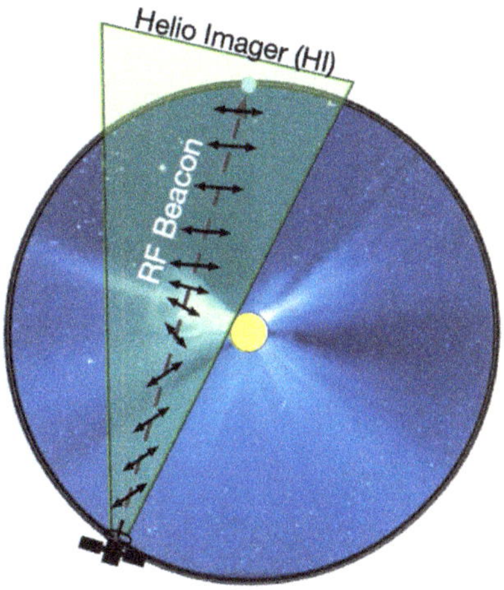

Figure 4. The Faraday Rotation Experiment (FRE) will probe the internal magnetic structure of Earth-directed transients while the Helio Imager provides the large-scale context and density information. FRE transmits an S-band linearly polarized signal that can be received by modestly sized (~15 m) ground stations around the world and used for SWx research and operations.

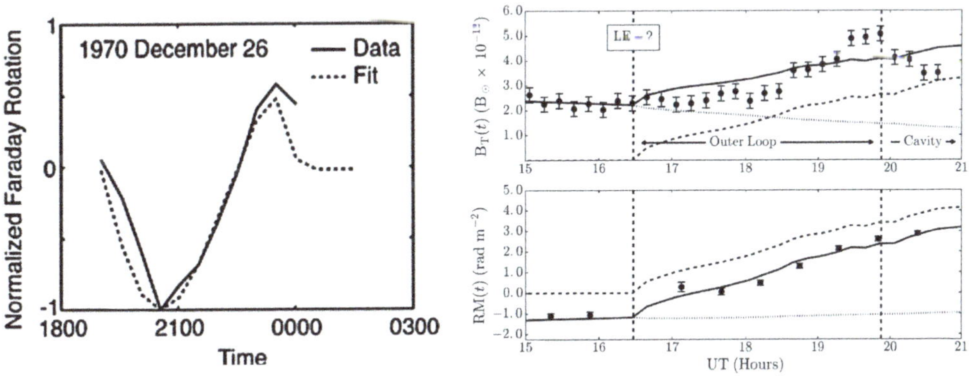

Figure 5. The FRE technique has been proven with both spacecraft transmission (**left**: Pioneer 4, from Jensen & Russell 2008 [12] ©AGU) and astronomical sources (**right**: 0842, from Kooi et al. 2017 [9] © AAS).

SRM offers an excellent opportunity to demonstrate the SWx value of the FR technique by deploying a radio experiment specifically designed for FR studies. We baseline a linearly polarized beacon at the X-Band (8.1 GHz) that transmits a prescribed signal pattern (i.e., square wave) through the 75 cm high-gain dual-feed antenna (HGA). The dual-feed system can accommodate both the FRE and downlink from the rest of the payload. The signal pattern facilitates calibration, while the relatively large HGA lowers the power requirement for the transmitter. The system requires a straightforward linear polarimeter at the ground station and may not require the use of the DSN. In any case, Ohlson et al. [13] have already demonstrated space-to-ground FR measurements across 1.8 AU with Pioneer 4 using 1970's technology and thus the technical challenges are minimal. During the 3.5-year cruise phase, the HGA is mostly pointed to Earth, except for burns and occasional interruptions for other tests. The FRE beams continuously during this time and allows us to test various techniques for deriving the heliospheric density and magnetic field with the assistance of the HTI observations that provide both context and a line-of-sight density estimate. We will also test and develop methods for ingesting the FRE information into research and operational models of the heliosphere and for developing forecasting workflows with this novel information.

3.2.3. Solar Energetic Particles Suite (SEPS)

The SRM trajectory is ideal for measuring SEPs directed to cislunar space. The data are critical for improving SEP forecasting schemes [14]. The required coverage is given in the NASA Gap Analysis Report (0.1–1 MeV electrons, 0.02–700 MeV protons, and heavy ion composition). Such an SEP package can be formed by spares of recently built instruments on Solar Orbiter and Parker Solar Probe but can also use adaptations of instruments from other missions. An example package that meets these requirements is shown in Figure 6. See [15] for details.

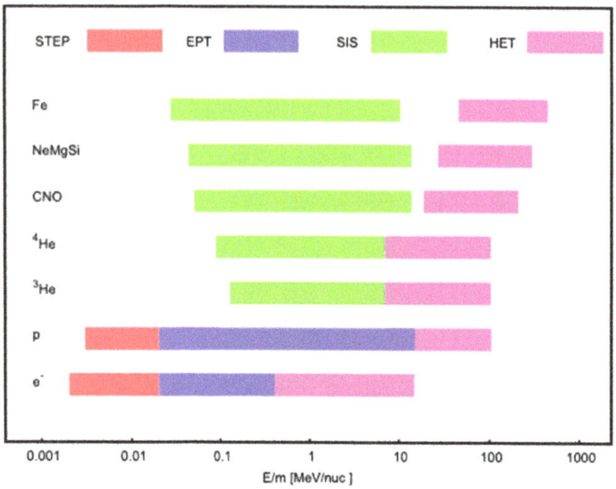

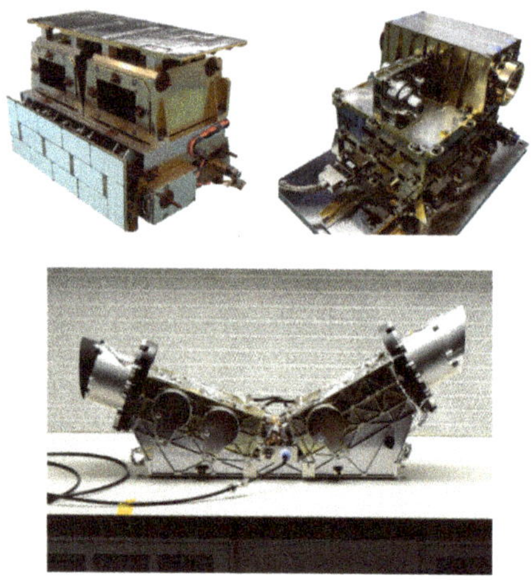

Figure 6. Candidate instrumentation for the SEP suite on SRM. **Left**: energy and ion species coverage of the EPD sensors. **Right**: Pictures of the flight units of STEP (upper left ©A&A), EPT-HET (upper right ©A&A), and SIS (bottom ©A&A) [15].

3.3. Concept of Operations

Resurrector operations are highly autonomous, utilizing onboard optical navigation to fully eliminate radiometric interference and allow the use of non-DSN ground stations like APL's Satellite Communications Facility for data downlink and FRE beacon reception. The spacecraft determines its orbit and trajectory from images taken by its navigation camera of Earth, the Moon, and a variety of asteroids while making its way to the vicinity of L3. This data is fed through a Kalman filter inside the commercialized AutoNav Mark 4 software onboard, solving for the orbit and maneuver modifications to keep its burn on track with its desired trajectory. Uplink is needed only occasionally, on the order of once per month, to load new background sequences for science gathering and housekeeping.

Upon reaching the vicinity of L3, the spacecraft autonomously completes a pop-up maneuver to place it in the desired location to use its helioimager at a nominal cadence of 30 min and continue transmitting the RF beacon to Earth. The autonomous sequences run the instruments according to a standard plan with observation modes and transmit data to Earth, eliminating the need for most uplink activities. Spacecraft station-keeping is performed using AutoNav Mark 4 software, which also performs onboard navigation relative to the nearby Spitzer Space Telescope.

The Resurrector's second mission of reactivating Spitzer by providing a crosslink and an Earth relay as the telescope comes out of solar conjunction occurs simultaneously with the space weather science operations.

3.4. Technology Development

The Resurrector requires no major technology development for its space weather mission. HTI is a repackaging of the WISPR instrument. The RFE is modeled after the Pioneer 4 beacon but updated with current space-qualified RF components. The SEPS uses TRL-9 design and spares from recently flown instruments. The spacecraft itself uses all TRL-9 off-the-shelf components, from its avionics and power subsystems to its attitude control and propulsion subsystems. Two key advanced software technologies and a key propulsion technology are described below.

3.4.1. Virtual Machine Language Sequencing 3.0 [TRL-9] and AutoNav Mark 4 [TRL-6]

VML 3 [2,16–19] is a commercialized version of the advanced spacecraft sequencing software with heritage back to fifteen missions, including the Spitzer Space Telescope, Mars Reconnaissance Orbiter, Dawn, Mars Phoenix, and OSIRIS-REx. The latest version enables highly advanced autonomous expert systems to operate the spacecraft, including navigation and maneuvering activities produced by AutoNav Mark 4. AN4 is a commercialized version (NASA SBIR 80NSSC18C0043) of the autonomous navigation software AutoNav [20,21] with heritage back to Deep Space 1, Stardust, and Deep Impact.

3.4.2. High Delta-V Maneuvering

Resurrector features a baselined Δv capability of approximately 6 km/s with a 30% margin on board. This is achieved using eight Enpulsion Micro R3 indium-ion propulsion modules, arranged to allow for differential thrust vectoring. Extra Δv allows mission trajectory options and backup for failed thruster modules. Xenon-ion propulsion has also been analyzed as a possible baseline using commercially available thrusters.

3.5. Programmatics

3.5.1. Operational Readiness of Spitzer

At the end of the mission (EOM), Spitzer was fully capable of continuing the Warm Mission [5], with substantial power and propellant margins and a healthy IRAC instrument. The spacecraft is anticipated to be fully operational upon the arrival of the Resurrector in health, power margin, available propellant, thermal stability, and IRAC functionality. Of the two thruster strings, String 2 was nominal. Inertial Reference Unit 1 was slightly degraded but functional, Unit 2 was nominal. The 3.6 μm and 4.5 μm channels of the IRAC instrument used during the Warm Mission were nominal. The transponder was set to carrier-only mode. The spacecraft was left on Side B of the avionics after a solar event, with Side A in reserve. All four Reaction Wheel Assemblies (RWAs) were fully operational.

At EOM, Spitzer was placed into indefinite safe mode in a two-hour rotation about the sunline, pointing its sunshield and solar cells directly normal to the Sun. In this configuration, the only nominal influences on the spacecraft that would affect the need to desaturate the RWAs are solar radiation pressure and precession of the sunline during an orbit, both minor in magnitude. At EOM, nitrogen propellant reserves were 6.674 kg remaining of an original 15.59 kg load (42%). Power generation at EOM was 12.2 A at 26.3 V for 345 W, with a total of 2015 battery cycles out of a minimum 25,000-lifetime capacity (90% remaining). Expected degradation in power generation by the time the Resurrector arrives places the generated current at 11.9 A (336 W). The Warm Mission is estimated to need 11 A (311 W) for nominal thermal zone profiles, yielding a margin of approximately 8% before taking any power-saving steps for an additional 10% margin.

Power generation is one limiting factor for the lifespan of the resurrected Spitzer. Current models show an expected reduction of 0.0377 A/year, giving Spitzer an additional 24 years of operations before power-saving measures such as turning off heaters in the spacecraft bus would be required. Unexpected incidents like CME damage and micrometeoroid impacts would decrease this timespan.

Another limiting factor is the propellant necessary to desaturate the PCS during nominal operations. In later years, sophisticated planning for pointing the spacecraft in ways that remove momentum inputs from prior turns was adopted to considerably reduce the rate of propellant usage. Assuming a similar conservative approach, the telescope can be expected to operate for an additional ten years or more.

3.5.2. Schedule

SRM is timed for a 2027 launch to rendezvous with the Spitzer Space Telescope in 2030 as the telescope comes out of solar conjunction. The extensive use of off-the-shelf TRL-9 technology and the use of instruments with few changes from previously flown versions help support this schedule (Figure 7). The orbital geometry and nature of the trajectory also easily support later launch dates and arrival times, reducing mission risk due to potential build and launch delays.

Q2 2026	Start of spacecraft integration
Q1 2028	Launch on rideshare to lunar transfer orbit, spacecraft commissioning, EP-1 burn start
L+20 days	Start of space weather mission
L+313 days	Descent to 0.875 AU complete, EP-1 end, continuing space weather observations
L+762 days	Start EP-2 burn
Q4 2030	Arrive at orbit 1.023 × 1.003 AU, end EP-2, resurrect Spitzer, continue space weather observations
Q4 2032	Complete two years of space weather mission, Spitzer characterization of asteroids, other deep space observations
Q4 2034	Complete additional two years of space weather/asteroid characterization/observations

Figure 7. Spitzer-Resurrector mission timeline.

3.5.3. Rideshare

SRM's inherent mission timing flexibility, the ESPA-sized spacecraft, and the compatibility of the trajectory with an LTO throw all combine to make SRM an ideal candidate for rideshare opportunities. Rideshares available on NASA, DoD, and commercial launches would dramatically expand opportunities for launch. The potential of in-kind support from interested agencies using this launch mechanism should also be explored.

3.5.4. Cost

Resurrector is designed to be a compact, low-cost, high-capability spacecraft. The entire mission, including launch costs, is targeted at USD 275 M–USD 375 M with 30% reserves. A rideshare arrangement covered by outside sources would reduce this amount to roughly USD 130 M–USD 180 M.

3.5.5. Public/Private Partnership

The currently funded Phase-I of the Spitzer-Resurrector concept has enabled rapid collaboration between space industry stakeholders. The mission as designed achieves advances in space weather experimentation and warning, on-orbit servicing in deep space, planetary defense, and astronomy. This success in leveraging the unique opportunities provided by the mission location can be attributed to mission leadership by a private company, responding to defined requirements and desires of public organizations such as the USSF, NASA, NOAA, and SAO (the Smithsonian Astrophysical Observatory, a partner in the Center for Astrophysics | Harvard and Smithsonian). Rather than any individual public organization proposing and developing a mission in isolation, the private-led effort can move quickly in coordinating stakeholders and find compromises to de-risk future standalone missions. The SRM team already has begun planning for making at least some of the data public and enabling some level of participation by outside experts. Public education and outreach and DEI are important components of the mission.

4. Conclusions

We presented the SRM concept to raise awareness in the community of a non-traditional opportunity to fill in observational gaps in Heliophysics. A tech demo opportunity from the USSF has brought together industry engineers, astrophysicists, and heliophysicists who devised an 'out-of-the-box' response to the call. The resulting concept goes far beyond the USSF's interest in exploring novel on-orbit servicing ideas. It aims for deep-space spacecraft recovery and operations, demonstrates propulsion and on-board autonomy, provides novel SWx measurements for improved forecasting around the Artemis missions, and resurrects an astrophysics mission that could provide critical synergies to JWST and perform planetary protection observations. Yet, for all these exciting aspects, there are equally challenging concerns. For example,

- This is a 'short fuse' opportunity. To provide SWx observations for Artemis III, SRM needs to launch by 2027. This timeline is too short to maneuver through the usual NASA proposal cycle. Our proposed cross-agency government/public/private collaboration has the additional advantage of facilitating rapid procurement of scientific instrumentation, as well as providing quick access to technical, scientific, and public resources;
- SRM-SWx should be of interest to NOAA or others since its operational structure naturally enables innovative participation via 'data buy,' guest observer, and/or other options.

We close with the following suggestions for consideration in the upcoming Decadal Survey discussions, based on our experience with this project:

1. Facilitate information exchange among the various Program Officers of government organizations involved in space projects. For example, being aware of potentially useful calls (or selections) to Heliophysics in other agencies should help with planning across the Division;
2. Consider a 'rapid response' space hardware development program. It could be an agile version of the Mission-of-Opportunity program able to respond to launch/rideshare/host opportunities from other agencies with a short proposal/evaluation cycle (1–2 months, instead of years). DoD experience may be particularly useful for this type of program;
3. Consider co-funding missions/hardware calls with DoD and NOAA so that successful Heliophysics concepts can be developed more efficiently and rapidly;
4. Finalize a 'data-buy' policy and implementation document to engage the space industry in closing the SWx infrastructure gaps in the next decade and beyond.

Author Contributions: Conceptualization, S.M.U., G.G.F., C.A.G., C.L. and W.B.R.; methodology, G.G.F., C.A.G., W.B.R., D.M.S., C.L., H.A.S., J.L.H., G.J.M., M.A., V.T. and S.W.; software, C.A.G.; validation, W.B.R., D.M.S. and R.C.H.; formal analysis, G.G.F., C.A.G., H.A.S., J.L.H., G.J.M., M.A., V.T., S.W. and A.V.; writing—original draft preparation, S.M.U., G.G.F., C.A.G., R.C.H., C.L., W.B.R., D.M.S., H.A.S., J.L.H., G.J.M., M.A., V.T., S.W. and A.V.; writing—review and editing, S.M.U., C.A.G., S.W. and S.B.; visualization, C.A.G., W.B.R., D.M.S. and A.V.; supervision, S.M.U., C.A.G. and C.L.; project administration, S.M.U., C.L. and S.B.; funding acquisition, S.M.U., C.A.G., C.L. and W.B.R. All authors have read and agreed to the published version of the manuscript.

Funding: This material is based upon work supported by the AFRL/RVK under Contract No. FA945322CA034. Any opinions, findings, conclusions, or recommendations expressed in this material are those of the author(s) and do not necessarily reflect the views of the United States Air Force Research Laboratory. A.V. was supported by APL B&P funding.

Data Availability Statement: The original contributions presented in the study are included in the article, further inquiries can be directed to the corresponding author.

Acknowledgments: G.F., H.A.S. acknowledge support from the Smithsonian Astrophysical Observatory. Some of the work described in this paper was carried out by Blue Sun Enterprises, Inc., under an agreement with the National Aeronautics and Space Administration and administered by the Office of Chief Technologist as two Small Business Innovation Research grants under contracts NNX17CP77P and 80NSSC18C0043. Part of this work was carried out at the Jet Propulsion Laboratory, California Institute of Technology, under a contract with the National Aeronautics and Space Administration.

Conflicts of Interest: Authors Shawn M. Usman, Cameo Lance, and Salma Benitez are employed by the company Rhea Space Activity, Inc. Author Christopher A. Grasso is employed by the company Blue Sun Enterprises, Inc. Authors Daveanand M. Singh and William B. Rideout have previously performed work for Rhea Space Activity, Inc. The remaining authors declare that the research was conducted in the absence of any commercial or financial relationships that could be construed as a potential conflict of interest.

References

1. Grasso, C.A. *Virtual Machine Language (VML) NASA Board Award*; NASA Technical Report 40365; NASA Inventions and Contributions Board: Pasadena, CA, USA, 7 September 2006.
2. United States Government. Space Weather Research-To-Operations and Operations-to-Research Framework. March 2022. Available online: https://testbed.swpc.noaa.gov/r2o2r/space-weather-r2o2r-framework (accessed on 31 May 2024).
3. National Academies of Sciences, Engineering, and Medicine. *Planning the Future Space Weather Operations and Research Infrastructure: Proceedings of a Workshop*; The National Academies Press: Washington, DC, USA, 2021.
4. Lock, P.D.; Grasso, C.A. Flight-ground integration: The future of operability. In Proceedings of the SpaceOps 2014 Conference, Pasadena, CA, USA, 5–9 May 2014.
5. Werner, M.W.; Roellig, T.L.; Low, F.J.; Rieke, G.H.; Rieke, M.; Hoffmann, W.F.; Young, E.; Houck, J.R.; Brandl, B.; Fazio, G.G.; et al. The Spitzer Space Telescope Mission. *Astrophys. J. Suppl. Ser.* **2004**, *154*, 1. [CrossRef]
6. Grasso, C.A. Spitzer-Resurrector: Reviving the Last Great Observatory. 2024; *Manuscript in preparation.*
7. Vourlidas, A.; Howard, R.; Plunkett, S.; Korendyke, C.; Thernisien, A.; Wang, D.; Rich, N.; Carter, M.; Chua, D.; Socker, D.; et al. The Wide-Field Imager for Solar Probe Plus (WISPR). *Space Sci. Rev.* **2015**, *204*, 83–130. [CrossRef]
8. Levy, G.S.; Sato, T.; Seidel, B.L.; Stelzried, C.T.; Ohlson, J.E.; Rusch, W.V. Pioneer 6: Measurement of Transient Faraday Rotation Phenomena Observed during Solar Occultation. *Science* **1969**, *166*, 596–598. [CrossRef] [PubMed]
9. Kooi, J.E.; Fischer, P.D.; Buffo, J.J.; Spangler, S.R. VLA Measurements of Faraday Rotation through Coronal Mass Ejections. *Sol. Phys.* **2017**, *292*, 56. [CrossRef]
10. Kooi, J.E.; Wexler, D.B.; Jensen, E.A.; Kenny, M.N.; Nieves-Chinchilla, T.; Wilson, L.B.; Wood, B.E.; Jian, L.K.; Fung, S.F.; Pevtsov, A.; et al. Modern Faraday Rotation Studies to Probe the Solar Wind. *Front. Astron. Space Sci.* **2022**, *9*, 841866. [CrossRef]
11. Kooi, J.E.; Ascione, M.L.; Reyes-Rosa, L.V.; Rier, S.K.; Ashas, M. VLA Measurements of Faraday Rotation Through a Coronal Mass Ejection Using Multiple Lines of Sight. *Sol. Phys.* **2021**, *296*, 11. [CrossRef]
12. Jensen, E.A.; Russell, C.T. Faraday rotation observations of CMES. *Geophys. Res. Lett.* **2008**, *35*, L02103. [CrossRef]
13. Ohlson, J.E.; Levy, G.S.; Stelzried, C.T. A Tracking Polarimeter for Measuring Solar and Ionospheric Faraday Rotation of Signals from Deep Space Probes. *IEEE Trans. Instrum. Meas.* **1974**, *23*, 167–177. [CrossRef]
14. Posner, A.; Arge, C.N.; Staub, J.; StCyr, O.C.; Folta, D.; Solanki, S.K.; Strauss, R.D.T.; Effenberger, F.; Gandorfer, A.; Heber, B.; et al. A multi-Purpose Heliophysics L4 Mission. *Space Weather* **2021**, *19*, e2021SW002777. [CrossRef]
15. Rodríguez-Pacheco, J.; Wimmer-Schweingruber, R.F.; Mason, G.M.; Ho, G.C.; Sánchez-Prieto, S.; Prieto, M.; Martín, C.; Seifert, H.; Andrews, G.B.; Kulkarni, S.R.; et al. The Energetic Particle Detector: Energetic particle instrument suite for the Solar Orbiter mission. *Astronomy Astrophys.* **2020**, *642*, A7. [CrossRef]

16. Grasso, C.A. Techniques for simplifying operations using vml (virtual machine language) sequencing on mars odyssey and SIRTF. In Proceedings of the 2003 IEEE Aerospace Conference Proceedings (Cat. No.03TH8652), Big Sky, MT, USA, 8–15 March 2003; pp. 8_3603–8_3613. [CrossRef]
17. Grasso, C.A.; Riedel, J.E.; Vaughn, A.T. Reactive Sequencing for Autonomous Navigation Evolving from Phoenix Entry, Descent, and Landing. In Proceedings of the AIAA Space Operations Conference Proceedings, Huntsville, AL, USA, 25–30 April 2010.
18. Grasso, C.A. VML 3.0 reactive rendezvous and Docking Sequencer for Mars Sample Return. In Proceedings of the SpaceOps 2014 Conference, Pasadena, CA, USA, 5–9 May 2014. [CrossRef]
19. Peer, S.; Grasso, C.A. Spitzer Space Telescope Use of Virtual Machine Language. In Proceedings of the 2005 IEEE Aerospace Conference, Big Sky, MT, USA, 5–12 March 2005.
20. Riedel, J.; Eldred, D.; Kennedy, B.; Kubitscheck, D.; Vaughan, A.; Werner, R.; Bhaskaran, S.; Synnott, S. AutoNav Mark 3: Engineering the Next Generation of Autonomous Onboard Navigation and Guidance. In Proceedings of the AIAA Guidance, Navigation, and Control Conference, Keystone, CO, USA, 21–24 August 2006.
21. Riedel, J.; Werner, R.; Vaughan, A.; Mastrodemos, N.; Huntington, G.; Grasso, C.; Wang, T.-C.; Myers, D.; Gaskell, R.; Bayard, D. Configuring the Deep Impact AutoNav System for Lunar, Comet and Mars Landing. In Proceedings of the AIAA Astrodynamics Specialist Conference, Honolulu, HI, USA, 18–21 August 2008.

aerospace

MDPI

Article

Ground-Based Characterisation of a Compact Instrument for Gamma-ray Burst Detection on a CubeSat Platform

Rachel Dunwoody [1,2,*], David Murphy [1,2], Alexey Uliyanov [1,2], Joseph Mangan [1,2], Maeve Doyle [1,2], Joseph Thompson [2,3], Cuan de Barra [1,2], Lorraine Hanlon [1,2], David McKeown [2,3], Brian Shortt [4] and Sheila McBreen [1,2]

1 School of Physics, University College Dublin, D04 C1P1 Dublin, Ireland; david.murphy@ucd.ie (D.M.); alexey.uliyanov@ucd.ie (A.U.); josephmangan42@gmail.com (J.M.); maeveedoyle@gmail.com (M.D.); cuan.debarra@ucdconnect.ie (C.d.B.); lorraine.hanlon@ucd.ie (L.H.); sheila.mcbreen@ucd.ie (S.M.)

2 Centre for Space Research, University College Dublin, D04 C1P1 Dublin, Ireland; joseph.thompson@ucd.ie (J.T.); david.mckeown@ucd.ie (D.M.)

3 School of Mechanical and Materials Engineering, University College Dublin, D04 C1P1 Dublin, Ireland

4 European Space Agency, ESTEC, 2201 Noordwijk, The Netherlands; brian.shortt@esa.int

* Correspondence: racheldunwoody2@gmail.com

Abstract: Gamma-ray bursts (GRBs) are intense and short-lived cosmic explosions. Miniaturised CubeSat-compatible instruments for the study of GRBs are being developed to help bridge the gap in large missions and assist in achieving full sky coverage. CubeSats are small, compact satellites conforming to a design standard and have transformed the space industry. They are relatively low-cost and are developed on fast timescales, which has provided unparalleled access to space. This paper focuses on GMOD, the gamma-ray module, onboard the 2U CubeSat EIRSAT-1, launched on December 1st 2023. GMOD is a scintillation-based instrument with a cerium bromide crystal coupled to an array of sixteen silicon photomultipliers, designed for the detection of GRBs. The characterisation of GMOD in the spacecraft, along with the validation of an updated spacecraft MEGAlib model is presented and this approach can be followed by other CubeSats with similar science goals. The energy resolution of the flight model is 7.07% at 662 keV and the effective area peaks in the tens to hundreds of keV, making it a suitable instrument for the detection of GRBs. An investigation into the instrument's angular response is also detailed. The results from this characterisation campaign are a benchmark for the instrument's performance pre-launch and will be used to compare with the detector's performance in orbit.

Keywords: CubeSat; gamma-ray instrumentation; gamma-ray bursts; characterisation; EIRSAT-1

1. Introduction

Gamma-ray bursts (GRBs) are violent, bright and brief explosions in the universe resulting from the death of a massive star or the collision of two compact objects, such as neutron star–neutron star or neutron star–black hole mergers [1,2]. The breakthrough coincident detection of electromagnetic signals (from radio to gamma rays) and gravitational waves in 2017 facilitated a new way to probe these complex events [3,4]. A major challenge to the future of gamma-ray astronomy and the progression of GRB science is the lack of new missions. The legacy and current gamma-ray missions have excelled in advancing the understanding of these objects but the majority are operating beyond their nominal lifetimes [5–8]. Additionally, current gamma-ray observatories cannot provide full-sky coverage in a single observation for these brief isotropic explosions, which is critical for future coincident detections. Obtaining a population of GRB/gravitational wave joint detections will be important in determining the characteristics of the different types of GRBs, confirming other theorised progenitors and the extracted properties can be used to search archival data (e.g., [9]) for similar candidates. Furthermore, increasing the number of joint detections could constrain the neutron star equation of state, advance the understanding of the origin of the heaviest elements in the universe and provide new insights into fundamental physics such as investigating the difference between the speed of gravity and the speed of light [3,10].

CubeSats are small-scale satellites, usually conforming to a design standardisation, defined by Cal Poly specifications [11]. CubeSats have grown in popularity in recent years due to their science and technology demonstration capabilities [12–15]. CubeSats were originally developed as an educational tool [16,17], resulting in the production of commercial-off-the-shelf (COTS) components. COTS components can be used for the fundamental subsystems of the satellite, e.g., power, communications and onboard computer (OBC), allowing for more resources to be applied to the science objectives of the mission [18]. CubeSats are relatively low-cost and have short launch timescales [19], making them ideal candidates to bridge the gap in gamma-ray missions during the multi-messenger era [20]. In addition to the current large-scale gamma-ray missions, a fleet of GRB-detecting CubeSats could allow for the adequate detection and localization of GRBs, and increased sky coverage. SmallSats with GRB detection and localisation capabilities are in development, e.g., BurstCube [21,22], MoonBEAM [23], HERMES [24], GRBAlpha [25] and CAMELOT [26].

The Educational Irish Research Satellite (EIRSAT-1) is a 2U CubeSat with a primary payload, the gamma-ray module, GMOD, developed for the detection of GRBs [27,28] at energies 60 keV to $\approx$2 MeV, as presented in Figure 1. GMOD is a $<$1 U scintillation-based detector and consists of two main components: the detector assembly and the GMOD Motherboard (MB) [29–31]. The detector assembly holds the $CeBr_3$ scintillation crystal (25 mm $\times$ 25 mm $\times$ 40 mm), which interacts with gamma-ray photons from GRBs to produce optical photons, which are collected by a tiled array of 16 silicon photomultipliers (SiPMs). The outputs from the SiPMs are read out and digitised by an application-specific integrated circuit (ASIC), produced by Integrated Detector Electronics AS (IDEAS), Oslo, Norway called SIPHRA (Silicon Photomultiplier Readout ASIC) [32]. The GMOD motherboard was developed in-house [33,34] and supports the electronics for power and the readout of the detector assembly. The GMOD MB converts the channel data into time-tagged events (TTEs). TTEs are generated for the 16 individual channels of SIPHRA and for the sum channel (sum of 16 channels). Both sum and 16 channel TTEs are sent in pages to OBC. GMOD is based on in-house instrument heritage and will be used to scale up to a 6U mission with detection and localisation capabilities, called Gamma-ray Investigation of the Full Transient Sky (GIFTS) [29,30,35–37].

The characterisation of an instrument in the spacecraft is necessary to investigate its performance and ensure that it is capable of achieving the mission's science goals [38]. Instrument characterisation as presented in this paper provides an in-depth understanding of the behaviour of the instrument that could not be achieved via functional testing. Performing the characterisation of the flight model of GMOD in EIRSAT-1 is essential for interpreting the downlinked science data throughout the mission and identifying intrinsic features in the data products. This characterisation campaign also provides a benchmark of the experiment's performance pre-launch. The results from this campaign can be used to compare with the detector's performance in-orbit, during commissioning and nominal operations.

Initially, the ground-based energy calibration of GMOD was performed to map the analog-to-digital (ADC) values to energy. Once the data were in energy space, the properties of the detector could be investigated, including the instrument's resolution. A calibration of the instrument will be performed post-launch using the signature of galactic positron annihilation at 511 keV to determine the shift in the energy spectrum [39]. Additionally, lines from solar activity, e.g., the 2.2 MeV line from neutron capture on hydrogen [40], can be used for in-orbit calibration. In addition to the ground characterisation, in-orbit cross-calibration will be performed when GRBs are detected in coincidence with other missions (e.g., Ref. [41]).

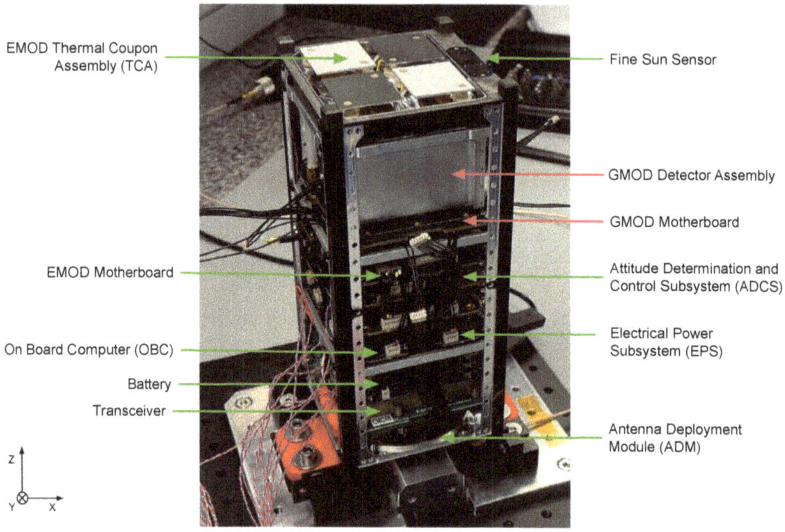

Figure 1. The flight model of EIRSAT-1 including the location of major subsystems and the GMOD detector assembly and motherboard. The coordinate system of the spacecraft is also detailed.

The aim of these tests was, first, to evaluate the effective area of the GMOD detector as a function of energy and source incident angle to correctly generate the detector response files [5]. This gave an understanding of the response of the detector and allowed an interpretation of the mission's science data when analysing potential GRB triggers from different sky positions (on- and off-axis). Some triggers were also simulated by bringing a source close to the spacecraft for a short period of time. The aim of these tests was to further validate the correct operation of the trigger algorithm and data processing and protection in the onboard software (OBSW) v18.3; see Refs. [30,42] for more details. Simulations were performed in MEGAlib, a collection of programs to simulate gamma-ray instrumentation [43], to correlate with measurements in the lab and validate a Geomega EIRSAT-1 model, with the original model detailed in Ref. [29].

Section 2 presents the experimental set-up in the cleanroom, including the angular measurements on a motorised turntable, where all the characterisation tests were performed with the EIRSAT-1 flight model (FM). The methods for producing the calibration and energy resolution are outlined in Sections 3 and 4. The effective area of GMOD as a function of energy and the angular response of GMOD are presented in Section 5. The trigger data analysis is presented in Section 6. The simulations performed in MEGALib are discussed in Section 7 and compared to the experimental data to assess the accuracy of the model.

2. Characterisation Test Set-Up

2.1. Energy Calibration Hardware Set-Up

Initially, the measurements to determine the channel–energy relation of GMOD were performed. The main goal of these measurements was to maximise the number of counts during the integration interval. The spacecraft was secured in a vertical support stand for the energy calibration source measurements. The EIRSAT-1 FM was connected to the power and data acquisition ground-support equipment (GSE) via the Umbilical Interface Unit (UIU), as labelled in Figure 2 (Panel B). The UIU interfaced with a bench power supply unit (PSU) to provide a battery voltage of 8.35 V for the spacecraft. The UIU also provided an interface between the spacecraft and the laptop used to command the satellite via the MCS v21.2 and Python v3.7 scripts and retrieve the experimental data stored on the OBC. On the laptop, a terminal was used to monitor the debug log from the spacecraft throughout the test activities.

In the first test, different radiative sources (enumerated 1–6, see Table 1) were placed close to GMOD. As in Figure 2, sources 1–5 were positioned on the cross-hairs, which identified the centre of GMOD's $CeBr_3$ crystal, on a cap on the +Z face of the spacecraft (Panels A and C of Figure 2). The distance from the end cap to the centre of the crystal (CoC) is 49.6 mm. The position of source 6 is centred on the X face of the crystal, 222 mm from the +Y face and 119 mm from the -Z face (see Panel D of Figure 2). It was found that, with some of the sources in this position, the OBC could not manage the rate at which GMOD was generating data for long durations. To determine the GMOD data rate that the OBC could successfully handle, the sources were placed at different distances from the CoC.

For each source position, the GMOD data production rate, in pages/min, was monitored via the MCS. The address of the GMOD sum page received by the OBC was determined at the same time as the address of the most recent page of generated summed data multiple times over a few minutes. If the difference between these was increasing as time progressed, the rate at which GMOD was producing data was greater than the rate at which the OBC was reading data. The source was moved until the difference between these two parameters remained the same over time, indicating that the OBC could manage the data rate produced by GMOD. This was found to be $\sim$250 pages/min. Each page contains 31 TTEs, so the TTE rate that the OBC can keep up with for long-duration experiments ($\sim$hours) is $\sim$130 TTEs/s or 130 counts/s. As GRBs are on much shorter timescales, this issue will not affect GMOD's ability to detect these events. Table 2 contains the exact sources used for the energy calibration measurements, their location with respect to the CoC in GMOD and the integration times.

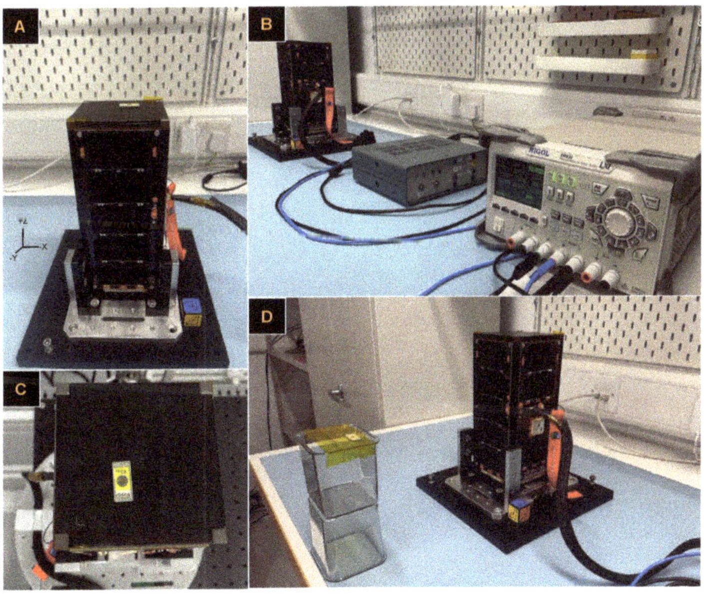

Figure 2. (**A**) EIRSAT-1, including co-ordinate system, on vertical integration stand for energy calibration measurements with a source positioned on the +Z endcap. (**B**) GSE for measurements; on the left is the FM, in the centre is the UIU and on the right is the PSU powering the spacecraft. (**C**) Source positioned on the +Z endcap on cross-hairs marking the centre of the $CeBr_3$ crystal. (**D**) Position of source 6 in Table 2, the source is centred on X of the crystal, 222 mm from the +Y face and 119 mm from the −Z face.

Table 1. Radioactive sources used for GMOD characterisation. The peak energies and transition probabilities for sources marked with an * are from [38] and those marked with † are from [44]. The source activity for sources 1, 2, 3 and 6 are in μCi and quoted as of 12:00 GMT on 1 December 1979. The activity of source 5 is quoted as of 15 May 2013.

Source No.	Source Name	Nuclide	Line Origin	Line Energy (keV)	Transition Probability	Half-Life (Years)	Activity (μCi)
1	Americium	^{241}Am	γ	59.40 *	0.359 *	432.2	11.16
2	Cobalt	^{60}Co	γ	1173.23 *	0.998 *	5.26	12.28
		^{60}Co	γ	1332.49 *	0.999 *	5.26	12.28
3	Barium	^{133}Ba	γ	80.99 †	0.901 †	10.5	11.42
		^{133}Ba	γ	276.39 †	0.075 †	10.5	11.42
		^{133}Ba	γ	302.85 †	0.191 †	10.5	11.42
		^{133}Ba	γ	356.01 †	0.636 †	10.5	11.42
		^{133}Ba	γ	383.85 †	0.091 †	10.5	11.42
4	Mixed Source	^{241}Am	γ	59.40 *	0.359 *	432.2	-
		^{137}Cs	Ba-137m	661.66 *	0.850 *	30.17	-
		^{60}Co	γ	1173.23 *	0.998 *	5.26	-
		^{60}Co	γ	1332.49 *	0.999 *	5.26	-
5	Sodium	^{22}Na	annihilation	511 *	1.798 *	2.6	10
		^{22}Na	γ	1274.54 *	0.999 *	2.6	10
6	Caesium	^{137}Cs	Ba-137m	661.66 *	0.850 *	30.17	11.16

Table 2. Energy calibration source positions and the duration of spectral integrations. Sources 1–5 were placed above the +Z endcap of the spacecraft at various distances from the centre of the GMOD crystal (CoC). Source 6 was aligned with the CoC in X, placed 222 mm from the +Y face and 119 mm from the −Z face on the table.

Source No.	Source Name	Location with Respect to Centre of EIRSAT-1 (mm)			Location Relative to CoC (mm)			Int. (Hours)
		X	Y	Z	X	Y	Z	
1	Americium	−10.57	−14.50	298.5	0.00	0.00	234.6	12
2	Cobalt	−10.57	−14.50	113.5	0.00	0.00	49.6	12
3	Barium	−10.57	−14.50	216.5	0.00	0.00	152.6	12
4	Mixed Source	−10.57	−14.50	113.5	0.00	0.00	49.6	12
5	Sodium	−10.57	−14.50	252.5	0.00	0.00	188.6	12
6	Cesium	−10.57	272.0	5.50	0.00	−286.5	−58.4	12

2.2. Angular Measurements Hardware Set-Up

For the determination of the angular response of the FM, two sources with known activities, ^{241}Am and ^{137}Cs, were used. Due to the limited time with the FM, integrations with these sources at various angles were performed at the same time, with the two sources taped together. They were selected as the gamma-ray peaks from each were at sufficiently different energies so as to not interfere with each other. Figure 3 presents the hardware set-up used, where the spacecraft was mounted on a turntable with an attached moveable appendage. The turntable had a slip ring and interface printed circuit boards (PCBs), which facilitated powering and commanding the spacecraft via the UIU. The UIU provided an interface with a bench PSU to provide power for the spacecraft. It also provided the connection to the MCS on the laptop, which facilitated setting up the experiment and downlinking the data. This hardware was designed for attitude control and mission testing and utilised for GMOD characterisation.

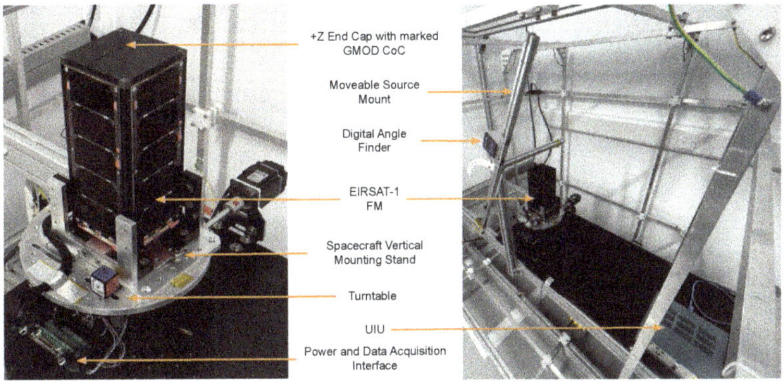

Figure 3. The test set-up for the measurements made to determine the angular response of GMOD. The FM was secured on a motorised turntable and the source was mounted on a moveable appendage, to vary the polar and azimuthal angles, respectively.

The sources were placed at a variety of different positions relative to the spacecraft, with some example orientations displayed in Figure 4. This was achieved by varying the polar and azimuthal angles. The polar angle was varied by mounting the source on a movable arm that could be rotated about an axis that went through the centre of the detector. The azimuthal angle was varied by securing the spacecraft on the controllable turntable, and moving it to the required positions. The turntable was also controlled via the Python test scripts, and a separate script was written for these measurements containing the set-up commands for GMOD and the commands for the turntable to move the spacecraft to a specified range of angles.

Table 3 shows the locations tested, in both polar and azimuthal angles with respect to the crystal centre and in Cartesian coordinates with respect to the centre of the spacecraft. These Cartesian coordinates are included, as they were required for the MEGAlib simulations discussed in Section 7. The final column of Table 3 shows the integration time (in seconds). During the test runs, all the scripts were run for 3.5 h; however, for locations where GMOD was most sensitive (e.g., +Y face, polar 90°, Az 270°) the number of TTEs downlinked from the OBC was not representative of the number of TTEs generated by GMOD. This issue was due to the OBC being unable to manage the rate at which GMOD was producing data; this was accounted for in the data analysis. The analysis and results of these off-axis measurements are presented in Section 5.

Table 3. Positions at which the source was placed for the angular response investigation for GMOD FM. These positions are provided in polar and azimuthal angles with respect to the crystal centre and in Cartesian coordinates with respect to the centre of the spacecraft. The integration time for each measurement is provided in the final column.

Polar Angle (°)	Azimuthal Angle (°)	Distance to Centre of Spacecraft (mm)			Integration (s)
		X	Y	Z	
0	0	−1.057	−1.450	28.890	3112.56
30	0	10.193	−1.450	25.876	8815.00
	90	−1.057	9.800	25.876	6226.38
	180	−12.307	−1.450	25.876	6137.81
	270	−1.057	−12.700	25.876	6201.48
60	0	18.429	−1.450	17.640	9337.74
	60	8.686	15.425	17.640	7057.31
	120	−10.799	15.425	17.640	6720.92
	180	−20.543	−1.450	17.640	6895.48
	240	−10.799	−18.325	17.640	8116.63
	300	8.686	−18.325	17.640	8021.03

Table 3. *Cont.*

Polar Angle (°)	Azimuthal Angle (°)	Distance to Centre of Spacecraft (mm)			Integration (s)
90	0	21.443	−1.450	6.390	12,597.20
	30	18.429	9.800	6.390	8079.04
	60	10.193	18.036	6.390	6063.91
	90	−1.057	21.050	6.390	5528.77
	120	−12.307	18.036	6.390	5759.64
	150	−20.543	9.800	6.390	6256.82
	180	−23.557	−1.450	6.390	6549.98
	210	−20.543	−12.700	6.390	6635.26
	240	−12.307	−20.936	6.390	5773.07
	270	−1.057	−23.950	6.390	5364.51
	300	10.193	−20.936	6.390	5981.63
	330	18.429	−12.700	6.390	7814.70
120	0	18.429	−1.450	−4.860	8490.85
	60	8.686	15.4250	−4.860	6611.60
	120	−10.799	15.425	−4.860	8862.90
	180	−20.543	−1.450	−4.860	6699.69
	240	−10.799	−18.325	−4.860	7760.87
	300	8.686	−18.325	−4.860	9409.55
150	0	10.193	−1.450	−13.096	13,941.38
	90	−1.057	9.800	−13.096	15,991.81
	180	−12.307	−1.450	−13.096	3324.10
	270	−1.057	−12.700	−13.096	9837.67
180	0	−1.0570	−1.450	−16.110	50,397.94

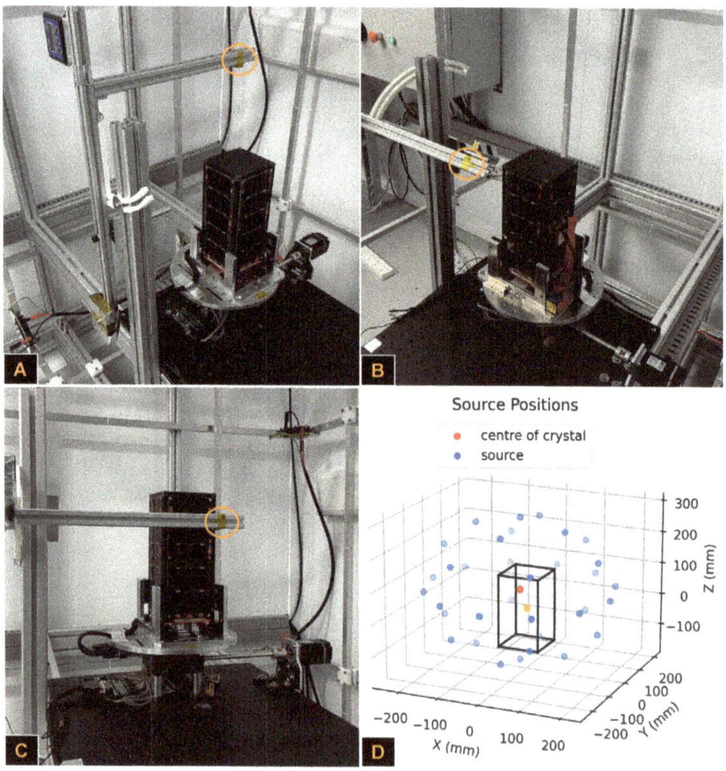

Figure 4. (**A**) EIRSAT-1 FM on vertical integration stand mounted to the turntable for polar 0° measurement, i.e., the sources, circled in orange, mounted at the crystal centre in X, Y and 22.5 cm

above the +Z face of the spacecraft. (**B**) FM on turntable with the sources, circled in orange, positioned at the polar 60°, azimuthal 270° measurements. (**C**) FM on turntable with the sources, circled in orange, positioned at the polar 90°, azimuthal 0° measurements. (**D**) Source positions (blue points) for all off-axis measurements. The red point marks the centre of the scintillator (40 mm in length along Y and 25 mm along X and Z). The orange point marks the centre of the spacecraft.

2.3. Data Acquisition and Scripting

Due to time constraints on the availability of the EIRSAT-1 FM, the ability to execute tests overnight and without an operator always present was paramount. Python scripts were composed to configure the ~40 parameters to set up GMOD and then allow for the specified amount of time for spectral integration. The scripts interfaced with the mission control software (MCS v21.2) and commanded the experiment as if it were in flight. The main interface with the spacecraft was via Bright Ascension's MCS [45]. The Python scripts performed the following tasks: turning on GMOD and experiment configuration including firmware version and operational mode. The script then synchronised the time on GMOD's microcontroller to the OBC time, and initiated GMOD generating data and streaming it to the OBC. The script recorded the storage location of each run on the OBC and, once a number of integrations had been completed, a second script could be used to downlink the data. The data were downlinked through the satellite telemetry interface using the MCS connected to the OBSW. This campaign provided the opportunity to test the robustness of the chain of data acquisition that will be followed during the actual mission.

3. Energy Calibration

3.1. Single-Source Spectra

The FM of GMOD was configured to take a 12 h measurement for each of the sources from 1 to 6 in Table 2. The summed channel TTEs stored on the OBC corresponding to each measurement were downlinked and analysed in Python. A 24 h background integration was also performed and downlinked. All of the data sets were divided by their respective integration times, and the background spectrum was subtracted from the raw source spectra. Figure 5 presents the raw spectrum (purple), background spectrum (grey) and background-subtracted spectrum (teal) for each source.

3.2. Channel–Energy Relation

All the source spectra produced with GMOD followed a primarily Gaussian shape, which varied depending on the underlying interaction with the $CeBr_3$ crystal. The photopeaks all fit with the following Gaussian function:

$$G(x) = \frac{N}{\sigma\sqrt{2\pi}} e^{\frac{-(x-\mu)^2}{2\sigma^2}}, \tag{1}$$

where N is the height of the peak, σ is the standard deviation and μ is the position of the centre of the peak [46]. In order to translate the ADC channels to energy values, a channel–energy relationship was first determined. Each photopeak from the sources listed in Table 2 was fitted with a Gaussian function, and the channel number for the centre of the photopeak, μ, was extracted. Figure 6 presents four examples of the functions fitted to energy peaks in the spectra to determine the channel number of the peak centre.

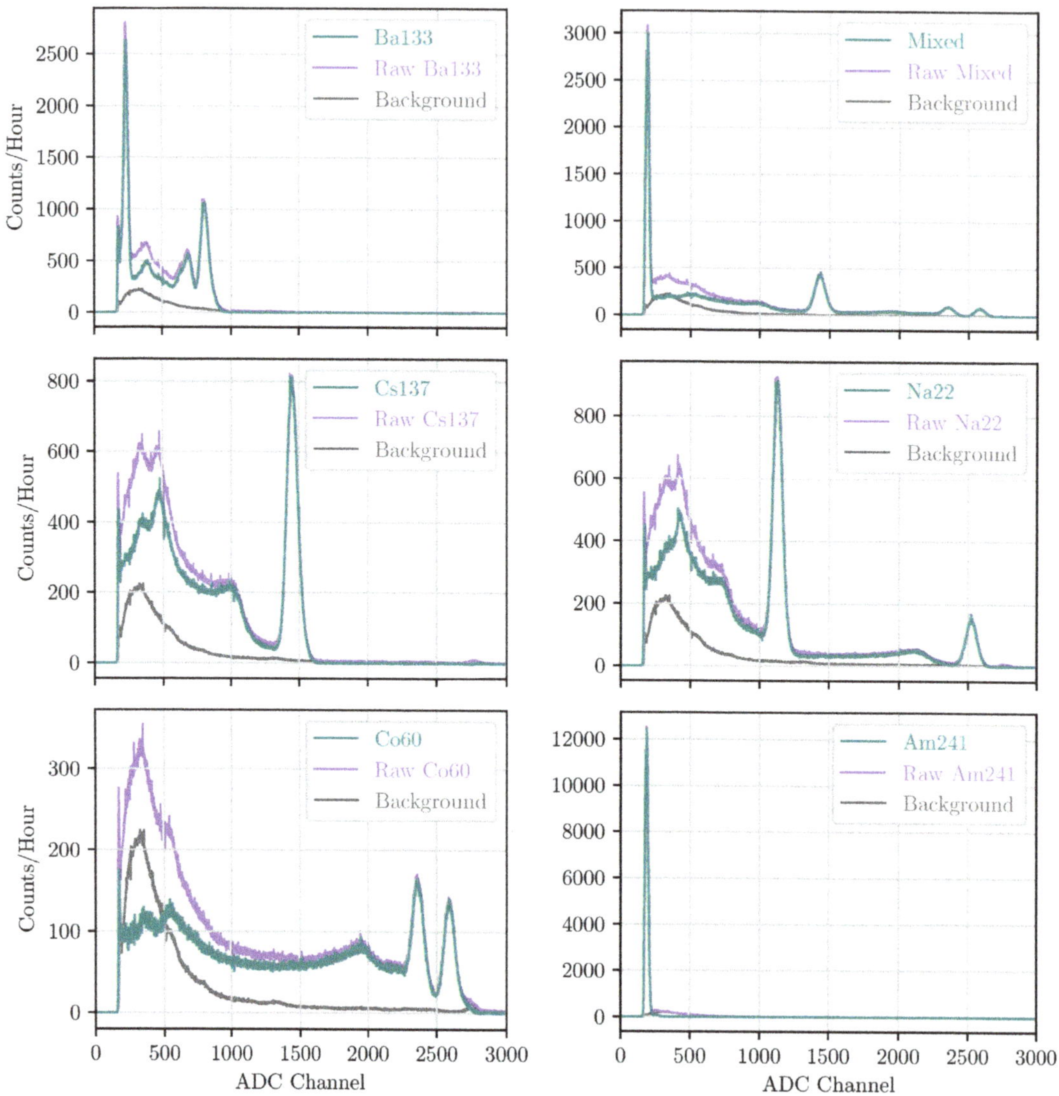

Figure 5. GMOD FM background-subtracted spectra (teal) in counts/hour for the six sources in Table 2. In each panel, the background spectrum (grey) and the raw source spectrum (purple) are presented.

In the analysis, the gamma-ray photopeaks could not be fitted with a simple Gaussian, as there are other contributions in the photopeak such as from back-scattering off of nearby structures, background contributions and Compton distributions (as seen in the spectra in Figure 5). In general, the peaks were fitted with a combination of a Gaussian function and a polynomial function to account for other contributions to the photopeak. The final fit is plotted as a dashed line and the composite functions are plotted separately as solid lines in Figure 6.

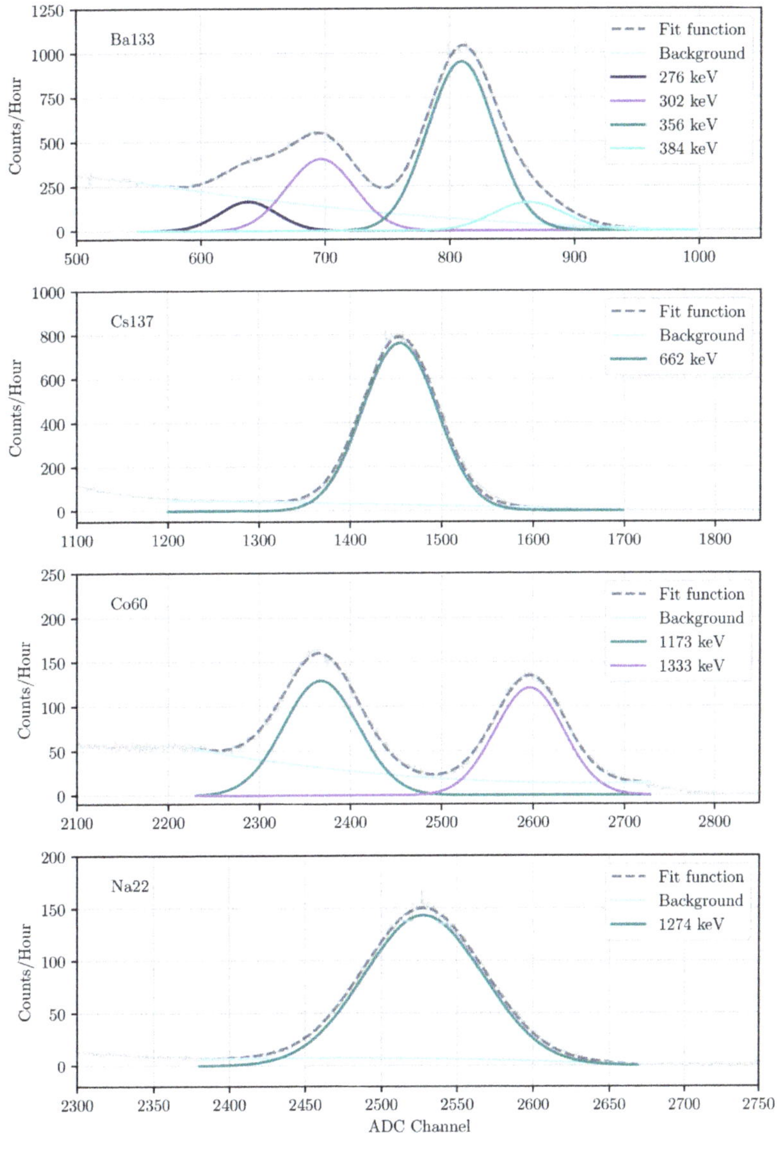

Figure 6. A sample of four fits to GMOD spectra in terms of ADC channel number. Each panel shows the fit (dashed grey line) to the background-subtracted spectrum. The composite functions of the fit are plotted as solid lines.

The ADC channel numbers associated with each peak centre, μ, were then plotted against the known energy of the photopeaks, as listed in Table 1, to map the channel numbers to energy. Figure 7 presents the channel–energy relation for GMOD FM. The relationship between the ADC channel number and energy, E, was best fitted with a quadratic of the form

$$E = a(\text{Channel Number})^2 + b(\text{Channel Number}) + c \quad (2)$$

where a, b and c are the fit coefficients. The coefficients of the fit to the GMOD channel energy relation are $a = 3.56 \times 10^{-5}$, $b = 0.42$ and $c = -17.30$.

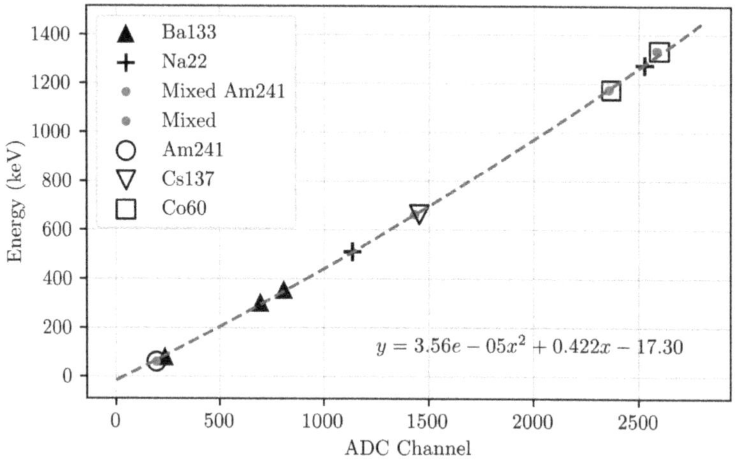

Figure 7. Energy–channel calibration relation for GMOD FM. The relationship was fit with a quadratic (dashed line) with the coefficients $a = 3.56 \times 10^{-5}$, $b = 0.42$ and $c = -17.30$. The teal points representing the mixed source were not used in the final fit due to the lab temperature being different for these measurements. Measurement errors for gamma-ray line centres are on the order of 0.3 ADC channel.

4. Energy Resolution

The energy resolution represents the broadening of a gamma-ray line due to photoelectron statistics and detector effects, it quantifies how well the detector can define a gamma-ray line for a specific energy.

The energy resolution, R, of a detector is defined as

$$R(\%) \equiv \frac{\text{FWHM}}{\mu} \times 100\,, \tag{3}$$

where the full-width at half-maximum (FWHM) of the energy pulse is divided by the location of the peak's centre, μ [38]. The FWHM is calculated using the standard deviation, σ, of the Gaussian fit function, as follows:

$$FWHM = 2\sqrt{2ln2}\sigma\,. \tag{4}$$

There are a number of contributing factors resulting in the increase of a detector's energy resolution such as the aforementioned photoelectron statistics, changes to the configuration of GMOD and non-uniformity of the scintillator's response or the SiPMs' performance. The sources used in this testing were only a sample of the full range of energies that could be emitted by astrophysical sources. The resolution of the detector for this larger energy range can be extrapolated by fitting the following function to the physical measurements [38]:

$$\sigma = \sqrt{a^2 + b^2E + c^2E^2}\,. \tag{5}$$

The constant, a, is negligible because SiPMs are high-gain devices and the noise is not an issue for the measurement of gamma rays above 50 keV. The $\sqrt{b^2E}$ term accounts for the fluctuations in the number of scintillated photons and photoelectrons produced by the SiPMs. The final term, $\sqrt{c^2E^2}$, is attributed to the non-ideal transfer of the scintillated photon energy from their creation locations to the SiPMs.

The energy calibration was first applied to the spectra produced by GMOD for the sources listed in Table 2, such that the spectral counts/keV/hour were plotted against energy in keV, rather than channel number, as in Figure 8.

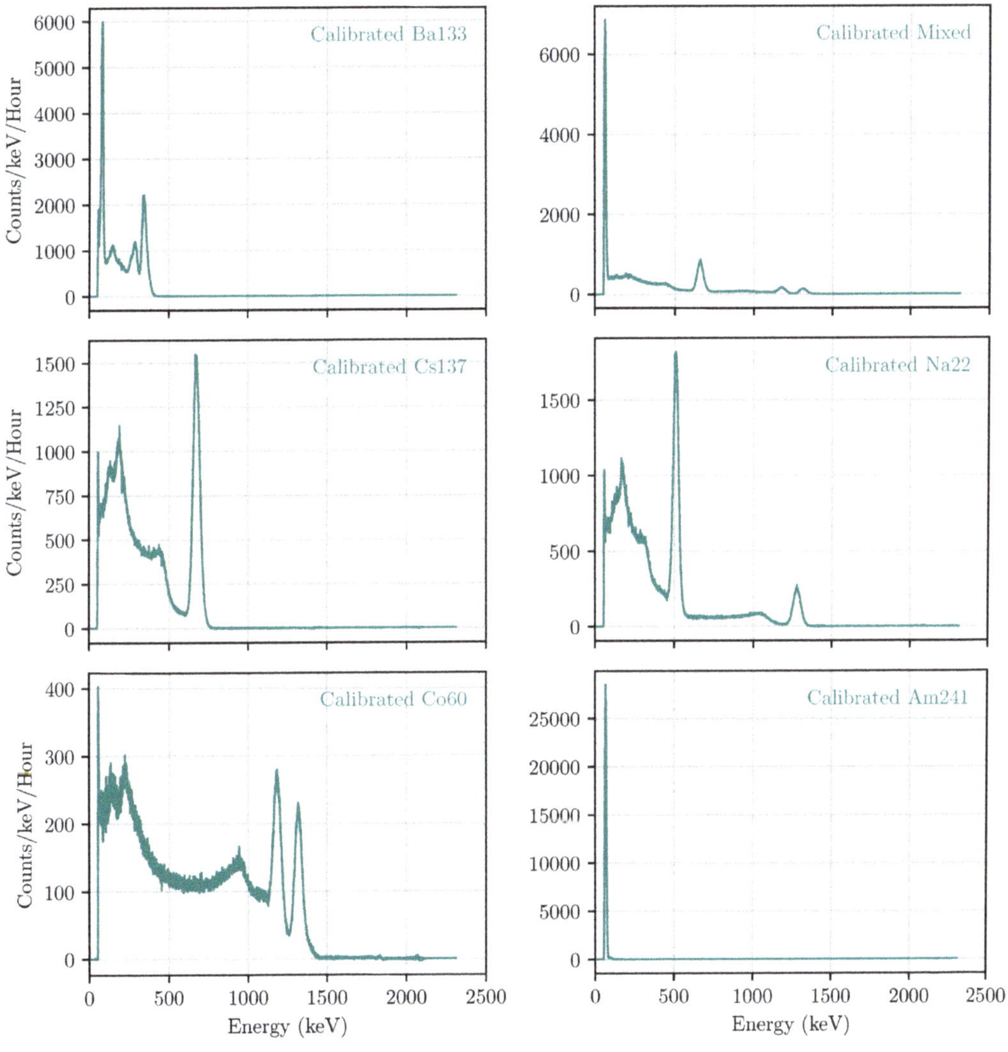

Figure 8. GMOD FM-calibrated spectra (teal) in counts/keV/hour for the six sources in Table 2.

The photopeaks were then re-fitted with a combination of a Gaussian(s) and a polynomial function to account for other contributions to the photopeak to determine σ and, using Equation (4), the FWHM in terms of energy. Figure 9 presents four examples of fitting the photopeaks of the sources in energy space. Figure 10 presents the σ values for the photopeaks as a function of energy. Equation (5) was then fitted to these points with coefficients $b = 0.6761$ and $c = 8.9754 \times 10^{-3}$.

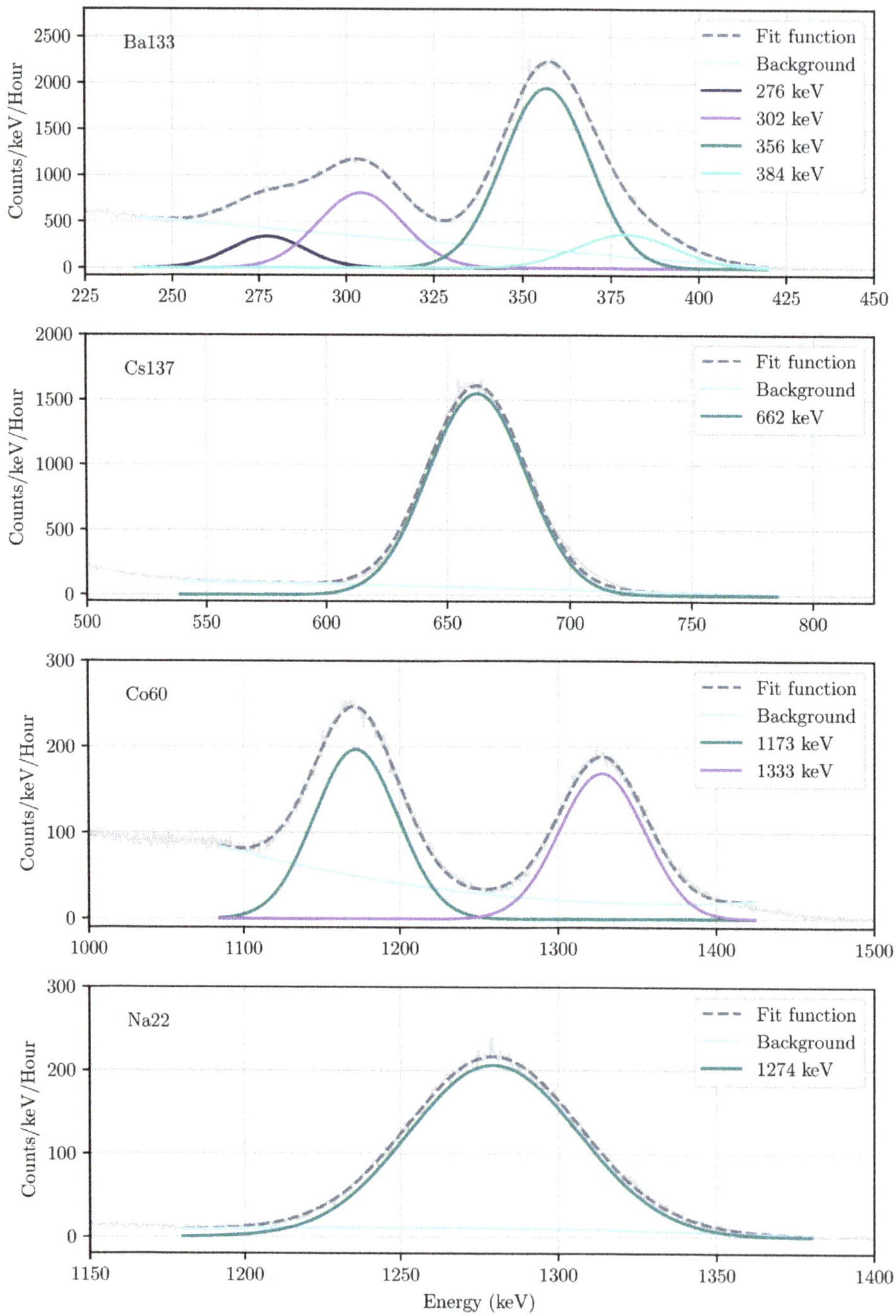

Figure 9. A sample of four fits to GMOD spectra in terms of energy (keV). Each panel shows the fit (dashed grey line) to the energy-calibrated spectrum. The composite functions of the fit are plotted as solid lines.

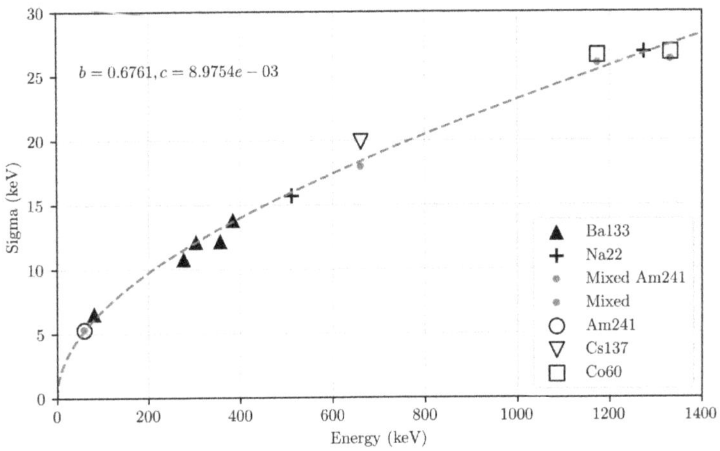

Figure 10. GMOD FM σ as a function of energy, fit (dashed line) with Equation (3) and coefficients $a = 0$, $b = 0.6761$ and $c = 8.9754 \times 10^{-3}$. Relative statistical errors in the sigma values are under 4%.

The energy resolution of the photopeaks were determined using Equation (3). This energy resolution was plotted as a function of energy, as in Figure 11. Equation (5) can be expressed in terms of the FWHM (%), as follows:

$$\text{FWHM (\%)} = 235.5 \times \sqrt{\frac{b^2}{E} + c^2}. \quad (6)$$

and can be fitted to the data in Figure 11 to determine the resolution of the GMOD FM over a wide range of energies. The obtained coefficients of the fit are $b = 0.6294$ and $c = 6.8994 \times 10^{-3}$. The energy resolution of GMOD FM improves for incoming photons with higher energies and was found to be 7.07% at 662 keV.

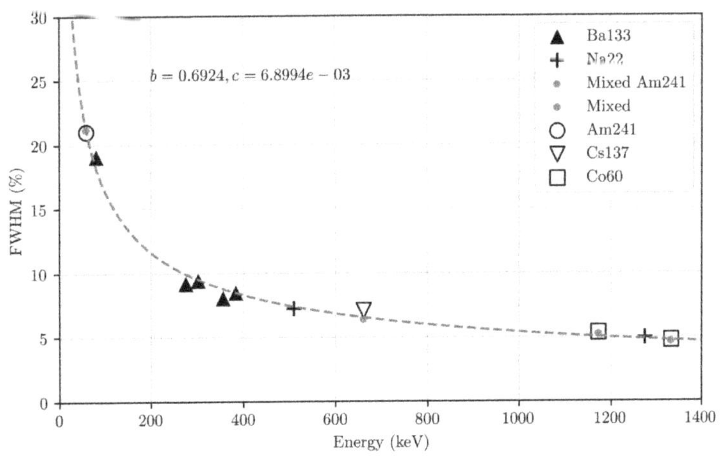

Figure 11. Energy resolution in terms of FWHM (%) fit with Equation (6) and coefficients $a = 0$, $b = 0.6294$ and $c = 6.8994 \times 10^{-3}$. Relative statistical errors in the resolution values are under 4%.

5. Effective Area

The effective area of a gamma-ray detector accounts for the detection efficiency of the instrument and the observable area. The full-energy peak effective area, A_{eff} (cm^2), for a detector is determined as [38]

$$A_{Eff} = \frac{C}{a_c P_T} 4\pi s^2 \,, \tag{7}$$

where C is the number of counts in the photopeak, i.e., the area under the Gaussian, P_T is the transition probability, s is the distance (cm) between the source and the detector and a_c is the activity of the radioactive source (1/s). The activity is given by

$$a_c = a_0 e^{-\lambda t} = a_0 e^{-\frac{ln(2)t}{T_{\frac{1}{2}}}} \,, \tag{8}$$

where a_0 is the activity of the source (s^{-1}) at a specified time, t is the time (s) between the determination of the activity and the performance of the test campaign and $T_{\frac{1}{2}}$ is the half-life of the source (s).

As the source is emitting gamma rays in all directions but only a fraction is observable to the detector, the term $4\pi s^2$ accounts for the fraction of photons emitted by the source that will actually come in contact with the detector.

Figure 12 presents the effective area of the flight model of GMOD as a function of energy, where all the sources were positioned above the +Z face of the spacecraft at the CoC. The source activity must be known for the determination of an absolute effective area, which limited the sources in Table 1 that could be used to determine GMOD's effective area. Sources 1–3, 5 and 6 were used to determine the effective area of GMOD as a function of energy. Table 2 provides the position of these sources relative to the crystal centre, the s term in Equation (7). In Figure 12, the previously simulated effective area (blue) from [29] is in good agreement with the physical measurements (see Section 7.2 for the updates made to the model to represent the FM including the addition of the GMOD top reinforcement bracket (GTRB), the structural bracket the detector assembly is mounted on). The instrument's sensitivity appears to peak in the range of tens to a few hundred keV, as is desired for the detection of the prompt emission of GRBs. The effective area is lower for smaller energies as a result of gamma-ray absorption by the detector housing and spacecraft.

The effective area of FM GMOD as a function of source incident angle was also investigated. As described in Section 2.2, due to time constraints, two sources, ^{137}Cs and ^{241}Am, were used together for the angular measurements. The effective area of GMOD was determined at 34 different incident angles, as listed in Table 3. Figure 13 presents the angular response of GMOD at 662 keV and Figure 14 at 59.4 keV. The angular response of GMOD at both these energies follows similar profiles. As shown in Figures 13 and 14, the calculated values of the effective area have an uncertainty of 6–7%, which is due to an uncertainty of the source activity (5% for the ^{241}Am source and 3.5% for the ^{137}Cs source) and errors in the source-to-crystal distance (about 5 mm).

The measurements at polar angle 90° show the most diversity and the impact of the crystal geometry on the detector's performance. The area of the Y and Z faces of the GMOD crystal is 10 cm^2, and the X faces are 6.25 cm^2. The effective area of GMOD peaks at the −Y and +Y faces, which correspond to the longer faces of the crystal. The only obstacle between the source and crystal at these faces are the spacecraft panels with mounted solar arrays. At the −X face, the effective area drops by $\sim$0.2 cm^2, as this face of the crystal is smaller. The effective area is at a minimum at the +X face as the gamma-ray photons are attenuated by the PCBs (SIPHRA and adapter boards) before reaching the crystal. For the other polar angles, the effective area tends to be at a minimum at azimuthal angle 0° (+X face). The effect of the spacecraft is also evident by the difference in effective area between polar angles 0° and 180°. At 59.4 keV, the effective area at polar angle 0° (+Z face) is similar to the −Y and +Y faces at >5 cm^2. The spacecraft components beneath GMOD attenuate the incoming photons as, at polar angle 180°, the effective area is ∼0.05 cm^2. As expected, at higher energies (662 keV), the photons are capable of reaching the crystal through the spacecraft and there is a smaller difference in the effective area between the −Z and +Z of ∼1 cm^2.

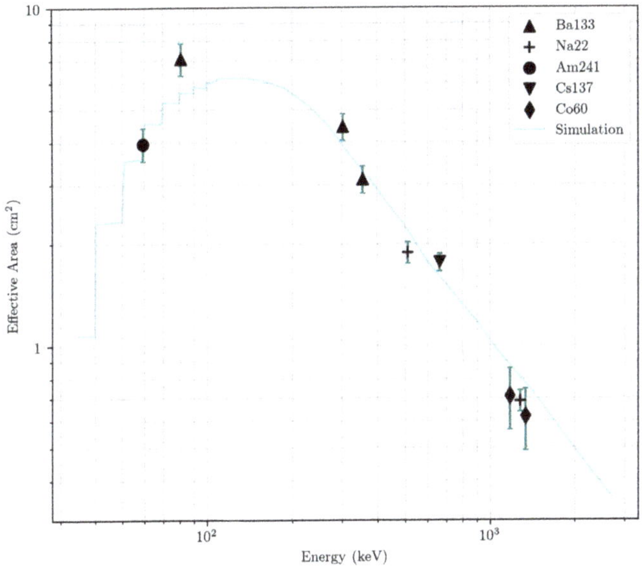

Figure 12. GMOD FM effective area as a function of energy, where all the sources were positioned above the +Z face of the spacecraft at the CoC. Simulated photopeak effective area from [29] plotted in teal for comparison to physical measurements.

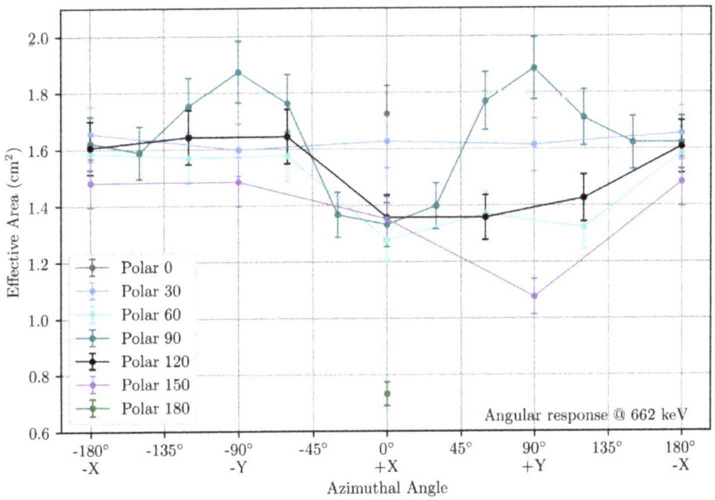

Figure 13. Angular effective area of GMOD FM at 662 keV.

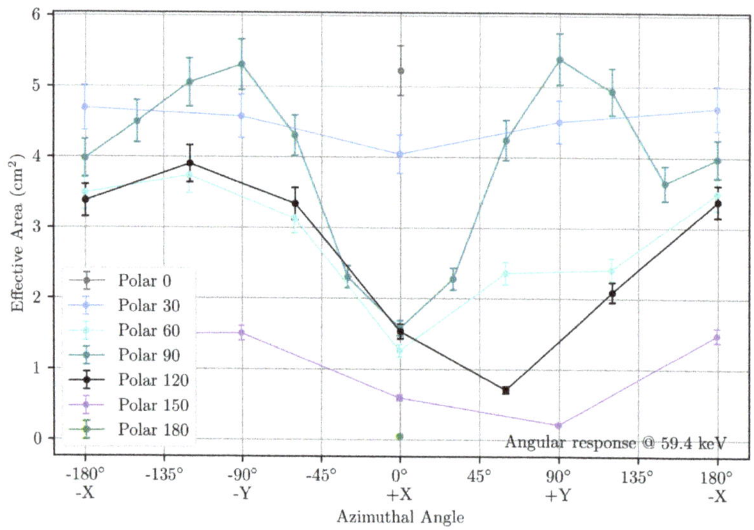

Figure 14. Angular effective area of GMOD FM at 59.4 keV.

6. Trigger Algorithm Testing

Additionally, the correct operation of the OBC-triggering algorithm and the processing and protection of the TTEs around the time of the trigger were tested during the characterisation campaign. GMOD was configured for experiment-running and the OBC's triggering algorithm was enabled. During these tests, the source was brought close to each face of the spacecraft (without touching the satellite) for a fixed amount of time, i.e., 10 s. GMOD triggered on the source at multiple positions, and it was confirmed that the trigger time and duration were correct, contributing to the verification of the functionality of the onboard triggering algorithm. The trigger data processing and protection in the OBSW around the time of the trigger were also validated as the data analysed from these triggers were downlinked from the protected trigger data storage channels on the OBC. The OBC successfully recorded 21 triggers. Figure 15 presents six triggers, where each trigger had the source located at a different face of the spacecraft (+X, −X, +Y, −Y, +Z, −Z) for ten seconds. Panel A of Figure 2 presents the EIRSAT-1 coordinate system.

In Figure 15, the sources were approximately the same distance from each face of the spacecraft for the same duration, so the intensity of the triggers reflects the impact of the crystal geometry and surrounding spacecraft on the performance of GMOD. The highest number of counts in a trigger ($\sim$3500) was when the source was at the $+Z$ face of the spacecraft; the only components between the source and GMOD are EMOD and the GTRB. As expected, the counts in the trigger are at a minimum when the source is at the $-Z$ face of the spacecraft. The trigger only creates $\sim$230 counts, as the CubeSat stack of PCBs attenuates the gamma-ray photons before they arrive at the GMOD detector. Similarly, the $+X$ trigger has relatively low counts, as the gamma-ray photons need to pass through the components in the detector assembly (SIPHRA, Interface PCB) before reaching the scintillator. The $+Y$ trigger has $\sim$2000 counts, as the source is further away from the GMOD detector. The $-Y$ and $-X$ triggers have a similar number of $\sim$3000 counts in the trigger, as there was only a solar panel between them and the source.

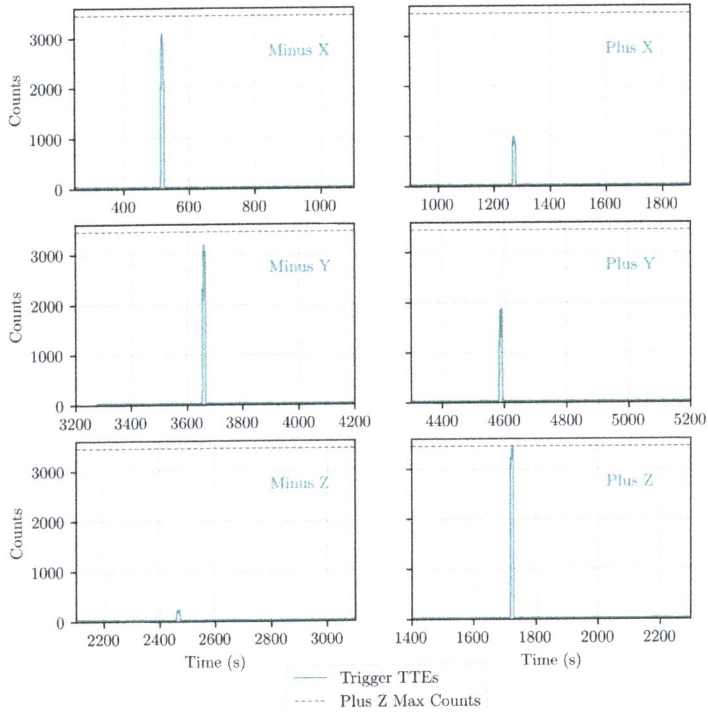

Figure 15. Each panel shows the counts recorded in a GMOD trigger as a function of time. A ^{137}Cs source was used to trigger GMOD; the source was briefly positioned at each face of the spacecraft for ~10 s. The protected trigger data was plotted. The maximum counts were recorded when the source was at the +Z face; each panel shows this max count value as a dashed green line to facilitate trigger comparison

7. Characterisation Simulations

In order to understand and predict the performance of GMOD as a GRB detector and ensure that the correct response is used if the event location is known, it is important to have an accurate spacecraft and detector model. The engineering qualification model (EQM) of EIRSAT-1, developed using MEGAlib, is described in [29]. This model needed to be updated to reflect the changes to the configuration of GMOD in the flight model. The first hardware design change was in the mounting of GMOD in EIRSAT-1. In the EQM, GMOD was mounted to the GMOD motherboard. In the FM, GMOD is suspended above the motherboard, and mounted in the PCB stack via, on a structural bracket, the GTRB. The second hardware change was that ceramic capacitors were used on the GMOD EQM MB. The GMOD FM motherboard has metal framed capacitors, as these are more robust against PCB flexure. Simulations were then performed with this updated model to correlate with measurements in the lab and validate the updated EIRSAT-1 model.

Section 7.1 provides an overview of the MEGAlib software v4.0 used to update the model and run the simulations. Section 7.2 outlines the updates made to the spacecraft geometry model. Section 7.3 presents the results of the simulations with a comparison to the physical measurements to confirm that the model was representative of the performance of the physical spacecraft.

7.1. MEGALib

The Medium-Energy Gamma-ray Astronomy library (MEGALib) v4.0 is a collection of programs for simulating gamma-ray instrumentation [43]). The software library was initially developed for the telescope MEGA [47]), which combined Compton scattering and pair production instrumentation; therefore, the higher-level data analysis tools concentrate on Compton telescope processes. However, the flexible design of the tools makes it applicable to many detector types, including scintillation-based instruments like GMOD on EIRSAT-1.

The simulations performed with MEGALib for EIRSAT-1 used the following three MEGAlib packages [48]:

1. Geometry for MEGAlib (Geomega);
2. A Cosmic Simulator for MEGAlib (Cosima);
3. Real Event Analyser (Revan).

The geometry of the detector and the satellite were first defined using Geomega, as seen in Figure 16; then, the gamma-ray source(s) were defined and, using Cosima and Revan tools, the response of GMOD in EIRSAT-1 to the sources was simulated.

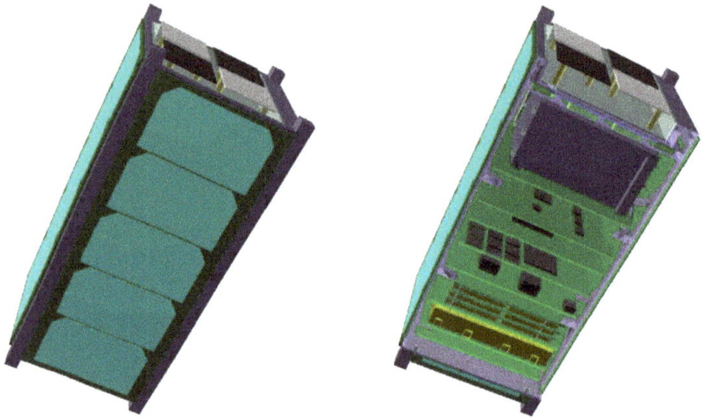

Figure 16. Updated EIRSAT-1 MEGAlib mass model including the GTRB with one solar panel cut out to view the internal components.

The main input file of Geomega includes the list of materials, the volumes of all the components of the spacecraft, the characteristics of the detector and any trigger criteria. In MEGAlib, the lengths of all objects are defined in half lengths and, throughout the programme, all the units are in cm, keV, g and deg. The materials that comprise the different components of the detector and the satellite, their densities and chemical compositions are defined in a separate file in Geomega and included in the main input file. The volumes can be a variety of different shapes including boxes, spheres, tubes and trapezoids.

The gamma-ray detector also needs to be defined in the geometry file in MEGAlib. Various properties of the detector are defined in the geometry file including trigger thresholds, noise thresholds and energy resolution in the case of scintillation-based detectors. The definition of the detector identifies which volume(s) in the model in which the positions and energies of interactions should be recorded.

The second tool used in MEGAlib is Cosima. Cosima is a Geant-4-based [49] simulator of the detection of gamma rays in the keV to a few hundreds of GeV range via photo-electric absorption, Compton scattering and pair production by gamma-ray telescopes. The Cosima simulation is directed by an input "source" file. The first input into the source file is the Geomega geometry definition of the satellite and the detector, as shown in the sample code below. The second section of the source file describes the electromagnetic interactions package that will be used to describe the interactions in the simulation. The Livermore low-energy EM package describes gamma-ray interactions in the 250 eV to 100 GeV range [50]. The output format of the simulated data is then defined (e.g., storage of simulation history and ionisation information) and the final section describes a simulated run. One source file can define multiple simulation runs that will be performed in succession. Each run must have a defined output file name, stop criterion (time) , definition of the radioactive source including the particle type, beam type and position, source spectrum and the average total flux integrated over beam and spectrum. The flux (particles/s) is calculated by multiplying the activity of the source (Bq) by the transition probability.

The output of the Cosima simulation contains the interaction history of all the primary and secondary particles and all of the hits in the detector. Along with each hit, the associated position in world coordinates and energy are recorded. Revan is the final MEGAlib package used in this chapter, to combine the Cosima simulation results and the Geomega geometry together to generate a simulated detector response. Revan is used to generate the spectrum recorded by the simulated detector of the gamma-ray source. This spectrum can then be analysed and compared to the physical measurements to confirm that the simulated model of the spacecraft is in close agreement.

7.2. EIRSAT-1 MEGAlib Model

In order to determine the performance of GMOD in orbit, it is important to consider the impact of the remainder of the spacecraft on its ability to detect incoming gamma rays. The initial geometry of EIRSAT-1 in MEGAlib is described in detail in [29]; however, due to design changes, the model required an update. The major addition to the model was the creation and inclusion of the GMOD GTRB.

In Geomega, to create the GTRB, a plate of aluminium was defined. The structure of the GTRB was then modelled by making vacuum cut-outs in the shapes of cuboids and trapezoids. Figure 17 shows the GTRB model in Geomega, the shapes highlighted in green are all vacuum cut-outs. The GTRB is 5 mm thick at the edge and then there is a 1 mm indent, as shown by the dark blue sections which are large vacuum cut-outs.

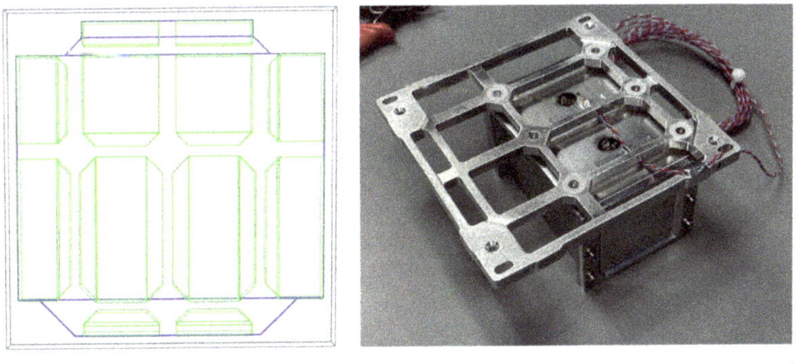

Figure 17. The left panel presents GMOD GTRB modelled in MEGAlib Geomega. The shapes highlighted in green are all vacuum cut-outs. The GTRB is 5 mm thick at the edge and then there is a 1 mm indent, as shown by the dark blue sections, which are large vacuum cut-outs. The right panel is the physical GMOD GTRB attached to the GMOD detector assembly.

The GTRB was inserted above the GMOD detector assembly, as shown in the left panel of Figure 18, and the right panel shows the new GMOD assembly included in the full spacecraft.

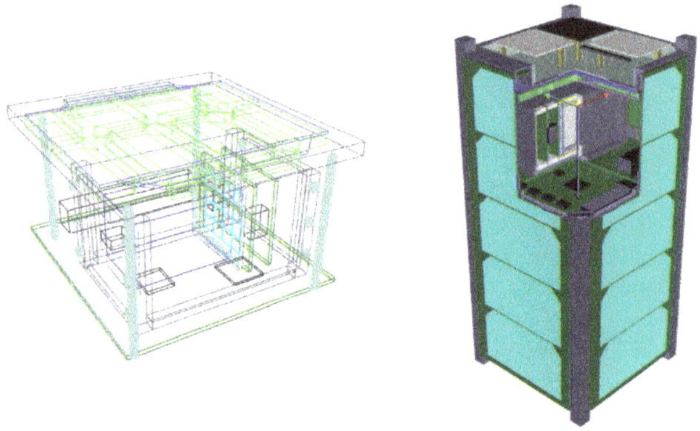

Figure 18. The GMOD GTRB modelled in MEGAlib Geomega included above the detector assembly (LHS) and then within the full spacecraft model (RHS) where the position of the detector assembly has also been updated to reflect the flight model.

The position of the detector assembly also required modification as, in the FM, the aluminium housing was inverted and secured to the GTRB (in the EQM, the aluminium housing was mounted to the GMOD MB below).

7.3. Simulations and Comparison with Physical Measurements

Two groups of simulations were performed to compare with the GMOD FM physical measurements and investigate the accuracy of the EIRSAT-1 MEGAlib model. Initially, the 12 h integrations were simulated using MEGAlib. The activity of the source is required as a simulation input; therefore, the response of GMOD to the five sources with known activities from Table 1 were simulated, i.e., ^{241}Am, ^{137}Cs, ^{133}Ba, ^{60}Co and ^{22}Na. The measurement duration, source activity and position relative to the centre of the spacecraft were input into the Cosima source file. Cosima was run to generate all the particle interactions. Revan was then used to process the Cosima output file, along with the spacecraft and instrument geometry, to produce a spectrum recorded by the simulated detector. Figure 19 presents a comparison of the simulated spectra (teal) and the spectra produced by the physical FM (grey) for each source. Both spectra are displayed in counts/keV/hour. As expected, the number of counts in the simulated photopeaks does not exactly match that of the physical measurements; these differences can be accounted for with uncertainties in the activities and positions of the sources.

As most evident in the ^{137}Cs spectrum, the additional low energy counts in the region of $\sim$50 keV to 250 keV in the physical measurements can be attributed to scattering from the surrounding environment. The MEGAlib simulation assumes that the spacecraft is in a vacuum, so there are no testing mounts or walls to scatter the gamma rays.

The photopeak of the gamma-ray line at 59.4 keV in the measured ^{241}Am spectrum (grey spectrum in the last panel of Figure 19) has an unusual profile, as this peak is at the energy threshold of the physical flight model of GMOD. Prior to FM vibration, GMOD was capable of detecting the full peak at 59.4 keV. After FM vibration testing, a shift in energies down in the channel number was observed in the GMOD spectra due to a lower current being produced by the SiPMs. A lower current could be caused by a lower light collection or an electronic issue with the SiPM bias voltage. As the bias voltage has been measured and confirmed to be configured correctly, the issue is attributed to the lower light collection. An optical coupling issue would cause lower light collection, e.g., trapped air between the scintillator and SiPMs. However, as the spacecraft had been qualified after the FM environmental testing, it could not be disassembled to confirm this possible cause or rectify it.

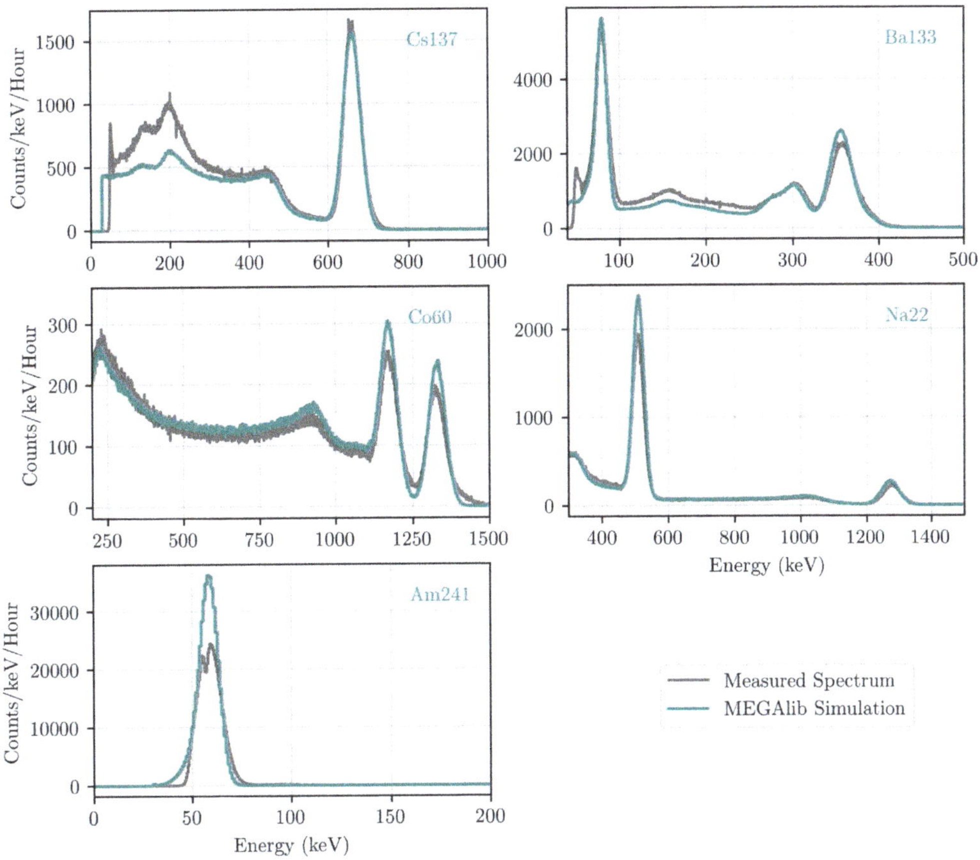

Figure 19. Spectra (counts/keV/hour) produced by the simulated MEGAlib GMOD detector (teal) for the five sources with known activities in Table 1. In each panel, the spectra produced by the physical GMOD FM are plotted for comparison.

The peaks of the simulated spectra were fitted using the same method as followed for the measured spectra to determine the relationship between σ and energy, as shown in Figure 20. The relationship described by Equation (5) was fitted to the simulated data points (teal) with the fit coefficients $b = 0.6810$ and $c = 9.6791 \times 10^{-3}$. The model is in good agreement with the relationship fit to the measured spectra (grey solid line).

The energy resolution, in terms of FWHM (%), of the simulated detector was then determined using these σ values and Equation (3). The energy resolutions of the simulated spectra were fitted with the relationship described by Equation (6) and fit coefficients $b = 0.6274$ and $c = 1.3940 \times 10^{-2}$, as in Figure 21. The MEGAlib model results (teal) reflect the measured energy resolution of GMOD FM (grey solid line).

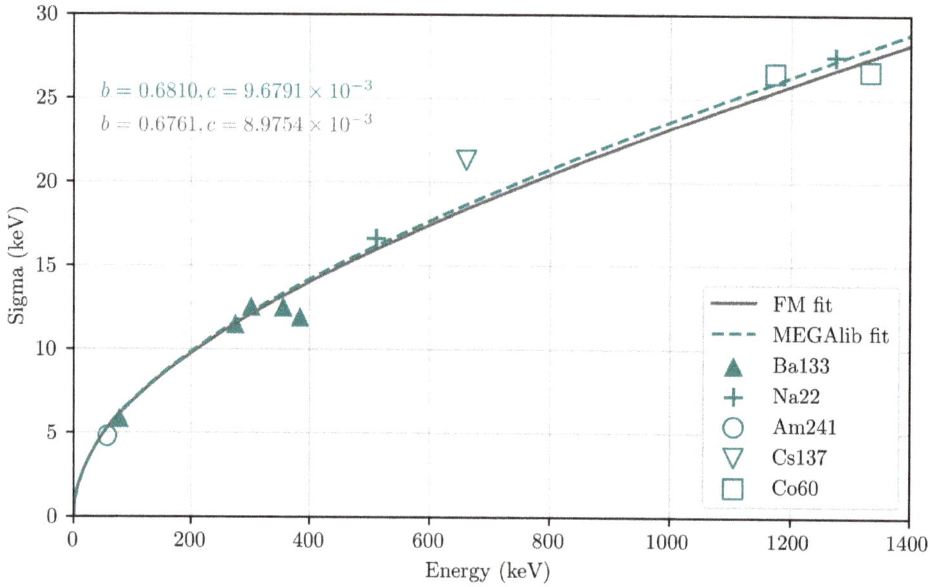

Figure 20. GMOD simulated σ as a function of energy, fitted with Equation (5) and coefficients $a = 0$, $b = 0.6810$ and $c = 9.6791 \times 10^{-3}$. The model is in good agreement with the relationship fit to the measured spectra (grey solid line). Relative statistical errors in the sigma values are under 4%.

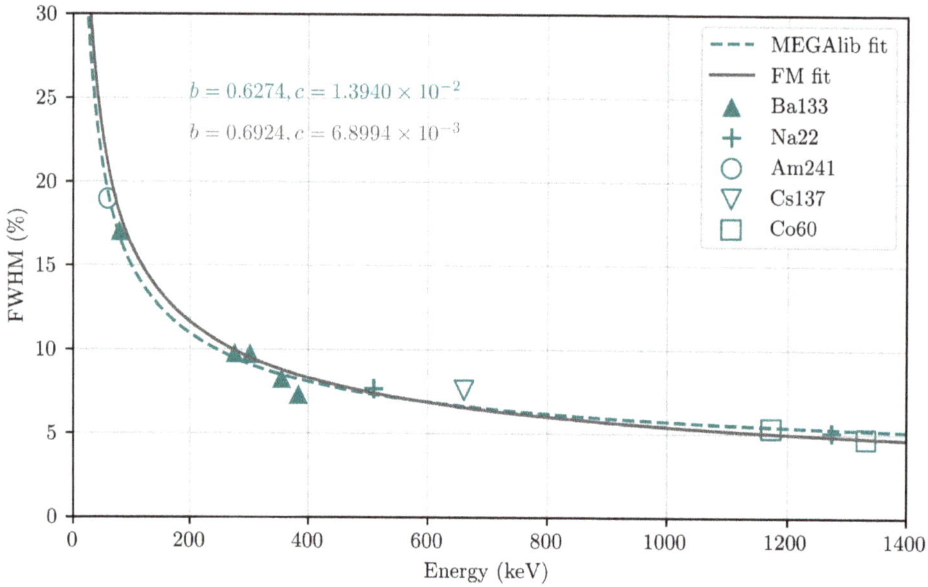

Figure 21. Simulated energy resolution in terms of FWHM (%) fit with Equation (6) and coefficients $a = 0$, $b = 0.6274$ and $c = 1.3940 \times 10^{-2}$. The model is in good agreement with the relationship fit to the measured spectra (grey solid line). Relative statistical errors in the resolution values are under 4%.

The effective area as a function of energy of the simulated detector was determined, as seen in Figure 22. As expected, the effective area of the model is slightly higher across all energies, with the average difference between the measured and simulated energies being 0.5 cm^2. This is due to the previously explained difference in photopeak counts of the simulated spectra in comparison to the measured spectra.

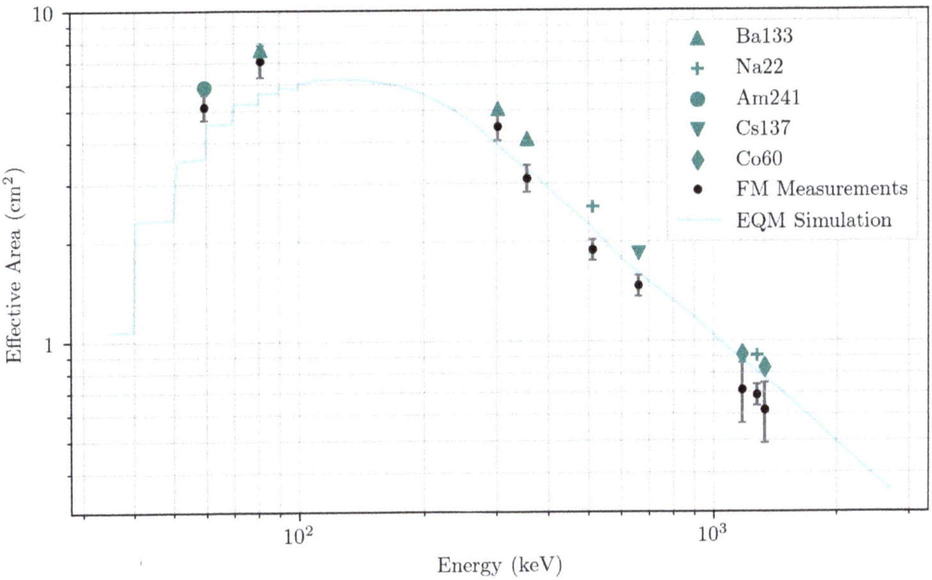

Figure 22. GMOD FM simulated effective area as a function of energy, where all the simulated source positions were above the +Z face of the spacecraft at the CoC.

The second set of simulations performed was to investigate the accuracy of the MEGAlib GMOD model's angular effective area. Source files for the measurements at the 34 angles were run with Cosima. The spectra were processed using Revan and analysed to calculate the effective areas. Table 4 compares the calculated effective areas at polar angle 0° and polar angle 180°, which correspond to the source being 22.5 cm from the +Z and −Z of the spacecraft, respectively. Figures 23 and 24 compare the measured (grey) and simulated (teal) angular response of GMOD FM at 662 keV and 59.4 keV, respectively. The updated FM model produces the expected angular response from GMOD in agreement with the measured results. Across all angles, the simulated effective area is higher due to there being more counts in the simulated photopeaks. There is also a larger difference in values for the 59.4 keV peak due to the physical FM not fully detecting this peak in ^{241}Am.

Table 4. Comparison of effective area for measured and simulated GMOD FM at 662 keV and 59.4 keV at +Z (polar angle 0°) and −Z (polar angle 180°).

Polar Angle	Peak Energy (keV)	Measured A_{eff} (cm^2)	Simulated A_{eff} (cm^2)
0	662	1.72	1.86
	59.4	5.23	5.33
180	662	0.73	1.15
	59.4	0.05	0.74

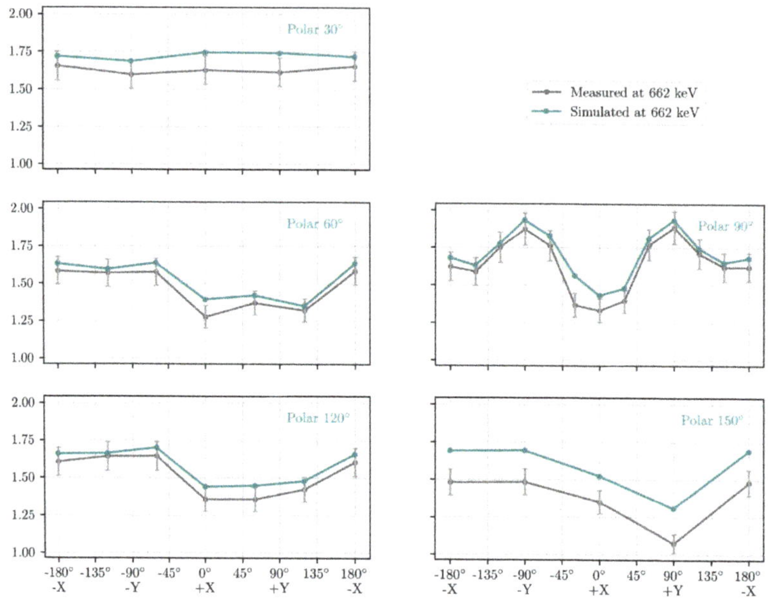

Figure 23. GMOD FM simulated angular response compared with the measured FM response at 662 keV. Errors in measured values are mainly due to systematic uncertainties. Errors for simulated values are purely statistical and under 1%.

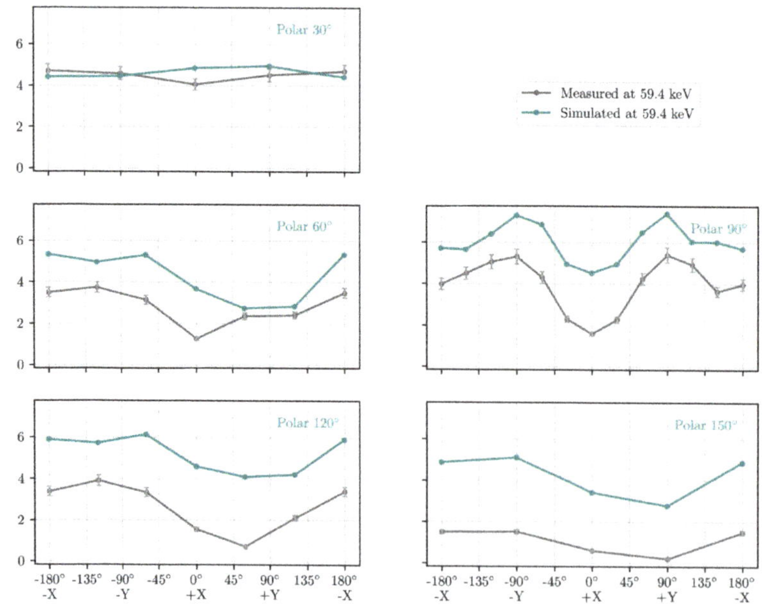

Figure 24. GMOD FM simulated angular response compared with the measured FM response at 59.4 keV. Errors in measured values are mainly due to systematic uncertainties. Errors for simulated values are purely statistical and under 1%.

8. Conclusions

The characterisation of the flight model instrument provides a comprehensive understanding of the instrument's behaviour pre-launch. Initially, an energy calibration was performed on GMOD to map the ADC channel values to energy. Then, the properties of the detector were determined, with the FM having an energy resolution of 7.07% at 662 keV. The effective area as a function of energy and incident angle was investigated, which is important for interpreting the downlinked science data during the mission. The effective area of GMOD peaks in the energy range of tens to a few hundred keV, as is required by the specifications of our scientific case for GRB detection. As expected, the detector is most sensitive on the faces with minimal obstruction and the least sensitive faces are those where gamma-ray photons must first pass through the CubeSat stack or the PCBs in the detector assembly. In addition to these measurements with the physical flight model, the EIRSAT-1 MEGAlib model was updated to reflect the new configuration of GMOD in the flight model. This updated model was validated by simulating the measurements made in the lab and confirming consistency between the simulated and physical results. This approach to characterising an instrument for GRB science can be followed by other CubeSats with similar science goals. The results from this characterisation campaign can be used for comparison with the detector's performance in orbit. In reference [29], simulations performed with the original MEGAlib model predicted that GMOD will detect 11–14 GRBs at a significance greater than 10σ during a nominal one-year mission. Future work would include performing these simulations with the updated MEGAlib model presented in this paper and producing the response files for spectral fitting.

Author Contributions: Conceptualization, D.M. (David Murphy), A.U., J.M., M.D., B.S. and S.M.; Methodology, R.D. and J.T.; Software, M.D.; Validation, A.U.; Formal analysis, R.D.; Investigation, R.D. and D.M. (David Murphy); Resources, J.M. and J.T.; Writing—original draft, R.D.; Writing—review & editing, R.D., A.U., J.M., M.D., J.T., C.d.B., L.H., D.M. (David McKeown), B.S. and S.M.; Supervision, D.M. (David Murphy), A.U., L.H. and S.M.; Project administration, L.H., D.M. (David McKeown) and S.M.; Funding acquisition, L.H., D.M. (David McKeown) and S.M. All authors have read and agreed to the published version of the manuscript.

Funding: The EIRSAT-1 project is carried out with the support of ESA's Education Office under the Fly Your Satellite! 2 programme. The authors acknowledge the guidance from Jean-Philippe Halain of the ESA PRODEX Office. The authors acknowledge all students who have contributed to EIRSAT-1 and support from Parameter Space Ltd. R.D., M.D., D.M. and J.T. acknowledge support from the Irish Research Council (IRC) under grants GOIPG/2019/2033, GOIP/2018/2564, GOIPG/2014/453 and GOIPG/2014/684, respectively. S.M.B., D.M., A.U. and J.M. acknowledge support from Science Foundation Ireland under grant 17/CDA/4723. S.M.B. and C.D.B. acknowledge support of Science Foundation Ireland grant 21/FFP-A/9043. L.H. acknowledges support from SFI under grant 19/FFP/6777 and support from the EU H2020 AHEAD2020 project (grant agreement 871158). This study was supported by the European Space Agency's Science Programme under contract 4000104771/11/NL/CBi and by the European Space Agency's PRODEX Programme under contract number C 4000124425.

Data Availability Statement: The original contributions presented in the study are included in the article, further inquiries can be directed to the corresponding author. The datasets presented in this article are not readily available because the data are part of an ongoing study. Requests to access the datasets should be directed to the corresponding author.

Conflicts of Interest: The authors declare no conflict of interest. The funders had no role in the design of the study; in the collection, analyses, or interpretation of data; in the writing of the manuscript, or in the decision to publish the results.

References

1. Vedrenne, G.; Atteia, JL. Gamma-ray Bursts: The Brightest Explosions in the Universe. *Phys. Today* **2010**, *63*. 56–57. [CrossRef]
2. Luongo, O.; Muccino, M. A Roadmap to Gamma-ray Bursts: New Developments and Applications to Cosmology. *Galaxies* **2021**, *9*, 77. [CrossRef]

3. Abbott, B.P.; Abbott, R.; Abbott, T.D.; Acernese, F.; Ackley, K.; Adams, C.; Adams, T.; Addesso, P.; Adhikari, R. X.; Cahillane, C.; et al. GW170817: Observation of Gravitational Waves from a Binary Neutron Star Inspiral. *Phys. Rev. Lett.* **2017**, *119*, 161101. [CrossRef] [PubMed]
4. Abbott, B.P.; Abbott, R.; Abbott, T.D.; Acernese, F. ; Ackley, K. ; Adams, C. ; Adams, T. ; Addesso, P. ; Adhikari, R. X.; Adya, V. B. et al. Gravitational Waves and Gamma-rays from a Binary Neutron Star Merger: GW170817 and GRB 170817A. *Astrophys. J.* **2017**, *848*, L13. [CrossRef]
5. Meegan. C.; Lichti, G.; Bhat, P. N.; Bissaldi, E.; Briggs, M.S.; Connaughton, V.; Diehl, R.; Fishman, G.; Greiner, J.; Hoover, A.S.; et al. The Fermi Gamma-ray Burst Monitor. *Astrophys. J.* **2009**, *702*, 791–804. [CrossRef]
6. Atwood, W. B.; Abdo, A. A.; Ackermann, M.; Althouse, W.; Anderson, B.; Axelsson, M.; Baldini, L.; Ballet, J.; Band, D. L.; Barbiellini, G.; et al. The Large Area Telescope on the Fermi Gamma-ray Space Telescope Mission. *Astrophys. J.* **2009**, *697*, 1071–1102. [CrossRef]
7. Gehrels, N.; Chincarini, G.; Giommi, P.; Mason, K.O.; Nousek, J.A.; Wells, A.A.; White, N.E.; Barthelmy, S.D.; Burrows, D.N.; Cominsky, L.R.; et al. The Swift Gamma-ray Burst Mission. *Astrophys. J.* **2004**, *611*, 1005–1020. [CrossRef]
8. Winkler, C.; Courvoisier, T.J.-L.; Di Cocco, G.; Gehrels, N.; Giménez, A.; Grebenev, S.; Hermsen, W.; Mas-Hesse, J.M.; Lebrun, F.; Lund, N.; et al. The INTEGRAL mission. *Astron. Astrophys.* **2003**, *411*, L1–L6. [CrossRef]
9. von Kienlin, A.; Meegan, C.A.; Paciesas, W.S.; Bhat, P.N.; Bissaldi, E.; Briggs, M.S.; Burns, E.; Cleveland, W.H.; Gibby, M.H.; Giles, M.M.; et al. The Fourth Fermi-GBM Gamma-ray Burst Catalog: A Decade of Data. *Astrophys. J.* **2020**, *893*, 46. [CrossRef]
10. Bailes, M.; Berger, B.K.; Brady, P.R.; Branchesi, M.; Danzmann, K.; Evans, M.; Holley-Bockelmann, K.; Iyer, B.R.; Kajita, T.; Katsa-nevas, S.; et al. Gravitational-wave physics and astronomy in the 2020s and 2030s. *Nat. Rev. Phys. 3* **2021**, *3*, 344–366. [CrossRef]
11. The CubeSat Program. CubeSat Design Specification Rev. 14. California Polytechnic State University. In *Technical Report CP-CDS-R14;* California Polytechnic State University (Cal Poly): San Luis Obispo, CA, USA, 2020.
12. Bloser, P.; Murphy, D.; Fiore, F.; Perkins, J. CubeSats for Gamma-ray Astronomy. In *Handbook of X-ray and Gamma-ray Astrophysics;* Springer Nature: Singapore, 2022; pp. 1–33. [CrossRef]
13. Serjeant, S.; Elvis, M.; Tinetti, G. The future of astronomy with small satellites. *Nat. Astron.* **2020**, *4*, 1031–1038. [CrossRef]
14. Mero, B.; Quillien, K.; McRobb, M.; Chesi, S.; Marshall, R.; Gow, A.; Clark, C.; Anciaux, M.; Cardoen, P.; Keyser, J.D.; et al. PICASSO: A State of the Art CubeSat. In Proceedings of the 29th Annual AIAA/USU Small Satellite Conference, Logan, UT, USA, 8–13 August 2015; Technical Session III: Next on the Pad: SSC15-III-2.
15. Liddle, J.D.; Holt, A.P.; Jason, S.J.; O'Donnell, K.A.; Stevens, E.J. Space science with CubeSats and nanosatellites. *Nat. Astron.* **2020**, *4*, 1026–1030. [CrossRef]
16. Nieto-Peroy, C.; Emami, M.R. CubeSat Mission: From Design to Operation. *Appl. Sci.* **2019**, *9*, 3110. [CrossRef]
17. Twiggs, R. Origin of CubeSat. In *Small Satellite: Past, Present and Future;* Helvajian, H., Janson, S.W., Eds.; The Aerospace Press: El Segundo, CA, USA, 2008; Chapter 5, pp. 151–173.
18. Straub, J.; Villela, T.; Costa, C.A.; Brandão, A.M.; Bueno, F.T.; Leonardi, R. Towards the Thousandth CubeSat: A Statistical Overview. *Int. J. Aerosp. Eng.* **2019**, *2019*, 5063145. [CrossRef]
19. Shkolnik, E.L. On the Verge of an Astronomy CubeSat Revolution. *Nat. Astron.* **2018**, *2*, 374–378. [CrossRef]
20. Engel, K.; Lewis, T.; Stein, M.; Venters, T.; Ahlers, M.; Albert, A.; Allen, A.; Solares, H.; Anandagoda, S.; Andersen, T. et al. Advancing the Landscape of Multimessenger Science in the Next Decade. *arXiv* **2022**, arXiv:2203.10074. [CrossRef]
21. Racusin, J.; Perkins, J.S.; Briggs, M.S.; de Nolfo, G.; Krizmanic, J.; Caputo, R.; McEnery, J.E.; Shawhan, P.; Morris, D.; Connaughton, V.; et al. BurstCube: A CubeSat for Gravitational Wave Counterparts. *arXiv* **2017**, arXiv:1708.09292. [CrossRef]
22. Joens, A.; Brewer, I.; Briggs, M.; Bruno, A.; Burns, E.; Caputo, R.; Cenko, B.; de Nolfo, G.; Goldstein, A.; Griffin, S.; et al. BurstCube: A CubeSat for gravitational wave counterparts. In Proceedings of the Space Telescopes and Instrumentation 2022: Ultraviolet to Gamma Ray, Montréal, QC, Canada, 17–22 July 2022; Volume 12181, p. 121811N.
23. Hui, C.M.; Briggs, M.S.; Goldstein, A.; Jenke, P.; Kocevski, D.; Wilson-Hodge, C.A. MoonBEAM: A Beyond-LEO Gamma-ray Burst Detector for Gravitational-Wave Astronomy. *Deep Space Gatew. Concept Sci. Workshop* **2018**, *2063*, 3060.
24. Fuschino, F.; Campana, R.; Labanti, C.; Evangelista, Y.; Feroci, M.; Burderi, L.; Fiore, F.; Ambrosino, F.; Baldazzi, G.; Bellutti, P.; et al. HERMES: An ultra-wide band X and gamma-ray transient monitor on board a nano-satellite constellation. *Nucl. Instruments Methods Phys. Res. Sect. A Accel. Spectrometers Detect. Assoc. Equip.* **2019**, *936*, 199–203. [CrossRef]
25. Pál, A.; Ohno, M.; Mészáros, L.; Werner, N.; Ripa, J.; Frajt, M.; Hirade, N.; Hudec, J. GRBAlpha: A 1U CubeSat mission for validating timing-based gamma-ray burst localization. In Proceedings of the Space Telescopes and Instrumentation 2020: Ultraviolet to Gamma Ray, Online, 14–18 December 2020; Volume 11444, pp. 825–833. [CrossRef]
26. Werner, N.; Řípa, J.; Pál, A.; Ohno, M.; Tarcai, N.; Torigoe, K.; Tanaka, K.; Uchida, N.; Mészáros, L.; Galgóczi, G.; et al. CAMELOT: Cubesats Applied for MEasuring and LOcalising Transients mission overview. In Proceedings of the Space Telescopes and Instrumentation 2018: Ultraviolet to Gamma Ray, Austin, TX, USA, 10–15 June 2018; Volume 10699, pp. 672–686. [CrossRef]
27. David, M.; Flanagan, J.; Joseph W.T.; Maeve, D.; Jessica, E.; Andrew, G.; Conor, O.; Lana, S.; Daire, S.; Sarah, W.; et al. EIRSAT-1—The Educational Irish Research Satellite. In Proceedings of the 2nd Symposium on Space Educational Activities, Barcelona, Spain, 27–28 April 2018; pp. 201–205.

28. Maeve, D.; David, M.; Jack, R.; Joseph, T.; Sarah, W.; Reddy, A.S.K.; Rachel, D.; Jessica, E.; Gabriel, F.; Joseph, M.; et al. Update on the status of the Educational Irish Research Satellite (EIRSAT-1). In Proceedings of the 4th Symposium on Space Educational Activities, Barcelona, Spain, 27–29 April 2022; pp. 294–299. [CrossRef]
29. Murphy, D.; Ulyanov, A.; McBreen, S.; Doyle, M.; Dunwoody, R.; Mangan, J.; Thompson, J.; Shortt, B.; Martin-Carrillo, A.; Hanlon, L. A compact instrument for gamma-ray burst detection on a Cubesat platform I: Design drivers and expected performance. *Exp. Astron.* **2021**, *52*, 59–84. [CrossRef]
30. Murphy, D.; Ulyanov, A.; McBreen, S.; Mangan, J.; Dunwoody, R.; Doyle, M.; O'Toole, C.; Thompson, J.; Reilly, J.; Walsh, S.; et al. A compact instrument for gamma-ray burst detection on a CubeSat platform II. *Exp. Astron.* **2022**, *53*, 961–990. [CrossRef] [PubMed]
31. Dunwoody, R.; Doyle, M.; Murphy, D.; Finneran, G.; O'Callaghan, D.; Reilly, J.; Thompson, J.; Akarapu, S.K.R.; de Barra, C.; Cotter, L.; et al. Development, description, and validation of the operations manual for EIRSAT-1, a 2U CubeSat with a gamma-ray burst detector. *J. Astron. Telesc. Instruments Syst.* **2023**, *9*, 037001. [CrossRef]
32. Dirk, M.; Jörg, A.; Alf, O.; Hans, K.; Hans, B.; Amir, H.; Akif, A.; Suleyman, A.; Bahram, N.; Jahanzad, T.; et al. SIPHRA 16-Channel Silicon Photomultiplier Readout ASIC. In Proceedings of the 6th International Workshop on Analogue and Mixed-Signal Integrated Circuits for Space Applications, Gothenburg, Sweden, 12–16 June 2016. [CrossRef]
33. Mangan, J.; Murphy, D.; Dunwoody, R.; Doyle, M.; Ulyanov, A.; Hanlon, L.; Shortt, B.; McBreen, S. Embedded Firmware Development for a Novel CubeSat Gamma-ray Detector. In Proceedings of the 2021 IEEE 8th International Conference on Space Mission Challenges for Information Technology (SMC-IT), Pasadena, CA, USA, 26–30 July 2021; pp. 14–22. [CrossRef]
34. Joseph, M.; David, M.; Rachel, D.; Maeve, D.; Alexey, U.; Mike, H.; Reddy, A.S.K.; Jessica, E.; Gabriel, F.; Fergal, M.; et al. Experiences in firmware development for a CubeSat instrument payload. In Proceedings of the 4th Symposium on Space Educational Activities 2022, Barcelona, Spain, 27–29 April 2022. [CrossRef]
35. Ulyanov, A.; Morris, O.; Hanlon, L.; McBreen, S.; Foley, S.; Roberts, O.J.; Tobin, I.; Murphy, D.; Wade, C.; Nelms, N.; et al. Performance of a monolithic $LaBr_3$:Ce crystal coupled to an array of silicon photomultipliers. *Nucl. Instruments Methods Phys. Res. A* **2016**, *810*, 107–119. [CrossRef]
36. Ulyanov, A.; Morris, O.; Roberts, O.J.; Tobin, I.; Hanlon, L.; McBreen, S.; Murphy, D.; Nelms, N.; Shortt, B. Localisation of gamma-ray interaction points in thick monolithic $CeBr_3$ and $LaBr_3$:Ce scintillators. *Nucl. Instruments Methods Phys. Res. A* **2017**, *844*, 81–89. [CrossRef]
37. Murphy, D.; Mangan, J.; Ulyanov, A.; Walsh, S.; Dunwoody, R.; Hanlon, L.; Shortt, B.; McBreen, S. Balloon flight test of a CeBr3 detector with silicon photomultiplier readout. *Exp. Astron.* **2021**, *52*, 1–34. [CrossRef] [PubMed]
38. Bissaldi, E.; von Kienlin, A.; Lichti, G.; Steinle, H.; Bhat, P.N.; Briggs, M.S.; Fishman, G.J.; Hoover, A.S.; Kippen, R.M.; Krumrey, M. et al. Ground-based calibration and characterization of the Fermi gamma-ray burst monitor detectors. *Exp. Astron.* **2009**, *24*, 47–88. [CrossRef]
39. Bandyopadhyay, R.M.; Silk, J.; Taylor, J.E.; Maccarone, T.J. On the origin of the 511-keV emission in the Galactic Centre. *Mon. Not. R. Astron. Soc.* **2009**, *392*, 1115–1123. [CrossRef]
40. Ramaty, R.; Lingenfelter, R.E. Gamma-ray Lines from Solar Flares. In *Solar Gamma-, X-, and EUV Radiation*; Kane, S.R., Ed.; Springer: Amsterdam, The Netherlands, 1975; Volume 68, p. 363.
41. Tierney, D.; McBreen, S.; Hanlon, L.; Foley, S.; Martin-Carrillo, A.; Topinka, M.; Meehan, S. Spectral Cross-Calibration of Fermi-GBM and INTEGRAL-ISGRI using Gamma-ray Bursts. In Proceedings of the Gamma Ray Bursts 2010, Annapolis, MD, USA, 1–4 November 2010; Volume 1358, pp. 397–400. [CrossRef]
42. Doyle, M.; Gloster, A.; Griffin, M.; Hibbett, M.; Kyle, J.; O'Toole, C.; Mangan, J.; Murphy, D.; Wong, N.L.; Akarapu, S.K.R.; et al. Design, development, and testing of flight software for EIRSAT-1: A university-class CubeSat enabling astronomical research. In Proceedings of the Software and Cyberinfrastructure for Astronomy VII, Montreal, QC, Canada, 17–23 July 2022; Volume 12189, p. 1218915. [CrossRef]
43. Zoglauer, A.; Andritschke, R.; Schopper, F. MEGAlib – The Medium Energy Gamma-ray Astronomy Library. *New Astron. Rev.* **2006**, *50*, 629–632. [CrossRef]
44. Be, M.-M.; Chiste, V.; Dulieu, C.; Kellett, M.A.; Mougeot, X.; Arinc, A.; Chechev, V.P.; Kuzmenko, N.K.; Kibedi, T.; Luca, A.; et al. LNE - LNHB/CEA Table de Radionucleides, 2016. Bureau International des Poids et Mesures, Pavillon de Breteuil, F-92310 Sevres, France. Available online: http://www.bipm.org/utils/common/pdf/monographieRI/Monographie_BIPM-5_Tables_Vol8.pdf (accessed on 9 July 2024).
45. Bright Ascension Ltd. User Manual—COAST MCS, BAL-COAST-MCS-UM - 21.2. 2021. Available online: https://brightascension.com/portal/knowledge-base/mcs-user-manual/ (accessed on 23 June 2023).
46. Knoll, G. *Radiation Detection and Measurement*; Wiley: Hoboken, NJ, USA, 2010.
47. Andritschke, R.; Zoglauer, A.; Kanbach, G.; Bloser, P.F.; Schopper, F. The Compton and pair creation telescope MEGA. *Exp. Astron.* **2005**, *20*, 395–403. [CrossRef]
48. Zoglauer, A.; Weidenspointner, G.; Galloway, M.; Boggs, S.E.; Wunderer Cosima, C.B. The cosmic simulator of MEGAlib. In Proceedings of the 2009 IEEE Nuclear Science Symposium Conference Record (NSS/MIC), Orlando, FL, USA, 24 October–1 November 2009; pp. 2053–2059. [CrossRef]

49. Agostinelli, S.; Allison, J.; Amako, K.; Apostolakis, J.; Araujo, H.; Arce, P.; Asai, M.; Axen, D.; Banerjee, S.; Barrand, G.; et al. GEANT4—A simulation toolkit. *Nucl. Instrum. Meth. A* **2003**, *506*, 250–303. [CrossRef]
50. Livermore Low-Energy Package (2023). Available online: https://twiki.cern.ch/twiki/bin/view/Geant4/LoweMigratedLivermore (accessed on 20 June 2023).

aerospace MDPI

Review

Overview of Space-Based Laser Communication Missions and Payloads: Insights from the Autonomous Laser Inter-Satellite Gigabit Network (ALIGN)

Othman I. Younus [1,*], Amna Riaz [1], Richard Binns [1], Eamon Scullion [1], Robert Wicks [1], Jethro Vernon [1], Chris Graham [2], David Bramall [2], Jurgen Schmoll [2] and Cyril Bourgenot [2]

1 Department of Mathematics, Physics and Electrical Engineering, Northumbria University, Newcastle upon Tyne NE1 8ST, UK; robert.wicks@northumbria.ac.uk (R.W.)
2 NETPark Research Institute, Centre for Advanced Instrumentation, Durham University, Joseph Swan Road, Sedgefield TS21 3FB, UK
* Correspondence: othman.younus@ieee.org

Abstract: This paper examines the growing adoption of laser communication (lasercom) in space missions and payloads for identifying emerging trends and key technology drivers of future optical communications satellite systems. It also presents a comprehensive overview of commercially available and custom-designed lasercom terminals, outlining their characteristics and specifications to meet the evolving demands of global satellite networks. The analysis explores the technical considerations and challenges associated with integrating lasercom terminals into LEO constellations and the Inter-satellite communications service provision in LEO due to their power, size, and weight constraints. By analyzing advancements in CubeSat lasercom technology designed to cater for the emergence of future mega constellations of interacting small satellites, the paper underscores its promising role in establishing high-performance satellite communication networks for future space exploration and data transmission. In addition, a brief overview of our ALIGN planned mission is provided, which highlights the main key operational features in terms of PAT and link budget analysis.

Keywords: free-space optical communication; satellites; low-earth orbit; geostationary orbit

1. Introduction

Wireless communications have recently undergone an exceptional surge in various areas. From cellular networks to satellite connections, wireless telecommunications have given rise to a multitude of novel services to provide communication and internet connectivity worldwide. Despite this, the "World Internet Development Report 2024" states that the global average internet penetration rate is approximately 67.1 percent, implying that approximately 32.9% of the global population lacks internet access. This means billions of people are still unconnected [1]. Traditional communication methods that utilize radio frequencies (RF) face data transmission capacity constraints due to technological and regulatory challenges [2]. From a technological perspective, these systems are mainly constrained by limited available bandwidth, signal attenuation, interferences, and hardware limitations in terms of weight and energy efficiency. On the regulatory front, they are bound by spectrum allocation, the necessity for licenses and coordination, and the international agreements governing the use of the RF spectrum and orbital slots [3]. Terrestrial networks, for example, make it impossible to deliver telecommunications services in isolated areas, the North and South Poles, and impoverished settlements due to high building costs. Likewise, for certain scenarios, such as high-speed trains and planes, terrestrial wireless communication networks deliver poor quality of service (QoS) due to a high call drop rate caused by high mobility and frequent cell handovers.

Satellite communications have received a lot of attention for offering abundant radio frequency resources, wide coverage areas, long communication distances, rapid deployment capabilities, and minimum interference from the ground network. However, the dependence on satellite communication highlights the inherent limits of present RF technologies. As an alternative to addressing these limits, the use of optical technology has demonstrably pushed the RF boundaries to enable data transmission with high bandwidth and secure channels.

Free-space optical (FSO) communication emerges as a subset of optical communication technology that is used to address the growing demand for high-speed data transmission in space applications. FSO, with its inherent advantages of large capacity, low detection likelihood, and minimal interception risk, provides an alternative solution to standard RF communication. It may also contribute to the deployment requirements of mega-constellations composed of multiple small-to-medium-sized satellites, thereby mitigating the issues related to spectrum congestion in RF communication [4]. The global space-based laser communication market, mostly commercial applications, is expected to reach USD 5 billion by 2031, i.e., a growth of about 26% per year since 2022 [5]. FSO links can be established between fixed or dynamic point-to-point systems or in interconnections between satellites, aircraft, maritime vessels, and ground terminals (stationary and mobile).

FSO links for space communication require beam tracking and locking prior to data transmission. In inter-satellite communications, where there is no need for ground stations, FSO links with wavelength division multiplexing (WDM) and relays can be deployed [6]. Recent space missions have demonstrated in orbit the core principles of space-based optical communication, which paved the way for its real-time deployment in both academic and business sectors [7]. These missions have demonstrated a range of optical communication links that are classified based on their separation from the Earth's surface and have investigated both the dynamic and stationary conditions of the transmitters and/or receivers. For instance, in geostationary Earth orbit (GEO), the satellite is typically deployed at about 36,000 km above the Earth's equator. GEO was the preferred technology used by satellite operators to execute their commercial satellite communication systems due to the advantages of fixed position relative to the Earth, coverage, and practically constant latency [8]. However, communication systems encounter multiple issues. The first issue is a bad user experience caused by high latency and channel attenuation. Second, it is inappropriate for mobile communications at high latitudes, not least because the elevation angle (the angle between the user's satellite and the horizon) of the user equipment is so narrow that terrain shelter should not be disregarded. Also, the system's coverage and capacity are constrained by a lack of orbital and frequency resources. Finally, the costs of satellite launch and operation are relatively high. Few missions were established between the GEO to ground. A Laser Communication Experiment (LCE) aboard the Japanese Engineering Test Satellite-VI (ETS-VI) was an example of the bi-directional laser communications link between the GEO and ground station [8,9]. Even though medium Earth orbit (MEO) satellite communications have a shorter propagation delay than GEO (closer to the Earth than GEO satellites, with distances ranging from 2000 km to 20,000 km), more MEO satellites are required to continually cover the region. It also requires a higher transmit power than low Earth orbit (LEO) systems to overcome path loss and other atmospheric losses [10]. O3b/SES "other 3 billion" has launched over 15 MEO satellites to provide Ka-band communication services [11], while Laser Light Communications plans to create an all-optical worldwide MEO communications network [10,11].

Unlike their GEO and MEO predecessors, LEO satellites orbit between 500 and 2000 km above the Earth's surface [12]. This low altitude translates into a variety of advantages, making LEO a suitable candidate for the satellite communication business, which include lower latency, reduced signal path loss, and lower production and launch costs. Prior demonstrations have successfully established FSO communication links between LEO to Ground Station. Examples include the Laser Utilizing Communications Equipment (LUCE) [13], Optical Payload for Lasercomm Science (OPALS), and Tesat LCT links facilitate optical communication capabilities from entities like the International Space Station (ISS) or LEO satellites to terrestrial stations [12]. The next section will further discuss these demonstrations and what they mean for future FSO communication installations employing LEO satellites.

The potential applications of FSO communication systems extend beyond traditional point-to-point links and established orbital altitudes. As shown in Figure 1, FSO can facilitate various array of interconnected links across different communication paths, including (i) Satellite-to-Satellite Links, as it may involves connections between LEO to GEO Satellite, which is established through the Semiconductor-laser Inter-satellite Link Experiment (SILEX) and the European Data Relay System (EDRS) [14,15]. Inter-satellite Links (ISLs) between satellites positioned in various orbits, including LEO, MEO, and GEO, are demonstrated by the Optical Inter-Orbit Communications Engineering Test Satellite (OICETS) and SILEX, with the Laser Communications Relay Demonstration (LCRD) showcasing optical inter-satellite links bridging LEO and geosynchronous orbit [16,17]. In addition ground to deep space connections were demonstrated by the Lunar Laser Communication Demonstration (LLCD) and Laser Communications Relay Demonstration (LCRD), showcasing optical communication capabilities from terrestrial stations to lunar orbit and deeper space regions [7,18,19]. (ii) Air-to-Ground/Ground-to-Air Link establishes a bidirectional laser communication between aircraft and LEO satellites, enabled by the Optical Communications and Sensor Demonstration (OCSD) [20]. Similarly, an optical downlink from high-speed aircraft to ground stations has been demonstrated, displaying integrated laser terminals suitable for free-space optical communications even under jet fighter vibrations conditions [21,22]. In addition, airborne free-space experimental laser terminals, such as those used in the stratospheric Optical Payload Experiment (STROPEX) [23], show data transmission from balloons to LEO satellites. CAPANINA, a project focused on broadband technology development from High Altitude Platforms (HAPs), extends coverage to users potentially marginalized by geography and infrastructural limitations, achieving burst data rates of up to 120 Mbit/s within a 60 km coverage area [24]. Stratobus and other projects have shown broadband internet connectivity via optical lines from a HAP balloon to ground [25]. (iii) The deep-sea-based OWC link was demonstrated by Japan Agency for Marine-Earth Science and Technology (JAMSTEC) with Trimatiz Limited (Trimatiz), which demonstrates an underwater vehicle at a depth of 900 m [26]. Table 1 illustrates instances of optical space communication links, organized across a variety of platforms.

A number of survey papers have discussed in detail the potential of FSO links in space [4,6,27]. For instance, the requirements to establish positioning, navigation, and timing (PNT) systems within LEO as a part of the FSO systems were analyzed [6]. The survey covered several areas of the signal design across space, ground, and user segments. The merits and limits of various instruments and techniques were evaluated to identify viable solutions for deploying LEO-based navigation systems. Thus, addressing the limitations of PNT systems in LEO is crucial for resolving current challenges in navigation, positioning, and timing, which can be directly reflected on the improvement of urban and indoors navigation. Likewise, the recent mega-constellation systems planned and launched from various countries were addressed in [28], in which the authors predict that the use of CubeSat laser communication will expand and that the deployment of the mega-constellation systems is subject to having sufficient number of operational in-orbit satellites. These papers highlight the significant advantages of optical communication over traditional radio frequency (RF) communication, in terms of higher bandwidth and data rates [4,27].

Similar to previous works, this paper presents a comparative analysis of all FSO links deployed in space both historically, considering commercial and non-commercial missions, and in the context of future missions specifically addressing LEO-to-LEO challenges. We also produced a comparative analysis in terms of latency reliability, data rate, and power consumption. The work emphasizes the development of compact and power-efficient laser communication terminals that are suitable for small satellite platforms and CubeSats and demonstrates how the emerging trend towards mega-constellations of small, interacting satellites, a topic that has not been extensively covered in earlier surveys. Included in the discussion about increased interest and future LEO missions, we will present a brief overview of our planned mission Autonomous Laser Inter-satellite Gigabit NetWork (ALIGN) and the Free-space Optical Communications Unit for Satellites (FOCUS) payload.

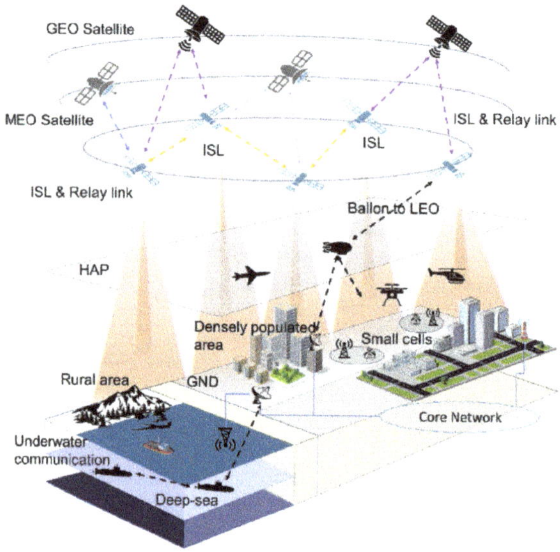

Figure 1. Architecture of satellite communication systems via different communication link types.

Table 1. An example of optical space communication across various platforms.

Platform	Project/Experiment	Maximum Link Span (km)	Maximum Data Rates
LEO to Ground Station	LUCE [13], OPALS, Tesat LCT links [12]	5000	1.8 Gbps
LEO to GEO Satellite	SILEX [14], EDRS [15]	45,000	50 Mbps to 1.8 Gbps
LEO to Aircraft	OCSD [9]	200	100 Mbps
Aircraft to Ground Station	[21]	50–79	1.25 Gbps
Balloon to LEO Satellite	STROPEX [23], CAPANINA [24]	200	2.5 Gbps
GEO to Ground Station	LCE aboard ETS-VI [8,9]	-	1 Mbps
HAP to Ground	Stratobus [2,29]	20 km above earth surface	-
Ground to Deep Space	LLCD, LCRD [7,18,19]	380,000–400,000	620 Mbps to 20 Gbps
Deep Sea	JAMSTEC [26]	0.9	1 Gbps
Inter-satellite Link	OICETS, SILEX [16], LCRD [17]	45,000	50 Mbps

The remainder of the paper is organized as follows. Section 2 describes all previous space-based laser communication missions. CubeSats and inter-satellite links are covered in Section 3. Finally, Section 4 presents the conclusions and future work.

2. Previous Space-Based Laser Communication Missions

In recent decades, a variety of space missions utilizing optical terminals have been launched for diverse purposes. For instance, the National Aeronautics and Space Administration (NASA) of the United States demonstrated a laser transmission link to the Galileo probe in outer space in 1992, as a part of Galileo Optical Experiment (GOPEX). The communication link used Earth-emitted pulses with megawatt power at a 532 nm wavelength, captured by a camera onboard the Galileo probe around 6 million kilometers away [30]. The first space-to-ground (i.e., GEO to the optical GNS) bi-directional laser communications link was demonstrated by LCE aboard the Japanese Engineering Test

Satellite-VI (ETS-VI) satellite in 1994, with a data rate of 1 Mbps, weight of 22 kg, and power consumption of 60 W [8,9].

Similarly, another milestone in creating the world's first inter-satellite laser communication link was developed, named the Advanced Relay and Technology Mission Satellite (ARTEMIS) by the European Space Agency (ESA). The communication link provided mobile services with a data transfer rate of 50 Mbps and a data relay from French Earth Observation satellite SPOT-4, positioned in LEO, through an optical channel to the geostationary ARTEMIS satellite. Then, the ARTEMIS satellite transmitted the information between the LEO and the GND via radio waves [14,31]. Likewise, the Optical Inter-orbit Communication Engineering Test Satellite (OICETS) demonstrated LEO to GND inter-orbit laser communication experiments in conjunction with ESA's ARTEMIS, known as "Kirari", and was a product of the collaborative efforts of the Japan Aerospace Exploration Agency (JAXA) [32]. It showcased an advanced on-orbit pointing, acquisition, and tracking methodologies. Structurally, Kirari was designed with a three-axis stabilization mechanism, weighing ~570 kg and having a power consumption of 1200 W, which was measured at the end of the satellite's operational life. The optical link can achieve a downlink data rate of 50 Mbps and uplink rates of 2 Mbps at a wavelength of 800 nm. The Kirari's orbital path is sun-synchronous, characterized by an inclination of 97.8° that occurs due to its altitude variations, ranging from 550 km at its operational end to 610 km at its commencement [13]. Alternatively, a HAP inter-connected with optical links was initiated by EADS Astrium (currently known as Airbus Group) in collaboration with the French Ministry of Defense's arms procurement division (DGA), in which a seamless laser communication link was established between an optical terminal on-board an aircraft flying at 9 km and the SILEX terminal on the ARTEMIS geostationary satellite. The link characteristics are dictated by the SILEX system, in terms of data rate, which is recorded as 50 and 2 Mbps for downlink and uplink channel, respectively, laser wavelength of ~800 nm and OOK (On–Off Keying) modulation [33].

The primary challenges encompassed: (i) ensuring quasi-error-free (bit error rate $< 10^{-9}$) data transmission, especially given the fast signal fading caused by laser propagation in the atmosphere; (ii) technical demands of instruments with µrad pointing requirements needed highly stable structures; (iii) optical communications introduce further complexities, particularly in optical design and detection; (iv) robust isolation between emission and reception, considering the significant power disparity of a factor of ~109 in laser power levels that range from 150 mW down to 20 to 200 pW; and (v) efficient stray light protection, especially during communication between satellites, with the Sun at angles nearing 5 degrees from the field of view, coupled with the need for exact alignment between emission and reception paths [20].

Furthermore, a coherent optical inter-satellite link in LEO was demonstrated in collaborative efforts between the German TerraSAR-X radar satellite and the U.S. Missile Defense Agency's NFIRE satellite [34]. The Laser-Communication-Terminal-on-TerraSAR-X (LCTSX) was established using a space-borne Nd:YAG laser with a wavelength of 1064 nm and two redundant 808 nm laser diode pump modules to attain higher reliability [35]. A mass of less than 30 kg and data rate of up to 5.6 Gbps was achieved over a distance of up to 80,000 km and a binary phase shift keying (BPSK) modulation scheme. The link provides sun exposure capability with a radiation tolerance greater than 80 Krad and a power consumption of less than 130 W [36]. In addition, China developed another LEO-GND communication terminal named the Haiyang-2 HY-2 satellite. The satellite was equipped with on-orbit micro-vibration measurement unit to assist in laser beam pointing prediction and correction. A data rate of 504 Mbps was achieved using a laser wavelength of 1.5 µm and an Intensity Modulation/Direct Detection (IM/DD) scheme [37,38].

In addition, the Onboard Terminal of a Laser Communication System (BTLS) on the Russian Segment demonstrated data rates of up to 1.8 Gbps between GEO-LEO and GEO-GND [39]. In addition, the USA continued its endeavours with LLCD (NASA GSFC) in 2013, which focused on Lunar-GND communication and demonstrated a 622 Mbps [40]. Table 2 summarizes the collective global effort to push the boundaries of space communication technology, each contributing unique capabilities and insights.

Table 2. Collective programs of space-based laser communication.

Ref.	Year	Mission	Link Type	Wavelength	Modulation	Data Rate
[9]	1994 1995 2000 2001	ETS-VI (NICT, Japan) GOLD (NASA JPL, USA) STRV-2 (BMDO, USA)-fail GOLITE (NRO, USA)	GEO-GND GEO-GND LEO-GND GEO-GND	0.8/0.5 μm 0.8/0.5 μm 0.8 -	IMDD IMDD IMDD -	1 Mbps 1 Mbps 1.2 Gbps -
[14]	2001	SILEX (ESA)	GEO-LEO, GEO-GND, GEO-Air	0.8 μm	IMDD	50 Mbps
[13]	2005	OICETS (JAXA/NICT)	GEO-LEO, LEO-GND	0.8 μm	IMDD	50 Mbps
[36]	2003	NFIRE (MDA, USA)	LEO-LEO	1.06 μm	BPSK	5.6 Gbps
[34,41]	2008	TerraSAR-X (DLR)	LEO-LEO, LEO-GND	1.06 μm	BPSK	5.6 Gbps
[37,38]	2011	HY-2 (China)	LEO-GND	1.5 μm	IMDD	504 Mbps
[39]	2011	BTLS (Russia)	LEO-GND	1.55/0.85 μm	IMDD	125 Mbps
[40]	2013	LLCD (NASA GSFC)	Lunar-GND	1.5 μm	PPM	622 Mbps
[42]	2013	EDRS/Copernics (ESA)	GEO-LEO, GEO-GND	1.06 μm	BPSK	1.8 Gbps
[43,44]	2014	SOCRATES/SOTA (NICT)	LEO-GND	0.98/1.5 μm	IMDD	10 Mbps
[45]	2014	OPALS (NASA JPL)	LEO-GND	1.5 μm	IMDD	30–50 Mbps
[46]	2015	OCSD-A (Aero. Corp., USA)-fail	LEO-GND	1.5 μm	IMDD	5–50 Mbps
[47]	2016	OSIRISv1-2 (DLR)	LEO-GND	1.5 μm	IMDD	20–100 Mbps
	2016	CAS SIOM (China)	LEO-GND	1.5 μm	DPSK	5.12 Gbps
[48]	2017	Micius (China)	LEO-GND	0.85/0.532/0.671 μm	-	-
	2018	OCSD-B/ AeroCube-7B (Aero. Corp.)	LEO-GND	1.064 μm	IMDD	50/100 Mbps
[49]	2019	Starlink/Space-X (USA)	-	-	-	1 Gbps
[50]	2019	RISESAT/VSOTA (NICT)	LEO-GND	0.98/1.5 μm	IMDD	~1 kbps
[51]	2019	SOLISS (Sony)	LEO-GND	1.5 μm	IMDD	100 Mbps
[52]	2019	EDRS-C (ESA)	GEO-LEO	1.06 μm	BPSK	~1.8 Gbps
[53]	2019	OPS-SAT (TU Graz)	LEO-GND	-	PPM	2 kbps (uplink)
[13]	2020	JDRS (JAXA)	GEO-GND	1.5 μm	DPSK	1.8 Gbps
[54]	2020	ALOS-3 (JAXA)	GEO-LEO	1.5 μm	DPSK	1.8 Gbps
[47]	2020	OSIRIS v3, v4 (DLR)	LEO-GND	1.5 μm	IMDD	10 Gbps
[55]	2021	ALOS-4 (JAXA)	GEO-LEO	1.5 μm	DPSK	1.8 Gbps
[56]	2021	LCRD (NASA GSFC, USA)	GEO-LEO, GEO-GND	1.5 μm	DPSK/PPM	2.8 Gbps/622 Mbps
[57]	2021	DSOC (NASA JPL, USA)	Deep space-GND	-	PPM	264 Mbps
[58]	2022	ETS-9/HICALI (NICT)	-	1.5 μm	DPSK	10 Gbps
	2022	LEMNOS (NASA GSFC)	Lunar-GND	-	PPM	311 Mbps

3. CubeSats and Inter-Satellite Links (ISLs)

While ISLs play a crucial role in satellite communication, enabling direct and efficient data exchange in space, their deployment in LEO constellations is less common compared to LEO-GND or GEO-LEO links, as shown in Figure 2 and Table 2. This scarcity is primarily due to the technical and operational challenges associated with implementing ISLs in LEO satellites, which are often smaller and have limited resources compared to satellites in other orbital regimes [59].

The advent of the miniature satellites with a standardized dimension, known as Cubesat's, have paved the way for the space-based laser communication system (lasercom) to be deployed for ISLs [60]. Compared to their larger, traditionally designed counterparts, Cubesat's offer significant advantages in cost, development time, and launch opportunities. Their compact size and modular design enable faster development cycles and lower production costs, making them particularly suitable for technology demonstration, scientific discovery, and educational purposes. Additionally, their compatibility with secondary launch opportunities allows for more frequent access to space, further accelerating the pace of exploration and innovation. However, CubeSat's are currently limited by their size and weight constraints, impacting their power generation, processing capabilities, and payload capacity. Thus, this has imposed a strict requirement in terms of the size, weight, and power (SWaP) when it comes to design of laser communication systems for space.

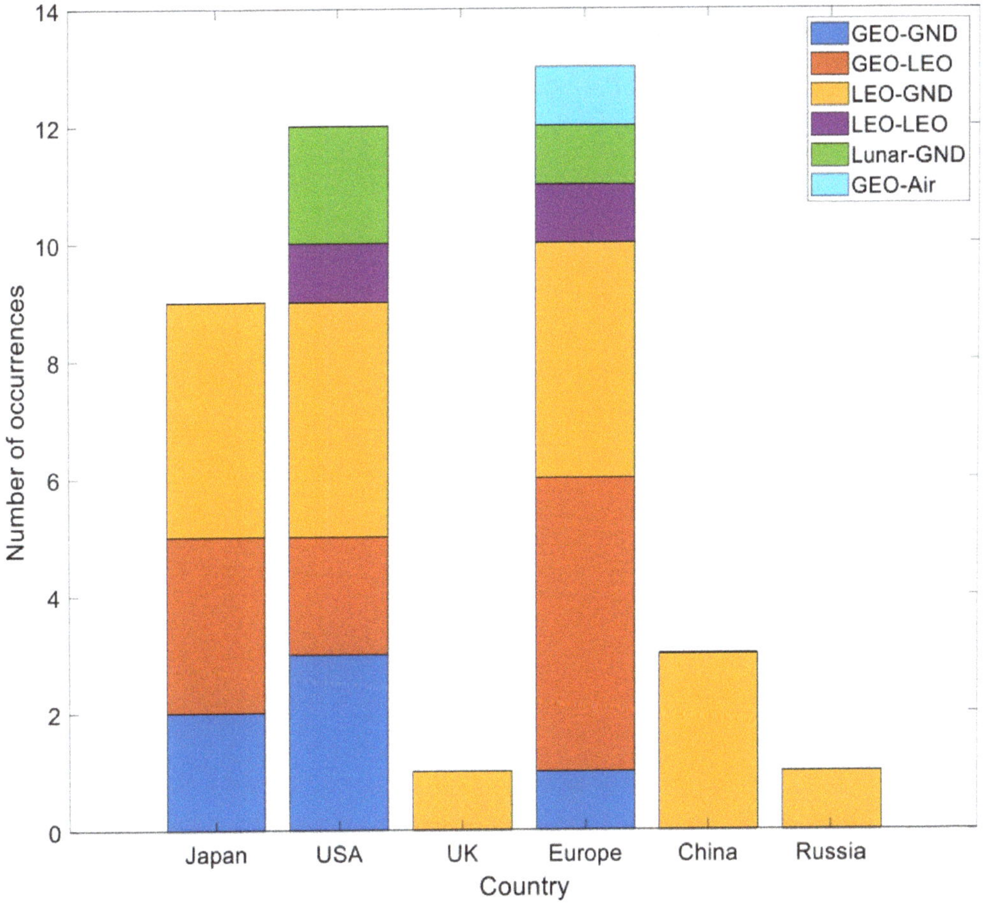

Figure 2. Distribution of link types in satellite communication systems by country.

3.1. Lasercom Terminals Payload Capacity

The analysis of commercially available and custom-designed laser communication (lasercom) systems suitable for CubeSats reveals a diverse landscape of technological capabilities, catering to the increasing demand for global satellite constellations. This paper surveyed leading corporations and research institutions involved in the development and deployment of CubeSat-compatible lasercom terminals, presenting a comprehensive overview of their characteristics and parameters.

In Table 3, a variety of commercial lasercom terminals from notable entities, such as TESAT, Mynaric, Thales Alenia Space, and Hyperion Technology, among others, are listed. These terminals exhibit a wide range of features including data rates, link types, size, power consumption, and release times. For instance, TESAT's TOSIRIS offers data rates ranging from 1.25 to 10 Gbps with a compact volume of 28.0 × 20.0 × 15.0 cm^3 and a power consumption of 40 W. Similarly, Mynaric's CONDOR Mk2 and Mk3 support data rates from 0.1 to 1.25 Gbps and 0.1 to 10 Gbps, respectively, showcasing the flexibility in bandwidth options.

Table 3. Commercial lasercoms terminals suitable for CubeSat or MicroSats.

Corporation	Name	Data Rate (D:Down; U:Up)	Link Type	Volume (cm^3)	Power Consumption (W)	Mass (kg)	Release (Year)
TESAT [12]	TOSIRIS	D: 1.25–10 Gbps, U: 1 Mbps	LEO-GND	20.0 × 20.0 × 15.0	40	8	2019
	SmartLCT	1.8 Gbps	GEO-GEO, GEO-LEO	<35.0 × 35.0 × 20.0	150	30	2020
	CubeLCT	D: 100 Mbps, U: 1 Mbps	LEO-GND	9.0 × 9.5 × 3.5	10	0.397	2021
	SCOT80	10 Gbps	LEO-LEO	2U	60–86	15	2023
	ConLCT	10 Gbps	GEO-LEO	OU:50.0 × 18.0 × 26.0, EU:26.0 × 11.0 × 17.5	80	15	2021
	LCT-135	1.8 Gbps	GEO-LEO	60.0 × 60.0 × 70.0	150	53	-
Mynaric	CONDOR Mk2	0.1–1.25 Gbps	GEO-LEO	OU:57.3 × 27.1 × 23.0, EU:34.0 × 25.9 × 16.3	-	-	2014
	CONDOR Mk3	0.1–2.5 Gbps	GEO-LEO	OU:35.1 × 21.1 × 17.0, EU:16.1 × 33.6× 25.5	-	-	2017
Thales Alenia Space	OPTEL-μ	2 Gbps	LEO-GND	8 U	43	8	2015
MOSTCOM	SOT-90	10 Gbps	GEO-LEO	45.0 × 30.0 × 38.0	60	16	2020
	SOT-150	1.25 Gbps	GEO-GND	60.0 × 40.0 × 48.0	100	50	2020
NICT [61]	SOTA [62]	1 Mbps–10 Mbps	LEO-GND	11.7 × 17.7 × 27.8	40	5.9	2014
	SOLISS [63]	80 Mbps	LEO-GND	All:29.5 × 35.6 × 43.6, OU:9.0 × 10.0 × 18.0	36	1.2	2019
	VSOTA	1 kbps–1 Mbps	LEO-GND	-	4.33	<1	2018
	HICALI	10 Gbps	GEO-GND	-	340	6	2021
	CubeSOTA [64]	10 Gbps	GEO-LEO	3 U	-	<14	2024
Hyperion Tech. & TNO [65]	CubeCat	D:100 M/300 M/1 G-bps, U:200 kbps	LEO-GND	1 U	15	1.33	2022
Aerospace Corporation [46]	AeroCube-7a	-	LEO-GEO	-	-	-	2015
	AeroCube-7b/7c	100 Mbps	-	1.5 U	-	3	2018
Astrogate Labs	ASTRO-LINK	1 Gbps	GEO-GND, LEO-GND	1 U	-	3	2022
HENSOLDT	Lasercom	U: 1.8 Gbps	Air-GEO	-	-	20–100	2015
CACI	CrossBeam-ST0	1.25 Gbps	LEO-LEO	-	50	10	2021
	CrossBeam-ST1	2.5 Gbps	LEO-LEO	-	75	10	2021
SONY [63]	SOLISS	80 Mbps	-	-	-	1.2	2019

Moreover, Table 4 presents customized (R&D) lasercom systems developed for CubeSats by research and development institutions like TESAT, NEC/JAXA, NASA/MIT/University of Florida, and DLR/University of Stuttgart. These systems offer unique transmission rates and link types tailored to specific mission requirements. For instance, TESAT's EDRS-A LCT provides a transmission rate of 1.8 Gbps for GEO-GND links, whereas NASA/MIT/University of Florida's CLICK-A focuses on LEO-LEO communication at a modest 10 Mbps.

The significant progress and innovation in CubeSat laser communication technology has offered a spectrum of solutions, encompassing varying data rates, link types, and spectral wavelengths. These advancements not only meet the increasing demand for global satellite constellations but also drive the miniaturization and efficiency enhancement of CubeSat payloads, ushering in a new era of space-based communication and exploration. However, despite these technological advancements, significant challenges remain. The earlier generation of terminals accentuates the pressing need for continued efforts towards miniaturization and decreased power consumption. Furthermore, factors including reliability, security, and the impact of atmospheric conditions necessitate careful consideration to ensure the dependable operation of laser communication systems within LEO constellations.

Table 4. R&D customized lasercom systems developed for CubeSats.

Corporation	Name	Data Rate (D:Down; U:Up)	Link Type	Volume (cm^3)	Power Consumption (W)	Mass (kg)	Release (Year)
NEC/JAXA	LUCAS	1.8 Gbps	LEO-GEO, LEO-LEO	-	-	-	2020
NASA, MIT, Uni. of Florida [67]	CLICK-A	10 Mbps	LEO-LEO	1.2 U	-	1.2	2018
	TBIRD [66]	D:200 Gbps, U:5 kbps	LEO-GND	1.8 U	-	2.25	2022
DLR Uni. of Stuttgart	OSIRISv1	200 Mbps	LEO-GND	-	26	1.3	2017
	OSIRISv2 [68]	1 Gbps	LEO-GND	-	37	1.65	2016
	OSIRISv3	10 Gbps	LEO-GND	-	150	9	2021
	OSIRIS 4 PIXL-1 [69]	100 Mbps	LEO-GND	0.3 U	10	0.39	2021
ESA (ARTES) Spire	OCT	-	LEO-LEO	-	-	-	2023

In conclusion, the evaluation of laser communication terminals presents a promising trajectory for their integration within LEO constellations. The escalating data rates, optimization of SWaP, diverse link options, and the advancing maturity of technology collectively position laser communication as a viable solution for future high-performance satellite networks.

3.2. LEO Exploration

Historically, LEO constellations have not utilized frequent LEO-to-LEO laser communication links due to the combined effect of several factors. LEO satellites, with their lower altitudes and higher orbital velocities (7–11 km/s) compared to MEO or GEO satellites, experience shorter communication windows and require more frequent handovers between satellites to maintain a connection. This handover process introduces complexity and increases the risk of disruption [70]. Additionally, atmospheric drag necessitates frequent orbital corrections for LEO satellites, further complicating link maintenance [71]. Finally, achieving comprehensive global coverage with LEO constellations demands a large number of satellites, and the challenge of maintaining continuous links between these rapidly moving spacecraft has historically favored communication link implementation in MEO and GEO constellations with fewer and slower-moving satellites [72].

3.3. Future Constellations

In Table 5, a list of known planned mega-constellations of small satellites that indicate an ambition to incorporate optical communication links in LEO (i.e., catering for LEO-LEO or LEO-ground or both), are presented (NB This list may not be complete and subject to modification). The emergence of low-cost, fast turnaround launch capability to LEO is heralding a significant growth in the development of optical communication terminals for small satellites such as CubeSats and nanosats. The mass, power, and volume envelope restrictions now force manufacturers of optical communications terminals to adapt their payloads, applying novel approaches to optical design, while maintaining competitive data rates that greatly outstrip RF communications systems.

Table 5. (Known) planned future constellations of small satellites in mesh networks.

Company	Origin	Constellation Type	Objective	Achievements
Amazon (Kuiper Systems)	USA	LEO	Provides broadband internet services with low latency	Plans to launch 3236 LEO satellites as part of the Kuiper Systems project. Intends to employ laser communication technology for enhanced connectivity.
Facebook (Athena)	USA	E-Band (71–86 GHz)	Conducts a satellite communication experiment focused on high-speed data.	Filed an application for E-band frequency usage with the Federal Communications Commission (FCC) and aims to deliver data 10 times faster than Space-X's Starlink internet satellites using laser communications.
Laser Light Communications	USA	MEO	Create sa 12-MEO satellite constellation with high capacity and data rates.	Collaborated with Australian telescommunications company Optus to prepare satellites and ground stations. Aims to establish a global network connecting terrestrial fiber systems with satellite systems.
ILLUMA-T	USA	LEO	Laser Communication Terminals for use in the ISS and Orion manned space vehicle utilizing the LCRD data relay with a data rate of 1.2 Gbps	-
BridgeCom (formerly BridgeSat)	USA	LEO	Create a 10-Gbps LEO constellation with laser communication services.	Plans to distribute 50 optical ground stations with software-defined modems worldwide to provide laser communication services, including support for small satellites and unmanned aerial vehicles. Also invested in governmental and 5G network communication needs.
Kaskilo (KLEO Connect GmbH)	Germany	LEO	Build a 288-satellite LEO constellation primarily for IoT services under Germany's Industry 4.0.	Will utilize laser communication for inter-satellite links to support IoT connectivity.
Huawei (Massive VLEO)	China	LEO	Construct a 10,000-satellite LEO constellation (Massive VLEO) for beyond-5G.	Plans to use low satellite altitudes of 300 km for ultra-reliable, low-latency communications, covering massive machine-type communications. Broadband communication will be achieved using terahertz and laser communications.

Table 5. *Cont.*

Company	Origin	Constellation Type	Objective	Achievements
Transcelestial Technologies	China	Various	Provide space laser communication.	Planning to offer space laser communication services using CubeSats and micro-satellites.
Golbriak Space	Estonia	Various	Provide space laser communication.	Intending to provide space laser communication services with CubeSats and micro-satellites.
Starlink/Space-X [49]	USA	LEO	-	-
Analytical Space Fast Pixel Network	USA	LEO	Six CubeSats, a constellation of 36 satellites across three orbital planes	Hybrid RF-laser to relay data for satellite operators and customers based on the technologies capability for data increases through RF.
CONDOR Mk3 Optical Communication Terminal	Germany	-	Optical Communication Terminal is capable of high-performance, high-bandwidth, and secure and reliable satellite communications.	It supports link distances greater than 10,000 km with a flexible data rate coverage of up to 100 Gbps, fast acquisition time, configurable Laser Ethernet Terminals (LETs), and seamless link configuration and interoperability across various optical communication terminals. It has a highly modularized design, and an option for Dual or Quad configuration for reduced power consumption and reduced mass.
Telesat's Lightspeed	USA	LEO	+300 satellites	

3.4. Future OISL Payloads

In Table 6, a list of known planned payloads for small satellites that indicate an ambition to provide optical communication links in LEO, are presented (NB: this list may not be complete and subject to modification). The next generation of optical communication terminals that will emerge over the next few years will push the limits of optical communications technology, not only to lower mass, power, and volume requirements, but also to more adaptable designs that allow for scalability at cost and can address the most challenging problem for LEO-to-LEO communications, which is PAT (Pointing, Acquisition, and Tracking). Table 6 is ordered in terms of mass from high to low. As can be shown, one of the most competitive products to emerge over the next few years is the FOCUS (Free Space Optical Communications Unit for Satellites) terminal being built by Northumbria University (UK) and partners and funded by the UK Space Agency. The FOCUS terminal is the most competitive because it will adequately address the growing requirement for lower mass, lower power consumption, and smaller volume envelope, while still achieving greater than 1 Gbps data transfer rates for LEO-to-LEO communications. It is only one of two (together with the Spire OCT) that is UK sourced that will address the LEO-to-LEO challenges that are agency funded (rather than commercially funded).

Table 6. Planned future payloads that provide laser communications capability in LEO, ordered by mass (kg) from high to low.

Corporation	Payload Name	Max. Data Rate (Gbps)	Link Type	Volume (cm^3)	Power Consumption (W)	Mass (kg)	Release (O:Operational, D:Development)
TESAT	LCT 135	1.8	GEO	252,000	~150	~53	O
Mynaric	CONDOR Mk3	2.5	LEO	32,422.5	-	-	O
TESAT	SCOT-80	10	LEO	54,109	60–86	~12.5	O
OneWeb [73]	Gen 2 Terminal	20	LEO	3750	70	25	D
UKSA, Northumbria Uni.	FOCUS	>1	LEO	3 U	40	3.6	D
NASA, MIT, Uni. of Florida	CLICK B/C	>0.02	LEO	1.3 U	-	-	D
TESAT [12]	SCOT20	2.5, 0.1	LEO	1135 (1U)	<35	<1.6	D

3.5. *Autonomous Laser Inter-Satellite Gigabit Network (ALIGN) and the FOCUS (Free-Space Optical Communications Unit for Satellites) Solution for LEO-to-LEO Platforms*

The ALIGN mission is designed to carry out an in-orbit demonstration of a novel laser communication system in LEO using two 6U CubeSats. We refer to the lasercom system herein as the Free-space Optical Communications Unit for Satellites (FOCUS) as the primary payload for each CubeSat. The aim of the mission is to establish Gbps full-duplex communication links autonomously, with a ranging capability of up to 1000 km and with SWaP requirements that are suitable for small- to nano-satellites in data-intensive constellations in LEO. The novel aspect of the FOCUS design includes (i) incorporation of a PAT approach that adheres to the Space Development Agency (SDA) standards; (ii) a powerful System-on-chip (SoC) communication device control board; and (iii) compact telescope integration. The proposed system offers efficient scalability, allowing for potential applications involving LEO-LEO, LEO-GEO, and LEO-ground resilient communications capable of networking mega-constellations.

Figure 3 represents the CAD design of FOCUS payload. The payload contains three main sections: (i) the top section contains the transmitter and receiver telescopes supported by a chassis (red structure); (ii) the middle section contains the amplifiers of the laser and the controller for the fast-steering mirror, held together by a cradle (grey structure); and (iii) the bottom section is where the Optical Interface Assembly (OIA) PCB stack is located. The primary function of this section is to manage the payload electronics, which include important components such as power management modules, data processing boards, and control electronics. The stack also controls the telescope actuation in addition to driving the high-speed communications. The actuators provides line-of-sight stabilization, micro-scanning, tracking, and fine pointing, essential for the PAT process.

3.5.1. Key Operational Features of ALIGN

Figure 4 depicts the concept of operation for the link acquisition process of ALIGN in LEO. The PAT process is highly challenging when considering two moving satellites with the relatively large uncertainties in real time, which progressively worsen due to minor variations in atmospheric drag. The uncertainties are represented by yellow circles along the orbital path (black dash line) with respect to transmitter Tx and receiver Rx satellite. As shown in Figure 4, the transmitter satellite must spiral the laser beam to search for the receiver and at the same time the receiver must be correctly orientated such that the transmitter is within the field of view (FoV) of the receiver's detector (green and blue cones).

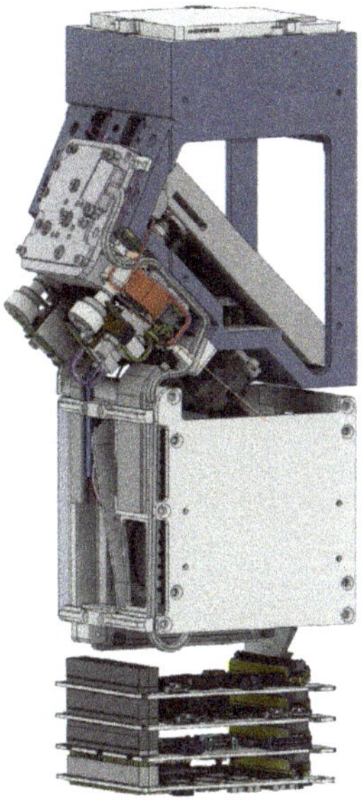

Figure 3. Current design of FOCUS.

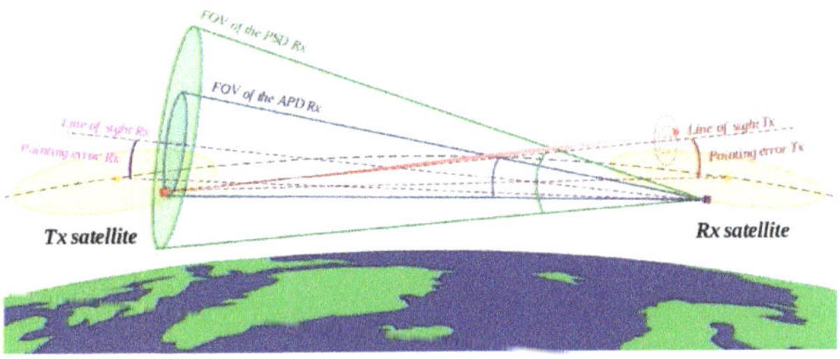

Figure 4. Mission concept in space, showing positional uncertainties (yellow circles) along the dashed path, with Tx and Rx alignment in green and blue cones.

The PAT system of ALIGN adopt spiral scans to ensure optimal coverage of search areas, which reduces the link acquisition time. Attitude Determination and Control System (ADCS) errors, which can cause pointing inaccuracies due to satellite movement, are mitigated by the system's adaptability, improving PAT precision. This adaptability is crucial for maintaining stable links between fast-moving satellites, even in environments with fluctuating signal-to-noise ratios or external interferences.

3.5.2. Link Budget

Table 7 presents a performance comparison between the proposed FOCUS link budget and other laser communication systems used on application scenarios [13,74]. The results indicate that FOCUS is comparable in terms of size, mass, data rates, pointing accuracy, and optical power requirements. Also, the FOCUS link-budget analysis confirms that by adjusting two fundamental parameters, i.e., the beam divergence and the transmit power, the same terminal design could also be adapted to serve a variety of communication scenarios, as also demonstrated by the different models of [13]. The market trend for next-generation laser communication terminals shows a shift toward smaller with extended link acquisition capabilities. However, the key factor for any commercially viable product will be its adaptability to various communication scenarios. The primary distinction between FOCUS and the other models discussed is that FOCUS is specifically designed for inter-satellite communication, which involves a significantly more complex PAT process. FOCUS's ability to autonomously establish a link between two fast-moving satellites is a groundbreaking technological advancement, critical for large satellite constellations that support high-speed global communication networks.

Table 7. Link budgets of the application scenarios.

Parameter	HAPS → OGS	LEO → GEO	LEO → OGS	Drone → OGS	LEO → LEO
		Referenced in [13]			ALIGN
Tx Power (W)	3	2	2	0.1	0.75
Transm. aperture (cm)	9	9	9	3	1.5
Pointing accuracy (µrad)	5	5	5	5	5
Beam divergence (µrad)	200	30	30	500	150
Distance (km)	100	37,000	1200	10	900
Footprint (m)	20	1110	36	5	112.5
Atmospheric attenuation (dB)	6	-	6	6	-
Pointing loss (dB)	1	2.2	2.2	1	<1
Tx loss (dB)	3	3	3	3	1.690
Rx aperture (cm)	9	15	30	9	9.57
Rx loss (dB)	10	6	12	10	1.690
Rx power (dBm)	−32	−55.6	−31.8	−34.9	−35.895
Bit rate (Gbps)	10	0.1	10	10	1.25
Required power (dBm)	−38.9	−58.9	−38.9	−38.9	−35.895
Margin (dB)	6.8	3.3	7.1	4	-

4. Conclusions

Space-based FSO systems have emerged as a viable solution for high-speed, long-distance, and secure communication compared to traditional RF communication systems used on satellites. While longer-distance links such as GEO-GND and GEO-to-LEO have been preferred in previous missions due to their less demanding pointing requirements, the deployment of FSO-based LEO-to-LEO communication systems is gaining traction. These systems offer reduced signal latency thanks to their proximity to Earth compared to GEO satellites. Hence, they provide high-speed, secure, and reliable data transmission, addressing the limitations of current communication technologies and meeting the future demands of global connectivity and space exploration.

This paper summarized the global effort in terms of current and future missions and payloads associated with FSO links deployed in space from commercial and non-commercial perspectives. In addition, a comparative analysis in terms of latency reliability, data rate, and power consumption was introduced. The work emphasized on the development of compact and power-efficient laser communication terminals that are suitable for small satellite platforms and CubeSats and demonstrates how the emerging trend towards mega-constellations of small, interacting satellites, a topic that has not been extensively covered in earlier surveys.

As part of the discussion on the growing interest in future LEO missions, we provided a brief overview of our ALIGN planned mission, along with a brief overview on the main key operational features in terms of PAT and link budget analysis.

Author Contributions: Conceptualization, O.I.Y. and E.S.; literature review, O.I.Y., A.R., E.S. and R.W.; resources, E.S.; data creation, O.I.Y., E.S. and R.W.; writing—original draft preparation, O.I.Y., A.R., E.S., R.B. and J.V.; writing—review and editing, all; visualization, O.I.Y.; supervision, E.S., R.W. and R.B.; project administration: E.S., R.W., R.B. and C.B.; funding acquisition, E.S., R.W., R.B. and C.B. All authors have read and agreed to the published version of the manuscript.

Funding: This research was funded by the UK Space Agency, grant number UKSAG22_0042.

Data Availability Statement: Not applicable.

Acknowledgments: The authors would like to thank e2E Services Ltd./Telespazio-UK and SMS Electronics Ltd., partners in the NSIP Phase 3 Project, "ALIGN Laser Optical Communications for CubeSats" for their support.

Conflicts of Interest: The authors declare no conflicts of interest.

Abbreviations

The following abbreviations are used in this manuscript:

FSO	Free-Space Optical
GEO	Geostationary orbit
GOPEX	Galileo Optical Experiment
HAP	High Altitude Platform
IMDD	Intensity Modulation with Direct Detection
JAXA	Japan Aerospace Exploration Agency
LCE	Laser Communication Equipment
LEO	Low Earth Orbit
LUCE	Laser Utilizing Communication Equipment
MEO	Medium Earth Orbit
NASA	National Aeronautics and Space Administration
NFIRE	U.S. Missile Defense Agency satellite
OCSD	Optical Communications and Sensor Demonstration
OGS	Optical Ground Station
OICETS	Optical Inter-orbit Communication Engineering Test Satellite
OOK	On-Off Keying
OPALS	Optical Payload for Lasercomm Science
OIA	optical interface assembly
PAT	Pointing Acquisition and Tracking
PCB	Printed Circuit Board
PSD	Position-sensitive detector
SDA	Space Development Agency
SILEX	Semiconductor-laser Intersatellite Link Experiment
SoC	System-on-chip
SOTA	Small Optical Transponder
SSM	Slow Steering Mirror
STROPEX	Stratospheric Optical Payload Experiment
SWaP	Size, Weight, and Power

References

1. DataReportal, We Are Social, M. Worldwide Internet User Penetration from 2014 to July 2024. Available online: https://www.statista.com/statistics/325706/global-internet-user-penetration/ (accessed on 3 October 2024).
2. Son, I.K.; Mao, S. A survey of free space optical networks. *Digit. Commun. Netw.* **2017**, *3*, 67–77. [CrossRef]
3. Rose, T.S.; Rowen, D.W.; LaLumondiere, S.D.; Werner, N.I.; Linares, R.; Faler, A.C.; Wicker, J.M.; Coffman, C.M.; Maul, G.A.; Chien, D.H.; et al. Optical communications downlink from a low-earth orbiting 1.5U CubeSat. *Opt. Express* **2019**, *27*, 24382–24392. [CrossRef] [PubMed]
4. Luo, X.; Chen, H.H.; Guo, Q. LEO/VLEO Satellite Communications in 6G and Beyond Networks—Technologies, Applications and Challenges. *IEEE Netw.* **2024**, *38*, 273–285. [CrossRef]
5. Straits, R. Space-Based Laser Communication Market Growth, Trends and Forecast to 2031. 2024. Available online: https://www.linkedin.com/pulse/space-based-laser-communication-market-size-trends-h8y7f (accessed on 25 March 2024).

6. Khalighi, M.A.; Uysal, M. Survey on Free Space Optical Communication: A Communication Theory Perspective. *IEEE Commun. Surv. Tutor.* **2014**, *16*, 2231–2258. [CrossRef]
7. Boroson, D.M.; Robinson, B.S.; Murphy, D.V.; Burianek, D.A.; Khatri, F.; Kovalik, J.M.; Sodnik, Z.; Cornwell, D.M. Overview and results of the Lunar Laser Communication Demonstration. In Proceedings of the SPIE 8971, Free-Space Laser Communication and Atmospheric Propagation XXVI, 89710S, San Francisco, CA, USA, 2–3 February 2014. [CrossRef]
8. Munemasa, Y.; Saito, Y.; Carrasco-Casado, A.; Trinh, P.V.; Takenaka, H.; Kubo-oka, T.; Shiratama, K.; Toyoshima, M. Feasibility study of a scalable laser communication terminal in NICT for next-generation space networks. In *International Conference on Space Optics—ICSO 2018; SPIE: Bellingham, WA, USA, 2019;* p. 111805W. [CrossRef]
9. Toyoshima, M.; Carrasco-Casado, A. NICT's optical communication projects and ground station development. In Proceedings of the KISS Workshop: Optical Communication on SmallSats, Online, 11–14 July 2016.
10. Lyras, N.K.; Efrem, C.N.; Kourogiorgas, C.I.; Panagopoulos, A.D.; Arapoglou, P.D. Optimizing the Ground Network of Optical MEO Satellite Communication Systems. *IEEE Syst. J.* **2020**, *14*, 3968–3976. [CrossRef]
11. Luini, L.; Riva, C.; Capelletti, F.; Comisso, A.; Rocha, A.; Mota, S.; Brás, M.; Biscarini, M.; Barbieri, S.; Consalvi, F.; et al. Ka-Band Rain Attenuation Derived from a MEO Satellite Constellation. In Proceedings of the 2024 18th European Conference on Antennas and Propagation (EuCAP), Glasgow, UK, 17–22 March 2024; IEEE: Piscataway, NJ, USA, 2024; pp. 1–5.
12. Products | TESAT. Available online: https://www.tesat.de/products (accessed on 26 September 2023).
13. Carrasco-Casado, A.; Shiratama, K.; Trinh, P.V.; Kolev, D.; Munemasa, Y.; Fuse, T.; Tsuji, H.; Toyoshima, M. Development of a miniaturized laser-communication terminal for small satellites. *Acta Astronaut.* **2022**, *197*, 1–5. [CrossRef]
14. Sodnik, Z.; Furch, B.; Lutz, H. Optical Intersatellite Communication. *IEEE J. Sel. Top. Quantum Electron.* **2010**, *16*, 1051–1057. [CrossRef]
15. ESA—Kernel (20). Available online: https://www.esa.int/Enabling_Support/Space_Engineering_Technology/Shining_light_on_the_future_of_space_communications (accessed on 26 September 2023).
16. Jono, T.; Takayama, Y.; Ohinata, K.; Kura, N.; Koyama, Y.; Arai, K.; Shiratama, K.; Sodnik, Z.; Bird, A.; Demelenne, B. Demonstrations of ARTEMIS-OICETS inter-satellite laser communications. In Proceedings of the 24th AIAA International Communications Satellite Systems Conference, San Diego, CA, USA, 11–14 June 2006; p. 5461.
17. Mohon, L. Laser Communications Relay Demonstration (LCRD). 2023. Available online: http://www.nasa.gov/mission_pages/tdm/lcrd/index.html (accessed on 26 September 2023).
18. Edwards, B. Overview of the laser communications relay demonstration project, 2012. *SpaceOps* **2012**, *2012*, 1261897.
19. Benna, M.; Mahaffy, P.; Halekas, J.; Elphic, R.; Delory, G. Variability of helium, neon, and argon in the lunar exosphere as observed by the LADEE NMS instrument. *Geophys. Res. Lett.* **2015**, *42*, 3723–3729. [CrossRef]
20. Cazaubiel, V.; Planche, G.; Chorvalli, V.; Hors, L.L.; Roy, B.; Giraud, E.; Vaillon, L.; Carré, F.; Decourbey, E. LOLA: A 40000 km optical link between an aircraft and a geostationary satellite. In Proceedings of the Sixth International Conference on Space Optics, Proceedings of ESA/CNES ICSO 2006, Noordwijk, The Netherlands, 27–30 June 2006.
21. Moll, F.; Horwath, J.; Shrestha, A.; Brechtelsbauer, M.; Fuchs, C.; Navajas, L.A.M.; Souto, A.M.L.; Gonzalez, D.D. Demonstration of high-rate laser communications from a fast airborne platform. *IEEE J. Sel. Areas Commun.* **2015**, *33*, 1985–1995.
22. Horwath, J.; Fuchs, C. Aircraft to ground unidirectional laser-communications terminal for high-resolution sensors. In *Free-Space Laser Communication Technologies XXI; SPIE: Bellingham, WA, USA,* 2009; Volume 719909. [CrossRef]
23. Horwath, J.; Knapek, M.; Epple, B.; Brechtelsbauer, M.; Wilkerson, B. Broadband backhaul communication for stratospheric platforms: The stratospheric optical payload experiment (STROPEX). In *Free-Space Laser Communications VI*; SPIE: Pune, India, 2006; pp. 436–447.
24. Grace, D.; Oodo, M.; Mitchell, P. An overview of the capanina project and its proposed radio regulatory strategy for aerial platforms. CAPANINA Consortium. 2005. Available online: https://capanina.com/ (accessed on 25 March 2024).
25. Karapantazis, S.; Pavlidou, F. Broadband communications via high-altitude platforms: A survey. *IEEE Commun. Surv. Tutor.* **2005**, *7*, 2–31. [CrossRef]
26. Ishibashi, S.; Susuki, K.I. 1Gbps x 100m Underwater Optical Wireless Communication Using Laser Module in Deep Sea. In *OCEANS 2022, Hampton Roads;* IEEE: Piscataway, NJ, USA, 2022; pp. 1–7.
27. Liu, S.; Gao, Z.; Wu, Y.; Kwan Ng, D.W.; Gao, X.; Wong, K.K.; Chatzinotas, S.; Ottersten, B. LEO Satellite Constellations for 5G and Beyond: How Will They Reshape Vertical Domains? *IEEE Commun. Mag.* **2021**, *59*, 30–36. [CrossRef]
28. Toyoshima, M. Recent trends in space laser communications for small satellites and constellations. *J. Light. Technol.* **2021**, *39*, 693–699. [CrossRef]
29. Giggenbach, D.; Epple, B.; Horwath, J.; Moll, F., Optical Satellite Downlinks to Optical Ground Stations and High-Altitude Platforms. In *Advances in Mobile and Wireless Communications: Views of the 16th IST Mobile and Wireless Communication Summit;* István, F., Bitó, J., Bakki, P., Eds.; Springer: Berlin/Heidelberg, Germany, 2008; pp. 331–349. [CrossRef]
30. Wilson, K.E.; Lesh, J.R.; Yan, T.Y. GOPEX: A laser uplink to the Galileo spacecraft on its way to Jupiter. In *Free-Space Laser Communication Technologies V; SPIE: Bellingham, WA, USA,* 1993; pp. 138–146. [CrossRef]
31. Perlot, N.; Knapek, M.; Giggenbach, D.; Horwath, J.; Brechtelsbauer, M.; Takayama, Y.; Jono, T. Results of the optical downlink experiment KIODO from OICETS satellite to optical ground station Oberpfaffenhofen (OGS-OP). *Proc. SPIE* **2007**, *6457*, 1. [CrossRef]

32. Fujiwara, Y.; Mokuno, M.; Jono, T.; Yamawaki, T.; Arai, K.; Toyoshima, M.; Kunimori, H.; Sodnik, Z.; Bird, A.; Demelenne, B. Optical inter-orbit communications engineering test satellite (OICETS). *Acta Astronaut.* **2007**, *61*, 163–175. [CrossRef]
33. Vaillon, L.; Planche, G.; Chorvalli, V.; Hors, L.L. Optical communications between an aircraft and a geo relay satellite: Design and flight results of the LOLA demonstrator. In *International Conference on Space Optics—ICSO 2008; SPIE: Bellingham, WA, USA*, 2017; Volume 1056619. [CrossRef]
34. Backes, N. TESAT Celebrates 10 Years of Laser Communications in Space. Available online: https://www.tesat.de/news/press/623-pi1286-tesat-celebrates-10-years-of-laser-communications-in-space (accessed on 25 March 2024).
35. Lange, R.; Smutny, B. BPSK laser communication terminals to be verified in space. In Proceedings of the IEEE MILCOM 2004. Military Communications Conference, Monterey, CA, USA, 31 October–3 November 2004; Volume 1, pp. 441–444. [CrossRef]
36. Pimentel, P.M.; Rödiger, B.; Schmidt, C.; Fuchs, C.; Rochow, C.; Hiemstra, T.; Zager, A.; Wertz, P.; Knopp, M.; Lehmann, M.; et al. Cube laser communication terminal (CubeLCT) state of the art. *Acta Astronaut.* **2023**, *211*, 326–332. [CrossRef]
37. Zhang, Q.; Wang, G. Influence of HY-2 Satellite Platform Vibration on Laser Communication Equipment: Analysis and On-Orbit Experiment. In *3rd International Symposium of Space Optical Instruments and Applications*; Urbach, H.P., Zhang, G., Eds.; Springer International Publishing: Berlin/Heidelberg, Germany, 2017; pp. 95–111.
38. Mertikas, S.P.; Zhou, X.; Qiao, F.; Daskalakis, A.; Lin, M.; Peng, H.; Tziavos, I.N.; Vergos, G.; Tripolitsiotis, A.; Frantzis, X. First preliminary results for the absolute calibration of the Chinese HY-2 altimetric mission using the CRS1 calibration facilities in West Crete, Greece. *Adv. Space Res.* **2016**, *57*, 78–95. [CrossRef]
39. Grigoryev, V.; Kovalev, V.; Shargorodskiy, V.; Sumerin, V. High-bit-rate laser space communication technology and results of on-board experiment. In Proceedings of the International Conference on Space Optics, Tenerife, Spain, 6–10 October 2014.
40. Singh, U.; Horan, S.; Antill, C.; Petros, M.; Joplin, M.; Spellmeyer, N.; Wright, M.; Zucker, E.; Yu, A.; Numata, K. Independent reliability assessment of the NASA GSFC laser transmitter for the LISA program. In Proceedings of the International Conference on Space Optics—ICSO 2022, Dubrovnik, Croatia, 3–7 October 2022; SPIE: Bellingham, WA, USA; pp. 1409–1415.
41. Schwander, T.; Lange, R.; Kämpfner, H.; Smutny, B. LCTSX: First on-orbit verification of a coherent optical link. *Proc. SPIE* **2004**, *554*, 419–421.
42. Tröndle, D.; Pimentel, P.M.; Rochow, C.; Zech, H.; Muehlnikel, G.; Heine, F.; Meyer, R.; Philipp-May, S.; Lutzer, M.; Benzi, E. Alphasat-Sentinel-1A optical inter-satellite links: Run-up for the European data relay satellite system. In Proceedings of the Free-Space Laser Communication and Atmospheric Propagation XXVIII, San Francisco, CA, USA, 15–16 February 2016; SPIE: Bellingham, WA, USA; p. 973902.
43. Takenaka, H.; Carrasco-Casado, A.; Fujiwara, M.; Kitamura, M.; Sasaki, M.; Toyoshima, M. Satellite-to-ground quantum-limited communication using a 50-kg-class microsatellite. *Nat. Photonics* **2017**, *11*, 502–508. [CrossRef]
44. Carrasco-Casado, A.; Kunimori, H.; Takenaka, H.; Kubo-Oka, T.; Akioka, M.; Fuse, T.; Koyama, Y.; Kolev, D.; Munemasa, Y.; Toyoshima, M. LEO-to-ground polarization measurements aiming for space QKD using Small Optical TrAnsponder (SOTA). *Opt. Express* **2016**, *24*, 12254–12266. [CrossRef]
45. Abrahamson, M.J.; Oaida, B.V.; Sindiy, O.; Biswas, A. Achieving operational two-way laser acquisition for OPALS payload on the International Space Station. In *Free-Space Laser Communication and Atmospheric Propagation XXVII*; SPIE: Bellingham, WA, USA, 2015; Volume 935408. [CrossRef]
46. Rose, T.S.; Rowen, D.W.; LaLumondiere, S.; Werner, N.I.; Linares, R.; Faler, A.; Wicker, J.; Coffman, C.M.; Maul, G.A.; Chien, D.H.; et al. Optical communications downlink from a 1.5U Cubesat: OCSD program. In Proceedings of the International Conference on Space Optics—ICSO 2018, Chania, Greece, 9–12 October 2018; SPIE: Bellingham, WA, USA, 2019; pp. 201–212. [CrossRef]
47. Schmidt, C.; Brechtelsbauer, M.; Rein, F.; Fuchs, C. OSIRIS Payload for DLR's BiROS Satellite. In Proceedings of the International Conference on Space Optical Systems and Applications 2014, Kobe, Japan, 7–9 May 2014.
48. Yin, J.; Li, Y.H.; Liao, S.K.; Yang, M.; Cao, Y.; Zhang, L.; Ren, J.-G.; Cai, W.-Q.; Liu, W.-Y.; Li, S.-L.; et al. Entanglement-based secure quantum cryptography over 1120 kilometres. *Nature* **2020**, *582*, 501–505. [CrossRef] [PubMed]
49. Mróz, P.; Otarola, A.; Prince, T.A.; Dekany, R.; Duev, D.A.; Graham, M.J.; Groom, S.L.; Masci, F.J.; Medford, M.S. Impact of the SpaceX Starlink Satellites on the Zwicky Transient Facility Survey Observations. *Astrophys. J. Lett.* **2022**, *924*, L30. [CrossRef]
50. Strotkamp, M.; Munk, A.; Jungbluth, B.; Hoffmann, H.D.; Höffner, J. Diode-pumped Alexandrite laser instrument for next generation satellite-based Earth observation. In Proceedings of the International Conference on Space Optics—ICSO 2018, Chania, Greece, 9–12 October 2018; SPIE: Bellingham, WA, USA; pp. 160–173. [CrossRef]
51. Komatsu, H.; Ohta, S.; Yamazoe, H.; Kubo, Y.; Nakao, T.; Ito, T.; Koda, D.; Sawada, H.; Ikeda, T.; Munemasa, Y.; et al. The pointing performance of the optical communication terminal, SOLISS in the experimentation of bidirectional laser communication with an optical ground station. In Proceedings of the Free-Space Laser Communications XXXIII, SPIE, Online, 6–11 March 2021; pp. 69–82. [CrossRef]
52. Calzolaio, D.; Curreli, F.; Duncan, J.; Moorhouse, A.; Perez, G.; Voegt, S. EDRS-C—The second node of the European Data Relay System is in orbit. *Acta Astronaut.* **2020**, *177*, 537–544. [CrossRef]
53. Zeif, R.; Kubicka, M.; Hörmer, A.J.; Henkel, M. OPS-SAT mission: An ESA Nanosatellite Mission. In Proceedings of the Lessons We Have Learned: 71st International Astronautical Congress, Online 12–14 October 2020.

54. Tadono, T.; Mizukami, Y.; Watarai, H.; Takaku, J.; Ohgushi, F.; Kai, H. Calibration and Validation of the Advanced Land Observing Satellite-3 "ALOS-3". *Int. Arch. Photogramm. Remote Sens. Spat. Inf. Sci.* **2020**, *XLIII-B1-2020*, 135–140. [CrossRef]
55. Motohka, T.; Kankaku, Y.; Miura, S.; Suzuki, S. Alos-4 L-Band SAR Mission and Observation. In Proceedings of the IGARSS 2019—2019 IEEE International Geoscience and Remote Sensing Symposium, Yokohama, Japan, 28 July–2 August 2019; pp. 5271–5273. [CrossRef]
56. Edwards, B.; Randazzo, T.; Babu, N.; Murphy, K.; Albright, S.; Cummings, N.; Ocasio-Perez, J.; Potter, W.; Roder, R.; Zehner, S.A.; et al. Challenges, Lessons Learned, and Methodologies from the LCRD Optical Communication System AI&T. In Proceedings of the 2022 IEEE International Conference on Space Optical Systems and Applications (ICSOS), Online, 29–31 March 2022; pp. 22–31. [CrossRef]
57. Rieländer, D.; Mira, A.D.; Alaluf, D.; Daddato, R.; Mejri, S.; Piris, J.; Alves, J.; Antsos, D.; Biswas, A.; Karafolas, N.; et al. ESA ground infrastructure for the NASA/JPL PSYCHE Deep-Space Optical Communication demonstration. In Proceedings of the International Conference on Space Optics—ICSO 2022, Dubrovnik, Croatia, 3–7 October 2022; SPIE: Bellingham, WA, USA; pp. 159–169. [CrossRef]
58. Awano, J.; Ogata, T.; Yajima, M. Development of wireless communication demonstration unit on intra-satellite. In Proceedings of the 39th International Communications Satellite Systems Conference (ICSSC 2022), Stresa, Italy, 18–21 October 2022; pp. 33–37. [CrossRef]
59. Chaudhry, A.U.; Lamontagne, G.; Yanikomeroglu, H. Laser Intersatellite Link Range in Free-Space Optical Satellite Networks: Impact on Latency. *IEEE Aerosp. Electron. Syst. Mag.* **2023**, *38*, 4–13. [CrossRef]
60. Toorian, A.; Diaz, K.; Lee, S. The CubeSat Approach to Space Access. In Proceedings of the 2008 IEEE Aerospace Conference, Big Sky, Montana, 1–8 March 2008; pp. 1–14. [CrossRef]
61. Kolev, D.; Takenaka, H.; Munemasa, Y.; Akioka, M.; Iwakiri, N.; Koyama, Y.; Kunimori, H.; Toyoshima, M.; Artaud, G.; Issler, J.L.; et al. Overview of international experiment campaign with small optical transponder (SOTA). In Proceedings of the 2015 IEEE International Conference on Space Optical Systems and Applications (ICSOS), New Orleans, LA, USA, 26–28 October 2015; pp. 1–6. [CrossRef]
62. Carrasco-Casado, A.; Takenaka, H.; Kolev, D.; Munemasa, Y.; Kunimori, H.; Suzuki, K.; Fuse, T.; Kubo-Oka, T.; Akioka, M.; Koyama, Y. LEO-to-ground optical communications using SOTA (Small Optical TrAnsponder)–Payload verification results and experiments on space quantum communications. *Acta Astronaut.* **2017**, *139*, 377–384. [CrossRef]
63. Iwamoto, K.; Komatsu, H.; Ohta, S.; Kubo, Y.; Nakao, T.; Yamazoe, H.; Kamata, T.; Munemasa, Y.; Kunimori, H.; Toyoshima, M.; et al. Experimental results on in-orbit technology demonstration of SOLISS. In *Free-Space Laser Communications XXXIII*; Hemmati, H., Boroson, D.M., Eds.; International Society for Optics and Photonics; SPIE: Bellingham, WA, USA, 2021; Volume 11678, p. 116780D. [CrossRef]
64. Carrasco-Casado, A.; Do, P.X.; Kolev, D.; Hosonuma, T.; Shiratama, K.; Kunimori, H.; Trinh, P.V.; Abe, Y.; Nakasuka, S.; Toyoshima, M. Intersatellite-Link Demonstration Mission between CubeSOTA (LEO CubeSat) and ETS9-HICALI (GEO Satellite). In Proceedings of the 2019 IEEE International Conference on Space Optical Systems and Applications (ICSOS), Portland, OR, USA, 14–16 October 2019; pp. 1–5. [CrossRef]
65. Dresscher, M.; Korevaar, C.W.; van der Valk, N.C.J.; de Lange, T.J.; Saathof, R.; Doelman, N.; Crowcombe, W.E.; Duque, C.M.; de Man, H.; Human, J.D.; et al. Key Challenges and Results in the Design of Cubesat Laser Terminals, Optical Heads and Coarse Pointing Assemblies. In Proceedings of the 2019 IEEE International Conference on Space Optical Systems and Applications (ICSOS), Portland, OR, USA, 14–16 October 2019; pp. 1–6. [CrossRef]
66. Schieler, C.M.; Riesing, K.M.; Bilyeu, B.C.; Robinson, B.S.; Wang, J.P.; Roberts, W.T.; Piazzolla, S. TBIRD 200-Gbps CubeSat Downlink: System Architecture and Mission Plan. In Proceedings of the 2022 IEEE International Conference on Space Optical Systems and Applications (ICSOS), Kyoto City, Japan, 28–31 March 2022; pp. 181–185. [CrossRef]
67. Tomio, H.; Grenfell, P.; Kammerer, W.; Serra, P.; Čierny, O.; Lindsay, C.; Garcia, M.; Cahoy, K.; Clark, M.; Coogan, D.; et al. Development and Testing of the Laser Transmitter and Pointing, Acquisition, and Tracking System for the CubeSat Laser Infrared CrosslinK (CLICK) B/C Mission. In Proceedings of the 2022 IEEE International Conference on Space Optical Systems and Applications (ICSOS), Kyoto City, Japan, 28–31 March 2022; pp. 224–231. [CrossRef]
68. Fuchs, C.; Schmidt, C. Update on DLR's OSIRIS program. In *International Conference on Space Optics—ICSO 2018*; Sodnik, Z., Karafolas, N., Cugny, B., Eds.; International Society for Optics and Photonics; SPIE: Bellingham, WA, USA, 2019; Volume 11180, p. 111800I. [CrossRef]
69. Schmidt, C.; Rödiger, B.; Rosano, J.; Papadopoulos, C.; Hahn, M.T.; Moll, F.; Fuchs, C. DLR's Optical Communication Terminals for CubeSats. In Proceedings of the 2022 IEEE International Conference on Space Optical Systems and Applications (ICSOS), Kyoto City, Japan, 28–31 March 2022; pp. 175–180. [CrossRef]
70. Alfriend, K.T.; Vadali, S.R.; Gurfil, P.; How, J.P.; Breger, L.S. Chapter 1—Introduction. In *Spacecraft Formation Flying*; Alfriend, K.T., Vadali, S.R., Gurfil, P., How, J.P., Breger, L.S., Eds.; Butterworth-Heinemann: Oxford, UK, 2010; pp. 1–11. [CrossRef]
71. Xu, X.; Wang, C.; Jin, Z. Perturbed ISL Analysis in LEO Satellite Constellation. In Proceedings of the 2020 IEEE 3rd International Conference on Electronics and Communication Engineering (ICECE), Xi'An, China, 14–16 December 2020; pp. 12–17. [CrossRef]
72. Yang, H.; Guo, B.; Xue, X.; Deng, X.; Zhao, Y.; Cui, X.; Pang, C.; Ren, H.; Huang, S. Interruption Tolerance Strategy for LEO Constellation With Optical Inter-Satellite Link. *IEEE Trans. Netw. Serv. Manag.* **2023**, *20*, 4815–4830. [CrossRef]

73. Kozhaya, S.; Kassas, Z.M. A First Look at the OneWeb LEO Constellation: Beacons, Beams, and Positioning. *IEEE Trans. Aerosp. Electron. Syst.* **2024**, *60*, 7528–7534. [CrossRef]
74. Rosano Nonay, J.; Rüddenklau, R.; Sinn, A.; Rödiger, B.; Schmidt, C.; Schitter, G. Horizontal Link Demonstration Over 143 km With CubeISL: The World's Smallest Commercial Optical Communication Payload for Inter-Satellite Links. In Proceedings of the Small Satellite Conference 2024, Logan, Utah, 3–8 August 2024.

MDPI AG
Grosspeteranlage 5
4052 Basel
Switzerland
Tel.: +41 61 683 77 34

Aerospace Editorial Office
E-mail: aerospace@mdpi.com
www.mdpi.com/journal/aerospace